Next-Generation Maize Research

Mukesh Choudhary • Pardeep Kumar
Bhupender Kumar • Shankar Lal Jat
P. Lakshmi Soujanya • Hanuman Sahay Jat
Editors

Next-Generation Maize Research

Springer

Editors
Mukesh Choudhary
ICAR-Indian Institute of Maize Research
Ludhiana, Punjab, India

Bhupender Kumar
ICAR-Indian Institute of Maize Research
Ludhiana, Punjab, India

P. Lakshmi Soujanya
Winter Nursery Centre
ICAR-Indian Institute of Maize Research
Hyderabad, Telangana, India

Pardeep Kumar
ICAR-Indian Institute of Maize Research
Ludhiana, Punjab, India

Shankar Lal Jat
ICAR-Indian Institute of Maize Research
Ludhiana, Punjab, India

Hanuman Sahay Jat
ICAR-Indian Institute of Maize Research
Ludhiana, Punjab, India

ISBN 978-981-92-0721-3 ISBN 978-981-92-0722-0 (eBook)
https://doi.org/10.1007/978-981-92-0722-0

This Springer imprint is published by the registered company Springer Nature Singapore Pte Ltd.
The registered company address is: 152 Beach Road, #21-01/04 Gateway East, Singapore 189721, Singapore

Foreword

Maize has emerged as a crop of growing strategic importance across the world, supporting food and nutritional security, livestock-based livelihoods, industrial growth, and renewable energy initiatives. In diverse agroecological regions, maize plays a vital role in both rainfed and irrigated farming systems, offering farmers flexibility, resilience, and market opportunities. As global agriculture confronts the compounded challenges of climate variability, natural resource degradation, and rising demand for food, feed, and energy, the advancement of maize research assumes renewed global significance.

The increasing relevance of maize as an energy crop is particularly noteworthy in the Indian context. With the Government of India's ambitious ethanol blending program and increasing demand for sustainable biofuels, maize offers significant potential to contribute to energy security while creating new income avenues for farmers.

Next-Generation Maize Research: Shaping the Future is a timely and valuable contribution that captures the breadth and depth of contemporary maize science. This volume brings together the collective wisdom of experienced scientists and the innovative perspectives of young researchers, presenting a comprehensive account of maize research across disciplines. Covering genetic resources, breeding innovations, genomics, crop management, mechanization, value addition, and technology transfer, the book reflects the expanding role of maize within India's evolving agricultural landscape.

The forward-looking emphasis on climate-resilient germplasm, accelerated breeding approaches, molecular and genomic tools, and data-driven innovations such as artificial intelligence aligns well with global research priorities for climate-smart and technology-enabled agriculture. The discussions on nutritional enhancement, crop physiology, and integrated pest and disease management underscore the need to enhance productivity while safeguarding environmental sustainability and human health.

I am confident that this publication will serve as an authoritative reference for researchers, students, extension professionals, policymakers, and stakeholders across the maize value chain. By promoting innovation, strengthening collaboration, and guiding evidence-based decision-making, this volume will contribute meaningfully to building resilient, inclusive, and future-ready maize systems in India and beyond.

New Delhi, India M. L. Jat
February 17, 2026

Preface

Maize has long occupied a central place in global agriculture, serving not only as a vital source of food and feed but also as a pillar of cultural heritage and economic stability across continents. As humanity navigates the complexities of the twenty-first century—defined by climate uncertainty, population growth, and increasing pressure on natural resources—the responsibility to sustain and advance this indispensable crop has never been greater. *Next-Generation Maize Research: Shaping the Future* emerges from this imperative, offering a comprehensive and forward-looking perspective on maize science and its evolving role in global food systems. In recent decades, maize has also gained strategic importance as an energy crop, particularly as a primary feedstock for bioethanol production, contributing to renewable energy initiatives, reduced dependence on fossil fuels, and the global pursuit of low-carbon agricultural economies.

This volume brings together the collective wisdom of distinguished scientists who have devoted decades to maize research, alongside the fresh perspectives of young researchers. Their shared experiences reflect the remarkable adaptability of maize across diverse agroecological environments and underscore the power of scientific progress when grounded in traditional knowledge. By weaving together established practices with cutting-edge advancements, the book highlights pathways toward resilient, productive, and sustainable maize systems.

The chapters span the full breadth of maize research. The journey begins with its historical evolution, production dynamics, and maize-based cropping systems, and progresses through the biological and reproductive foundations that govern growth and yield. Special attention is given to the potential of wild relatives in enhancing climate resilience and to the enduring influence of conventional breeding in shaping modern improvement strategies. Advances in molecular markers, accelerated breeding techniques, and artificial intelligence–driven approaches are explored as transformative tools for developing next-generation maize varieties suited to future challenges.

Beyond genetic improvement, the book addresses critical dimensions of maize production and utilization, including nutritional enhancement, crop physiology, biotechnological interventions, and integrated disease and pest management. Equally significant are the discussions on agronomic practices, soil health, farm mechanization, and precision agriculture, all of which contribute to sustainable and efficient production systems. The scope further extends to fodder maize, silage, biofuels, and value-added products, emphasizing maize's expanding role in food security, renewable energy, and evolving consumer demands.

Recognizing maize as more than a crop as a socioeconomic lifeline, the book also examines market dynamics, farmer–industry linkages, and public–private partnerships, with particular emphasis on India's success in hybrid maize seed development. The concluding chapter reflects on future trends, emerging challenges, and new opportunities, inviting readers to envision maize as a continuing force for nourishment, innovation, and resilience.

As editors, we hope this book serves as a meaningful resource for farmers, researchers, students, policymakers, and all those invested in the future of agriculture. Each seed sown carries a promise of productivity, sustainability, and shared prosperity. Through informed research and collective action, maize will continue to shape a resilient agricultural future for generations to come.

Ludhiana, Punjab, India Mukesh Choudhary
Ludhiana, Punjab, India Pardeep Kumar
Ludhiana, Punjab, India Bhupender Kumar
Ludhiana, Punjab, India Shankar Lal Jat
Hyderabad, Telangana, India P. Lakshmi Soujanya
Ludhiana, Punjab, India Hanuman Sahay Jat

Contents

Contributors

T. V. Anirudh College of Agriculture, Govind Ballabh Pant University of Agriculture & Technology, Pantnagar, U.S. Nagar, Uttarakhand, India

Prakash Athare ICAR-Indian Institute of Maize Research, Ludhiana, Punjab, India

Pravin Kumar Bagaria ICAR-Indian Institute of Maize Research, Ludhiana, Punjab, India

Shanti Devi Bamboriya ICAR-Indian Institute of Maize Research, Ludhiana, Punjab, India

Sumitra Devi Bamboriya SKNAU, Jobner, Rajasthan, India

R. K. Bhavyasree Punjab Agricultural University, Regional Research Station, Faridkot, Punjab, India

Bharat Bhushan ICAR-Indian Institute of Maize Research, Ludhiana, Punjab, India

Arpit Chaudhary ICAR-Indian Institute of Maize Research, Ludhiana, Punjab, India

Mukesh Choudhary ICAR-Indian Institute of Maize Research, Ludhiana, Punjab, India

Poonam Choudhary ICAR-Central Institute of Post-Harvest Engineering and Technology, Ludhiana, Punjab, India

Abhijit Kumar Das ICAR-Indian Institute of Maize Research, Ludhiana, Punjab, India

Manesh Chander Dagla ICAR-Indian Institute of Maize Research, Ludhiana, Punjab, India

Sayantan Das ICAR-Indian Institute of Maize Research, Ludhiana, India

Sain Dass Directorate of Maize Research, New Delhi (Presently ICAR-IIMR, Ludhiana), Ludhiana, India

Deepak ICAR-Indian Institute of Maize Research, Ludhiana, Punjab, India

Jomika Devi ICAR-Indian Institute of Maize Research, Ludhiana, Punjab, India

Veena Devi ICAR-Indian Institute of Maize Research, Ludhiana, Punjab, India

Sukhmani Dhaliwal ICAR-Indian Institute of Maize Research, Ludhiana, Punjab, India

Subhank Dixit ICAR-Indian Institute of Maize Research, Ludhiana, Punjab, India

Shourat Fatma ICAR-Indian Institute of Maize Research, Ludhiana, Punjab, India

Priya Garkoti College of Agriculture, Govind Ballabh Pant University of Agriculture & Technology, Pantnagar, U.S. Nagar, Uttarakhand, India

ICAR-Indian Institute of Maize Research, Ludhiana, Punjab, India

Gulshan Gaur ICAR-Indian Institute of Maize Research, Ludhiana, India

Mamta Gupta ICAR-Indian Institute of Maize Research, Ludhiana, Punjab, India

Bahadur Singh Jat ICAR-Indian Institute of Maize Research, Ludhiana, Punjab, India

Hanuman Sahay Jat ICAR-Indian Institute of Maize Research, Ludhiana, Punjab, India

Shankar Lal Jat ICAR-Indian Institute of Maize Research, Ludhiana, Punjab, India

Jawala Jindal Punjab Agricultural University, Ludhiana, Punjab, India

B. Jyothi Winter Nursery Centre, ICAR-Indian Institute of Maize Research, Rajendranagar, Hyderabad, India

Manish Kakraliya ICAR-Indian Institute of Maize Research, Ludhiana, Punjab, India

Chikkappa Gangadhar Karjagi ICAR-Indian Institute of Maize Research, Ludhiana, India

Regional Maize Research and Seed Production Centre, ICAR-Indian Institute of Maize Research, Begusarai, India

Ankita Kataria ICAR-Indian Institute of Maize Research, Ludhiana, Punjab, India

K. Kamesh Krishnamoorthy ICAR-Indian Institute of Maize Research, Ludhiana, Punjab, India

Bhupender Kumar ICAR-Indian Institute of Maize Research, Ludhiana, Punjab, India

Krishan Kumar ICAR-Indian Institute of Maize Research, Ludhiana, Punjab, India

Pardeep Kumar ICAR-Indian Institute of Maize Research, Ludhiana, Punjab, India

Ramesh Kumar ICAR-Indian Institute of Maize Research, Ludhiana, Punjab, India

Santosh Kumar ICAR-Indian Agricultural Research Institute, Hazaribagh, Jharkhand, India

Manish Kumar International Maize and Improvement Centre (CIMMYT), BISA Farm, Ludhiana, Punjab, India

Chayanika Lahkar Assam Agricultural University – Zonal Research Station (AAU-ZRS), Gossaigaon, Assam, India

Deep Mohan Mahala ICAR-Indian Institute of Maize Research, Ludhiana, Punjab, India

Shreya Mehindiratta ICAR-Indian Institute of Maize Research, Ludhiana, Punjab, India

Kajal Mehra College of Agriculture, Govind Ballabh Pant University of Agriculture & Technology, Pantnagar, U.S. Nagar, Uttarakhand, India

Nikhil Mehta ICAR-Indian Institute of Maize Research, Ludhiana, Punjab, India

Sandesh Kumar Mishra ICAR-Indian Institute of Maize Research, Ludhiana, Punjab, India

Sudip Nandi ICAR-Indian Institute of Maize Research, Ludhiana, India

Thotla Naresh College of Agriculture, Govind Ballabh Pant University of Agriculture & Technology, Pantnagar, U.S. Nagar, Uttarakhand, India

Pankaj Pandey ICAR-Indian Institute of Maize Research, Ludhiana, Punjab, India

Laishram Priscilla Punjab Agricultural University, Ludhiana, Punjab, India

Pushpendra ICAR-Indian Institute of Maize Research, Ludhiana, Punjab, India

Komal Rani ICAR-Indian Institute of Maize Research, Ludhiana, Punjab, India

Rumesh Ranjan Punjab Agricultural University, Ludhiana, Punjab, India

U. Srinivasulu Reddy Machine Learning and Data Analytics Lab, Department of Computer Applications, National Institute of Technology, Tiruchirappalli, India

Shaik Rehana ICAR-Indian Institute of Maize Research, Ludhiana, Punjab, India

Anirban Roy Ramakrishna Mission Vivekananda Educational and Research Institute, Kolkata, West Bengal, India

Alok Kumar Sahoo ICAR-Indian Institute of Maize Research, Ludhiana, Punjab, India

Ankush Sharma ICAR-Indian Institute of Maize Research, Ludhiana, Punjab, India

Ph. Romen Sharma ICAR-Indian Institute of Maize Research, Ludhiana, Punjab, India

H. S. Sidhu Punjab Agricultural University, Ludhiana, Punjab, India

Alla Singh ICAR-Indian Institute of Maize Research, Ludhiana, Punjab, India

Gurpinder Singh ICAR-Indian Institute of Maize Research, Ludhiana, Punjab, India

Manpreet Singh Punjab Agricultural University (PAU), Ludhiana, Punjab, India

N. K. Singh College of Agriculture, Govind Ballabh Pant University of Agriculture & Technology, Pantnagar, U.S. Nagar, Uttarakhand, India

Navdeep Singh ICAR-Indian Institute of Maize Research, Ludhiana, Punjab, India

Rakhi Singh ICAR-Indian Institute of Maize Research, Ludhiana, Punjab, India

Vishal Singh ICAR-Indian Institute of Maize Research, Ludhiana, Punjab, India

Aayushi Singla ICAR-Indian Institute of Maize Research, Ludhiana, Punjab, India

P. Lakshmi Soujanya Winter Nursery Centre, ICAR-Indian Institute of Maize Research, Hyderabad, Telangana, India

S. B. Suby ICAR-Indian Institute of Maize Research, Ludhiana, Punjab, India

N. Sunil Winter Nursery Centre, ICAR-Indian Institute of Maize Research, Rajendranagar, Hyderabad, India

C. Susmita Winter Nursery Centre, ICAR-Indian Institute of Maize Research, Hyderabad, India

Sunandan Swain Punjab Agricultural University, Ludhiana, India

Mayank Tiwari College of Agriculture, Govind Ballabh Pant University of Agriculture & Technology, Pantnagar, U.S. Nagar, Uttarakhand, India

Thalari Vasanthrao Professor Jayashankar Telangana Agricultural University, Hyderabad, India

Shrikant Yankanchi ICAR-Indian Institute of Maize Research, Ludhiana, Punjab, India

Kondaji Rangappa Yathish ICAR-Indian Institute of Maize Research, Ludhiana, India

Winter Nursery Centre, ICAR-Indian Institute of Maize Research, Hyderabad, India

Maize History: Production Trends and Maize-Based Systems in India

1

Manish Kakraliya, Mukesh Choudhary, Shankar Lal Jat, and Hanuman Sahay Jat

Abstract

Maize (*Zea mays* L.) is a highly versatile and strategic cereal crop supporting food, feed, fodder, and bio-industrial sectors in India. Cultivated across diverse agro-climatic zones, national maize production exceeds 43 million tonnes, with over 85% utilized by poultry, starch, and ethanol industries, reflecting a major shift in production trends and end-use diversification. The widespread adoption of improved hybrids, specialty maize types, and climate-smart management practices has significantly enhanced system productivity, resulting in higher grain yield (25–45%), greater economic returns (30–60%), and improved nutrient-use efficiency (15–35%) compared to conventional farming. However, productivity remains constrained by climate variability, drought, heat stress, declining soil fertility, and emerging pests such as fall armyworm (FAW), causing maximum yield losses (20–35%) under farmers' practices. Integration of conservation agriculture (CA), mechanization, precision nutrient management (PNM), and diversified maize-based cropping systems enhances system performance, achieving yield improvement (20–40%), water savings (30–45%), reduction in labour input (25–50%) with higher profitability (30–65%), while legume intercropping further improves soil organic carbon (10–25%) and enhances system sustainability. The mechanized operations and site-specific nutrient management (SSNM) also reduce production costs (15–30%) and improve energy productivity (20–35%) than traditional systems. Expansion of maize into non-traditional regions using climate-resilient hybrids has achieved higher yield gains (15–40%) over existing local cropping systems. Additionally, maize-based ethanol offers higher energy efficiency 20–30% and lower greenhouse gas

M. Kakraliya · M. Choudhary · S. L. Jat (✉) · H. S. Jat
ICAR-Indian Institute of Maize Research, Ludhiana, Punjab, India
e-mail: manishkakraliya719@gmail.com; mukesh.choudhary1@icar.org.in; sl.jat@icar.org.in; hs.jat@icar.org.in

M. Choudhary et al. (eds.), *Next-Generation Maize Research*,
https://doi.org/10.1007/978-981-92-0722-0_1

(GHG) emissions (35–50%) than fossil fuels, reinforcing its role in sustainable bioenergy development. Overall, integrated technological innovations and policy interventions are essential for sustainably intensifying maize production and strengthening its contribution to India's food, feed, and energy security.

Keywords

Production trends · Maize-based cropping systems · Climate-smart agriculture · Mechanization and sustainable intensification

1.1 Introduction

Maize (*Zea mays* L) stands as one of the most versatile and widely cultivated cereal crops globally. The crop's significance spans diverse sectors, including food, feed, fodder, ethanol, and industrial applications. In India, maize is cultivated on approximately 12.1 million hectares of land, demonstrating its vital role in the nation's agricultural economy (MoA and FW 2025; https://upag.gov.in). Maize is grown across varied agro-climatic zones, has the ability to thrive in multiple seasons; *kharif*, *rabi*, and spring, making it adaptable and resilient to different environmental conditions. Among the cereal crops, maize exhibits the highest yield potential, with yields exceeding 5.0 tonnes per hectare during the *rabi* and spring seasons. This remarkable productivity makes maize a key crop for ensuring food security, supporting livestock nutrition, and providing raw materials for industrial sectors. Its use in the industrial sector has significantly reshaped its utilization pattern in recent years, making maize's role no longer confined to traditional food usage alone. Over 85% of India's maize production is now channelled toward industrial uses such as poultry and animal feed, starch, processed food products, and ethanol production (FICCI 2024; MoA and FW 2025).

Maize's history in India dates back to the sixteenth century when it was introduced by Portuguese traders (Watt 1893), just after trade routes were discovered by Vasco da Gama in 1498. Initially grown in isolated regions, particularly in the southern and western parts of the country, its cultivation expanded rapidly following the Green Revolution in the 1960s. The introduction of hybrid varieties, improved irrigation practices, and the adoption of modern farming technologies boosted maize production significantly, making it a major staple crop across the country (Yadav et al. 2015). In the past 5 years, India has witnessed a steady increase in maize production, which is growing at an annual growth rate (AGR) of 6.5%, reaching 33.73 million tonnes in 2020–21, with projections indicating it will exceed 50 million tonnes in the 2025–26 period (FICCI 2024). This surge in production is largely attributed to the expanding demand from the poultry, starch, and biofuel industries, positioning maize as a critical crop in India's bio-economy. Its high-yield potential, combined with its suitability for bioethanol production, underscores its importance as a sustainable source of energy in India's energy transition. India ranks as the 5th largest producer of maize globally, yet its productivity is relatively low compared to top maize-producing nations like the United States and China. This disparity highlights the untapped potential of India's diverse agro-climatic conditions, which could support further expansion of maize production. However, being mainly a rainfed crop, the growth of maize production is not without challenges. Climate change, water scarcity, soil health issues, and the need for efficient input management are crucial concerns to be addressed to ensure sustainable growth. The increasing demand for maize, especially in the biofuel and industrial sectors, necessitates innovations in agricultural practices to boost yields while maintaining environmental sustainability. Given its broad range of uses and critical role in India's agricultural and industrial sectors, understanding maize's production trends, cropping systems, and the challenges is essential for shaping future agricultural strategies and ensuring the crop's continued success in meeting the country's evolving needs.

1.1.1 Importance of Maize in Global and Indian Agriculture

Since its domestication nearly 9000 years ago, maize (*Zea mays* L.) has emerged as a cornerstone of global agri-food systems. It is currently the most produced cereal crop globally, cultivated across diverse agro-ecological regions owing to its remarkable adaptability. According to the Food and Agriculture Organization (FAOSTAT) in 2025, the global maize area reached 208.2 million hectares (Mha) with a total production of 1241.6 million tonnes (Mt), reflecting an increase of area by two times and an increase in production by six times as compared to 1961 (105 Mha area and 205 Mt production).

The average global productivity rose to 6.0 tonnes per hectare, a significant improvement from 1.9 tonnes per hectare in 1961, marking a 207% increase (Fig. 1.1). The rapid growth in global maize production has been driven by a combination of rising demand, technological innovations, improved hybrid varieties,

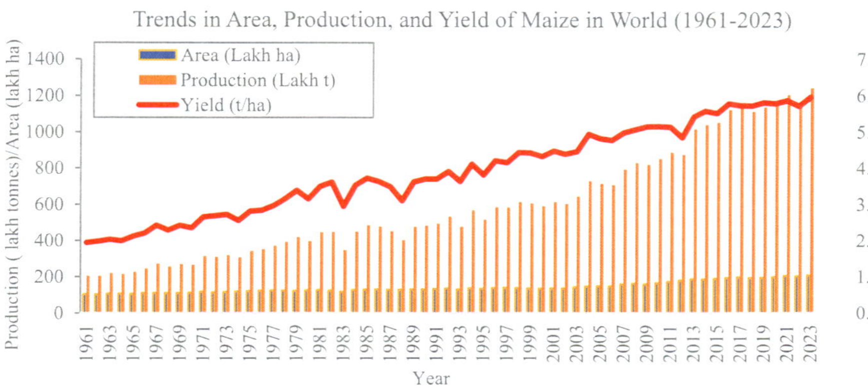

Fig. 1.1 Trends in area, production, and yield of maize in the world from 1961 to 2023. (Data Source: FAOSTAT 2025)

and expansion into new areas. Maize is a highly versatile crop used predominantly as animal feed, but also serving as a staple food in several developing regions, particularly in sub-Saharan Africa and Latin America, where it forms a major source of dietary energy. Its high photosynthetic efficiency (C4 pathway), short growth duration, and suitability for hybridization make it an ideal choice as a high-yielding and climate-resilient crop. Furthermore, maize is increasingly utilized for bioenergy production, especially ethanol, and plays a significant role in international trade, with the United States alone contributing nearly 30% of global maize exports (FAOSTAT 2025). In the Indian context, maize holds immense significance, both as a food crop and a key driver of the agricultural economy. As the seventh-largest maize producer globally, India recorded 43 million metric tonnes (Mt) of maize production in 2025 (MoA and FW 2025; https://upag.gov.in). Maize is grown across multiple states and agro-climatic zones, from the hills of Himachal Pradesh and North Eastern India to the plains of Bihar, Karnataka, and Telangana. Traditionally, maize has been consumed as a staple food in many parts of Southern and Eastern India, where it features in local cuisines as maize flour, boiled cobs, or regional dishes like *khakra*, *bhutte ka khees*, and maize porridge (Singh 1977). Beyond human consumption, maize has become an indispensable input in the livestock and poultry feed industry, which drives much of its demand, particularly in the dairy and poultry sectors. A notable emerging use of maize in India is its role in ethanol production, aligning with national biofuel policies aimed at reducing fossil fuel dependence and enhancing climate resilience. The use of maize as a feedstock for ethanol not only offers environmental benefits but also opens new market opportunities for farmers (Choudhary et al. 2020). Maize also contributes significantly to crop diversification and is increasingly being integrated into rice-maize (RM), maize-wheat (MW), and maize-pulse cropping systems. These systems support soil health improvement, pest and disease management, and reduce the risks associated with monocropping. Its short duration, wide adaptability, and ability to fit into multiple cropping systems make maize a vital crop for achieving food, feed, fodder, and fuel security in India.

1.2 History of Maize Cultivation in India

Maize, originally domesticated around 7000 years ago in the Balsas River Valley of present-day Mexico, has had a fascinating journey into the Indian subcontinent. Today, maize is India's third most important cereal after rice and wheat, with applications across food, feed, and industrial sectors. The timeline of maize's arrival in India is debated, but the dominant view credits Portuguese traders with introducing it during their colonial expansion in the sixteenth century (Watt 1893). However, some linguistic and archaeological evidence suggests an earlier presence through Arab trade or Central Asian overland routes, possibly even before the European connection. Linguistic analysis supports the theory of Arab mediation. The word "makka" or "makka jola," used in many Indian languages to describe maize, appears to derive from "Mecca," indicating a route via Islamic trade networks. Additionally, temple sculptures from the twelfth to thirteenth century in South India, particularly, at Somnathpur, depict cob-like structures resembling maize. Genetic studies of traditional maize varieties in the Himalayan and Northeastern regions show unique traits, lending weight to the hypothesis of multiple possibly pre-Columbian introductions through Southeast Asia or the Pacific (Table 1.1). In pre-modern India, maize found a niche within the traditional rainfed agricultural systems. Farmers practiced mixed cropping, crop rotation, and used organic fertilizers like compost and farmyard manure in maize. Water was managed through community structures such as *baolis* (step wells) and *kunds* (rainwater reservoirs). Landraces are locally adapted maize varieties that played a vital role in these systems and are valued for their genetic diversity, stress tolerance, and low input requirements. These practices promoted sustainability, food security, and ecological balance. During post-independence, maize began receiving systematic scientific attention. A landmark event was the establishment of the All India Coordinated Maize Improvement Project (AICMIP) in 1957 by the Indian Council of Agricultural Research (ICAR). This initiative aimed to develop high-yielding and disease-resistant varieties adapted to different agro-climatic zones, improve agronomic practices, and ensure seed

Table 1.1 Timeline of maize development and adoption in India

Period	Key developments
Pre-Colonial Era	Maize was likely introduced to India by Portuguese traders in the sixteenth century. Initially grown in isolated pockets, especially in tribal and hilly regions
1940s–50s	Grown as a traditional crop using landraces and local varieties. Low productivity (~1 t/ha), mainly used for food and fodder. Limited research infrastructure
1957	Establishment of the All India Coordinated Maize Improvement Project (AICMIP) by ICAR to initiate systematic maize research and development
1960s–70s	Introduction of composite varieties (e.g., Vijay, Ranjit) through mass selection and population improvement. National seed programs started promoting these varieties
1980s	Release and popularization of double-cross hybrids. Expansion of maize into non-traditional areas like eastern India began. Efforts to promote quality seed supply
1990s	Private sector participation grew in hybrid seed development. Single-cross hybrid research intensified. Adoption of QPM (Quality Protein Maize) started for nutritional security
2000–2010	Single-cross hybrids (SCH) gained dominance, offering 20–30% higher yields than composites. Significant increase in productivity (up to 3 t/ha nationally)
2015	Establishment of ICAR–Indian Institute of Maize Research (IIMR), Ludhiana, to lead maize R&D in India. Greater emphasis on climate-smart and mechanized maize
2010s–Present	Focus on biotic/abiotic stress tolerance, nutritional security (biofortification), industrial use (starch, ethanol), and expansion to non-traditional areas (e.g., Northeastern states, tribal belts)
2020s	Maize is positioned as a climate-resilient and industrially relevant crop under schemes like the Ethanol Blending Program (EBP). Research focuses on genomics, conservation agriculture, disease and pest resistance, genome editing, and digitization of maize value chains.

availability. AICMIP coordinated research across national and regional centres, significantly contributing to the structured advancement of maize breeding and cultivation in India. The Green Revolution of the 1960s, although centred on rice and wheat, indirectly benefited maize through the promotion of input-intensive agriculture. Hybrid maize research was boosted by taking inspiration from the success of high-yielding varieties (HYVs) in other cereals by ICAR scientists and global partners. However, unlike rice and wheat, maize's cross-pollination nature made it especially suitable for hybrid breeding. This led to the introduction and spread of hybrid maize in the 1980s and 1990s. The real transformation began in the late 1990s and early 2000s with the significant entry of the private seed sector. Seed companies like Monsanto (now Bayer), DuPont Pioneer, Syngenta, and numerous Indian firms played a pivotal role in developing and marketing hybrid seeds. As a result, currently over 80% of the maize grown in India is hybrid seed-based, leading to significant yield improvements.

Between 1999–2000 and 2024–25, maize production in India surged from around 11.5 million tonnes to more than 43.41 million tonnes, with productivity improving from 1.9 to 3.6 tonnes per hectare (Fig. 1.1). By 2025, maize had transformed from a minor, introduced crop into a central pillar of India's agriculture and industry. Madhya Pradesh, Karnataka, Bihar, Maharashtra, Telangana, West Bengal, and Rajasthan are major maize-producing states with year-round cultivation in *kharif*, *rabi*, and spring seasons, adding to its versatility. Maize became essential not only as a human food grain but also as a key component of poultry feed, animal fodder, starch industries, pharmaceuticals, and, most recently bioethanol. The government's Ethanol Blending Programme (EBP), targeting 20% ethanol blending by 2025 and 30% by 2030, has further boosted the strategic importance of maize as a feedstock for sustainable fuel. This has triggered fresh policy interest and private investment in maize cultivation, especially in rainfed and non-traditional regions.

1.2.1 Technological Advancements and Research on Maize in India

Since its inception in 1998, the ICAR-Indian Institute of Maize Research (IIMR) has played a pivotal role in transforming maize cultivation across India. As of 2024, IIMR has developed and released over 550 maize hybrids and varieties, including high-yielding single cross hybrids, Quality Protein Maize (QPM), and nutri-rich speciality maize types such as baby corn, sweet corn, popcorn, and waxy maize (IIMR 2024). These cultivars exhibit enhanced resistance to major biotic stresses like *Turcicum* leaf blight, maydis leaf blight, stem borers, and abiotic stresses, including drought and heat, making them suitable for diverse agro-climatic zones. Beyond varietal development, IIMR has significantly advanced climate-smart and precision agriculture through the promotion of mechanized sowing and harvesting, micro-irrigation systems (drip and sprinkler), zero-tillage, and nutrient management strategies. The integration of molecular breeding, genomic selection, and biotechnological tools has accelerated the development of climate-resilient and nutrient-enriched lines, particularly QPM hybrids that address protein deficiency in human diets. IIMR has also emphasized the role of maize in sustainable cropping systems, especially maize–wheat (MW) and maize-legume rotations, to enhance soil fertility, reduce input dependency, and minimize the impact of monoculture. Extensive technology dissemination is achieved through a robust network of Krishi Vigyan Kendras, state agricultural universities, public-private partnerships, and frontline demonstrations under national initiatives. These efforts collectively contribute to the adoption of climate-resilient, resource-efficient, and economically viable maize-based production systems in India.

1.2.2 Trends of Area, Production, and Yield of Maize in India

1.2.2.1 Area of Maize

In the 1960s, the area under maize cultivation in India was around 4.6 million hectares, primarily grown in traditional maize belts such as Uttar Pradesh, Madhya Pradesh, and Bihar, mostly as a subsistence crop for local consumption, and maize contributed modestly to the national food basket. However, as India's population grew and food security became a national focus, the importance of maize as a multipurpose crop gained attention not only as human food but also as a vital component in animal feed, poultry rations, starch, and industrial processing.

During the 1970s and 1980s, though maize was not at the forefront of the Green Revolution, it still benefited from gradual improvements in farming techniques, fertilizer use, and improved seed varieties, leading to steady growth in cultivated area. By the early 1990s, the area had reached approximately 9.0 million hectares, marking a 95% increase from the 1960s, indicating maize's growing role in Indian agriculture. The early 2000s marked a turning point, with rising demand from poultry, feed, starch, and ethanol sectors, along with the introduction of hybrid maize varieties by both public institutions and private companies. States like Karnataka, Andhra Pradesh, and Rajasthan emerged as major maize producers, while non-traditional regions such as the Northeast, Tamil Nadu, Chhattisgarh, and parts of Maharashtra witnessed rapid adoption due to maize's adaptability and profitability. By the 2010s, the area had crossed the 10 million ha mark, an over 115% increase compared to the 1960s (Figs. 1.2 and 1.3). As of 2024–25, the area under maize cultivation is 12.09 million hectares, representing a ~220% increase from the 1960s baseline, with a slight rise expected in 2024–25 (MoA and FW 2025).

1.2.2.2 Maize Production

India's maize production has shown a remarkable upward trajectory over the past six decades, evolving from a marginal cereal crop into a commercially important component of the country's agricultural landscape. In the early 1960s, maize production was approximately 5.5 million tonnes, which was largely cultivated under rainfed conditions with traditional varieties and limited input use. During the following decades, modest gains were made, and by the 1990s, production had increased to about ten million tonnes, marking an ~105% rise. This growth was

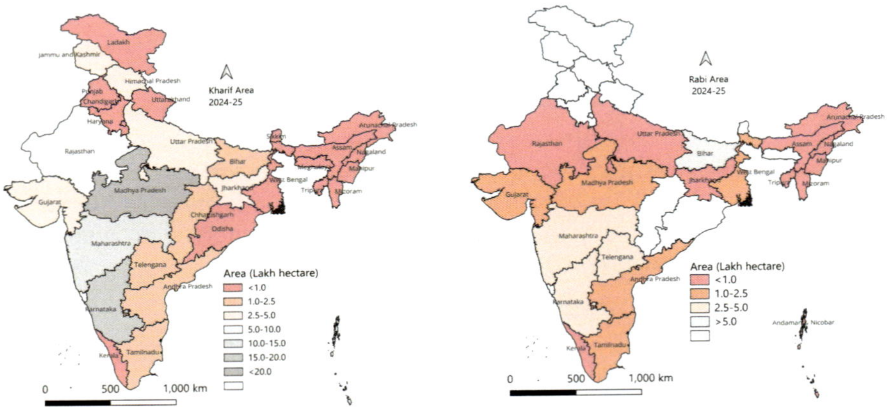

Fig. 1.2 State-wise area of maize in India during (**a**) *kharif* and (**b**) *rabi* seasons of 2024–25

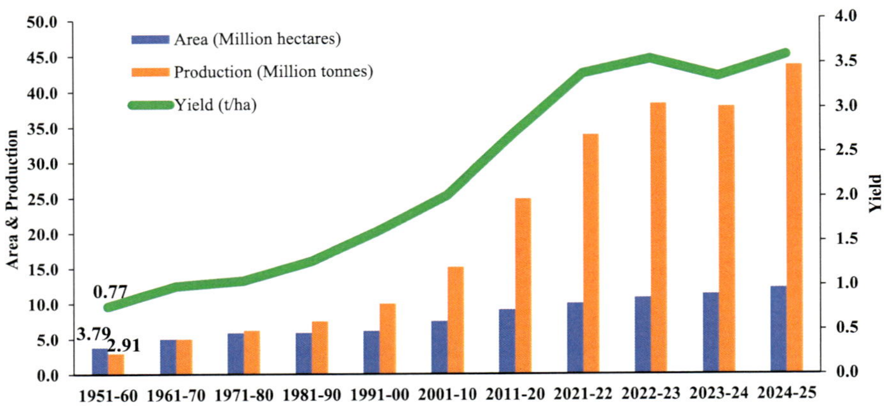

Fig. 1.3 Area, production, and yield of maize in India from 1951 to 2025

spurred by increasing demand from the poultry and starch industries, coupled with early technological interventions. The real transformation, however, began in the 2000s with the introduction of single-cross hybrids (SCH) and the active participation of the private seed sector. In 2000–01, maize production rose to 12–13 million tonnes and continued to climb rapidly, reaching 21.7 million tonnes by 2010–11, by a ~80% increase within a decade. This period also saw expansion into non-traditional regions supported by better irrigation, mechanization, and pest management. By 2019–20, India produced 28.8 million tonnes of maize, which further surged to 43.4 million tonnes in 2024–25, recording a ~51% growth in just 5 years (Figs. 1.3 and 1.4). Overall, maize production in India has increased by nearly 780% since the 1960s, driven by scientific crop management, cereal diversification strategies, and the growing role of maize in ethanol production and climate-resilient agriculture in India.

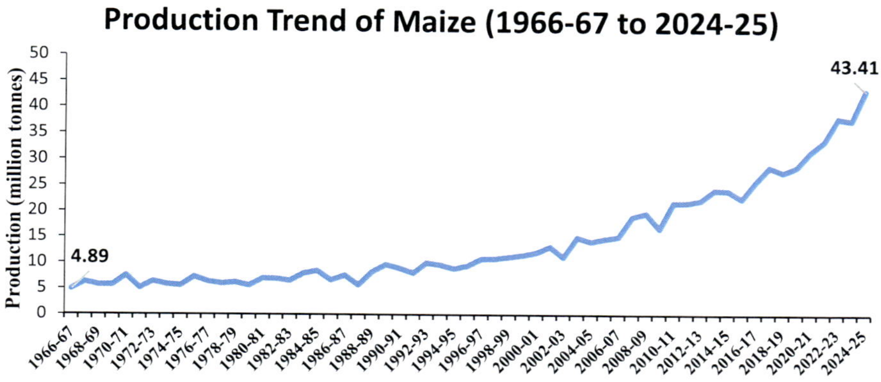

Fig. 1.4 Production trend of maize in India from 1966–67 to 2024–25

1.2.2.3 State-Wise Production Trends

The state-wise analysis of maize production in India from 1966–67 to 2024–25 reveals a remarkable growth trajectory, with national production increasing from approximately 48.94 lakh tonnes to 43.41 lakh tonnes, a more than sevenfold rise. Madhya Pradesh stands out with an extraordinary surge from just 4.54 lakh tonnes in 1966–67 to 66.42 lakh tonnes in 2024–25, and now contributes 15.3% of the country's total maize production, making it one of the top maize-producing states (Table 1.2; Fig. 1.5).

Karnataka also demonstrated substantial growth, rising from 0.11 to 61.90 lakh tonnes, driven largely by the expansion of *kharif* maize, contributing 14.26% to national production. Other leading states include Bihar (12.60%), Maharashtra (10.94%), Telangana (7.81%), West Bengal (6.53%), Rajasthan (6.17%), Tamil Nadu (6.13%), Andhra Pradesh (4.95%), and Uttar Pradesh (4.52%), all of which have significantly contributed to the surge, reflecting the widespread adoption of high-yielding hybrids and market-driven cultivation (Table 1.2; Fig. 1.5).

1.2.2.4 Season-Wise Production Trends

The production of *kharif* maize in India has shown a remarkable upward trend from 1966-67 to the figures for 2024–25 (Figs. 1.6 and 1.7). In 1966–67, the production was approximately 5.5 million tonnes, which remained relatively stagnant throughout the 1970s and early 1980s. A gradual increase was observed in the 1990s, with production reaching around 8.5 million tonnes by the mid-1990s. The significant growth began in the early 2000s, coinciding with the adoption of hybrid varieties and improved agronomic practices, pushing production to approximately 15 million tonnes by 2010–11. This marked a 172% increase from the 1966 to 67 level. The upward trend continued, and by 2021–22, production peaked near 23 million tonnes, registering an overall increase of over 318% compared to 1966–67. The production for 2024–25 stands at around 25 million tonnes, which represents a 354% increase over the base year of 1966–67.

Table 1.2 State-wise production (Lakh tonnes) share of maize from 1966–67 to 2023–24

State	1966–67	1970–71	1980–81	1990–91	2000–01	2010–11	2020–21	2024–25
Madhya Pradesh	4.54	6.27	7.13	12.37	12.18	10.52	38.82	66.42
Karnataka	0.11	2.05	3.71	6.37	21.36	44.44	53.62	61.90
Bihar	9.49	11.13	8.70	10.38	14.97	14.40	20.84	54.68
Maharashtra	0.47	0.42	1.52	1.35	3.03	26.02	35.20	47.49
Telangana	0.00	0.00	0.00	0.00	0.00	20.71	17.57	33.92
West Bengal	0.40	0.48	0.56	0.82	0.88	3.52	24.36	28.36
Rajasthan	6.14	9.30	7.86	13.03	10.16	20.53	22.70	26.78
Tamil Nadu	0.03	0.15	0.21	0.44	1.40	10.28	25.65	26.63
Andhra Pradesh	1.73	3.44	7.25	6.46	15.81	18.85	17.81	21.48
Uttar Pradesh	10.76	17.96	8.94	14.32	14.73	11.14	18.18	19.61
Gujarat	1.45	4.44	4.03	5.30	2.89	8.20	6.68	7.95
Jharkhand	0.00	0.00	0.00	0.00	1.14	2.62	6.52	6.98
Himachal Pradesh	3.64	4.83	5.23	6.55	6.84	6.71	7.29	6.93
Jammu & Kashmir	2.22	3.66	4.93	4.46	5.26	5.28	5.09	5.88
Assam	0.13	0.07	0.13	0.13	0.15	0.14	1.48	4.75
Chhattisgarh	0.00	0.00	0.00	0.00	1.26	1.86	3.41	3.61
NEH Region	0.34	0.14	0.74	1.73	2.15	3.51	4.44	3.66
Punjab	6.14	8.61	6.05	3.36	4.61	4.91	3.95	3.70
Odisha	0.49	0.59	1.74	2.07	0.69	2.99	2.27	2.65
Uttarakhand	0.00	0.00	0.00	0.00	0.60	0.43	0.41	0.51
Haryana	0.86	1.30	0.83	0.49	0.34	0.19	0.20	0.16
All India	**48.94**	**74.86**	**69.57**	**89.62**	**120.43**	**217.26**	**316.47**	**434.09**

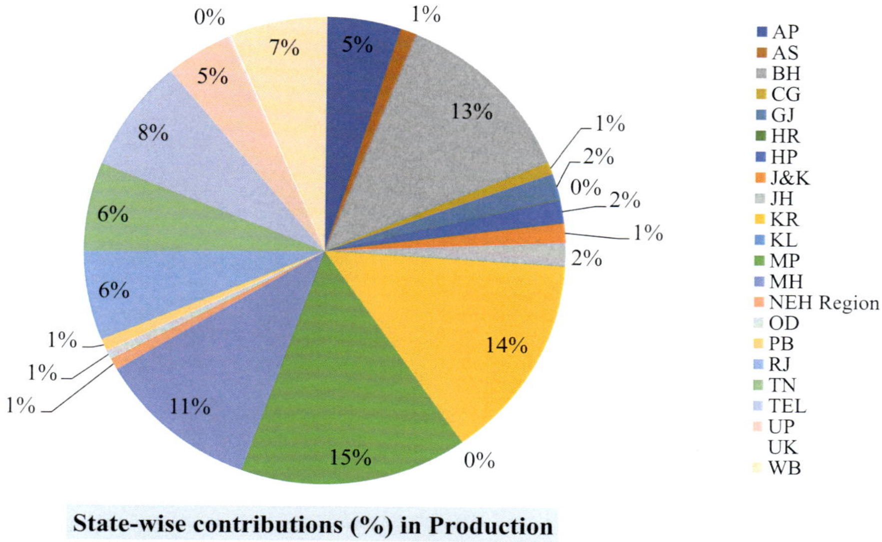

Fig. 1.5 State-wise contributions in maize production during 2024–25

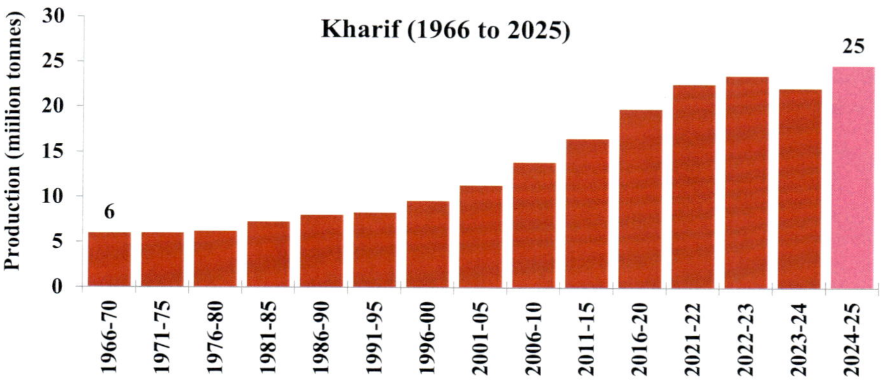

Fig. 1.6 Production trends of maize during *kharif* season of 1966 to 2025

In the *rabi* season, the production in 1982–83 was merely around 0.5 million tonnes, reflecting the limited adoption of Rabi maize during that period. However, over the years, especially after the early 2000s, the area and production began to expand significantly due to improvements in irrigation facilities, short-duration hybrids, and greater demand for maize in the feed and processing sectors.

By 2002–03, the production rose to about 1.2 million tonnes, and by 2012–13 it had reached approximately 4.5 million tonnes, recording an ~800% increase over three decades. This upward trend continued, with rabi maize production crossing 7.0 million tonnes around 2020–21. In 2023–24, production touched 13.8 million

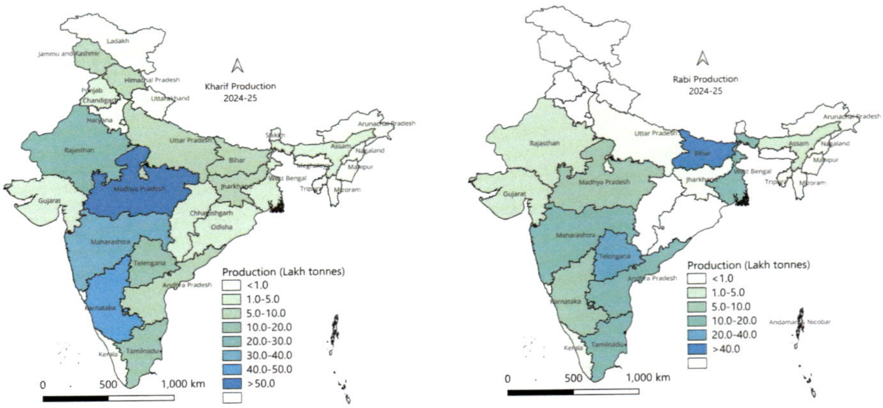

Fig. 1.7 State-wise production of maize during *kharif* and *rabi* seasons of 2024–25

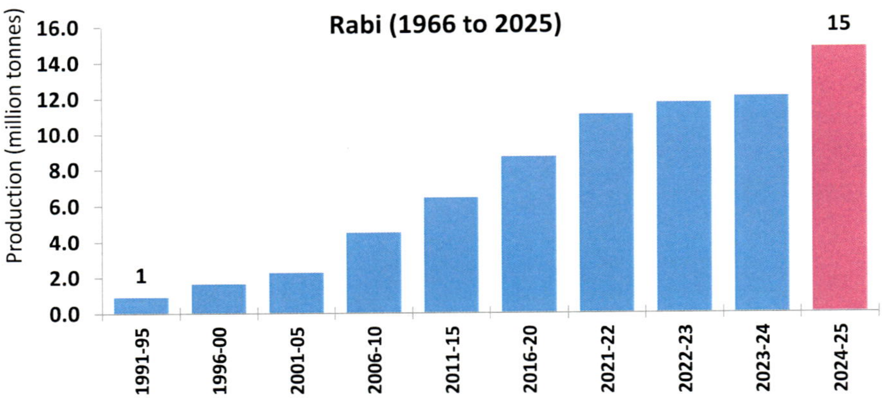

Fig. 1.8 Production trends of maize during *rabi* season of 1966 to 2025

tonnes, and the production for 2024–25 is around 14.75 million tonnes, marking an astonishing ~2850% increase compared to the 1982–83 level (Figs. 1.7 and 1.8).

1.2.2.5 Major Maize-Growing States with Productivity

Alongside the impressive rise in production, maize productivity in India has improved significantly over the past six decades. In the early 1960s, the average yield was just 1100 kilograms per hectare, reflecting limited use of inputs and traditional farming methods. By 1990–91, yields had increased to 1500 kg/ha, a ~36% improvement. The upward trend continued, with productivity reaching 1900 kg/ha in 2000–01 and 2500 kg/ha by 2010–11, driven by the adoption of hybrid seeds, improved agronomic practices, and increased use of inputs. In the following decade, average yields rose to 3100 kg/ha in 2020–21 and touched around 3590 kg/ha in 2024–25, representing a ~226% increase from the 1960s. However, despite these

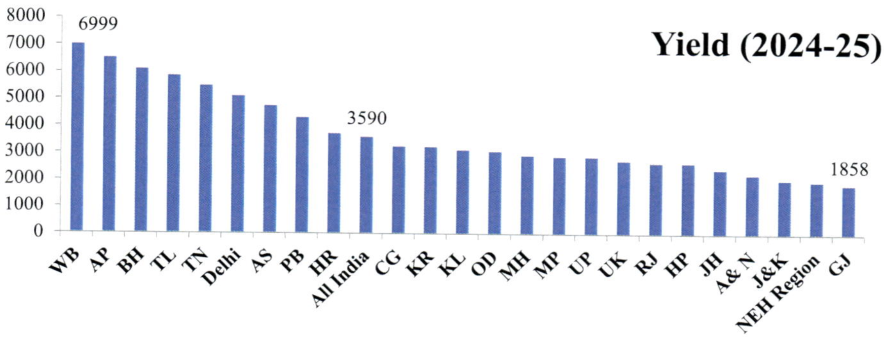

Fig. 1.9 State-wise maize productivity of 2024–25

gains, India's maize productivity remains significantly lower than that of global leaders like the United States (11,200 kg/ha) and China (6300 kg/ha). During 2024–25, West Bengal leads the country with the highest yield of 6999 kg/ha, followed closely by Andhra Pradesh, Bihar, and Telangana, each exceeding 5500 kg/ha (Fig. 1.9). In contrast, Gujarat records the lowest maize yield among major states at 1858 kg/ha, which is 48% below the national average of 3590 kg/ha. The North Eastern Hill (NEH) region (1982 kg/ha), Jammu & Kashmir (2041 kg/ha), Jharkhand (2418 kg/ha), and many others also fall significantly below the national average, which indicates challenges. States like Punjab, Assam, and Delhi show moderate yields (around 4300–5100 kg/ha), while major maize producers such as Madhya Pradesh, Karnataka, and Chhattisgarh are seen with yields ranging between 2900 and 3200 kg/ha (MoA and FW 2025).

1.3 Factors Influencing Production Trends

Maize production trends are shaped by climate variability, supportive policies, and technological advancements. Rising temperatures, especially a 1 °C increase, can reduce maize yields by around 5% in tropical regions, while erratic rainfall patterns influence sowing time and crop growth (Tran et al. 2025). Government interventions such as the National Food Security Mission and input subsidies have significantly improved maize productivity by 20–25% in targeted areas (MoA and FW 2023). Technological progress, notably the adoption of single-cross hybrids, has led to 30–50% higher yields compared to traditional open-pollinated varieties, while conservation agriculture and precision farming further enhance resource use efficiency and climate resilience (Table 1.3).

Table 1.3 Climatic, policy, and technological factors and their influence on production

Factors	Aspect/policy/ technology	Influence on production
Climatic factors	Rainfall variability	Maize is sensitive to rainfall at the silking and grain filling stages. Droughts, especially in kharif, reduce yields
	Temperature	Optimal range: 18–27 °C. High temperatures during flowering cause poor grain set. Heat stress affects rabi maize in north India
	Climate change	Erratic rainfall, shortened growing seasons, and increased pest pressure (e.g., FAW) are emerging threats
	Seasonal shift	Rabi maize is gaining importance due to better water and pest control, especially in Bihar and Eastern UP
Policy and Institutional Factors	National Food Security Mission	Financial support for seed, IPM kits, and farm machinery helped boost productivity
	Minimum Support Price (MSP)	MSP declared, but procurement is weak, except in select states (MP, Bihar). Limits assured market access
	Ethanol Blending Policy (20% by 2025)	Boosts demand for starch-grade maize, leading to acreage expansion near distilleries
	Pradhan Mantri Krishi Sinchayee Yojana (PMKSY)	Promotes drip irrigation, crucial for rabi maize in water-scarce areas
	Public–Private Partnerships (PPPs)	Widespread adoption of hybrid seeds and input packages. Private sector is key to hybrid seed penetration
Technological Factors	Single-cross hybrids (SCH)	Major yield breakthrough (20–30% over composites). Over 80% maize area now under hybrids
	Quality Protein Maize (QPM)	Addresses nutritional security, especially in tribal and mid-day meal schemes
	Mechanization	Zero till, permanent raised beds, and planters improving labour efficiency and precision
	Precision Nutrient Management	Tools like Leaf Color Chart (LCC), Nutrient Expert®, and Soil Health Cards boost NUE and yields
	ICT and Mobile Advisory	Farmers now access real-time advisories (e.g., FAW management) via apps and WhatsApp
	Conservation Agriculture (CA)	Promotes residue retention, minimal tillage, and crop rotation, enhancing soil health and productivity in maize–wheat systems
	Integrated Pest Management (IPM)	Crucial for combating Fall Armyworm (FAW), especially in kharif maize

1.4 Maize-Based Cropping Systems in India

Maize (*Zea mays* L) is highly adaptable to diverse soil and climatic conditions, making it suitable for cultivation either sequentially or in combination with various crops across multiple agroecological zones. Its compatibility with crops exhibiting different growth habits has led to the development of numerous intercropping techniques. For enhanced productivity in such systems, nutrient management should cater to the combined requirements of both crops. Additionally, when applying herbicides, it's crucial to ensure they are selective and safe for all crops involved. Maize's versatility in intercropping positions it as a key contributor to crop diversification in various agricultural contexts. Common intercropping partners for maize include legumes such as cowpea (*Vigna unguiculata*), green gram (*Vigna radiata*), and groundnut (*Arachis hypogaea*) (Table 1.4). Studies have demonstrated that intercropping maize with these legumes not only boosts productivity and profitability but also enhances forage quality and nutrient uptake. For instance, intercropping maize with cowpea in a 1:1 ratio, combined with integrated nutrient management practices, has shown significant improvements in forage yield and quality. Moreover, such intercropping systems contribute to better soil fertility and can reduce the need for nitrogen fertilizers in subsequent crops. In response to market-driven agriculture, high-value maize-based intercropping systems incorporating vegetables and flowers are gaining popularity, particularly in peri-urban areas. For example, the maize-potato-onion cropping system has been identified as a profitable alternative to the traditional rice-wheat system in the Indo-Gangetic Plains, offering higher system productivity and economic returns. Additionally, intensified maize-based systems like maize-wheat-mungbean and maize-mustard-mungbean have proven to be both productive and profitable. The rice-winter maize cropping system has emerged as a promising strategy in Asia, covering approximately 3.5 million hectares (Timsina et al. 2010). This system is rapidly expanding in southern and north-eastern India, driven by the increasing demand for maize from the poultry and fish industries. Studies have shown that maize-based cropping systems utilize resources more efficiently than traditional rice-wheat and rice-rice systems, leading to improved water productivity and soil health. Below are some of the major cropping systems involving maize:

1.4.1 Maize–Wheat (MW) System

The maize–wheat system is a dominant double-cropping model in many agro-climatic regions due to its adaptability, resilience to climatic fluctuations, and compatibility with conservation agriculture (CA). Practices like zero tillage and residue retention improve soil structure, increase soil organic carbon (SOC), and enhance water holding capacity. This system also supports mechanization and crop diversification. While it offers substantial environmental benefits, such as reduced greenhouse gas emissions and improved nutrient cycling, there may be increased nitrous oxide emissions under intensified nitrogen regimes due to high N inputs and

Table 1.4 Maize-based sequential cropping systems in different agro-climatic zones of India

Agro-climatic region	Irrigated cropping system	Rainfed cropping system
Western Himalayan	Maize–wheat Maize–potato–wheat Maize–mustard Maize–sugarcane	Maize–mustard Maize–legumes
Eastern Himalayan	Summer rice–maize–mustard Maize–maize Maize–maize–legumes	Sesame–rice + maize
Lower Gangetic Plain	Autumn rice–maize Jute–rice–maize	Maize–pulses
Middle Gangetic Plain	Maize–early potato–wheat–mungbean Rice–potato–maize Maize–wheat Maize–wheat–mungbean Maize–wheat–urdbean Maize–sugarcane–mungbean	Maize–wheat
Upper Gangetic Plain	Maize–wheat Maize–wheat–mungbean Rice–potato/pea–maize Maize–potato–wheat Maize–potato–sunflower Maize–potato–onion Maize–potato–sugarcane–ratoon Rice–potato–maize	Maize–wheat Maize–barley Maize–safflower
Trans Gangetic Plain	Maize–wheat Maize–wheat–mungbean Maize–potato–wheat Maize–potato–sunflower Maize–potato–onion Rice–potato/pea–maize Mungbean–maize–toria–wheat Maize–early potato–late potato–mungbean	–
Eastern Plateau and Hills	Maize–groundnut–vegetables Maize–wheat–vegetables	Rice–potato–maize Jute–maize–cowpea
Central Plateau and Hills	Maize–wheat	Maize–groundnut
Western Plateau and Hills	Sugarcane + Maize	
Southern Plateau and Hills	Maize–rice Rice–maize	Sorghum–maize Maize–sorghum–pulses Maize–potato–groundnut
East Coast Plain and Hills	Rice–maize–pearlmillet Maize–rice Rice–maize Rice–rice–maize	Maize–maize–pearlmillet Rice–maize + cowpea

(continued)

Table 1.4 (continued)

Agro-climatic region	Irrigated cropping system	Rainfed cropping system
West Coast Plain and Hills	Maize–pulses Rice–maize	Rice–maize Groundnut–maize
Gujarat Plains and Hills	Maize–wheat	Rice–maize
Western Dry Region	Maize–mustard Maize–chickpea	Maize + legumes
Island Region	Rice–maize	Maize–rice Rice–maize + cowpea Rice–maize–urdbean Rice–rice–maize

irrigation use (Jat et al. 2014; Gathala et al. 2011; Wang et al. 2020; Hussain et al. 2023). Compared to rice-wheat (RW), the MW system demonstrates higher system productivity (MEY: 12.6 t/ha), reduced irrigation water usage (84 cm vs. 112–205 cm), and improved WUE, leading to energy savings and lower greenhouse gas emissions (Kumar et al. 2013; Gill and Sharma 2005). Traditional maize–wheat systems yield ~14.1 t/ha of maize equivalent yield (MEY), underscoring their stability under input-limited conditions (Tables 1.4, 1.5, and 1.6).

1.4.2 Maize–Mustard System

This system involves sequential cropping with early-maturing mustard varieties that efficiently utilize the residual moisture and nutrients following maize harvest (Tables 1.5 and 1.6; Fig. 1.10). The contrasting root systems and residue characteristics of maize and mustard improve microbial activity and promote synchronized nutrient release. Incorporating mustard also enhances soil enzymatic activities, nitrogen-use efficiency, and helps reduce pest and weed pressure in cereal-dominated rotations (Pooniya et al. 2021; Sarkar et al. 2018; Yang et al. 2023). Maize equivalent yields exceeding 8 t/ha with a favourable B:C ratio (2.26) make it economically and ergonomically viable (Pooniya et al. 2021; Yang et al. 2023).

Advantages of the triple M cropping system include advancing mustard planting, increased mustard duration, reduced aphid infestation, enhanced mustard yield, timely planting of the third crop of longer-duration summer mungbean, increased mungbean yield, three crops with much less water than rice, and all these three crops can be grown under conservation agriculture/zero tillage.

1.4.3 Maize–Pulses System

Legume-based systems, particularly those involving pulses such as mungbean or chickpea, contribute significantly to nitrogen cycling through biological nitrogen fixation. Their inclusion improves soil fertility, enhances soil organic matter, and

Table 1.5 Key agronomic benefits, productivity, economics, soil health, and environmental impact of different maize-based cropping systems

Cropping system	Key agronomic benefits	Productivity (Yield)	Economics (B:C ratio/ net returns)	Soil health/ environmental impact	References
Maize–wheat	Climate-resilient, best suits in CA, supports intercropping	Maize: 5–7 t/ha Wheat: 4–5 t/ha	B:C ratio: 1.94 ↑ Net returns by 43.6%	↑ Soil organic carbon, Available Soil NPK, Water holding capacity (WHC) ↓ Greenhouse gases (15%) Saved energy by 91% ↑ N_2O (26%)	Jat et al. (2014), Gathala et al. (2011), Parihar et al. (2020), Wang et al. (2020), and Jain et al. (2015)
Maize–mustard	Short-duration mustard, synergistic residue use	MEY (maize eq. yield): 8.2 t/ha	Net returns: 1.35 lakh INR/ha B:C ratio: 2.26	↑ SOC (12%) ↑ Microbial biomass Improved residual N	Pooniya et al. (2021), Sarkar et al. (2018), and Yang et al. (2023)
Maize–pulses	Biological N fixation, resource sustainability	Land equivalent ratio: 1.27 ↑ Yield by 17–21%	Net returns: 1.95 lakh INR/ha/year B:C ratio: 2.12	↑ Nitrogen in soil (15–20%) ↑ Organic matter, P, Zn Improved soil structure	Prakash et al. (2024), Flavia et al. (2023), Tripathi et al. (2021), and Baishya et al. (2021)
Maize–vegetables	High-value crops, marketable diversity	MEY: Up to 24.1 t/ha	Net returns: 3.22 lakh INR/ha B:C: 6.04	↑ SOC, WUE Improved nutrient cycling LER > 1	Mauriya et al. (2024) and Dimande et al. (2024)
Maize–sugarcane	Staggered harvest, efficient space use	↑ Sugarcane eq. yield	Higher profitability over sole crops	↑ SOM Better input-use efficiency	Dwivedi et al. (2025), Anjaly and Ajith (2024)
Maize–potato	Double-cropping intensive model	↑ MGEY by 63.7% over sole maize	↑ Net returns by 69.08% vs. sole maize and 82% vs. rice–wheat	↑ N efficiency Saved 30 kg N/ha Improved tilth	Singh et al. (2015), Zhao et al. (2024), and Asaye et al. (2022)

promotes microbial biomass. The deep-rooted structure of pulses also improves soil porosity and structure. Systems like maize–wheat–greengram (MEY: 15.0 t/ha) and maize–mungbean–baby corn (MEY: 30.59 t/ha) illustrate the productivity gains achievable through strategic legume inclusion (Gill and Sharma 2005) (Tables 1.5

Table 1.6 Advantages of the major maize-based cropping system

S. No.	Cropping system	Advantages
1.	Maize–mustard–mungbean (MMM)	*Northern India*: Maize to be grown in areas receiving <1000 mm rainfall (rice to be replaced in such areas) Three major crops of cereal, oilseed, and pulse annually Reduced oil import Improved soil fertility Improved productivity of each crop with a long-duration variety Check lowering water table: three crops in much lesser water than one rice crop Save power and water by solving issues of residue burning In mustard, it gives an option for timely planting and reduced risk of aphid infestation
2.	Maize–potato–watermelon/ mungbean	*Northern India* Highly remunerative, productive system Suitable for small and marginal farmers Enhanced soil fertility
3.	Maize–potato–maize	*Northern India* High productivity and profitability potential Enhanced vegetable availability Increased cropping intensity and farm income
4.	Rice–maize	*South and eastern India* Zero-till planting in high soil moisture Highly productive and remunerative Saving of water over the rice–rice system

(continued)

Table 1.6 (continued)

S. No.	Cropping system	Advantages
5.	Maize–wheat–mungbean	*Northern and central India* Saving of water ~90% by maize cultivation over paddy Maize cultivation saves 70% power compared to paddy Increased wheat yield by 10–15% Maize takes less time than rice, which advances the wheat planting and thereby increases its productivity by ensuring increased crop duration and avoiding the risk of terminal heat stress. Third mungbean crop for enhancing soil fertility. Reduced infestation of *Phalaris minor* The stover of stay-green maize hybrids can also be directly used for livestock feeding.
6.	Silage/grain maize-vegetable based cropping system	*Peri-urban areas* Maize green fodder is of the best quality, and maize silage can enhance livestock productivity in the state and can be transported to neighbouring states, meeting green fodder demand. High productivity and profitability potential Enhanced vegetable availability

Fig. 1.10 Protein and edible oil security through maize–mustard–mungbean (triple MMM) cropping systems under conservation agriculture/zero tillage (Kadam et al. 2023)

and 1.6; Fig. 1.11). Moreover, WUE in maize-pea-wheat was recorded at 194 kg/ha-cm, indicating strong efficiency under water-limited conditions.

This system enhances resource-use efficiency (radiation, water, and nutrients), system productivity, and offers ecological advantages like pollinator support and pest suppression through increased biodiversity (Prakash et al. 2024; Flavia et al. 2023; Baishya et al. 2021). Growing mungbeans under maize-based cropping systems (maize–wheat/mustard) improved yield and net returns with lower environmental footprints and energy consumption in the Indo-Gangetic plains (IGP) and similar agro-ecologies (Kadam et al. 2023).

1.4.4 Maize–Vegetables System

This high-value system provides dietary diversity and income security while utilizing available resources intensively. Vegetables have rapid nutrient uptake and short growth cycles, which complements the nutrient release pattern following maize. The inclusion of shallow- and deep-rooted vegetable crops improves nutrient scavenging and water use efficiency. Systems such as maize–potato–onion (MEY; 22.35–27.6 t/ha) and maize–radish–onion (Wheat equivalent yield: 20.51 t/ha; Net returns: INR 75,909/ha) outperform cereal-only systems in profitability (Gill and Sharma 2005; Singh 2006) (Tables 1.5 and 1.6). The system also supports intensive land use and increases labour demand, offering rural employment opportunities (Mauriya et al. 2024; Dimande et al. 2024).

1.4.5 Maize–Sugarcane System

Sugarcane, being a long-duration and space-extensive crop, can be effectively intercropped or sequenced with maize to optimize resource utilization during its early growth stages (Tables 1.5 and 1.6). The system enhances input-use efficiency, particularly for water and nutrients, and utilizes land during otherwise underutilized periods. Crop diversification through this combination supports better pest and weed management and contributes to higher carbon sequestration due to biomass accumulation (Dwivedi et al. 2025; Anjaly and Ajith 2024).

Fig. 1.11 Maize–Mungbean–Baby corn Systems

1.4.6 Maize–Potato System

A highly intensive double-cropping model, maize–potato improves system productivity (MEY: 18.2 t/ha) through complementary growth patterns and efficient nutrient extraction and higher net returns than traditional cereal systems (Gill and Sharma 2005). Advanced variants such as maize–potato–onion reach even higher yields (MEY up to 27.6 t/ha). Potato's shallow rooting allows for minimal competition with maize residual roots, and its fast growth enhances soil aeration and structure (Tables 1.5 and 1.6). It is suitable for areas with adequate irrigation and infrastructure for timely planting and harvesting (Singh et al. 2015; Zhao et al. 2024; Asaye et al. 2022).

1.4.7 Intercropping with Maize

Integrating legumes like cowpea and soybean into maize-based systems is a proven strategy to boost soil fertility, productivity, and sustainability, especially under conservation agriculture and low-input conditions (Tables 1.7 and 1.8). Legumes enhance biological nitrogen fixation, improve soil structure, and raise organic carbon levels. For instance, intercropping maize with cowpea has been shown to significantly increase soil microbial biomass carbon and nitrogen compared to sole maize cropping, indicating enhanced soil fertility and microbial activity (Dare et al. 2014). Similarly, a maize–soybean rotation improved soil nitrogen and phosphorus by 12% and 15%, respectively, and raised maize yield by 1.9 t/ha over maize monoculture (Tables 1.7 and 1.8). Studies have reported that such rotations can enhance soil nitrogen and phosphorus levels, leading to increased maize yields compared to continuous maize monoculture (Zhang et al. 2024).

Comparative evaluation of maize-based intercropping systems reveals that intercropping significantly enhances system productivity, profitability, and resource-use efficiency over sole maize cultivation (Bhatnagar et al. 2012; Soujanya et al. 2024; Hussain et al. 2023; Table 1.7 Fig. 1.12). Among cereal-based systems, maize + wheat provides crop diversification but often shows lower economic returns due to resource competition and lower land-use efficiency (Tripathi et al. 2010). In contrast, pulse-based intercropping systems, particularly those involving urdbean, mungbean, cowpea, groundnut, soybean, and black gram, contribute positively to soil fertility through biological nitrogen fixation and consistently improve net

Table 1.7 The details of maize intercropping systems being practiced in different agro-ecological regions in India

Intercropping system	Agro-ecological regions
Maize + pigeon pea/cowpea/mungbean/urdbean/sugarcane/soybean; rice + maize	All maize-growing agro-ecological regions
Maize + high value vegetables; flowers; baby corn/sweet corn + vegetables	*Peri*-urban interface regions
Maize + turmeric/ginger/mungbean/French bean	Hilly regions

Table 1.8 Yield of maize and intercrop, gross margin, energy productivity, and net returns of different maize-based intercropping systems

I/C systems	Maize yield (t/ha)	Intercrop yield (t/ha)	Gross margin (INR/ha)	Energy productivity (g/MJ)	Net return (INR/ha)	References
Maize (sole)	3.48–6.99	–	55,952	1713	18,181–28,250	Bhatnagar et al. (2012), Ali et al. (2015), Soujanya et al. (2024), and Baishya et al. (2014)
Maize + wheat (1:2)	3.12	3.06	–	–	15,098	Tripathi et al. (2010)
Maize + linseed (1:2)	4.66	0.81	–	–	16,524	Tripathi et al. (2010)
Maize + urdbean	4.01–4.73	0.49–0.63	–	–	22,466–23,115	Bhatnagar et al. (2012)
Maize + mungbean	1.35–6.64	0.33–1.16	–	785–1569	18,715–47,120	Bhatnagar et al. (2012), Soujanya et al. (2024), Baishya et al. (2014), and Jat et al. (2025, "unpublished")
Maize + cowpea	4.13–8.58	0.60–0.77	–	–	24,023–32,568	Bhatnagar et al. (2012), Soujanya et al. (2024), and Jat et al. (2025, "unpublished")
Maize + groundnut	4.28–7.39	0.96–1.29	–	–	28,747–41,842	Bhatnagar et al. (2012), Soujanya et al. (2024), Jat et al. (2025, "unpublished")
Maize + soybean	4.18–7.81	0.74–0.97	–	–	21,475–28,492	Bhatnagar et al. (2012), Choudhary et al. (2014), and Jat et al. (2025, "unpublished")
Maize + potato	6.09–9.70	10.08–20.49	33,677	–	32,369	Ali et al. (2015) and Tripathi et al. (2010)
Maize + radish	9.92	8.12	51,702	–	–	Ali et al. (2015)
Maize + coriander	10.54	0.64	48,877	–	–	Ali et al. (2015)
Maize + Indian mustard (1:2)	3.38	1.25	–	–	14,693	Tripathi et al. (2010)
Maize + *Toria* (1:2)	2.76	0.76	–	–	5611	Tripathi et al. (2010)

(continued)

Table 1.8 (continued)

I/C systems	Maize yield (t/ha)	Intercrop yield (t/ha)	Gross margin (INR/ha)	Energy productivity (g/MJ)	Net return (INR/ha)	References
Maize + pea (1:2)	5.35	1.28	–	–	27,679	Tripathi et al. (2010)
Maize + bushbean	10.50	6.27	75,527	–	–	Ali et al. (2015)
Maize+ spinach	10.49	5.08	53,727	–	–	Ali et al. (2015)
Maize + red amaranthus	5.51–10.90	3.31	53,977	–	–	Ali et al. (2015), Soujanya et al. (2024), and Jat et al. (2025, "unpublished")
Maize + black gram	4.67–7.43	–	–	–	–	Soujanya et al. (2024) and Jat et al. (2025, "unpublished")
Maize + pigeon pea	6.71–8.07	–	–	–	–	Jat et al. (2025, "unpublished")
Maize + sesame	5.83–7.35	–	–	–	–	
Maize + lady finger	5.22–6.78	–	–	–	–	
Maize + rice bean	6.03–6.37	–	–	–	–	
Maize + sesbania	4.82–6.51	–	–	–	–	

Fig. 1.12 Cowpea, groundnut, and red amaranthus as an intercrop in the maize cropping system

returns and sustainability (Bhatnagar et al. 2012; Baishya et al. 2014; Soujanya et al. 2024; Jat et al. 2025, "unpublished"). Oilseed-based intercropping systems such as maize + linseed, mustard, and sesame offer moderate returns and crop diversification benefits, though they are generally less profitable compared to legumes and vegetables (Tripathi et al. 2010; Jat et al. 2025, "Unpublished""Own study"). Notably, vegetable-based intercropping systems like maize + bushbean, spinach, potato, coriander, radish, and red *amaranthus* are the most profitable and productive due to higher market demand, enhanced resource-use efficiency, and complementary growth habits (Ali et al. 2015; Soujanya et al. 2024; Jat et al. 2025, "unpublished").

These systems also contribute to nutritional security and short-duration income generation. Additionally, intercropping maize with pigeon pea, rice bean, or lady finger supports longer-duration systems suitable for specific agro-ecologies, while inclusion of green manure crops like *sesbania* improves soil organic matter and system sustainability (Jat et al., 2025, "unpublished"). Overall, intercropping maize with pulses, oilseeds, vegetables, or leafy greens significantly enhanced system productivity, profitability, and resource-use efficiency compared to sole maize cropping. The results underscore the importance of context-specific intercropping strategies to sustainably intensify maize-based systems across diverse agro-climatic zones.

1.4.8 Specialty Corn for Diversification

The cultivation of specialty corn such as baby corn, sweet corn, and popcorn in peri-urban regions of India is increasingly recognized as a strategic approach for agricultural diversification, value addition, and rural employment generation. These specialty crops align with the growing demand of the food processing industry and contribute to enhancing farm incomes, especially in areas close to urban markets. Farmers cultivating 2–3 crops of baby corn or sweet corn annually earn approximately 50,000–60,000 INR per acre, along with the added benefit of producing around 100 quintals of nutritious green fodder per crop per acre. This green biomass plays a crucial role in addressing India's significant fodder deficit, which is currently estimated at 11.24% for green fodder, 23.4% for dry fodder, and 28.9% for concentrate feed ingredients. Incorporation of this high-quality fodder into dairy animal diets has been reported to increase milk production by 15–20% without the need for additional land for fodder cultivation.

1.4.9 Maize Type Selection

Maize is a highly versatile crop cultivated in various forms to meet the diverse demands of food, feed, fodder, and industrial applications. The major types of maize grown in India include normal grain maize, sweet corn, baby corn, popcorn, Quality Protein Maize (QPM), waxy corn, high oil corn, and fodder maize. These speciality types support nutritional security, livestock productivity, food processing, and income diversification across different agro-ecologies:

(i) *Normal grain maize*: This includes both yellow and white maize, which is cultivated either for grain or green cob purposes. For grain production, maize is typically grown in hinterland areas where market proximity is less critical. Conversely, cultivation for green cob marketing should be located near urban or peri-urban markets to maximize profitability through reduced transportation costs and faster delivery to consumers.

(ii) *QPM*: QPM is nutritionally enriched maize with enhanced levels of lysine and tryptophan are essential amino acids deficient in normal maize. The biological value of protein in QPM is approximately 80%, nearly double that of conventional maize (40%) and close to that of milk (90%). More than nine QPM hybrids with diverse grain colours have been developed and released in India for different agro-climatic zones. While agronomic practices remain similar to normal maize, a spatial isolation of 150–200 m must be maintained to preserve genetic purity. QPM is particularly suitable for maize-consuming tribal populations in Madhya Pradesh, Rajasthan, Chhattisgarh, Gujarat, Maharashtra, Jharkhand, Himachal Pradesh, Jammu and Kashmir, Bihar, and particularly NEH region states. Additionally, due to its superior protein quality, it reduces the protein supplementation cost in poultry and piggery industries.

(iii) *Baby corn*: Baby corn is harvested as small, tender, unfertilized cobs within 1–3 days of silk emergence. It is an ideal crop for employment generation, engaging family labour in activities such as harvesting, peeling, packaging, and marketing. Agronomic practices are similar to grain maize but require higher plant population density, greater nitrogen input, use of early maturing single-cross (SC) hybrids, and timely harvesting. Its suitability for year-round cultivation under irrigated conditions allows for 3–4 crops annually, especially in peri-urban areas. The remaining plant biomass (green tassel, cob sheath, stalk) serves as nutritious fodder, especially useful during lean periods, thus also benefiting the livestock sector and boosting farm income from the same land.

(iv) *Sweet corn*: Sweet corn is highly valued for its taste and nutritional composition, being rich in energy, vitamin C, and vitamin A. Its popularity in urban diets makes it a profitable crop for peri-urban farmers. Sweet corn matures early, typically in 70–75 days during the *kharif* season, and cobs are harvested 18–20 days after pollination. Besides the economic return from fresh cobs, the green biomass also contributes to fodder availability.

(v) *Pop corn*: Popcorn is a flint maize type with a hard endosperm widely consumed as a snack. It is especially suited to hilly and remote regions under the "Popcorn Village" concept, which can promote local entrepreneurship and value addition in far-flung areas.

(vi) *Fodder maize*: Maize is an excellent fodder crop as it can be harvested at various growth stages without the risk of anti-nutritional factors commonly present in other fodder crops. However, fodder quality declines after anthesis due to nutrient remobilization. Cultivars selected for fodder should be tall, leafy, long-duration, and exhibit stay-green traits. Fodder maize can be cultivated year-round in most parts of the country, though December–January sowing is discouraged in North India due to low temperatures.

1.5 Challenges in Maize Production

1.5.1 Biotic Constraints

Maize production in India and globally is significantly constrained by biotic stresses, primarily caused by insect pests and fungal diseases, which reduce both yield and grain quality. Among the most destructive insect pests, the Fall Armyworm (*Spodoptera frugiperda*) has emerged as a serious invasive threat. It feeds voraciously on leaves, stems, and ears, leading to up to 60% yield losses in severe infestation, and is notoriously difficult to manage due to its rapid reproduction and pesticide resistance (FAO 2021). Another widespread pest, the Stem Borer (*Chilo partellus*), damages the plant by boring into the stalk, weakening the plant and making it prone to lodging, thereby compromising photosynthesis and nutrient flow, which further reduces productivity. In addition to insect pests, fungal diseases also pose major challenges. Common rust (*Puccinia sorghi*) manifests as orange pustules on leaves, impairing photosynthetic efficiency and reducing grain fill, especially under humid agro-climatic conditions. Downy Mildew (*Peronosclerospora sorghi*), causes leaf chlorosis, plant stunting, and death in young plants, severely impacting stand establishment and grain quality. This disease is particularly problematic in moist and poorly drained soils, making it prevalent during the *kharif* season in high-rainfall zones. The management of these biotic constraints often requires integrated pest and disease management (IPDM) strategies that combine resistant cultivars, biological control agents, seed treatment, and agro-ecological practices.

1.5.2 Abiotic Stresses

Abiotic stresses are among the most critical limitations to maize productivity, particularly in the context of climate variability and soil degradation. One of the most severe stresses is drought, which affects maize during key growth stages such as pollination and grain filling, leading to kernel abortion, poor seed set, and significant yield reductions. Maize is especially sensitive during the reproductive phase, where even short periods of water stress can drastically impact grain yield (IIMR 2023). Similarly, heat stress, particularly when temperatures exceed 35 °C, impairs pollen viability and fertilization, which in turn reduces grain number and size. Heat also negatively influences photosynthesis and enzyme activity, weakening the plant's metabolic balance and slowing development. These stresses are becoming increasingly frequent and intense due to global warming. Another major abiotic constraint is the decline in soil fertility, resulting from continuous cultivation without adequate nutrient replenishment, poor organic matter content, and nutrient imbalances. Maize, being a nutrient-exhaustive crop, demands high levels of nitrogen, phosphorus, and potassium, and deficiencies in these nutrients lead to chlorosis, stunted growth, and poor root development, ultimately reducing biomass and grain output. To mitigate the impacts of these stresses, integrated approaches are required, including the use of climate-resilient maize hybrids, site-specific nutrient management, micro-irrigation, and conservation agriculture practices.

1.5.3 Economic and Policy Issues

One major issue is price fluctuation in economic and policy issues. For example, the modal price of maize in Davangere (key market in Karnataka) dropped below the MSP to Rs. 1850 per quintal in October 2023 but rose to Rs. 2309 per quintal by January 2024, a rise of over 20% in 3 months (The Economic Times 2024). This was largely due to a 25–27% decline in arrivals triggered by erratic monsoons affecting *kharif* season yields. Simultaneously, rising demand from poultry, starch, and ethanol industries, bolstered by the government's ethanol blending program, intensified price pressure (ChiniMandi 2025). On the export front, India's maize exports fell by 58% in 2023–24 due to high domestic prices and low yields, causing importers like Bangladesh and Vietnam to switch to more affordable sources (The Economic Times 2024). Such market volatility creates economic uncertainty for farmers, complicating production planning. Another key constraint is the lack of mechanization in smallholder-dominated regions. Manual farming practices persist due to the unaffordability of machinery-tractors alone, which cost ₹3–6 lakh, not including maintenance, and limited access to custom hiring centres (CHCs), with less than 20% of villages having operational CHCs (MoA and FW 2023). Additionally, poor awareness hampers mechanization; over 60% of maize farmers in Bihar and Jharkhand have never received formal training on farm machinery. These gaps hinder timely operations like sowing and harvesting, directly affecting productivity and profitability. Further economic constraints include escalating input costs and poor access to credit. Prices of hybrid seeds, fertilizers, and pesticides have surged, with fertilizers like urea and DAP increasing by 20–30% over 3 years, while hybrid maize seeds cost Rs. 300–400/kg. Coupled with fluctuating prices, these costs make maize farming less profitable. Moreover, nearly 55% of small and marginal farmers rely on informal loans due to complicated banking systems and collateral requirements, exposing them to high interest rates and financial stress.

1.6 Future Prospects and Research Priorities

1.6.1 Adoption of Climate-Smart Agriculture (CSA) Practices

The adoption of CSA offers transformative potential in building resilient and sustainable agricultural systems. According to the FAO (2019), CSA can increase productivity by 15–25%, enhance resilience by 20–30%, and reduce emissions by up to 30% when appropriately implemented. Practices such as drought-tolerant maize hybrids, zero tillage, and precision irrigation have already shown success in countries like India and Kenya. For instance, adoption of drought-tolerant maize in sub-Saharan Africa has resulted in yield increases of up to 30% under water-limited conditions. CSA improves resource efficiency through conservation agriculture, integrated nutrient management, and agroforestry, all of which contribute to long-term sustainability. Globally, agriculture contributes about 21% of greenhouse gas emissions, and CSA-based interventions like agroforestry and reduced tillage could sequester up to 1.2–1.5 Gt CO_2 equivalent annually. Widespread adoption depends on enabling policies, institutional support, and investment in climate-resilient technologies. India's National Mission on Sustainable Agriculture under the National Action Plan on Climate Change is one such step promoting CSA through funding support and awareness campaigns. To scale CSA effectively, research must target innovations in crop genetics, soil health, water use, and socio-economic adoption. Developing climate-resilient maize varieties through CRISPR and genomic selection is essential, as maize yields decline by 4–5% for every 1 °C increase above optimum temperature during flowering (Tran et al. 2025). Soil degradation affects about 30% of global arable land, requiring research into bio-fertilizers, microbial consortia, and organic amendments to restore fertility. In India, precision irrigation techniques like drip and sprinkler systems have shown water savings of 30–50% and yield increases of 20–30% in maize and vegetables (Behuria et al. 2024) Agroforestry research highlights that systems integrating nitrogen-fixing trees enhance soil fertility and productivity, whereas mechanized farms, in contrast, can achieve 25–30% higher productivity (FAO 2022), and also enhance soil carbon stocks by 25–40% over a decade. Social science research is equally crucial, as only 28% of smallholders in South Asia are aware of CSA practices, and among them, less than 15% have access to climate information services. Research into gender-responsive approaches, mobile-based advisory tools, and farmer behaviour will support wider adoption. Future research must also evaluate the profitability, long-term benefits, and scalability of CSA innovations under diverse agro-ecological zones.

1.6.2 Expansion of Maize in Non-traditional Areas

The expansion of maize cultivation into non-traditional areas is a promising strategy to meet the increasing demand for maize, enhance food security, and boost rural economies. These non-traditional areas often include regions with less favourable climates, degraded soils, or those previously dominated by other crops. With proper interventions, such expansion can provide a sustainable solution to regional disparities in agricultural productivity. According to the IIMR, maize cultivation is steadily expanding in states like Jammu & Kashmir, Himachal Pradesh, Tamil Nadu, and parts of the North-East, where traditional crops are being replaced or supplemented by maize due to its relatively short duration and multiple uses. For instance, Assam witnessed a nearly 60% increase in maize area between 2015 and 2022, largely due to improved varieties and increased demand for poultry feed and ethanol production (IIMR 2023). Maize is a staple crop for over 900 million people globally and is used in more than 4000 food and industrial products. In India, it contributes about 9% to the total food grain production, and its demand is projected to grow by 8–9% annually due to increased consumption in the poultry, starch, and ethanol sectors (ChiniMandi 2025a). The key benefits and drivers of the expansion of maize in non-traditional areas:

- *Improved Food Security and Economic Growth*: Expanding maize in new areas strengthens local food systems and reduces dependency on imports. Rural areas benefit from job creation and enhanced income through value chains involving storage, processing, and transport.
- *Climate Adaptation*: Developing and adopting drought-tolerant and early-maturing maize hybrids allows cultivation in semi-arid and rainfed zones.
- *Market Diversification and Processing Opportunities*: New maize-producing belts enable decentralized procurement and processing hubs, boosting local agro-industries. With ethanol blending targets rising to 30% by 2030, demand for maize-based ethanol is expected to increase, particularly in states where sugarcane is not viable.
- *Technological Advancements*: Mechanization and climate-smart practices such as zero-tillage, precision planting, and fertigation enhance productivity even in marginal soils. For instance, the introduction of happy seeders and single-row planters in North-Eastern hill regions has improved maize yields by up to 25% (IIMR 2023).
- *Research Priorities for Maize Expansion in Non-Traditional Areas*

  Breeding programs must develop maize varieties tolerant to heat, drought, salinity, and poor soils using CRISPR and genomic tools; Site-specific nutrient management (SSNM), along with biofertilizers and mycorrhizal fungi, should be promoted to address nutrient imbalances in soils; Drip irrigation, alternate wetting and drying (AWD), and climate-based irrigation using soil moisture sensors can reduce water use by 30–40%; Lightweight, small-scale machinery suited to hilly terrains and AI-based tools for disease detection and nutrient application are needed for wider adoption; Cost–benefit analysis, market price trends, contract farming, and digital platforms like FPOs can help strengthen market linkages and economic viability; customized farmer training modules in regional languages and digital extension tools such as mobile apps can increase technology adoption; long-term monitoring is essential to assess the impacts of maize expansion on soil health, pollinator populations, and land-use changes; policies should consider gender, youth, and land access, with emphasis on subsidy frameworks, crop insurance, and schemes like PM-KISAN and PMFBY.

1.6.3 Development of Drought-Tolerant, Heat-Tolerant, and Biofortified Maize Varieties

With the increasing challenges posed by climate change, such as erratic rainfall, high temperatures, and deteriorating soil fertility, the development of climate-resilient and nutritionally enriched maize varieties has become imperative are given in Table 1.9.

1.6.4 Increasing Mechanized Maize Cultivation to Improve Efficiency

Mechanization plays a pivotal role in enhancing maize production efficiency, reducing labour dependency, and ensuring higher yields with minimal post-harvest losses. With the increasing demand for maize in food, fodder, and biofuel industries, adopting mechanized farming techniques is essential to sustain production growth and improve farm profitability. Mechanization enables timely sowing, precision

Table 1.9 Details of different drought- and heat-tolerant, and biofortified varieties in India

Variety name	Details of the varieties
DHM 117, DHM 121, Pusa Vivek Hybrid 27	Early vigour, deeper root architecture, stay-green phenotype, efficient stomatal regulation, and osmotic adjustment. Yield advantage of 15–30% over non-tolerant varieties under drought stress conditions
Shaktiman-4, HQPM-1, HQPM-5	Higher in protein (9.0–9.6%), provitamin-A, essential amino acids like tryptophan, lysine (2–3%), etc., and good yield
Pusa Vivek QPM9 Improved	An early maturing hybrid with high provitamin A (8.15 ppm), lysine (2.67%), and tryptophan (0.74%) content. Released in 2017 for various states, including Jammu and Kashmir, Himachal Pradesh
Pusa HM4 Improved	A medium-maturing hybrid rich in lysine (3.62%) and tryptophan (0.91%), released in 2017 for regions like Punjab, Haryana, and Delhi
Pusa HM8 Improved	This hybrid contains high levels of lysine (4.18%) and tryptophan (1.06%), suitable for cultivation in Maharashtra, Karnataka, and other states
Pusa HM9 Improved	Released for Bihar, Jharkhand, and neighbouring states, it possesses lysine (2.97%) and tryptophan (0.68%) with an average grain yield of 52.0 q/ha

planting, and uniform seed placement, leading to improved plant stand and higher yields. It also reduces reliance on manual labour, which is particularly important in regions facing labour shortages, while lowering production costs. Efficient land preparation, mechanized weed control, and nutrient management further contribute to optimal crop growth and productivity. For example, studies in Telangana have shown that mechanized maize cultivation significantly outperformed conventional methods in terms of yield (Sunitha et al. 2023). Precision agriculture tools such as GPS-guided tractors, drones, and automated planters enhance the efficiency of input use, including seed, fertilizers, and pesticides.

The integration of IoT-based sensors and AI-driven decision support systems helps optimize irrigation scheduling and nutrient application, improving overall resource-use efficiency. Mechanized soil and water management techniques, including conservation tillage, laser land levelling, and precision irrigation systems like drip and sprinklers, help retain soil moisture, reduce erosion, and ensure uniform water distribution across fields. Looking ahead, research priorities include developing automated precision planters, herbicide-tolerant hybrids suited for mechanized weed management, drone-based input applications, smart fertigation systems, and low-cost, multi-functional machinery for small farms (Fig. 1.13). Efforts must also focus on energy-efficient and climate-smart equipment, biodegradable inputs, and sustainable mechanization practices that minimize environmental impact while maintaining soil health. Mechanized harvesting using combine harvesters reduces field losses, improves grain quality, and enhances labour productivity, while mechanized threshing, drying, and storage infrastructure help minimize post-harvest losses and preserve grain quality (Fig. 1.14). Economically, mechanization reduces per-unit production costs and makes maize cultivation more profitable and scalable. It is particularly beneficial for smallholder farmers through custom hiring centres and cooperative models, as seen in Odisha, where mechanization has improved the economic viability of maize farming.

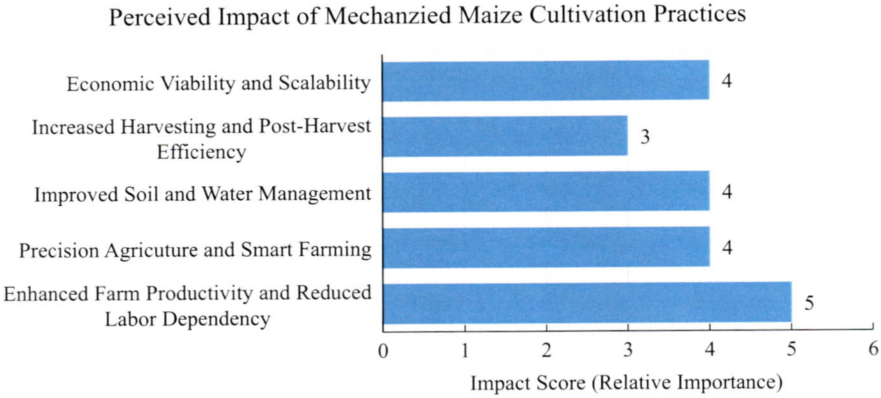

Fig. 1.13 Perceived impact of mechanized cultivation practices

Fig. 1.14 Mechanized maize cultivation to improve efficiency

1.6.5 Enhancing Ethanol Production from Maize to Support India's Biofuel Policy

India's National Biofuel Policy (2018) targets 20% ethanol blending in petrol (E20) by 2025 to enhance energy security and reduce dependence on fossil fuels, which was achieved in 2025. While sugarcane molasses remains the primary ethanol feedstock, its high-water requirement and regional limitations necessitate diversification. Maize has emerged as a viable alternative due to its high starch content, shorter crop duration, and adaptability across diverse agro-climatic zones. Future prospects for maize-based ethanol include the development of high-starch, drought-tolerant hybrids, expansion of cultivation into non-traditional and rainfed areas, and integration with decentralized bio-refineries to ensure year-round availability and processing efficiency. According to projected trends (2023–2036), maize production in India is expected to rise steadily from approximately 40 to 75 million metric tonnes, supported by a compound annual growth rate (CAGR) aligned with increasing demand from feed and fuel sectors (Fig. 1.15). Notably, while poultry sector demand is projected to surpass 50 million tonnes by 2035–36, demand from other traditional sectors is declining, creating strategic room for ethanol feedstock allocation without compromising food or feed security. This transition underlines the importance of efficient maize utilization, especially in the ethanol sector. Economic and environmental benefits include reduced greenhouse gas emissions, improved water-use efficiency, and enhanced farmer income through assured markets and value-added byproducts such as distillers' dried grains with soluble. Research priorities should focus on genetic improvement for starch quality, optimization of enzymatic hydrolysis and fermentation processes, life cycle assessment of maize-to-ethanol systems, and techno-economic evaluations comparing maize with other feedstocks. Policy interventions, including MSP assurance, production-linked incentives, and investment in ethanol infrastructure, are critical to mainstreaming maize as a sustainable and scalable feedstock for biofuel production in India.

ICAR-IIMR is disseminating best management practices and improved maize varieties across 15 catchment areas in 78 districts spanning 15 states. This initiative aims to sustainably boost maize production, ensuring a sufficient feedstock supply for India's growing ethanol demand.

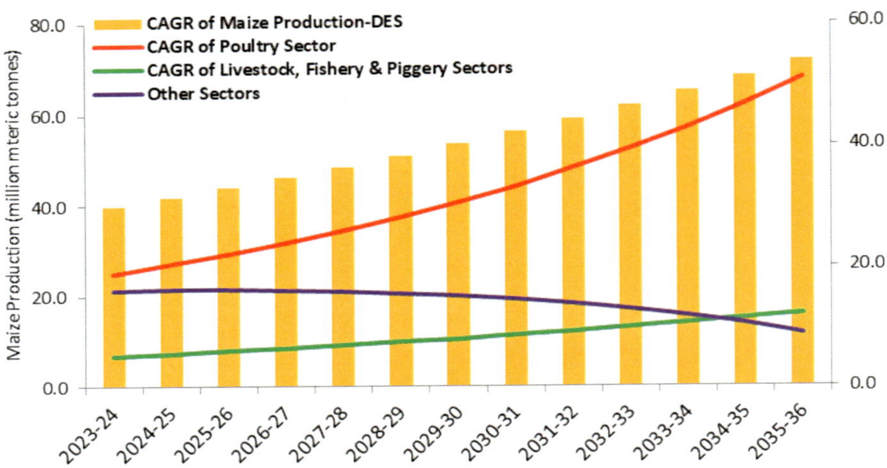

Fig. 1.15 Scenarios of production of maize, share of poultry, livestock, and other sectors for maize. (Source: NAAS 2024)

1.7 Conclusion

Maize has evolved from a relatively minor crop into a cornerstone of India's agricultural and industrial economy. Its journey, beginning with introductions through trade routes in the sixteenth century and later accelerated by scientific interventions, highlights the dynamic interplay between tradition, innovation, and policy. Over the decades, maize has demonstrated remarkable adaptability across diverse agro-climatic zones, thriving in kharif, rabi, and spring seasons, and steadily expanding its footprint from subsistence farming to large-scale commercial cultivation. The transformation of maize production in India has been driven by technological advancements, particularly the development of hybrids, precision farming practices, and climate-smart approaches. Institutions such as ICAR-IIMR, ICAR-IARI, ICAR-VPKAS, and AICRP Maize Centres have played pivotal roles in strengthening research, disseminating improved varieties, and promoting sustainable cropping systems. These efforts have not only enhanced productivity but also positioned maize as a vital contributor to food, feed, fodder, and fuel security. The crop's integration into rice–maize, maize–wheat, and maize–pulse systems underscores its role in diversification, soil health improvement, and resilience against climate variability.

Economically, maize has become indispensable to the poultry, starch, and ethanol industries, with more than 85% of production now directed toward industrial uses. Its emerging role in biofuel production aligns with India's energy transition goals, offering both environmental benefits and new market opportunities for farmers. Yet, challenges remain. Productivity levels, though rising, still lag behind global leaders, and issues such as water scarcity, climate change, and soil degradation demand innovative solutions and efficient resource management. Looking ahead, maize's importance will only deepen as India strives to balance food security with industrial demand and sustainability. Harnessing its genetic diversity, expanding hybrid adoption, and integrating modern technologies will be crucial. Ultimately, maize stands not just as a crop of historical significance, but as a strategic pillar for India's agricultural future and bio-economy.

References

Ali MR, Rahman MS, Asaduzzaman M, Hossain MM, Mannan MA (2015) Intercropping maize with different vegetables. Bangladesh Agron J 18(1):49–52

Anjaly V, Ajith A (2024) The sweet gains of diversification: a review of intercropping benefits in sugarcane farming. Agric Rev 46(4):585–593

Asaye Z, Kim D-G, Yimer F, Prost K, Obsa O, Tadesse M, Gebrehiwot M, Brüggemann N (2022) Effects of combined application of compost and mineral fertilizer on soil carbon and nutrient content, yield, and agronomic nitrogen use efficiency in maize–potato cropping systems in Southern Ethiopia. Land 11(6):784. https://doi.org/10.3390/land11060784

Baishya LK, Ansari MA, Walling I, Sarma PK, Prakash N (2014) Productivity, profitability and energy budgeting of maize (*Zea mays*)/greengram (*Vigna radiata*) intercropping system under rainfed conditions of Eastern Himalayan Region. Indian J Agric Sci 84(9):1073–1077

Baishya LK, Temjenna J, Walling N, Rajkhowa DJ (2021) Evaluation of maize (*Zea mays* L.) + legume intercropping system for productivity, profitability, energy budgeting and soil health in hill terraces of Eastern Himalayan region. Legum Res 44(11):1343–1347. https://doi.org/10.18805/LR-4190

Behuria PR, Ojha S, Nayak L, Kar JK, Gochhayat D, Jaiswal N, Rajvir A (2024) The transformative potential of precision farming in India: a comprehensive review. Int J Res Agron 7(10):732–741. https://doi.org/10.33545/2618060X.2024.v7.i10j.1887

Bhatnagar A, Pal MS, Singh V (2012) Productivity and profitability of maize-based intercropping systems. Madras Agric J 99(7–9):530–532. https://doi.org/10.29321/MAJ.10.100132

Choudhary VK, Dixit A, Kumar PS (2014) Productivity, competition behaviour and weed dynamics of various row proportions of maize (*Zea mays*)–legumes intercropping in Arunachal Pradesh. Indian J Agric Sci 84(11):1329–1334

Choudhary M, Singh A, Gupta M, Rakshit S (2020) Enabling technologies for utilization of maize as a bioenergy feedstock. Biofuels Bioprod Biorefin 14(2):402–416

ChiniMandi (2025). Rising ethanol demand drives up maize prices, impacting poultry feed costs: Poultry India President Uday Singh Bias. Available at https://www.chinimandi.com/rising-ethanoldemand-drives-maize-prices-impacting-poultry-feed-costs-poultry-india-president-uday-singh-bias/

ChiniMandi (2025a). Maize production must grow at 8-9 per cent annually to support India's ethanol blending target of E30. Available at https://www.chinimandi.com/maize-production-must-grow-at-8-9-per-cent-annually-to-support-indias-ethanol-blending-target-of-e30/?utm

Dare MO, Soremekun JA, Inana FO, Adenuga OS, Ajiboye GA (2014) Microbial biomass carbon and nitrogen under different maize cropping systems. In: Hartemink A, McSweeney K (eds) Soil carbon. Springer, Cham, pp 349–357. https://doi.org/10.1007/978-3-319-04084-4_32

Dimande P, Arrobas M, Rodrigues MÂ (2024) Intercropped maize and cowpea increased the land equivalent ratio and enhanced crop access to more nitrogen and phosphorus compared to sole crops. Sustainability 16(4):1440

Dwivedi AP, Shukla SK, Jaiswal VP, Singh VP, Singh AK, Tripathi MK, Yadav SK, Singh SR, Singh KK, Srivastava TK, Kumar R (2025) Soil quality parameters, crop yields and economics as influenced by intercropping with autumn sugarcane in subtropical India. Sugar Tech:1–13. https://doi.org/10.1007/s12355-025-01560-y

The Economic Times (2024). Cabinet approves MSP hike on Rabi crops, providing pre-Diwali relief to farmers. Available at https://economictimes.indiatimes.com/news/economy/agriculture/cab-inet-approves-msp-hike-on-rabi-crops-providing-pre-diwali-relief-to-farmers/articleshow/114279-546.cms?from=mdr
FAO (2019) The State of Food and Agriculture. Food and Agriculture Organization of the United Nations, Rome. Available at: https://www.fao.org/family-farming/detail/en/c/1245425/ (Accessed: date)
FAO (2021) Fall Armyworm: measuring damage and loss caused by a novel invasive pest as a guide to sustainable management. FAO statistics working paper series, issue 23–28. FAO, Rome
FAO (2022) The State of Food and Agriculture 2022: agricultural mechanization and sustainable agrifood system transformation. FAO, Rome
FAOSTAT (2025) Maize production statistics. Food and Agriculture Organization of the United Nations, Rome
FICCI (2024) The Indian maize sector: trends, challenges & imperatives for sustainable growth. FICCI/YES Bank Ltd, New Delhi/Mumbai
Flavia FTA, Rajendran K, Sharmili K, Kumar PD (2023) Evaluation of maize and pulses intercropping system under rainfed condition in Western Zone of Tamil Nadu, India. Int J Environ Climate Chang 13(8):678–683. https://doi.org/10.9734/IJECC/2023/v13i81996
Gathala MK, Ladha JK, Kumar V, Saharawat YS, Kumar V, Sharma PK, Sharma S, Pathak H (2011) Tillage and crop establishment affects sustainability of South Asian rice–wheat system. Soil Tillage Res Manage 103(4):961–971
Gill MS, Sharma GC (2005) Cropping systems diversification opportunities and conservation agriculture. J Farm Syst Res Dev 11(2):127–134
Hussain S, Naseer MA, Guo R, Han F, Ali B, Chen X, Ren X, Alamri S (2023) Nitrogen application enhances yield, yield-attributes, and physiological characteristics of dryland wheat/maize under strip intercropping. Front Plant Sci 14:1150225
IIMR (2023) Annual report 2023. Indian Institute of Maize Research, New Delhi
IIMR (2024) Annual report 2024. Indian Institute of Maize Research, New Delhi
Jat RK, Sapkota TB, Singh RG, Jat ML, Kumar M, Gupta RK (2014) Seven years of conservation agriculture in a rice–wheat rotation of Eastern Gangetic Plains of South Asia: yield trends and economic profitability. Field Crop Res 164(1):199–210
Jat SL, et al. (2025) Maize-based intercropping systems: productivity and yield analysis. Unpublished data, ICAR-Indian Institute of Maize Research, New Delhi, India.
Jain NK, Singh H, Dashora LN, Mundra SL (2015) Maize–wheat cropping system: intensification through introduction of pulses. Indian J Agron 60(3):347–351
Kadam PV, Jat SL, Mahala DM, Parihar CM, Singh AK, Kumar B, Gambhir G, Radheshyam, Chandra MS (2023) Residue retention and nitrogen management of preceding crops influences the productivity, profitability and environmental footprints of summer mungbean in maize-based cropping systems under conservation agriculture. Arch Agron Soil Sci 69(11):1990–2005
Kumar A, Jat SL, Kumar R, Yadav OP (2013) Maize production systems for improving resource-use efficiency and livelihood security. Directorate of Maize Research, New Delhi
Mauriya AK, Hashim M, Verma RB, Kumar P, Kumari M (2024) Evaluation of Rabi Maize-based intercropping system for augmenting the productivity and profitability of Rabi Maize under irrigated conditions of Bihar. Indian J Agric Res 1:5
Ministry of Agriculture and Farmers Welfare. (2023). Annual Report 2022-23. Government of India, Department of Agriculture and Farmers Welfare. agriwelfare.gov.in
Ministry of Agriculture & Farmers Welfare (2025) All India: crop-wise area, production & yield. Available at: https://upag.gov.in/dash-reports/allindiaapy. Accessed 7 Oct 2025
NAAS (2024) Maize for bioethanol production in India: prospects and strategy. Strategy paper no. 19. National Academy of Agricultural Sciences, New Delhi, p 20
Pooniya V, Biswakarma N, Parihar CM, Swarnalakshmi K, Lama A, Zhiipao RR, Nath A, Pal M, Jat SL, Satyanarayana T, Majumdar K, Jat RD, Shivay YS, Kumar D, Ghasal PC, Singh K (2021) Six years of conservation agriculture and nutrient management in maize–mustard

rotation: impact on soil properties, system productivity and profitability. Field Crops Res 260:108002

Prakash V, Ashraf AM, Sivagamy K, Begam SN (2024) Influence of pulses intercropping on growth and productivity of maize in Delta region of Tamil Nadu. Agric Sci Dig https://doi.org/10.18805/ag.D-6040

Parihar CM, Jat HS, Jat SL, Kakraliya SK, Nayak HS (2020) Precision nutrient management for higher nutrient use efficiency and farm profitability in irrigated cereal-based cropping systems. Indian J Fert 16(10):1000–1014

Soujanya PL, Vanisree K, Giri GS, Mahadik S, Jat SL, Sekhar JC, Jat HS (2024) Intercropping in maize reduces fall armyworm *Spodoptera frugiperda* (JE Smith) infestation, supports natural enemies, and enhances yield. Agric Ecosyst Environ 373:109130

Sunitha G, Sunandini GP, Lavanya T, Reddy DS, Meena A (2023) Comparative economic analysis of adopters and non-adopters of seed-to-seed mechanization in maize. Int J Plant Soil Sci 35(19):1722–1727

Singh B (1977) Races of maize in India. ICAR, New Delhi

Singh P (2006) Alternate cropping systems in peri-urban areas. PDCSR Annual Report 2005–06

Singh SK, Lal SS, Dua VK, Singh RK (2015) Nitrogen management in maize + potato intercropping system under Eastern Indo-Gangetic Plains of Bihar. J AgriSearch 2(4):244–250

Sarkar D, Baishya LK, Meitei CB, Naorem GC, Thokchom RC, Singh J, Bhuvaneswari S, Batabyal K, Das R, Padhan D, Prakash N, Rahman FH (2018) Sustainability of maize–mustard cropping system through balanced nutrient management. Field Crops Res 225:9–21

Timsina J, Jat ML, Majumdar K (2010) Rice–maize systems of South Asia: current status, future prospects and research priorities for nutrient management. Plant Soil 335:65–82

Tripathi SC, Venkatesh K, Meena RP, Chander S, Singh GP (2021) Sustainable intensification of maize and wheat cropping system through pulse intercropping. Sci Rep 11:18805

Tripathi AK, Kumar A, Nath S (2010) Production potential and monetary advantage of winter maize-based intercropping systems. Indian J Agric Sci 80(2):125–128

Tran BL, Tseng WC, Chen CC (2025) Climate change impacts on crop yields across temperature rise thresholds. Sci Rep 15:23424

Watt G (1893) Maize or Indian corn. Dictionary of the Economic Products of India

Wang X, He C, Liu B, Zhao X, Liu Y, Wang Q, Zhang H (2020) Effects of residue returning on soil organic carbon storage and sequestration rate in China's croplands: a meta-analysis. Agronomy 10(5):691

Yadav OP, Hossain F, Karjagi CG, Kumar BVKV, Zaidi PH, Jat SL, Chawla JS, Kaul J, Hooda KS, Kumar P, Yadava P (2015) Genetic improvement of maize in India: retrospect and prospects. Agric Res 4(4):325–338

Yang J, Jiang R, Zhang H, He W, Yang J, He P (2023) Modelling maize yield, soil nitrogen balance and organic carbon changes under longterm fertilization. J Environ Manag 325:116454

Zhang L, Yang Y, Zhao Z, Feng Y, Bate B, Wang H, Li Q, Cui J (2024) Maize–Soybean rotation and intercropping increase maize yield by influencing the structure and function of rhizosphere soil fungal communities. Microorganisms 12(8):1620

Zhao J, Sun X, Xue Y, Sher A, Ran J, Liu P, Zhao B, Ren B, Yu N, Ren H, Zhang J (2024) Optimizing nitrogen management for grain yield and nitrogen use efficiency in summer maize via coordinating the N supply–demand balance. J Integr Agric. https://doi.org/10.1016/j.jia.2024.12.028

Maize Biology and Reproductive Mechanisms

2

K. R. Yathish, C. Susmita, Thalari Vasanthrao, Chikkappa Gangadhar Karjagi, P. Lakshmi Soujanya, Abhijit Kumar Das, and Santosh Kumar

Abstract

Maize is a major cereal crop globally and serves as an exemplary model for investigating plant biology, reproductive development, and genome evolution during domestication. Approximately 10,000 years ago, maize was domesticated from teosinte in Mesoamerica. The crop was subsequently introduced to Asia and India in the sixteenth and seventeenth where it continued to spread. Farmers later selected landraces for cultivation in diverse agro-ecological regions. India, particularly the northeast and northwest Himalayas, emerged as a significant center of maize diversity. These farmer-preserved landraces represent a valuable yet underutilized reservoir of alleles for enhancing stress resistance and productivity. The biological basis for maize's reliance on human-mediated propagation is examined through key aspects of its growth and

K. R. Yathish (✉) · C. Susmita
Winter Nursery Centre, ICAR-Indian Institute of Maize Research, Hyderabad, India
e-mail: kr.yathish@icar.org.in

T. Vasanthrao
Professor Jayshankar Telangana Agricultural University, Hyderabad, India

C. G. Karjagi
Regional Maize Research and Seed Production Centre, ICAR-Indian Institute of Maize Research, Begusarai, India

P. L. Soujanya
Winter Nursery Centre, ICAR-Indian Institute of Maize Research, Hyderabad, Telangana, India

A. K. Das
ICAR-Indian Institute of Maize Research, Ludhiana, Punjab, India

S. Kumar
ICAR-Indian Agricultural Research Institute, Hazaribagh, Jharkhand, India

M. Choudhary et al. (eds.), *Next-Generation Maize Research*,
https://doi.org/10.1007/978-981-92-0722-0_2

reproduction, including monoecy, protandry, inflorescence architecture, pollination biology, and kernel structure. The roles of critical domestication genes, notably teosinte branched1 (*tb1*) and ramosa1 (*ra1*), are highlighted for their substantial impact on plant architecture and inflorescence development during the transition from teosinte to modern maize. Advances in maize genomics, particularly the B73 reference genome and the expanding pangenome framework, reveal extensive variation in gene content, structure, and regulation. The importance of male sterility systems for hybrid seed production and yield enhancement is discussed, as is the potential of apomixis to stabilize heterosis in future breeding. The chapter demonstrates how reproductive biology, genome architecture, and evolutionary processes continue to shape contemporary maize breeding and improvement strategies.

Keywords

Maize · *Tripsacum* · Teosinte · Domestication · Pollination · Transposable elements

2.1 Introduction

The word "maize" is believed to come from the Taino term "mahiz" used in the Caribbean, which was adapted to "maiz" in Spanish. The "Center of origin" for cultivated crops is defined as a geographical area where the majority of crop diversity exists, and where its wild relatives live or have lived. It is established that the center of origin for *Zea mays* is the Mesoamerican region, specifically present-day Mexico and Central America (Watson 1992), and it is thought to have evolved from wild grasses known as teosintes approximately 9000 years ago. Evidence from archaeology indicates that maize was domesticated at least 6000 years ago in regions like the southwestern United States, Mexico, and Central America, occurring independently (Mangelsdorf 2013). Table 2.1 indicates the centers of origin and the main geographical distribution of cultivated maize and its relatives. Following its origin, maize subsequently spread to various parts of the globe, including the Americas, Europe, Africa, and Asia. The Portuguese brought maize to Southeast Asia from the Americas in the sixteenth century. In India, maize was introduced by the Portuguese in the seventeenth century. From India, it made its way to China and was later brought to the Philippines and the East Indies. Over the years, each region has developed specific maize cultivars that are suited to their distinct environmental conditions. There is speculation that maize was first introduced to the Himalayan region of India. India remains a significant center of maize diversity, exhibiting extensive variation in traits, particularly in the northeastern and northwestern highlands. The country continues to harbor a substantial diversity of maize landraces, which serve as a valuable reservoir of beneficial genes. Landrace accessions containing unexplored alleles and genes are important donors for key agronomic traits. These landraces have been preserved by farmers over time and were selected to thrive under specific environmental conditions to meet the local needs.

Table 2.1 Centers of origin and primary geographical distribution of cultivated maize and its relatives

Species Name	Place of origin/distribution
North America	
Tripsacum floridanum	Florida, USA
Tripsacum dactyloides	Central and eastern United States
Tripsacum andersonii	Southern Mexico/northern Central America region
Central America	
Zea mays ssp. *diploperennis* (teosinte)	Western Mexico
Zea mays ssp. *mexicana* (teosinte)	Central Mexican Highlands
Zea mays ssp. *perennis* (teosinte)	Western Mexico
Zea mays	Mexico
Zea mays ssp. *huehuetenangensis* (teosinte)	Huehuetenango, Guatemala
Zea mays ssp. *mexicana* (teosinte)	Central Mexico
Zea mays ssp. *parviglumis* (teosinte)	Balsas River Basin, southwestern Mexico
Zea mays (cultivated maize)	Southwestern Mexico
Zea mays ssp. *luxurians* (teosinte)	Southern Mexico and Guatemala
Zea mays ssp. *nicaraguensis* (teosinte)	Northwestern Nicaragua
Tripsacum zopilotense	Southern Mexico
Tripsacum bravum	Southern Mexico
Tripsacum laxum	Southern Mexico to Central America
Tripsacum latifolium	Central America
Tripsacum maizar	Nicaragua–Costa Rica
Tripsacum pilosum	Southern Mexico and Central America
South America	
Tripsacum peruvianum	Peru
Tripsacum australe	Southern South America (Argentina–Chile region)
Tripsacum cundinamarcae	Northern South America (Colombia)

Modified from Mammadov et al. (2018)

Teosinte (*Zea mays* ssp. *parviglumis*) is the ancestral form of maize (Matsuoka et al. 2002) and is a wild grass that naturally occurs in Mexico and several Central American nations, such as Guatemala (Wilkes 1977), Nicaragua (Iltis and Benz 2000), and Honduras (Standley 1950). The genus *Zea* consists of various species, including the annual *Zea mays* ssp. *luxurians*, the diploid perennial *Zea mays* ssp. *diploperennis*, the tetraploid perennial *Zea mays* ssp. *perennis*, and the polytypic annual *Zea mays*. Within *Zea mays*, there are four subspecies: ssp. *mays* (maize), ssp. *mexicana*, ssp. *parviglumis*, and ssp. *huehuetanangensis* (Fukunaga et al. 2005). Teosinte (*Zea mays* ssp. *parviglumis*) (Iltis and Benz 2000), along with *Tripsacum*, are two crop wild relatives (CWR) that have been thoroughly studied. Because of its similar morphology and its ability to hybridize with *Zea* to create viable but typically sterile offspring, *Tripsacum* is regarded as closely related to *Zea* (Galinat 1961). The genus *Tripsacum* includes nine species of warm-season, perennial grasses found in North and South America (Mammadov et al. 2018).

2.2 Origin: Domestication of Genes and the Molecular Basis of Maize Evolution

Maize (*Zea mays* ssp. mays) represents a classic model of plant domestication, having originated from its wild progenitor teosinte (*Zea mays* ssp. *parviglumis*) approximately 9000 years ago in southern Mexico. This transition was marked by profound morphological and developmental changes driven by human selection acting on a limited number of domestication genes (Doebley et al. 2006; Piperno 2011). Advances in molecular genetics and genomics have enabled the identification of these genes and clarified the evolutionary history of maize. Among them, teosinte branched1 (*tb1*) is one of the most influential domestication genes. In teosinte, low tb1 expression promotes extensive lateral branching, whereas increased expression in domesticated maize suppresses axillary branches, resulting in a single dominant stalk (Doebley et al. 1997). Importantly, domestication of tb1 involved selection on upstream cis-regulatory elements rather than changes in the protein-coding sequence, underscoring the role of regulatory variation in crop evolution (Studer and Doebley 2011). The architectural transformation of maize was largely driven by selection on genes regulating branching at distinct developmental stages. Two major loci, tb1 and ramosa1 (*ra1*), function in complementary developmental contexts. While *tb1* primarily controls vegetative branching and apical dominance, ra1 regulates branching patterns in reproductive inflorescences, together shaping the compact, high-yielding maize ideotype. tb1 encodes a TCP family transcription factor that suppresses axillary bud growth. Elevated tb1 expression in maize enhances apical dominance, improving resource allocation to the primary ear and increasing harvest efficiency (Doebley et al. 1997; Studer and Doebley 2011). Inflorescence architecture is another key trait targeted during maize domestication, with ra1 playing a central role. ra1 encodes a C2H2 zinc-finger transcription factor that inhibits secondary branch meristem formation in tassels and ears. Loss-of-function ra1 mutants exhibit highly branched inflorescences resembling those of ancestral teosinte (Vollbrecht and Sigmon 2005). Strong ra1 activity in domesticated maize promotes compact, unbranched ears with paired spikelets, enhancing kernel density and harvest efficiency. Comparative analyses of maize ra1 and its teosinte ortholog (ta1) reveal conserved coding sequences but substantial divergence in upstream regulatory regions, leading to differences in spatial and temporal expression patterns (Studer and Doebley 2011). Collectively, tb1 and ra1 demonstrate how modest regulatory changes with large phenotypic effects facilitated rapid architectural evolution during maize domestication (Doebley 2004; Doebley et al. 2006). The domestication of maize from its wild ancestor teosinte required coordinated selection on genes controlling vegetative architecture, inflorescence determinacy, kernel development, and seed dispersal. Beyond classical loci tb1, recent genetic and genomic studies have identified additional domestication genes that collectively shaped the maize ideotype favored in early agriculture, including tru1, tga1, ZmSWEET4c, UB3, ids1/Ts6, ZmSh1, and ra1 (Table 2.2, Fig. 2.1).

Teosinte reduced upper internodes1 (*tru1*) regulates the elongation of upper stem internodes and the spatial relationship between tassel and ear. In teosinte, elongated internodes contribute to a dispersed reproductive structure, whereas domesticated maize exhibits shortened internodes that compact ear placement along the main stalk. Selection on tru1 altered auxin-responsive growth, reinforcing apical dominance established by tb1 and improving harvest efficiency (Dong et al. 2019). Kernel morphology was dramatically altered by teosinte glume architecture1 (*tga1*), where a single amino acid substitution reduced glume hardness and produced naked kernels, a key domestication trait facilitating threshing and consumption (Wang et al. 2005). Selection also reduced tillering and modified ear placement through genes such as gt1 and broader changes in meristem determinacy and inflorescence architecture (Vollbrecht and Sigmon 2005). ZmSWEET4c, a sugar transporter expressed in the basal endosperm transfer layer, enhanced carbohydrate flux into developing kernels, increasing kernel size and density in domesticated maize (Sosso et al. 2015). Unbranched3 (*UB3*), together with ids1 and Ts6, promoted determinacy of spikelets and stabilized floral development, increasing kernel number and uniformity (Chuck et al. 1998). Selection on ZmSh1 genes suppressed seed shattering by modifying the abscission layer, ensuring kernel retention on the cob (Lin et al. 2012; Stitzer and Ross-Ibarra 2018). Finally, ramosa1 (*ra1*) integrated with *tru1*, *UB3*, and ids1 to suppress excessive inflorescence branching, balancing determinacy and productivity during maize domestication (Vollbrecht and Sigmon 2005).

2.3 Phenology of Maize (Life Cycle)

Maize is a monoecious species with a determinate growth habit, and its growth phases are categorized into vegetative (V) and reproductive (R) stages, as depicted in Fig. 2.2. The vegetative stages are numbered as V1, V2, V3, and so forth up to V(n), where (n) indicates the final leaf stage prior to VT. These stages are determined by counting the number of collars present on the maize plant. The leaf collar is a light-colored, collar-like "band" found at the base of an extended leaf blade. Leaves that are not fully developed within the whorl and lack a visible leaf collar are excluded from this count. For instance, a plant with three collars is classified as a V3 plant, even if it has six leaves visible. The initial and concluding vegetative stages are referred to as VE (emergence) and VT (tasseling), respectively (Ritchie 2005).

2.3.1 VE—Emergence

Optimum temperature and soil moisture are essential for seed germination. The process of emergence happens when the coleoptile, which serves as a protective sheath for the young shoot, breaks through the soil surface just before the first leaf collar appears. The mesocotyl, a white internode located between the seed and the coleoptilar node, elongates to push the coleoptile upward through the soil. Under optimal conditions, emergence typically occurs within 4–5 days after planting. During this period, the embryonic leaves grow swiftly, pushing through the tip of the coleoptile, followed by the emergence of the first true leaf, at which point

Table 2.2 Major domestication genes involved in maize evolution and their functional significance

Gene	Full name/alias	Primary biological function	Domestication-related trait	Key evidence from domestication	Representative references
tb1	*Teosinte branched1*	Regulates apical dominance and lateral branching	Reduced tillering; single main stalk	Cis-regulatory changes increase *tb1* expression, suppressing axillary branches	Doebley et al. (1997), Studer and Doebley (2011)
tru1	*Teosinte reduced upper internodes1*	Controls internode elongation in upper stem	Compact ear placement	Shortened upper internodes improve harvest efficiency	Dong et al. (2017, 2019)
tga1	*Teosinte glume architecture1*	Controls glume hardness and kernel enclosure	Naked kernels	Single amino acid substitution softens glumes	Wang et al. (2005)
ra1	*ramosa1*	Suppresses secondary inflorescence branching	Compact ear and tassel	Regulatory divergence reduces inflorescence ramification	Vollbrecht et al. (2005)
UB3	*Unbranched3*	Regulates meristem identity and branching	Reduced tassel and ear branching	Alters spikelet fate and inflorescence determinacy	Chuck et al. (1998)
ids1/Ts6	*Indeterminate spikelet1/tasselseed6*	Controls spikelet determinacy	Increased kernel number	Promotes determinate spikelet development	Chuck et al. (1998)
ZmSWEET4c	SWEET sugar transporter	Hexose transport to developing endosperm	Increased kernel size and filling	Enhanced sugar import supports yield	Sosso et al. (2015)
ZmSh1–1	*ZmShattering1* (Chr 1)	Controls abscission layer formation	Reduced seed shattering	Loss or reduction of shattering phenotype	Lin et al. (2012)
ZmSh1–5.1	*ZmSh1* paralog (Chr 5)	Seed dispersal regulation	Kernel retention	Functional diversification after duplication	Lin et al. (2012)
ZmSh1–5.2	*ZmSh1* paralog (Chr 5)	Seed dispersal regulation	Stable ear integrity	Selection suppresses natural shattering	Lin et al. (2012), Stitzer and Ross-Ibarra (2018)

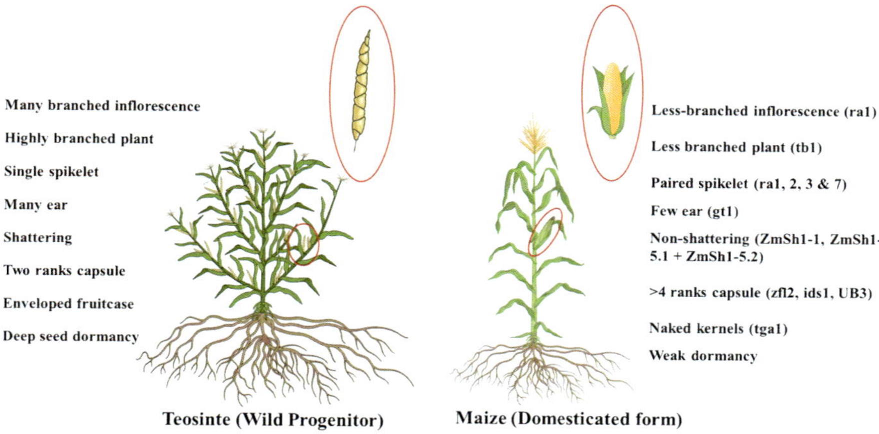

Fig. 2.1 Domestication-driven morphological changes in maize and their underlying genetic regulators

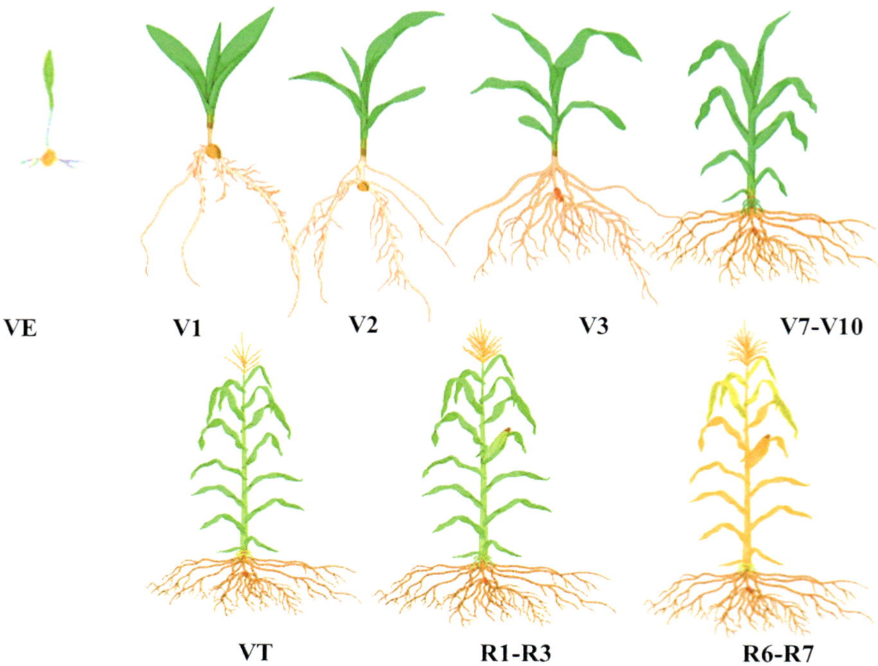

Fig. 2.2 Phenological stages of maize development

photosynthesis begins. The growth of the seminal roots starts to decrease, while nodal roots begin to develop at the crown. At this early stage, the seminal root system serves as the primary root system for the maize plant and provides support for the maize seedlings. If the seminal root system suffers damage before the establishment of nodal roots, it may lead to stunted growth or even death of the plant.

2.3.2 Vegetative Stage (V1)

Plants that have developed their first collared leaf are categorized as V1. The tip of this first leaf is rounded or oval, distinguishing it from all other leaves that have pointed tips, and nodal roots begin to elongate. Leaf initiation continues at the growing point located beneath the soil surface. Below the ground, one or two nodal roots may be visible while the growing point remains below the soil surface.

2.3.3 Vegetative Stage 2 (V2)

Plants that have produced their first two collared leaves are classified as V2. Leaf initiation persists, and the formation of the nodal root system commences. The nodal root system, which emerges above the seed approximately ½ to ¾ inch under the soil surface, can be distinguished from the seminal roots that originate from the seed.

2.3.4 Vegetative Stage 3 (V3)

Plants that have their first three collared leaves are designated as V3. At this point, the plant is capable of standing upright due to the strength provided by the overlapping leaf sheaths. The growing point is still beneath the soil, as minimal stalk elongation has taken place. Lateral roots initiate from the nodal roots, while the development of the seminal root system halts. All leaves and ear shoots that the plant will develop are formed during this phase. Since the growing point remains underground, the plant can often recover from damage caused by hail, wind, or frost; however, it cannot tolerate flooding at this stage.

2.3.5 Vegetative Stage 6–7 (V6–V7)

This stage is crucial for the growth of maize during the vegetative phase. In this stage, the process of rapid stem elongation starts, and the formation of ear shoots begins. The ear's circumference, meaning the number of rows around it, is also established during this phase. A new leaf comes out approximately every 3 days, while older leaves begin to wilt. The growing point is located above the soil, making it susceptible to damage from frost or hail. The plant's root system is robust and spread throughout the soil, enhancing its ability to take in nutrients. Checking for the need for additional fertilizer is vital during the V6 growth phase. Applying nitrogen (N) through top dressing is most effective between the V6 and V8 stages.

2.3.6 Vegetative Stage 8–11 (V8–V11)

In the V9 stage, numerous ear shoots become visible when analyzing the plant and ear shoots (potential ear) will form at every above-ground node except for the final six to eight nodes below the tassel. Ultimately, only one or two upper shoots develop into harvestable ears. Most lower stalk ear shoots cease growth, leading only the upper one or two ear shoots to become harvestable. Plants that produce multiple harvestable ears on the main stem are referred to as prolific. The likelihood of a plant exhibiting prolificness increases with lower plant densities. During this period, deficiencies in both macronutrients and micronutrients can begin to manifest. If these deficiencies are not addressed, they can significantly hinder leaf growth. By the V10 stage, the plant experiences rapid growth, with new leaves emerging every 2–3 days, requiring considerable amounts of water and nutrients to sustain this growth rate.

2.3.7 V12 Stage

A plant is classified as V12 when it has its 12th collared leaf. At this point, roughly 10% of the total dry matter of the plant has been accumulated. Although ear shoots (potential ears) were formed just prior to tassel development, the number of ovules (potential kernels) on each ear and the ear size are now being established at the V12 stage. The number of kernel rows per ear has already been determined, but the final count of kernels per row will not be completed until about a week post-silking. The growing point continues to rise as the lower internodes become fully elongated. Insufficient moisture or nutrient levels at this stage can drastically decrease both the potential number of seeds and ear size.

2.3.8 VT—Tasseling

A plant reaches the VT stage when all of its tassel branches are visible and protrude from the upper leaves. VT marks the conclusion of the vegetative phase, as the plant has achieved its maximum height and tassels begin to emerge. The VT stage commences when the terminal portion of the tassel becomes visible, and before the silks have emerged. The tasseling phase occurs 2–3 days prior to the silking phase. The duration between VT and R1 (the silk stage) varies among different maize varieties and under various weather conditions. The pollination period is distinctively timed based on whether it pertains to the entire field or an individual plant. Not every plant in a field begins or ceases pollen shedding simultaneously due to variability between plants and fields. Typically, pollen is released in the morning. Plants at the VT/R1 stage are particularly susceptible to stress caused by excessive water. If the tassel suffers damage or is entirely lost, it may fail to produce any grain.

2.3.9 R1—Silking

The appearance of silks (R1) signifies the initial phase of the reproductive cycle, and critical stages of reproduction as well as adverse environmental factors, can significantly impact yield. Each possible kernel (ovule) on the ear develops its own silk. Silks start to lengthen shortly after reaching the V12 stage. When the plant has silks that are visible outside the husks, it has entered the first reproductive stage (R1), Silks grow at an approximate rate of 1.5 inches per day. They continue to grow until they capture pollen and germinate or until they start to deteriorate with maturity. Environmental issues like drought stress can cause delays in silk elongation and emergence. Typically, silks remain capable of receiving pollen for up to 10 days following their emergence, although they begin to deteriorate just 5 days after they appear. Pollen grains that fall are captured by the silk, and they grow down the silk within a 24-h timeframe, ultimately fertilizing the ovule. Pollen release usually happens in the early to mid-morning when conditions of moisture and temperature are optimal. The ovule then transforms into a kernel. It takes nearly 3 days for all silks on one ear to be fully exposed and pollinated. The quantity of fertilized ovules is established at this stage. An ovule that does not get fertilized will not lead to kernel production and will eventually degenerate. Environmental stress during this period negatively affects pollination and seed development, with moisture stress leading to the drying out of silks and pollen grains. Potassium absorption is complete, while nitrogen and phosphorus absorption is happening quickly. At this point, symptoms of nitrogen and phosphorus deficiency can be seen in the lower leaves.

2.3.10 R2—Blister

Once pollination occurs, the formation of kernels begins. The kernels during the R2 stage are pale and resemble blisters. This stage is evident roughly 10–14 days post-silking. At this point, the silks start to brown and dry out quickly. As the plant starts a period of kernel fill, starch begins to build up within the kernel. During the R2 growth phase, the radicle, coleoptile, and the first embryonic leaf have developed within the embryo. The moisture content of the kernels at the R2 stage is approximately 85%. Significant stress during the pre-blister and blister stages may lead to the abortion of kernels and a decrease in the total number of kernels on the cob. At this stage, the plant requires 960 growing-degree days (GDD), also known as growing-degree units, to achieve physiological maturity. Extra water supplied at or after the R2 stage does not improve yield, delays the drying process, and can promote diseases in stalks and grains.

2.3.11 R3—Milking Stage

Approximately 22 days after silking, the kernel milk stage begins. At this stage, the outer surface of the kernels is primarily yellow, starch accumulates rapidly, the kernels contain a milky white fluid internally, and the endosperm cells have finished their division. The main factors contributing to noticeable kernel growth are cell expansion and starch accumulation. Severe stress may lead to kernel abortion. The kernel holds around 80% moisture content, and around 880 GDD is required to achieve physiological maturity. While this stage is not as critical as the R1 growth stage, stress during this period can diminish both kernel size and weight.

2.3.12 R4—Dough

When the kernels reach the dough (R4) stage, they transform from a milky state to being soft and sticky. At R4, the kernels have achieved nearly 50% of their maximum weight, and the cob exhibits a light red to pink hue. By this stage, four embryonic leaves have developed, and the kernel contains roughly 70% moisture. Kernel weight can decrease if the environmental conditions are poor or if there's an insufficient supply of nutrients. Stress can lead to kernels that are not fully developed or deep, resulting in "chaffy" ears.

2.3.13 R5—Dent

At the R5 growth stage, almost all of the kernel crowns are beginning to dent, the moisture level is around 55%, and a noticeable horizontal line known as the milky line can be observed between the yellow (starch solid) and white (milky liquid) regions on the kernel. As the kernel continues to mature and the starch solidifies, this line gradually moves towards the tip of the kernel. A severe frost during the R5 stage can harm the plant, leading to decreased yield and kernel development.

2.3.14 R6—Black Layer or Physiological Maturity

Approximately 45–50 days following silking, the maize plant reaches its physiological maturity stage (R6). At this point, the kernels are fully mature, with moisture levels ranging from 30% to 35%, the starch line has progressed to the tip of the kernel, and a black layer has developed at the base of the mature kernels. After this stage, any severe stress will not impact grain yield, unless there is damage caused by diseases such as stalk rots or by insect feeding on the stalk or ear. Although the stalk of the plant may still appear green, the leaves and husk tissue have lost their green coloration.

2.3.15 Botanical Features

Maize is a tall, annual C_4 cereal crop that grows in an organized way and has many different heights between different types of plants. The plant has a strong, solid stem and long, narrow leaves that grow in an arrangement along the culm. These leaves grow from different nodes and are usually arranged in pairs on opposite nodes, which gives the maize plant its unique shape. The shape and structure of each plant part, such as its roots, stem, leaves, flowers, and seeds, show how it has changed over time and how it has adapted to different types of agricultural and ecological environments. Below are the main botanical features of these plant parts.

2.3.15.1 Root

Maize roots initially do not occupy a large volume of soil during the early stages of growth, but expand quickly as the plant matures. Typically, maize plants exhibit three root types: (i) seminal roots—which emerge right after germination, stop growing at the V3 stage, but remain functional throughout the plant's lifespan; (ii) nodal roots—which begin to develop at the formation of the first node (V1) and continue to grow until the kernel reaches the blister stage. By the time the plant reaches the V6 growth stage, nodal roots serve as the primary source of water and nutrients; and (iii) brace or prop roots—which are generated by the lower two nodes. These roots exhibit rapid growth, spreading outwards and downwards almost equally. In suitable soils, corn roots can extend up to 60 cm laterally and in depth.

2.3.15.2 Stem

The stems are robust, solid, featuring distinct nodes and internodes, and typically reach a thickness of three to four centimeters. The internodes are relatively short and thick at the plant's base; they become longer and thicker as they ascend the stem before tapering again. The internode that supports the ear is longitudinally indented to facilitate the correct alignment of the ear head. The leaves are simple with smooth edges and are arranged in an alternating pattern, known as distichous. The upper leaves in corn play a crucial role in capturing light and significantly contribute to the production of photosynthates for the grain. A membranous ligule is located at the connection between the base and the blade.

2.4 Floral Biology

Maize is a monoecious species characterized by the presence of distinct male and female reproductive organs on separate inflorescences on the same plant. The pistillate inflorescence (ear) is the female flower, and the staminate inflorescence (tassel) is the male flower. The plant grows in an organized way, and the main stem stops growing when the terminal staminate tassel forms.

2.4.1 Male (Staminate) Structures

Maize typically exhibits protandry, meaning that the male flowers mature before the female flowers. The male flowers are arranged in a structure called a tassel located at the top of the stem. A tassel consists of a central stalk featuring branches (referred to as rames) along its sides. Each branch has multiple pairs of spikelets. Each spikelet pair contains one pedicellate (stalked) spikelet and one sessile (stalkless) spikelet. Every spikelet consists of two flowers and is encased by four glumes. The first two glumes, located at the bottom, are sterile, while the third glume, known as the flowering glume, and the fourth glume, which is the two-nerved palea, encase a flower. The flowers are distinctly unisexual. The perianth is represented by a pair of fleshy, scale-like structures called lodicules. Each male floret contains three stamens comprising a stalk-like filament and an anther, which produces pollen. The color of the anthers can differ among maize varieties, ranging from purple to red and yellow. Reports indicate that the number of pollen grains per anther can vary from 2000 to 7500 (Kiesselbach 1949). These pollen grains are very small, nearly imperceptible to the naked eye, lightweight, and easily carried by the wind. The combination of wind-driven pollen dispersal and maize's protandrous characteristics promotes cross-pollination, although around 5% of pollination may occur via self-pollination.

2.4.2 Female (Pistillate or Carpellate) Structures

The female flower structure of maize is a spadix that arises from the axil of a lower leaf, forming what is known as the ear. The female flowers are clustered in compact spikes located at the ends of sturdy lateral branches that emerge from the main stem. The cob serves as a fleshy central axis upon which many spikelets are tightly arranged in systematic rows. It is surrounded by several large, modified bracts referred to as spathes or husk leaves, which safeguard the developing ear until the silks emerge from the leaf whorl at the tip of the ear. Female spikelets appear in pairs, much like male spikelets, and are protected by two glumes. Each spikelet comprises two florets: an inferior sterile floret that has just a lemma and palea, and a superior fertile floret with a lemma, palea, and pistil. The pistil is made up of a single carpel featuring a superior, single-chambered ovary that contains one anatropous ovule. Although the style is short, the two stigmas are elongated and fused to create the silk. Lodicules are not present. Silks start off looking hair-like and subsequently change to yellow, green, or purple. The numerous trichomes on the surface of the silk capture pollen grains that are dispersed by wind. The base of each silk continues to grow until fertilization takes place. Following pollination and fertilization, the ovary transforms into a maize kernel. The ear remains firmly covered by husk leaves, which prevents natural seed dispersal and makes maize reliant on human actions for seed release and propagation.

2.5 Pollination and Fertilization

In maize, the process of pollen shedding, known as anthesis, occurs at specific times. Typically, it begins 2–3 days prior to silk emergence and continues for 5–8 days under favorable conditions. The release of pollen is highly responsive to environmental variations. Excessive or insufficient moisture can temporarily halt this process, although it will resume once the conditions of temperature and humidity are optimal. In ideal circumstances, individual pollen grains can survive for approximately 18–24 h. The anthesis–silking interval for well-watered, unstressed plants is generally 1–2 days. However, this interval tends to extend under stress conditions, resulting in less effective fertilization. The long, sticky strands on maize silks function like stigmas, capturing airborne pollen grains. When a pollen grain settles on a receptive silk, it quickly absorbs moisture and initiates growth, forming a pollen tube within just a few minutes. This pollen tube penetrates the silk tissue, typically through openings adjacent to the silk hairs, and grows toward the ovule from within. Fertilization occurs rapidly; the pollen tube reaches the embryo sac merely 12–28 h after pollination. Maize exhibits a distinct generation change; the diploid sporophyte produces haploid gametophytes via meiosis. The male gametophyte is represented by the pollen grain, while the female gametophyte is the embryo sac. Each pollen grain contains two male gametes. Double fertilization involves one sperm cell fertilizing the egg cell, resulting in a diploid embryo, while the other sperm cell fuses with the two polar nuclei to form the triploid endosperm. For fertilization to be successful, the silk needs to maintain adequate moisture. Conditions of low humidity or high temperatures can hinder pollen tube growth, potentially leading to unfertilized ovules, which may appear as absent or aborted kernels on the cob. Larger kernels neighboring the absent ones may compensate for this.

The design of maize ears prevents individual kernels from naturally dispersing their seeds. As a fully domesticated plant, maize relies on human assistance to spread its seeds (Stoskopf 1985). The structure of the cobs holds the kernels securely, preventing them from easily falling off and being carried away by animals, wind, or water. If corn cobs were to drop to the ground and the kernels dispersed, the seedlings that emerged would likely have to vie for resources, diminishing their chances of survival.

2.6 Grain

The distinct maize grain is scientifically classified as a caryopsis, which is a dry fruit that houses a single seed integrated with the inner tissues of the fruit case. This seed consists of two connected structures: the germ, which encompasses the plumule and radicle that will develop into a new plant, and the endosperm, which supplies nutrients to the germinating seedling until it can produce enough leaf area to become autotrophic. The germ serves as the origin of maize "vegetable oil" (with the total oil content in maize grain being 4% by weight). The endosperm makes up about two-thirds of a maize kernel's volume and contributes roughly

86% of its dry weight. Maize kernel endosperm may be either yellow or white. The main component of the endosperm is starch, accompanied by 10% bound protein (gluten), and this stored starch forms the foundation for the nutritional applications of the maize kernel. As stated by Goodman (1999), the various types of endosperms, along with their applications and the preferences of farmers, are linked to different phases in the maize domestication timeline. Some of the oldest maize kernels bear a resemblance to popcorn, indicating that ancient maize was likely popcorn, with other endosperm varieties, such as flint, floury, dent, and sweet, emerging later (Brieger et al. 1958). It was proposed by Brieger et al. (1958) that popcorn is associated with the initial and oldest domestication stage since it retains certain "wild type" traits, including small and rigid grains that are frequently pointed. These traits would necessitate the use of stones for grinding or heat for roasting or popping the grains. Flint endosperm may signify the second domestication phase, with the primary distinction between flint and popcorn types being size, which leads to a notable increase in the volume of the inner core that contains soft endosperm encased in a tough shell. Floury endosperm has been proposed as the third domestication stage and was highly valued by indigenous populations because of its ease of grinding and cooking. Dent endosperm represents a more contemporary maize variety and is advantageous for animal feed, a factor that held little significance for indigenous groups. It is believed that various maize endosperms could have originated at different times and in separate locations (Brieger et al. 1958; Goodman 1999; Staller et al. 2006).

2.7 Male Sterility and Apomixis in Maize

Male sterility in maize is a naturally occurring phenomenon that is controlled by genes. It has been very important in the development of modern hybrid breeding systems. It is defined by the failure to produce functional pollen while female fertility remains intact. A lot of cytological, genetic, and molecular research has shown that male sterility in maize is caused by complex interactions between the nuclear and cytoplasmic genomes that control the growth of pollen and anthers (Chase 2007; Kaul 1988). There are two main types: genetic male sterility (GMS) and cytoplasmic male sterility (CMS). Recessive nuclear genes that affect meiosis, tapetum development, or pollen wall formation control GMS. It is useful for research, but it can't be used to make hybrid seeds because you have to remove fertile plants manually (Bedinger and Fowler 2009). CMS happens when mitochondrial genes and the nuclear genome don't work well together. It has had a big effect on maize breeding. There are three main types of CMS: CMS-T, CMS-C, and CMS-S. Each type is linked to a different rearrangement of the mitochondrial genome (Levings III 1990). CMS-T was widely used because it was stable, but it rendered crops more vulnerable to Bipolaris maydis race T during the southern corn leaf blight epidemic of the 1970s (Ullstrup 1972). Chimeric mitochondrial open reading frames that tangle up pollen development are usually what causes CMS at the molecular level. Nuclear-encoded restorer-of-fertility (Rf) genes, many of which code for pentatricopeptide repeat proteins that are involved in mitochondrial RNA processing (Chase 2007; Hanson and Bentolila 2004), fix the problem. Researchers have found that these CMS–Rf systems are useful for studying how organelles' genomes change over time and how nuclear and cytoplasmic interactions work. Apomixis, the process of making seeds without meiosis or fertilization, is a game-changing chance to improve maize, even though maize is a strictly sexual species. Apomixis includes apomeiosis, parthenogenesis, and the development of functional endosperm. It is common in wild relatives like *Tripsacum* (Koltunow and Grossniklaus 2003). Genetic and genomic studies aimed at transferring apomictic traits have been possible because maize and *Tripsacum* are so similar. However, early attempts showed that there were strong barriers to reproduction. Recent progress in genomics and genome editing has shown that parts of apomixis can be engineered in sexual crops. This opens up the possibility of recreating apomixis in maize (Khanday et al. 2019). Successful implementation would allow fixation of hybrid vigor across generations, fundamentally reshaping maize breeding and seed systems.

2.8 The Genome of Maize

Maize has one of the most complex genomes among all cereal crops. This is because it has been polyploid for a long time, has many transposable elements (TEs) that keep growing, and its genome keeps changing. Its haploid genome is about 2.3 Gb, which is much bigger than those of related grasses like rice and sorghum (Schnable et al. 2011). This is mostly because it has a lot of long terminal repeat retrotransposons, which make up more than 80% of the genome. These elements are not evenly spread out, which makes big intergenic areas that separate gene-rich domains and make the chromosomal architecture very compartmentalized. A whole-genome duplication that only happened in maize, followed by biased fractionation and diploidization, led to sub-genome dominance, unequal gene retention, and regulatory divergence among paralogous genes. The B73 inbred line has been the main reference genome for finding genes, functionally annotating them, and mapping genes, but it only represents a small part of the genomic diversity in the species (Schnable et al. 2009). This limitation prompted the formulation of the maize pangenome concept, which includes the entire set of genes within the species, comprising conserved core genes and a significant dispensable genome characterized by presence–absence variation. Pangenome analyses indicate that maize possesses a gene content that surpasses that of any individual reference genome by a minimum of 1.5-fold, featuring numerous variable genes linked to stress responses, metabolism, and environmental adaptation. Comparative genomics shows that maize and other grasses have a lot of synteny, which shows that the genome is organized in a way that has stayed the same over time and that maize has evolved, become more diverse, and improved as a crop.

2.9 Transposable Elements as Architects of the Maize Genome

Transposable elements (TEs) are important elements of the maize genome and have a big impact on its size, structure, and evolution. Barbara McClintock's discovery of mobile "controlling elements" in maize was the first time people realized how important they were. This showed that DNA segments could move around in the genome and change gene expression, which led to visible changes in phenotypes (McClintock 1950). Subsequent molecular analyses validated that the genetic instability she characterized resulted from TE insertions, excisions, and related chromosomal rearrangements, thereby establishing maize as a fundamental model for TE biology (Wessler 1988). Sequencing of the B73 reference genome showed that about 75–85% of the maize genome is made up of TEs, which makes it one of the most TE-rich crop genomes (Schnable et al. 2009). Class I LTR retrotransposons, especially those from the Gypsy and Copia families, are the most common in this repetitive landscape. Their unique nested insertion pattern has led to large intergenic blocks of repetitive DNA, which has caused the genome to grow a lot without adding more genes (SanMiguel et al. 1996). TE activity changes the structure of DNA in many ways, such as by adding, deleting, flipping, and duplicating pieces. This also affects the size of the genome. These rearrangements result in extensive presence–absence variation among maize lines and represent a significant element of the dispensable genome, establishing the structural foundation of the maize pangenome (Stitzer and Ross-Ibarra 2018).

TEs have a big impact on gene regulation and phenotypic diversity, in addition to structural effects. Insertions close to or inside genes can add new regulatory elements like transcription factor binding sites, enhancers, or promoters. Many of these elements have been added to host regulatory networks. TE-associated regulatory variation is especially pronounced during abiotic stress, where certain TE families accumulate near stress-responsive genes and influence inducible gene expression (Makarevitch et al. 2015). Polymorphic TE insertions significantly influence the expression of quantitative trait loci and the variability of complex traits. TE insertions can mess up gene function and create classical mutants (Wessler 1988), but they can also lead to new evolutionary changes by shuffling exons, duplicating genes, and diversifying regulatory elements. Some TE-derived sequences even become functional parts of maize genes (Bennetzen 2005). Epigenetic mechanisms, such as DNA methylation, histone modifications, and small RNAs, tightly control TE activity. These mechanisms limit transposition but allow for some flexibility when stress or hybridization occurs (Lisch 2012). Comparative analyses with teosinte suggest that TE insertions have facilitated regulatory divergence at loci associated with domestication. TE-associated variation is the basis for many traits that are important for agriculture, even though it makes genomic analyses more difficult. This means that TE diversity must be included in maize pangenomes for QTL mapping, GWAS, and breeding (Springer et al. 2009; Wallace et al. 2014).

2.10 Conclusion

This chapter integrates maize biology and reproductive mechanisms into a comprehensive discussion on domestication and genome evolution. It describes the evolutionary change of maize from its wild ancestor, teosinte, with a focus on how selection on important domestication genes like *teosinte branched1* (*tb1*) and *ramosa1* (*ra1*) changed the way the plant was built and how the flowers were arranged. We discuss these developmental and evolutionary insights in light of modern genomic resources, especially the B73 reference genome and the growing maize pangenome. Together, these resources show that there is a lot of variation in gene content and genome structure across different germplasm. Comparative synteny studies with other grasses, coupled with assessments of transposable element dynamics, demonstrate the interaction between long-term genomic conservation and lineage-specific innovation. Moreover, the chapter emphasizes the practical importance of cytoplasmic and genic male sterility systems in facilitating efficient hybrid seed production, while also addressing apomixis as a promising yet formidable approach for the stabilization of hybrid vigor. These views work together to make a complete picture of how maize has changed over time and how to improve future crops.

References

Bedinger PA, Fowler JE (2009) The maize male gametophyte. In: Handbook of maize: its biology. Springer, pp 57–77

Bennetzen JL (2005) Transposable elements, gene creation and genome rearrangement in flowering plants. Curr Opin Genet Dev 15(6):621–627

Brieger FG, Gurgel J, Paterniani E, Blumenschein A, Alleoni M (1958) Races of maize in Brazil and other eastern South American countries. National Academy of Sciences, Washington, DC

Chase CD (2007) Cytoplasmic male sterility: a window to the world of plant mitochondrial–nuclear interactions. Trends Genet 23(2):81–90

Chuck G, Meeley RB, Hake S (1998) The control of maize spikelet meristem fate by the APETALA2-like gene indeterminate spikelet1. Genes Dev 12(8):1145–1154

Doebley J (2004) The genetics of maize evolution. Annu Rev Genet 38(1):37–59

Doebley J, Stec A, Hubbard L (1997) The evolution of apical dominance in maize. Nature 386(6624):485–488

Doebley JF, Gaut BS, Smith BD (2006) The molecular genetics of crop domestication. Cell 127(7):1309–1321

Dong Z, Li W, Unger-Wallace E, Yang J, Vollbrecht E, Chuck G (2017) Ideal crop plant architecture is mediated by tassels replace upper ears1, a BTB/POZ ankyrin repeat gene directly targeted by TEOSINTE BRANCHED1. Proc Natl Acad Sci 114(41):E8656–E8664

Dong Z, Xiao Y, Govindarajulu R, Feil R, Siddoway ML, Nielsen T, Lunn JE, Hawkins J, Whipple C, Chuck G (2019) The regulatory landscape of a core maize domestication module controlling bud dormancy and growth repression. Nat Commun 10(1):3810

Fukunaga K, Hill J, Vigouroux Y, Matsuoka Y, Sanchez G J, Liu K, Buckler ES, Doebley J (2005) Genetic diversity and population structure of teosinte. Genetics 169(4):2241–2254

Galinat WC (1961) Tripsacum floridanum crosses readily with corn. Maize Genet Coop Newslett 35:38–39

Goodman MM (1999) Broadening the genetic diversity in maize breeding by use of exotic germplasm. In: Genetics and exploitation of heterosis in crops, pp 139–148 https://doi.org/10.2134/1999.geneticsandexploitation.c13. https://acsess.onlinelibrary.wiley.com/series/5172
Hanson MR, Bentolila S (2004) Interactions of mitochondrial and nuclear genes that affect male gametophyte development. Plant Cell 16(Suppl. 1):S154–S169
Iltis HH, Benz BF (2000) Zea nicaraguensis (Poaceae), a new teosinte from Pacific coastal Nicaragua. Novon 10:382–390
Kaul MLH (1988) Male sterility in higher plants. Monogr Theor Appl Genet 10:335–336
Khanday I, Skinner D, Yang B, Mercier R, Sundaresan V (2019) A male-expressed rice embryogenic trigger redirected for asexual propagation through seeds. Nature 565(7737):91–95
Kiesselbach TA (1949) The structure and reproduction of corn, 50th Anniversary edn. University of Nebraska Press
Koltunow AM, Grossniklaus U (2003) Apomixis: a developmental perspective. Annu Rev Plant Biol 54(1):547–574
Levings CS III (1990) The Texas cytoplasm of maize: cytoplasmic male sterility and disease susceptibility. Science 250(4983):942–947
Lin Z, Li X, Shannon LM, Yeh C-T, Wang ML, Bai G, Peng Z, Li J, Trick HN, Clemente TE (2012) Parallel domestication of the Shattering1 genes in cereals. Nat Genet 44(6):720–724
Lisch D (2012) Regulation of transposable elements in maize. Curr Opin Plant Biol 15(5):511–516
Makarevitch I, Waters AJ, West PT, Stitzer M, Hirsch CN, Ross-Ibarra J, Springer NM (2015) Transposable elements contribute to activation of maize genes in response to abiotic stress. PLoS Genet 11(1):e1004915
Mammadov J, Buyyarapu R, Guttikonda SK, Parliament K, Abdurakhmonov IY, Kumpatla SP (2018) Wild relatives of maize, rice, cotton, and soybean: treasure troves for tolerance to biotic and abiotic stresses. Front Plant Sci 9:886
Mangelsdorf PC (2013) Corn: its origin, evolution and improvement. Harvard University Press
Matsuoka Y, Mitchell SE, Kresovich S, Goodman M, Doebley J (2002) Microsatellites in Zea—variability, patterns of mutations, and use for evolutionary studies. Theor Appl Genet 104(2):436–450
McClintock B (1950) The origin and behavior of mutable loci in maize. Proc Natl Acad Sci 36(6):344–355
Piperno DR (2011) The origins of plant cultivation and domestication in the New World tropics: patterns, process, and new developments. Curr Anthropol 52(S4):S453–S470
Ritchie SW (2005) How a corn plant develops. Iowa State University of Science and Technology, Cooperative Extension Service
SanMiguel P, Tikhonov A, Jin Y-K, Motchoulskaia N, Zakharov D, Melake-Berhan A, Springer PS, Edwards KJ, Lee M, Avramova Z (1996) Nested retrotransposons in the intergenic regions of the maize genome. Science 274(5288):765–768
Schnable PS, Ware D, Fulton RS, Stein JC, Wei F, Pasternak S, Liang C, Zhang J, Fulton L, Graves TA (2009) The B73 maize genome: complexity, diversity, and dynamics. Science 326(5956):1112–1115
Schnable JC, Springer NM, Freeling M (2011) Differentiation of the maize subgenomes by genome dominance and both ancient and ongoing gene loss. Proc Natl Acad Sci 108(10):4069–4074
Sosso D, Luo D, Li Q-B, Sasse J, Yang J, Gendrot G, Suzuki M, Koch KE, McCarty DR, Chourey PS (2015) Seed filling in domesticated maize and rice depends on SWEET-mediated hexose transport. Nat Genet 47(12):1489–1493
Springer NM, Ying K, Fu Y, Ji T, Yeh C-T, Jia Y, Wu W, Richmond T, Kitzman J, Rosenbaum H (2009) Maize inbreds exhibit high levels of copy number variation and presence/absence variation in genome content. PLoS Genet 5(11):e1000734
Staller J, Tykot R, Benz B (2006) Histories of maize: multidisciplinary approaches to the prehistory, linguistics, biogeography, domestication, and evolution of maize. Left Coast Press
Standley PC (1950) New plants from Honduras. Ceiba 1(1):38
Stitzer MC, Ross-Ibarra J (2018) Maize domestication and gene interaction. New Phytol 220(2):395–408

Stoskopf NC (1985) Cereal grain crops. Reston Publishing Co. Inc., Virginia, pp 429–443
Studer AJ, Doebley JF (2011) Do large effect QTL fractionate? A case study at the maize domestication QTL teosinte branched1. Genetics 188(3):673–681
Ullstrup AJ (1972) The impacts of the southern corn leaf blight epidemics of 1970–1971. Annu Rev Phytopathol 10(1):37–50
Vollbrecht E, Sigmon B (2005) Amazing grass: developmental genetics of maize domestication. Biochem Soc Trans 33(6):1502–1506
Vollbrecht E, Springer PS, Goh L, Buckler ES IV, Martienssen R (2005) Architecture of floral branch systems in maize and related grasses. Nature 436(7054):1119–1126
Wallace JG, Bradbury PJ, Zhang N, Gibon Y, Stitt M, Buckler ES (2014) Association mapping across numerous traits reveals patterns of functional variation in maize. PLoS Genet 10(12):e1004845
Wang H, Nussbaum-Wagler T, Li B, Zhao Q, Vigouroux Y, Faller M, Bomblies K, Lukens L, Doebley JF (2005) The origin of the naked grains of maize. Nature 436(7051):714–719
Watson L (1992) The grass genera of the world: descriptions, illustrations, identification, and information retrieval. http://delta-intkey.com
Wessler SR (1988) Phenotypic diversity mediated by the maize transposable elements Ac and Spm. Science 242(4877):399–405
Wilkes HG (1977) Hybridization of maize and teosinte in Mexico and Guatemala and the improvement of maize. Econ Bot 31:254–293

Harnessing Wild Relatives for Climate Resilience in Maize

3

T. V. Anirudh, Mayank Tiwari, Kajal Mehra, Thotla Naresh, Priya Garkoti, and N. K. Singh

Abstract

Maize has originated from Mexico and Central America, and is classified under the Maydae tribe within the Poaceae family. Being a staple crop of global significance, it provides essential nutrition, animal feed, biofuel, and industrial raw material. To fulfil global food requirements, the annual growth rate must increase from 1.6% to 2.4%. The domestication and selective breeding of maize have created a remarkable crop but have also led to the loss of numerous allelic variations in traits that underwent intensive selection. The wild relatives offer greater genetic diversity and are a reservoir of novel alleles that confer resistance/tolerance to various biotic and abiotic stresses, improve yield, and enhance nutritional quality. The wild relatives of maize include *Coix, Chionachne, Sclerachne, Trilobachne Polytoca*, *Tripsacum*, and teosinte. Among these teosintes (*Zea mays* ssp. *parviglumis*) is the direct ancestor of maize. Studies of teosinte-derived maize lines have shown diversification in resilient traits such as waterlogging tolerance, drought, diseases, yield, and adaptability. Despite their potential, wild relatives remain underutilized due to limited investigation and validation. This chapter provides insights into the different wild relatives, their characterisation, utilization, and explores strategies to leverage these genetic resources using traditional and modern breeding methods like backcrossing and hybridization, molecular marker studies alongside biotechnological approaches such as CRISPR/Cas9 and genomic selection. By integrating these approaches, we can accelerate the development of climate-resilient, high-yielding maize varieties to meet the growing global demand.

T. V. Anirudh · M. Tiwari · K. Mehra · T. Naresh · P. Garkoti · N. K. Singh (✉)
College of Agriculture, Govind Ballabh Pant University of Agriculture & Technology, Pantnagar, U.S. Nagar, Uttarakhand, India

M. Choudhary et al. (eds.), *Next-Generation Maize Research*,
https://doi.org/10.1007/978-981-92-0722-0_3

Keywords

Teosinte · *Tripsacum* · Maize · Wild relatives · Climate resilience · CWR

3.1 Introduction

Maize cultivation extends across tropical, subtropical, and temperate regions throughout Asia, Africa, Europe, and America. The crop's global significance has increased owing to its versatile applications in food, animal feed, biofuel, and various industrial processes (Cassidy et al. 2013). The escalating demand for maize stems from its diverse uses and is projected to rise further as the global population is expected to reach nine billion by 2050 (Sahoo and Adhikari 2021). To address the needs of this growing population, it is necessary to boost maize production at an annual rate of 2.2% (Foley et al. 2011). Maize is essential for global food and economic stability and has both historical and modern cultural significance.

Maize possesses a rich history of experimental breeding methods, with the potential for high yield through hybridization, as initially observed by Shull (1908, 1909). The primary aim of maize breeding programs is to create high-yielding varieties or hybrids that exhibit stress tolerance and nutritional enhancement, while adapting to diverse agro-ecological and socio-economic contexts. But, with the escalating uncertainties and unpredictability driven by climate change, breeding objectives must now prioritize resilience to both biotic and abiotic stresses alongside elevated productivity. Breeding processes should evaluate multiple traits to enhance economic value, including tolerance to environmental stresses such as moisture deficits, high- and low-temperature extremes, frost, and soil-related constraints like acidity, salinity, and alkalinity. Genetic variability for these traits is essential in breeding populations to facilitate the selection of superior genotypes. However, genomic reshuffling through selective breeding, along with the domestication bottleneck, has narrowed the genetic base in maize cultivars and their primary gene pool (Garkoti and Senthilkumar 2025). The erosion of adaptive, canopy architecture, and yield-related alleles stemming from domestication bottlenecks and intensive selection threatens the capacity of modern cultivars to withstand evolving biotic and abiotic stresses linked to climate change. Hence, identifying novel sources of genetic diversity from wild relatives and progenitors is imperative. Reintroducing key wild alleles into elite domesticated germplasm will expand the genetic base and mitigate genetic erosion.

3.2 Impacts of Climate Change on Maize Production

Climate change is increasingly threatening global food production, particularly for climate-sensitive crops such as maize. Rising temperatures, frequent heatwaves, drought, waterlogging, and salinity stress directly disrupt maize growth and productivity (Kumar and Choudhary 2022). Heat stress reduces photosynthesis, impairs reproductive development, and reduces kernel set, with flowering being the most vulnerable stage (Lobell and Field 2007). The IPCC has warned that recurring extreme heat events pose serious risks to food security, with projected yield losses of up to 25% if adaptation strategies are not adopted (Masson-Delmotte and Zhai 2018; IPCC 2014). Over the past three decades, climate variability has already reduced global maize production by 0.17 MMT per year, and yields are expected to decline by 7.4% for every 1 °C rise in global temperature (Ray and West 2019). Among the abiotic stresses, drought stress severely affects maize at all growth stages, especially during tassel emergence and grain formation, leading to reduced kernel number and size (Bolaños et al. 1993). In tropical and semi-arid regions of South and Southeast Asia, high temperatures often coincide with moisture stress, while waterlogging affects more than 18% of maize-growing areas and causes annual yield losses of 25–30% (Zaidi and Maniselvan 2010). Salinity further restricts maize productivity by disturbing nutrient uptake and reducing germination, vegetative growth, and reproductive performance (Szalai and Janda 2009; Kaya et al. 2013). Climate extremes also intensify pest and disease pressure, particularly stalk and ear rots caused by *Fusarium* spp. and *Aspergillus* flavus, increasing mycotoxin risks (Savary et al. 2011). Charcoal rot (*Macrophomina phaseolina*) is favoured by hot and dry conditions, further aggravating yield losses (Kumar and Shekhar 2005). To mitigate these challenges, maize breeding programs must expand the genetic base by conserving and utilizing diverse germplasm, including wild relatives, to develop climate-resilient varieties suited to future growing conditions.

3.3 Concept of Wild Relatives and Their Genetic Diversity

The domestication and selective breeding of maize have created a remarkable crop but have also led to the loss of numerous allelic variations in traits that underwent intensive selection. These lost alleles can be reintroduced by utilizing wild relatives like teosinte, which is rich in novel alleles that confer resistance to stresses, improve yield, and enhance quality. Studies of teosinte-derived maize lines have shown diversification in agronomical traits such as ear and kernel characteristics, yield, and adaptability (Kumar et al. 2024a, b, 2026). Despite their potential, wild relatives remain underutilized due to limited investigation and validation. Genetic and archaeological evidence suggests that cultivated species are shaped by concurrent selective forces on their progenitors, including wild relatives, through a process called domestication syndrome. This process results in prominent modifications like gigantism in harvested parts, reduced branching, and loss of shattering (Gepts 2012). Wild species of crops continue to thrive in their natural habitats, surviving and reproducing successfully in the face of diverse biotic and abiotic stresses without human intervention. Their ability to adapt to such harsh environments is largely attributed to inherent or evolved genetic mechanisms that enable them to withstand unfavourable conditions. In contrast, these adaptive traits are often absent or underrepresented in modern maize cultivars. This gap likely arose because traditional breeding programs largely emphasized yield improvement under favourable growing conditions, inadvertently overlooking traits associated with environmental resilience and survival. Recognizing the importance of allelic variation, the introgression or domestication of wild alleles presents a valuable opportunity to enrich the primary gene pool of maize. This genetic diversification ensures broader access to functionally diverse alleles that plant breeders can utilize to enhance both resilience and productivity in maize cultivars. Among the wild relatives, teosinte stands out as a particularly important genetic resource. As the wild progenitor of maize, teosinte has evolved under natural selection pressures, accumulating a wide range of alleles that confer tolerance to biotic and abiotic stresses, improve yield potential, and enhance grain quality (Kumar et al. 2025). The wild relatives harbour far greater genetic and allelic diversity than cultivated maize, making them invaluable in efforts to address existing and emerging agricultural challenges (Singh et al. 2024). Harnessing these unique allelic variants can help overcome yield stagnation, strengthen stress tolerance, and contribute to the sustainable improvement of maize productivity.

Comparative population genetic studies on maize and teosinte indicate a reduction in genetic diversity resulting from domestication, alongside beneficial gene flow from teosinte to maize, which has contributed to improved adaptation across diverse environments (Hufford et al. 2012). The ability of domesticated crops to crossbreed with their wild relatives facilitates gene transfer and enables genetic analysis of traits with agronomic significance. The current loss of biodiversity is largely due to climate change, which is negatively affecting species' geographic range, local abundance, and genetic diversity levels (Allendorf et al. 2010). According to the Food and Agriculture Organization (FAO), approximately 75% of genetic diversity in crop varieties has been lost over the past century (Khoury and Brush 2022), thereby restricting breeders' ability to enhance key agronomic traits. Conservation initiatives for teosinte, the wild progenitor of maize, remain limited, with protection efforts primarily concentrated in the Sierra de Manantlán Biosphere Reserve in Mexico and the Reserva de Recursos Genéticos de Apacunca in Nicaragua. Genetic diversity is essential to produce cultivars that are climate-resilient because it helps address difficulties presented by shifting environmental conditions, such as resistance to biotic and abiotic stresses, thereby ensuring food security and agricultural productivity (Li and Li 2024). According to Fustier (2017), genomic investigations of teosintes have revealed notable genetic variations among populations linked to different local climates, indicating that teosintes frequently adapt locally to opposing environments, such as warm and dry climates. For example, teosinte, particularly *Zea mays* subsp. *parviglumis*, closely related to maize, is highly adapted to local environments which harbours novel alleles for disease resistant and yield traits. *Zea nicaraguensis* found in the shallow water near riverbeds of Nicaragua has a higher ability for water logging tolerance. The frequency of climate-related features linked to single-nucleotide polymorphisms (SNPs) in two subspecies of teosinte, *Z. mays* ssp. *mexicana* and *Z. mays* ssp. *parviglumis*. Mexicana is indigenous to the drier and colder highlands of Central Mexico, whereas *Z. mays* ssp. *parviglumis* is indigenous to the more tropical lowlands of southwestern Mexico. These two teosintes flourish in different environments (Gonzalez-Segovia and Pérez-Limon 2019).

3.4 Maize Origin and History

The origin of maize remained controversial due to its highly specialized ear structure, which shows little similarity to its wild relatives. This led botanists to propose several hypotheses explaining maize domestication. The three major hypotheses are the Tripartite hypothesis, the Teosinte hypothesis, and the *Tripsacum–Z. diploperennis* hypothesis. The Tripartite hypothesis, proposed by Mangelsdorf and Reeves (1938, 1939), suggested that maize originated from an extinct wild ancestor, while teosinte resulted from hybridization between maize and *Tripsacum*. They attributed major morphological differences to linked chromosomal segments from *Tripsacum*. This hypothesis was widely accepted until the 1960s.

In contrast, Beadle (1939) proposed the Teosinte hypothesis, stating that maize is a domesticated form of teosinte selected by ancient humans. He suggested that a few major mutations controlled key traits such as ear structure and plant architecture. Later evidence identified *Zea mays* ssp. *parviglumis* (Balsas teosinte) as the progenitor of modern maize. QTL studies revealed major domestication genes such as *tga1* and *tb1* (Buckler and Stevens 2006), making this the most widely accepted hypothesis. The *Tripsacum–Z. diploperennis* hypothesis proposed by Eubanks (1995) suggested a hybrid origin of maize. However, cytological and molecular evidence showed closer similarity to teosinte than *Tripsacum*, providing limited support for this model. Evolutionary studies indicate that *Zea* diverged from *Tripsacum* 2.3–4.8 million years ago, with diversification within *Zea* occurring around 1.2–1.4 million years ago. The split between *Z. mays* and *Z. luxurians* occurred approximately 700,000 years ago (White and Doebley 1999).

3.5 Gene Pool Classification

Taxa are classified according to their ability to engage in genetic exchange, as proposed by Harlan and de Wet (1971). Maize (*Z. mays* ssp. *mays*) represents the primary gene pool, other *Zea* species (teosintes) form the secondary gene pool, and *Tripsacum* species constitute the tertiary gene pool. *Tripsacum* spp. have a base chromosome number of $n = 18$ and are perennial, presenting substantial obstacles to genetic exchange with *Zea*. Nevertheless, genomic analysis comparing maize with diploid *Tripsacum dactyloides* L. indicates a similar gene content despite significant differences in transposable element families (Chia et al. 2012). This suggests that utilization of *Tripsacum* for maize improvement may become feasible in the future via gene editing technologies. *Tripsacum* species exhibit considerable ploidy variation, from diploid to tetraploid. *T. andersonii*, a naturally occurring *Tripsacum* × *Zea* hybrid, exemplifies the genomic similarities between *Zea* and *Tripsacum* as well as the complexities of their reproductive biology. Diploid *Zea luxurians* serves as the *Zea* parent, whereas triploid *T. latifolium* ($2n = 3\times = 54$) is proposed as the *Tripsacum* parent, itself derived from a hybridization between *T. latifolium* Hitchc. (2×) and *T. maizar* Hern. & Randolph (2×) (Berthaud et al. 1997). *T. andersonii*, commonly known as Guatemala grass, has become a globally successful forage crop due to prolific vegetative growth. All reported wild maize relatives are presented in Table 3.1.

(i) *Primary gene pool*: Includes maize and related germplasm that hybridize freely, such as racial complexes, OPVs, inbreds, and diploid teosintes. Morphologically, teosinte often resembles maize, and hybrids are fertile with regular meiotic pairing (Iltis 1983). *Z. diploperennis*, however, shows minor fertilization barriers.
(ii) *Secondary gene pool*: *Tripsacum* species, perennials with chromosomes in multiples of $x = 18$, can cross with maize with difficulty, and the resulting hybrids show variable sterility. Chromosome morphology and knob positions differ greatly from maize (Goodman and Galinat 1988), though introgression is possible (Galinat 1988).
(iii) *Tertiary gene pool*: Distantly related genera such as *Coix*, *Sclerachne*, *Polytoca*, and *Chionachne* constitute this pool, with varying chromosome numbers (*Chionachne* and *Sclerachne*: 20; *Polytoca*: 40; *Coix*: 10, 20, 40). Other Andropogoneae genera are also included.

3.6 Utilisation of Wild Relatives

Wild maize relatives like teosintes and *Tripsacum* are valuable for improving maize tolerance to diseases, drought, and waterlogging (Table 3.1). Their stress-resilient genes can be used in breeding or through modern tools like marker-assisted selection and gene editing to develop hardier maize varieties.

3.7 Tolerance to Abiotic Stresses

Abiotic stresses have a significant impact on agricultural productivity, especially in situations where the climate is unpredictable and variable. Drought is the most important environmental conditions that restrict agricultural productivity. These adverse effects have been thoroughly investigated in the majority of major crops, and genes or alleles linked to similar stress tolerances were frequently identified in the crop's wild counterparts.

3.7.1 Drought Tolerance (DT)

A prolonged period of insufficient rainfall that causes soil water to be depleted through transpiration or evaporation, rendering it insufficient for crop demands, is

Table 3.1 Wild relatives of maize and their origin

S.no.	Genus	Species	Place of origin	2*n*	GP
1.	*Zea*	*mays* ssp. *mays*	Mexico and Central America	20	1
2		*mays* ssp. *mexicana*	Mexico	20	2
3.		*mays* ssp. *parviglumis*	Southern and Eastern Mexico	20	2
4.		*mays* ssp. *huehuetenangensis*	Guatemala	20	2
5.		*diploperennis*	Jalisco and Mexico	20	2
6.		*luxurians*	South Eastern Guatemala	20	2
7.		*nicaraguensis*	Nicaragua	20	2
8.		*perennis*	Jalisco and Mexico	40	2
9.	*Tripsacum*	*australe*	South America	36	3
10.		*cundinamarce*	Colombia	36	–
11.		*floridanum*	South Florida and Cuba	36	–
12.		*latifolium*	North and Central America	36	–
13.		*laxum*	Central America	36	–
14.		*bravum*	Mexico	36, 72	–
15.		*zopilotense*	Meso-America	36, 72	–
16.		*maizar*	Meso-America	36, 72	–
17.		*andersonii*	Central America	64	–
18.		*dactyloides*	Mexico and Guatemala	36, 72, 108	–
19.		*intermedium*	Guerrero, Chiapas, Guatemala, Honduras	72	–
20.		*manisuroides*	Tuxtla Gutierrez, Chiapas and Mexico	72	–
21.		*jalapense*	Oaxaca, Chiapas, Guatemala, El Salvador	72	–
22.		*lanceolatum*	South West USA	72	–
23.		*pilosum*	North and Central America	72	–
24.		*peruvianum*	South America	72, 90, 108	–

(continued)

Table 3.1 (continued)

S.no.	Genus	Species	Place of origin	$2n$	GP
25.	*Coix*	*aquatica*	South East Asia	10	3
26.		*poilanei*		10	–
27.		*lacryma-jobi*		20	–
28.		*gigantea*		20, 40	–
29.		*ouwehandii*		–	–
30.		*puellarum*		–	
31.		*gasteenii*		–	–
32.	*Chionachne*	*koenigii*	South East Asia	20, 40	–
33.		*semiteres*		20	–
34.		*hubberdiana*		–	–
35.		*biaurata*		–	–
36.	*Sclerachne*	*punctata*	South East Asia	20	–
37.	*Trilobachne*	*cookei*	South East Asia	20	–
38.	*Polytoca*	*macrophylla*	South East Asia	20, 40	–
39.		*massii*		–	–
40.		*digitata*		–	–
41.		*javanica*		–	–
42.		*wallichiona*		–	–
43.		*cyanthopoda*		–	–

Adopted with permission from Hossain et al. (2016), Springer Nature (6171180620449) known as a drought. The primary cause of crop loss in hybrid maize is drought, and its frequency and intensity are predicted to rise due to climate change brought on by the accumulation of greenhouse gases. A key component of the approach is to stabilize world production by application of genetics to increase drought tolerance and ensure yield stability (Sayadi Maazou et al. 2016). If a maize plant experiences drought stress during the 2–10-week period leading up to anthesis, the impact on yield can be significant, as this time is crucial for the development of the ears (Eubanks 2006). Numerous studies have been conducted in the past three decades to assess and improve DT in maize, with a particular focus on dominant and secondary characteristics like grain yield (GY) and the anthesis-silking interval (ASI), which is significantly correlated with GY, respectively (Campos et al. 2017). Due to its high correlation with grain output, increased flowering synchrony (e.g., reduced ASI) has been widely exploited in traditional breeding for drought tolerance. Aerenchyma tissue in roots, root penetration and enhanced biomass are some of the morphological and other characteristics of wild relatives, particularly *Tripsacum*, that contribute to its capacity to tolerate drought. According to physiological data, high photosynthesis and water use efficiency in leaf gas exchange analysis are the foundations of *Tripsacum*'s exceptional drought tolerance (Kemper and Alberts 1997). Additionally, Eubanks (2006) noted in another study that the *Tripsacum*-introgressed cultivar performs better than the control maize, even when drought stress is present. Larger, more resilient root systems and an overall rise in grain yield seem to be the results of *Tripsacum* introduction. A study conducted at Pantnagar identified three QTLs, which explained phenotypic variation of 14.67% (leaf rolling), 12.84% (plant height), both linked with the marker umc1303, 11.79% (anthesis-silking interval) linked with the marker phi113. Hence, traits linked and co-localized markers for drought-tolerance traits could be used for MAS for the development of drought-tolerant maize (Kumar et al. 2024a). Different researchers identified the various QTLs using the wild species (Table 3.2).

3.7.2 Waterlogging Tolerance (WT)

The developmental stage of the plant, the duration of the flooding period, and the soil–air temperatures all have a significant impact on the damage caused by flooding to maize. Early studies on temperate maize indicate that WT-related seedling traits show complex inheritance and expression (Subbaiah and Sachs 2003). Responses to waterlogging involve anaerobic protein synthesis, gene expression changes, metabolic shifts to fermentation, and structural modifications like aerenchyma formation. A CIMMYT study on tropical maize lines subjected to 7 days of waterlogging at V7–V8 stages also suggested polygenic inheritance of waterlogging tolerance, with tolerant alleles showing partial dominance over susceptible ones (Zaidi and Maniselvan 2010). Flooding damage is influenced by duration, soil–air temperature, and growth stage, causing severe early-stage yield losses. Teosintes from high-rainfall areas may provide useful genetic resources for flood tolerance, as some maize lines can form adventitious roots under flooding (Mano and Omori 2007).

Table 3.2 List of QTLs introgressed from wild relative in to maize for waterlogging

Crop wild relatives	Stress	QTL Location	Tolerance mechanism	References
Zea mays ssp. *huehuetenangensis*	Waterlogging	QTL (Chr 3, 7, and 8)	Increasing oxygen access through adventitious root	Mano et al. (2005)
Z. nicaraguensis	Waterlogging	QTL (Chr 1, 5, and 8)	Constitutive root aerenchyma and improves root and shoot oxygen availability	Mano and Omori (2008)
Z. nicaraguensis	Waterlogging/low oxygen in hydroponic conditions	QTL located at locus RBF1, (Chr 3 of teosinte)	Promotes oxygen diffusion inside adventitious and lateral roots and stops oxygen loss from roots to the surrounding soil.	Pedersen and Nakayama (2021)
Z. nicaraguensis	Waterlogging	QTL Qft-rd4.07–4.11 (Chr 4)	Minimizes damage to roots and leaves	Mano and Nakazono (2021)

Prominent QTLs for waterlogging have been identified over two locations, and their validation could support marker-assisted breeding for tolerant maize (Senthil 2024).

In comparison to certain maize inbreds, the teosinte *Z. nicaraguensis* (Iltis and Benz 2000), *Z. luxurians*, and *Z. mays* ssp. *huehuetenangensis* (Mano et al. 2005) have been found to have a greater capacity for adventitious root development. By growing adventitious roots above the soil surface, *Z. mays* ssp. *huehuetenangensis* seedlings were found to demonstrate a high degree of flooding tolerance (Mano and Omori 2007). This teosinte's adventitious roots are able to absorb oxygen and this property might be crucial to its ability to adapt to flooding circumstances. According to Ray et al. (1999), *Z. nicaraguensis* and *Z. luxurians* also generate well-formed aerenchyma in adult plants, which confers resistance to flooding circumstances. *Z. luxurians* was identified as a source of a possible waterlogging tolerance (Omori and Mano 2007). Because it offers wider root angles, promote shallow root systems that remain in better-aerated soil layers, thereby enhancing oxygen availability, reducing hypoxia-induced root damage, and improving plant survival under waterlogged conditions. The constitutive production of root aerenchyma, the tissue that permits oxygen to reach the distal areas of the roots (Gitz and Baker 2013), is the reason why eastern gamagrass can withstand waterlogging.

3.8 Tolerance to Biotic Stresses

3.8.1 Diseases Resistance

Plant diseases caused by diverse pathogens lead to significant reductions in crop yield and pose major threats to food security worldwide. A virulent pathogen, a vulnerable host, and a conducive environment are necessary for the development of a disease. The various wild species provide resistance against the different diseases (Table 3.3). Maize ear and stalk rots, particularly those caused by *Fusarium* spp., are the most common disease among the numerous biotic variables that cause significant losses in maize yield. *Zea diploperennis* and *Zea mays* spp. *parviglumis* had higher resistance towards the *F. verticilloids* which is a causal agent of *Fusarium* ear rot, was identified through QTL studies using the Teosinte introgressed RIL population (Feng et al. 2022). Using a population of near isogenic lines and traditional quantitative trait loci (QTL) mapping, Lennon and Krakowsky (2016) investigated the genetics of teosinte resistance to the fungal disease gray leaf spot (GLS) and were able to locate the teosinte-derived GLS resistance QTL in bin 4.07. Additionally, Teosinte (*Zea mays* ssp. *Parviglumis*) was found to exhibit resistance to corn smut disease (Chavan and Smith 2014). A study on BLSB resistance using teosinte (*Zea. mays* ssp. *Parviglumis*) introgressed lines had detected seven QTLs, among them a major QTL linked with phi10918 was co-localized with three minor QTLs for ear length, kernel rows per ear, and kernels per row (Adhikari et al. 2022).

Resistance to the maize chlorotic dwarf virus (MCDV) was introduced from *Z. diploperennis* through a backcross population by Findley et al. (1982). In fact, Ramirez (1997) used sexual hybridization to introduce resistant genes *from Z. mays*

Table 3.3 List of important wild species as source of maize disease resistance

S. no.	Wild species donor	Diseases	References
1	*Zea diploperennis*	Maize chlorotic mottle virus	Nault et al. (1982)
		Maize chlorotic dwarf virus	Findley et al. (1982)
		Maize streak virus	Nault et al. (1982)
		Northern corn leaf blight & Southern	Wei and Zhao (2003)
2	*Zea mays* ssp. .*parviglumis*	Gray leaf spot	Lennon and Krakowsky (2016)
		Banded leaf and sheath blight	Adhikari et al. (2022)
		Southern corn leaf blight	Lennon and Krakowsky (2017)
3	*Zea mexicana*	Downy mildew	Ramirez (1997)
		Gray leaf spot	Guerrero-Corona and Santacruz-Ruvalcaba (2021)
4	*Tripsacum dactyloides*	Common Rust	Bergquist (1981)
		Northern Corn Leaf Blight	Hoisington et al. (1999)
		Maize weevil	Throne and Eubanks (2015)

ssp. *mexicana, Z. diploperennis*, and *Z. perennis* into maize. Furthermore, it was noted that the F_1 and F_2 generations of maize-*Zea. mays* ssp. *mexicana* crosses showed the introduction of resistance against *Fusarium* (Pásztor and Borsos 1990). A study conducted by Joshi and Adhikari (2023) utilized 169 backcross inbred lines (BILs), derived using teosinte (*Zea mays* ssp. *parviglumis*), to map stable and reliable MLB-resistant QTLs and found four QTLs linked to four markers by various methods and considered reliable.

3.8.2 Insect Resistance

Insect pests are organisms that feed on plants and harm their tissues or spread plant diseases, which lowers crop yields, plant health, and growth. Leaf-chewing caterpillars, sap-sucking aphids, borers, thrips, whiteflies, beetles, and other herbivorous arthropods that engage with plant hosts through a variety of feeding techniques are examples of these pests. A major biotic stress in agriculture, insect pest infestations in crops cause large financial losses and pose a threat to global food security. Insect-pest resistance is another advantageous characteristic that wild relatives have possessed. In fact, by directly harming plants and making grains unsuitable for ingestion by humans and animals, insect pests have caused significant losses. Teosinte leaves are thought to be less appealing to herbivores than farmed maize because of physical obstacles such as increased tissue toughness, higher silica deposition, thicker cuticle and epidermis, greater trichome density, elevated lignin content, and enhanced chemical defences, collectively reducing herbivore feeding efficiency. Different sources of wild relatives for various insect resistance are mentioned in

Table 3.4 List of insect resistance traits in wild relatives of maize

Trait	Crop wild relative	Potential reason for resistance
Tolerance to fall armyworm (*Spodoptera frugiperda*)	*Z. mays* ssp. *parviglumis*	Leaf trichome and leaf toughness
	Z. diploperennis	Chemical composition of leaves
	Z. mays spp. *parviglumis*	The increased expression of wip1, RP1, and chitinase genes
	Teosinte (no information related to specific species)	Larval parasitoids, including *Cotesia marginiventris* and *Meteorus laphygmae*, are drawn to the emission of herbivore-induced volatiles, including indole and many mono- and sesquiterpenes produced by FAW leaf herbivory
Tolerance to maize spotted stalk borer	*Z. mays* ssp. *mexicana*	Increased concentration of benzoxazinoids (BXs)
	Z. mays spp. *parviglumis* *Z. perennis* *Z. mays* spp. *mexicana*	The egg oviposition of the maize spotted stalk borer (*Chilo partellus*), which draws parasitoids for its eggs (*Trichogramma boucheri*) and larvae (*Cotesia sesamiae*), led to the emission of (E)-4,8-Dimethyl-1,3,7-nonatriene
Tolerance to western corn rootworm (*Diabrotica v. virgifera*)	Teosinte (no information related to specific species)	Root herbivory that releases (E)-β-caryophyllene, which draws the entomopathogenic worm *Heterorhabditis megibis*

Table 3.4. In addition to physical barriers, tolerance to herbivore insect consumption may be produced by the chemical makeup of teosinte leaves, namely by poisonous secondary metabolites (Howe and Jander 2008). Farias-Rivera and Hernandez-Mendoza (2003) demonstrated that *Z. diploperennis* leaf extracts and residual fibre impacted FAW growth, resulting in decreased pupae length and width and a higher cumulative death rate. Eubanks (2001) created viable recombinants that were cross-fertile with maize by crossing *Tripsacum* with diploid perennial teosinte. According to insect bioassays and field root damage assessments, this made it possible to introduce *Tripsacum* genetic material into maize and create experimental lines, some of which showed resistance to rootworms. Similarly, in order to improve plant resistance or tolerance to corn rootworms, Prischmann and Dashiell (2009) created *Tripsacum*-introgressed maize germplasm in breeding programs.

3.9 Breeding Strategies for Incorporating Wild Relatives

The use of crop wild relatives in maize breeding is a powerful strategy to enhance genetic diversity and improve resilience to various biotic and abiotic stresses. Wild relatives such as Teosinte and *Tripsacum* offer untapped genetic resources that hold potential for improving stress tolerance, disease resistance, and yield stability. However, the incorporation of beneficial alleles from wild relatives into cultivated maize presents several challenges, including linkage drag and genetic incompatibility. Over time, both traditional and modern breeding strategies have evolved to address these issues and facilitate the successful introgression of favourable traits from wild germplasm.

3.9.1 Traditional Breeding Approaches

Traditional breeding methods for maize improvement have largely relied on hybridization, backcrossing, and selection to transfer beneficial traits from crop wild relatives into elite varieties. One of the earliest and most widely adopted approaches involved crossing maize with its closest wild relative, Teosinte. However, successful hybridization between maize and Teosinte is challenging due to cross-compatibility barriers. For instance, the *tcb1* (Teosinte crossing barrier 1) gene governs unidirectional compatibility, allowing Teosinte pollen to fertilize maize silk but not the reverse (Kermicle and Evans 2010). Early breeding efforts focused on incorporating genetic material from temperate, tropical, and subtropical regions to enhance maize's adaptability and genetic diversity. A notable example is the Germplasm Enhancement of Maize (GEM) project initiated in 1993, which successfully integrated exotic germplasm into U.S. maize breeding programs, resulting in breeding lines with improved grain yield and adaptability (Rogers and Bian 2022).

Traditional methods such as mass selection and Backcross have also played a foundational role. Pure-line selection, introduced by Johannsen, involves creating superior homozygous lines through repeated self-pollination (Poehlman 2013). Although effective, this method is limited by low heritability due to environmental influences. Mass selection improves upon this by focusing on highly heritable traits and combining genetic contributions from multiple parents through controlled gamete segregation. These methods, though slow, have been crucial in developing maize varieties better suited to specific environments (Lamichhane and Thapa 2022). Over the last century, plant breeders have leveraged wild relatives of maize to introduce resistance to biotic stresses Backcross breeding is another key traditional method that facilitates the introgression of desirable traits from wild relatives while retaining the elite genetic background. This method involves crossing a donor parent carrying the desired trait with a recurrent elite parent, followed by repeated backcrossing to recover nearly 98% of the elite parent's genome after five to six generations (Vogel 2009). It has been extensively used to improve resistance to pests and diseases in maize. The initial hybrids between maize and *Z. diploperennis* were sterile; subsequent backcrossing led to fertile progeny with resistance to MCDV (Findley et al. 1982). Recurrent selection, introduced by Hull, builds on these traditional methods by continuously crossing high-performing individuals over multiple generations to increase the frequency of favourable traits in a population (Ramya and Singh 2016).

3.9.2 Modern Breeding Approaches

Traditional breeding methods like backcrossing and recurrent selection have laid the foundation for utilizing wild germplasm; however, these methods are often time-consuming and less precise. Recent advancements in molecular biology, genomics, and genome editing have significantly improved the efficiency and precision of introgression breeding, allowing breeders to harness the genetic potential of CWR more effectively. Marker-assisted selection (MAS) has become a cornerstone of modern maize breeding by enabling precise identification and introgression of genes linked to desirable traits. Through gene pyramiding, MAS enhances durable pathogen resistance across generations, while marker-assisted backcrossing (MABC) accelerates the recovery of elite genetic backgrounds using molecular markers (Singh and Wang 2020). These approaches have reduced breeding cycle time and improved genetic gains, particularly for traits that are difficult to select phenotypically. Genomic selection (GS) and genome-wide association studies (GWAS) further strengthen modern breeding by enabling the prediction of hybrid performance based on genomic profiles, facilitating rapid incorporation of beneficial alleles from wild relatives. CIMMYT has demonstrated that GS can nearly halve breeding timelines while improving yield, ear traits, and stress tolerance, especially for complex polygenic traits such as drought and heat tolerance. Genomic tools, including QTL mapping and haplotype–trait association mapping, have also enhanced introgression breeding by identifying loci linked to stress tolerance, nutrient use efficiency, and disease resistance, as demonstrated by the GEM project (Rogers and Bian 2022).

Doubled haploid (DH) technology has further accelerated maize improvement by enabling rapid development of homozygous inbred lines. The discovery of natural haploid inducers such as Stock 6, along with genes like MATRILINEAL (MTL), has significantly increased haploid induction rates, making DH technology a powerful and widely adopted breeding tool (Rotarenco et al. 2010). Multiparent advanced generation intercross (MAGIC) and nested association mapping (NAM) populations have added further value by increasing genetic diversity and improving the mapping of complex traits such as disease resistance and yield stability in maize and other crops.

3.9.3 Biotechnological Approaches

Genome editing technologies, particularly CRISPR/Cas9, represent a major breakthrough in maize breeding by enabling precise modification of target genes without introducing large chromosomal segments. A notable example is the editing of the ARGOS8 promoter, which regulates ethylene responses, resulting in significant yield gains under drought stress without compromising performance under optimal conditions (Shi et al. 2017). CRISPR has also shown promise in increasing haploid induction rates, thereby accelerating inbred line development. Furthermore, genome editing supports de novo domestication by introducing domestication-related traits into wild species, allowing the development of crops better adapted to specific environments (Fernie and Yan 2019). This precision helps overcome linkage drag during wild introgressions, as demonstrated by the successful mapping and transfer of flooding tolerance genes from *Zea nicaraguensis* into defined chromosomal regions without undesirable traits (Mano and Omori 2013). Although other genome-editing tools such as zinc-finger nucleases (ZFNs) and TALENs have been applied in maize, CRISPR/Cas9 remains the most widely adopted due to its simplicity, efficiency, and cost-effectiveness (Altpeter et al. 2016). Despite certain limitations related to efficiency and feasibility, genome editing holds strong potential for future pre-breeding efforts involving wild relatives.

Omics technologies, including transcriptomics, proteomics, and metabolomics, further enhance the utilization of wild relatives in maize improvement. Transcriptomic studies have identified differentially expressed genes associated with drought tolerance in wild maize relatives, providing valuable targets for genome editing (Van Tassel and Tesdell 2020), while proteomic analyses of teosinte have revealed stress-responsive proteins linked to climate resilience. Advances in genetically modified (GM) maize also illustrate the potential of modern breeding approaches; although early GM varieties primarily targeted biotic stresses, improving abiotic stress tolerance remains challenging due to the polygenic nature of these traits. Long-term breeding efforts using wild relatives, including distant hybridization with species such as *Zea diploperennis*, have successfully produced inbred lines with enhanced resistance to diseases, pests, and environmental stresses, exhibiting strong heterosis

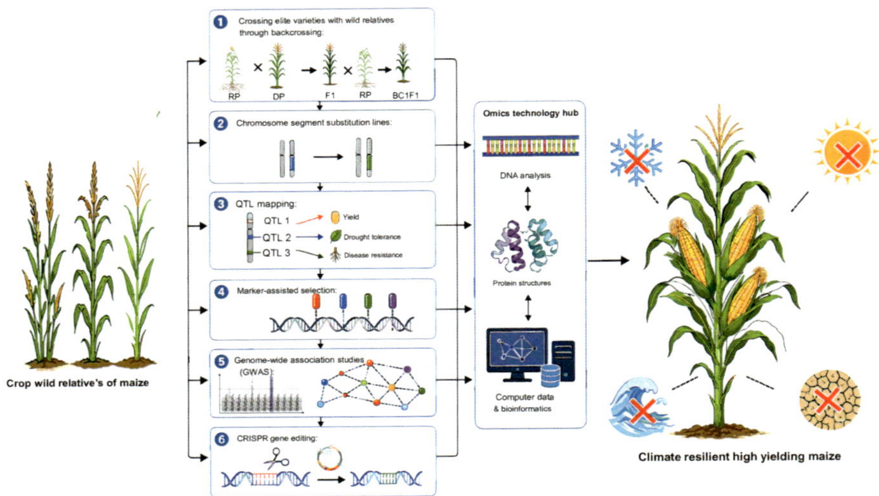

Fig. 3.1 Overview of genetic interventions for utilization of wild relatives in maize improvement

and yield stability when crossed with elite testers. An integrated breeding strategy combining traditional approaches such as backcrossing and recurrent selection with modern tools including MAS, GS, genome editing, and doubled haploid technology is essential for addressing emerging challenges in maize production (Fig. 3.1). This synergistic framework enhances both efficiency and precision, enabling the development of climate-resilient, high-yielding maize varieties and supporting future food security under changing environmental conditions.

3.10 Conclusion

Climate change poses major challenges to maize cultivation, including temperature fluctuations, extreme weather, soil degradation, and increased pest and disease pressures. Heat stress, droughts, floods, and erratic rainfall reduce yield and quality, while soil erosion and nutrient loss threaten long-term sustainability. Limited genetic diversity further restricts the development of resilient varieties. Climate-smart practices—such as minimum tillage, crop rotation, and cover cropping—improve soil health and water retention. Advanced breeding approaches, including genomic selection (GS), GWAS, marker-assisted selection (MAS), and speed breeding, help develop stress-tolerant and pest-resistant maize. Precision agriculture, multi-omics studies, and global germplasm exchange enhance trait discovery and breeding efficiency. Policies supporting research, extension services, and ethical deployment of GMOs are essential. Emerging technologies like gene editing, synthetic biology, and international collaborations will enable the development of region-specific, climate-resilient maize varieties, ensuring food security and sustainable agriculture.

References

Adhikari S, Joshi A, Kumar A (2022) Identification of QTL for banded leaf and sheath blight in teosinte-derived maize population. Agric Res 11:155–163. https://doi.org/10.1007/s40003-021-00567-7

Allendorf FW, Hohenlohe PA, Luikart G (2010) Genomics and the future of conservation genetics. Nat Rev Genet 11:697–709. https://doi.org/10.1038/nrg2844

Altpeter F, Springer NM, Bartley LE (2016) Advancing crop transformation in the era of genome editing. Plant Cell 28:1510–1520. https://doi.org/10.1105/tpc.16.00196

Beadle GW (1939) Teosinte and the origin of maize. J Hered 30:245–247

Bergquist RR (1981) Transfer from *Tripsacum dactyloides* to corn of a major gene locus conditioning resistance to *Puccinia sorghi*. Phytopathology 71:518–520

Berthaud J, Savidan Y, Barre M, Leblanc O (1997) Tripsacum. In: Fucillo D, Sears L, Stapleton P (eds) Biodiversity in trust. Cambridge University Press, pp 261–278

Bolaños J, Edmeades GO, Martinez L (1993) Eight cycles of selection for drought tolerance in lowland tropical maize: responses in drought-adaptive physiological and morphological traits. Field Crop Res 31:269–286. https://doi.org/10.1016/0378-4290(93)90066-V

Buckler ES, Stevens NM (2006) Maize origins, domestication, and selection. In: Darwin's harvest: new approaches to the origins, evolution, and conservation of crops. Columbia University Press, pp 67–90

Campos H, Caligari PDS, Edmeades GO (2017) Tropical maize (Zea mays L.). In: Genetic improvement of tropical crops. Springer, pp 57–109. https://doi.org/10.1007/978-1-4419-0293-1

Cassidy ES, West PC, Gerber JS, Foley JA (2013) Redefining agricultural yields: from tonnes to people nourished per hectare. Environ Res Lett 8:034015. https://doi.org/10.1088/1748-9326/8/3/034015

Chavan S, Smith SM (2014) A rapid and efficient method for assessing pathogenicity of *Ustilago maydis* on maize and teosinte lines. J Vis Exp 83:e50712. https://doi.org/10.3791/50712

Chia JM, Song C, Bradbury PJ (2012) Maize HapMap2 identifies extant variation from a genome in flux. Nat Genet 44:803–807. https://doi.org/10.1038/ng.2313

Eubanks MW (1995) A cross between two maize relatives: *Tripsacum dactyloides* and *Zea diploperennis*. Econ Bot 49:172–182. https://doi.org/10.1007/BF02862785

Eubanks MW (2001) The mysterious origin of maize. Econ Bot 55:492–514. https://doi.org/10.1007/s12231-001-0024-2

Eubanks MW (2006) A genetic bridge to utilize Tripsacum germplasm in maize improvement. Maydica 51:315–327

Farias-Rivera LA, Hernandez-Mendoza JL (2003) Effect of leaf extracts of teosinte (*Zea diploperennis* L.) and a Mexican maize variety on the growth and survival of the fall armyworm (*Lepidoptera*: Noctuidae). Fla Entomol 86:239–243. https://doi.org/10.2307/3495391

Feng X, Xiong H, Zheng D (2022) Identification of *Fusarium verticillioides* resistance alleles in three maize populations with teosinte gene introgression. Front Plant Sci 13:942397

Fernie AR, Yan J (2019) De novo domestication: an alternative route toward new crops for the future. Mol Plant 12:615–631. https://doi.org/10.1016/j.molp.2019.04.009

Findley WR, Nault LR, Styer WE, Gordon DT (1982) Inheritance of maize chlorotic dwarf virus resistance in maize × *Zea diploperennis* backcrosses. Maize Newsl 56:165–166

Foley JA, Ramankutty N, Brauman KA (2011) Solutions for a cultivated planet. Nature 478:337–342. https://doi.org/10.1038/nature10452

Fustier M-A (2017) Signatures of local adaptation in lowland and highland teosintes from whole-genome sequencing of pooled samples. Mol Ecol 26:2738–2756. https://doi.org/10.1111/mec.14082

Galinat WC (1988) The origin of corn. In: Sprague GF, Dudley JW (eds) Corn and corn improvement, 3rd edn. Crop Science Society of America, pp 1–31

Garkoti P, Senthilkumar V (2025) Deciphering the potential *Zea nicaraguensis* for addressing yield plateau in maize. Genet Resour Crop Evol 72:7601

Gepts P (2012) Biodiversity in agriculture: domestication, evolution, and sustainability. Cambridge University Press, Cambridge

Gitz DC, Baker JT (2013) Suitability of eastern gamagrass for in situ precipitation catchment forage production in playas. Agron J 105:907–914. https://doi.org/10.2134/agronj2012.0358

Gonzalez-Segovia E, Pérez-Limon S (2019) Characterization of introgression from the teosinte *Zea mays* ssp. *mexicana* to Mexican highland maize. PeerJ 7:e6815. https://doi.org/10.7717/peerj.6815

Goodman MM, Galinat WC (1988) The history and evolution of maize. Crit Rev Plant Sci 7(3):197–220

Guerrero-Corona A, Santacruz-Ruvalcaba F (2021) Phenotypic identification of near isogenic maize–teosinte lines (*Zea* ssp.) involved in resistance to gray leaf spot. Plant Breed 140(5):775–785

Harlan JR, de Wet JMJ (1971) Toward a rational classification of cultivated plants. Taxon 20:509–517. https://doi.org/10.2307/1218252

Hoisington D, Khairallah M, Reeves T (1999) Plant genetic resources: what can they contribute toward increased crop productivity. Proc Natl Acad Sci USA 96:5937–5943. https://doi.org/10.1073/pnas.96.11.5937

Hossain F et al (2016) Maize. In: Hossain F, Singh M, Kumar S (eds) Broadening the genetic base of grain cereals. Springer, pp 1–35. https://doi.org/10.1007/978-81-322-3613-9_4
Howe GA, Jander G (2008) Plant immunity to insect herbivores. Annu Rev Plant Biol 59:41–66. https://doi.org/10.1146/annurev.arplant.59.032607.092825
Hufford MB, Xu X, van Heerwaarden J (2012) Comparative population genomics of maize domestication and improvement. Nat Genet 44:808–818. https://doi.org/10.1038/ng.2309
Iltis HH (1983) From teosinte to maize: the catastrophic sexual transmutation. Science 222:886–894
Iltis HH, Benz BF (2000) Zea nicaraguensis (Poaceae), a new teosinte from Pacific coastal Nicaragua. Novon 382–390.
IPCC (2014) Climate change 2014: synthesis report. Contribution of Working Groups I, II and III to the fifth assessment report of the intergovernmental panel on climate change. In: Pachauri RK, Meyer LA (eds) IPCC, Geneva
Joshi A, Adhikari S (2023) Prospecting quantitative trait loci for maydis leaf blight resistance using a population of teosinte-introgressed maize (*Zea mays* ssp. *mays*) and in silico identification of candidate resistance genes. J Phytopathol 171:369–387. https://doi.org/10.1111/jph.13116
Kaya C, Ashraf M, Dikilitas M, Tuna AL (2013) Alleviation of salt stress-induced adverse effects on maize plants by exogenous application of indoleacetic acid and inorganic nutrients: a field trial. Aust J Crop Sci 7:249–254
Kemper WD, Alberts EE (1997) Aerenchyma, acid tolerance and associative nitrogen fixation enhance carbon sequestration in soil. In: Proceedings of management of carbon sequestration in soil. CRC Press, pp 221–234
Kermicle JL, Evans MMS (2010) The *Zea mays* L. sexual compatibility gene *ga2*: naturally occurring alleles, their distribution and role in reproductive isolation. J Hered 101:737–749. https://doi.org/10.1093/jhered/esq090
Khoury CK, Brush S (2022) Crop genetic erosion: understanding and responding to loss of crop diversity. New Phytol 233:84–118. https://doi.org/10.1111/nph.17733
Kumar P, Choudhary M (2022) Salinity stress tolerance and omics approaches: revisiting progress and achievements in major cereal crops. Heredity 128(6):497–518
Kumar S, Shekhar M (2005) Stress on maize in tropics. Directorate of Maize Research, New Delhi, pp 172–194
Kumar A, Adhikari S, Joshi A (2024a) Study of co-localization of QTL for drought tolerance of morphological traits in a teosinte-introgressed maize population. Cereal Res Commun 52:555–567. https://doi.org/10.1007/s42976-024
Kumar P, Choudhary M, Sheoran S, Longmei N, Kumar B, Jat BS, Dagla MC, Bhushan B, Aggarwal SK, Bagaria PK, Sharma A (2024b) Teosinte-derived advanced backcross population harbors genomic regions for grain yield attributing traits in maize. Int J Mol Sci 25(19):10300
Kumar P, Choudhary M, Sheoran S, Kumar B, Kumar S, Sharma A, Bhushan B, Jat BS, Paul D, Aggarwal SK, Singh SB (2025) Harnessing teosinte for quality traits enhancement and genetic diversity in maize. Cereal Res Commun:1–12. https://doi.org/10.1007/s42976-025-00734-4
Kumar P, Choudhary M, Kumar B, Longmei N, Jat BS, Sharma A (2026) Exploring teosinte (*Zea mays* ssp. *parviglumis*) for identification of genomic regions for root and shoot traits at seedling stage in maize. Plant Mol Biol Report 44(1):1–8
Lamichhane S, Thapa S (2022) Advances from conventional to modern plant breeding methodologies. Plant Breed Biotechnol 10:1–4
Lennon JR, Krakowsky M (2016) Identification of alleles conferring resistance to gray leaf spot in maize derived from its wild progenitor species teosinte. Crop Sci 56:209–218. https://doi.org/10.2135/cropsci2015.07.0457
Lennon JR, Krakowsky M (2017) Identification of teosinte alleles for resistance to southern leaf blight in near-isogenic maize lines. Crop Sci 57:1973–1983. https://doi.org/10.2135/cropsci2016.11.0937
Lobell DB, Field CB (2007) Global-scale climate–crop yield relationships and the impacts of recent warming. Environ Res Lett 2:014002. https://doi.org/10.1088/1748-9326/2/1/014002

Mammadov J, Buyyarapu R (2018) Wild relatives of maize, rice, cotton and soybean: treasure troves for tolerance to biotic and abiotic stresses. Front Plant Sci 9:886. https://doi.org/10.3389/fpls.2018.00886

Mangelsdorf PC, Reeves RG (1938) The origin of maize. Proc Natl Acad Sci USA 24:303–312. https://doi.org/10.1073/pnas.24.7.303

Mangelsdorf PC, Reeves RG (1939) The origin of Indian corn and its relatives. Texas Agricultural Experiment Station Bulletin No. 574, pp 1–315

Mano Y, Nakazono M (2021) Genetic regulation of root traits for soil flooding tolerance in genus Zea. Breed Sci 71:20117–20139. https://doi.org/10.1270/jsbbs.20117

Mano Y, Omori F (2007) Breeding for flooding-tolerant maize using teosinte as a germplasm resource. Plant Root 1:17–21. https://doi.org/10.3117/plantroot.1.17

Mano Y, Omori F (2008) Verification of QTL controlling root aerenchyma formation in a maize × teosinte (*Zea nicaraguensis*) advanced backcross population. Breed Sci 58:217–223. https://doi.org/10.1270/jsbbs.58.217

Mano Y, Omori F (2013) Relationship between constitutive root aerenchyma formation and flooding tolerance in *Zea nicaraguensis*. Plant Soil 370:447–455. https://doi.org/10.1007/s11104-013-1641-0

Mano Y, Omori F, Muraki M, Takamizo T (2005) QTL mapping of adventitious root formation under flooding conditions in tropical maize seedlings. Breed Sci 55:343–347. https://doi.org/10.1270/jsbbs.55.343

Masson-Delmotte V, Zhai P (2018) Global warming of 1.5°C: an IPCC special report. IPCC, Geneva

Nault LR, Gordon DT, Damsteegt VD, Iltis HH (1982) Response of annual and perennial teosintes (*Zea*) to six maize viruses. Plant Dis 66:61–62. https://doi.org/10.1094/PD-66-61

Omori F, Mano Y (2007) QTL mapping of root angle in F_2 populations from maize 'B73' × teosinte *Zea luxurians*. Plant Root 1:57–65. https://doi.org/10.3117/plantroot.1.57

Pásztor K, Borsos O (1990) Inheritance and chemical composition in inbred maize (*Zea mays* L.) × teosinte (*Zea mays* subsp. *mexicana* (Schräder) Iltis) hybrids. Növénytermelés 39:193–213

Pedersen O, Nakayama Y (2021) Lateral roots, in addition to adventitious roots, form a barrier to radial oxygen loss in *Zea nicaraguensis* and a chromosome segment introgression line in maize. New Phytol 229:94–105. https://doi.org/10.1111/nph.16870

Poehlman JM (2013) Breeding field crops. Springer Science & Business Media, Berlin

Prischmann DA, Dashiell KE (2009) Evaluating Tripsacum-introgressed maize germplasm after infestation with western corn rootworms (Coleoptera: Chrysomelidae). J Appl Entomol 133:10–20. https://doi.org/10.1111/j.1439-0418.2008.01341.x

Ramirez DA (1997) Gene introgression in maize (*Zea mays* ssp. *mays* L.). Philipp J Crop Sci 22:51–63

Ramya P, Singh GP (2016) Effect of recurrent selection on drought tolerance and related morpho-physiological traits in bread wheat. PLoS One 11:e0156869. https://doi.org/10.1371/journal.pone.0156869

Ray J, Sinclair T, Kindiger B (1999) Introgressing root aerenchyma into maize [Zea mays L.]. Maydica 44:113–117

Ray DK, West PC (2019) Climate change has likely already affected global food production. PLoS One 14:e0217148. https://doi.org/10.1371/journal.pone.0217148

Rogers A, Bian Y (2022) Genomic prediction for the germplasm enhancement of maize project. Plant Genome 15:e20267. https://doi.org/10.1002/tpg2.20267

Rotarenco VA, Dicu G, State D, Fuia S (2010) New inducers of maternal haploids in maize. Maize Genet Coop Newsl 84:1–7

Sahoo S, Adhikari S (2021) Use of wild progenitor teosinte in maize (*Zea mays* ssp. *mays*) improvement: present status and future prospects. Trop Plant Biol 14:156–179. https://doi.org/10.1007/s12042-021-09286-1

Savary S, Nelson A, Sparks AH, Willocquet L (2011) International agricultural research tackling the effects of global and climate changes on plant diseases in the developing world. Plant Dis 95:1204–1216. https://doi.org/10.1094/PDIS-07-11-0575

Sayadi Maazou A, Tu J, Qiu J, Liu Z (2016) Breeding for drought tolerance in maize (*Zea mays* L.). Am J Plant Sci 7:1858–1870. https://doi.org/10.4236/ajps.2016.712174
Senthil (2024) Identification of genomic regions for waterlogging tolerance and kernel traits in *Zea nicaraguensis* introgressed maize lines. PhD thesis, Govind Ballabh Pant University of Agriculture and Technology
Shi J, Gao H, Wang H, Lafitte HR, Archibald RL, Yang M (2017) ARGOS8 variants generated by CRISPR–Cas9 improve maize grain yield under field drought stress conditions. Plant Biotechnol J 15:207–216. https://doi.org/10.1111/pbi.12603
Shull GH (1908) The composition of a field of maize. In: American Breeders' Association Reports, vol 4. American Breeders' Association, pp 296–301
Shull GH (1909) A pure-line method in corn breeding. In: American Breeders' Association Reports, vol 5. American Breeders' Association, pp 51–59
Singh N, Wang DR (2020) A coordinated suite of wild-introgression lines in indica and japonica elite backgrounds. Front Plant Sci 12:564824. https://doi.org/10.3389/fpls.2020.564824
Singh AP, Pandey PK, Joshi RP (2024) Harnessing genetic diversity for climate-resilient maize: a comprehensive review. J Appl Biol Agric 1:21–28
Subbaiah CC, Sachs MM (2003) Molecular and cellular adaptations of maize to flooding stress. Ann Bot 91:119–127. https://doi.org/10.1093/aob/mcf057
Szalai G, Janda T (2009) Effect of salt stress on salicylic acid synthesis in young maize (*Zea mays* L.) plants. J Agron Crop Sci 195:165–171. https://doi.org/10.1111/j.1439-037X.2008.00352.x
Throne JE, Eubanks MW (2015) Resistance of Tripsacum-introgressed maize lines to *Sitophilus zeamais*. J Stored Prod Res 64:62–64. https://doi.org/10.1016/j.jspr.2015.08.007
Van Tassel DL, Tesdell O (2020) New food crop domestication in the age of gene editing: genetic, agronomic and cultural change remain co-evolutionarily entangled. Front Plant Sci 11:789. https://doi.org/10.3389/fpls.2020.00789
Vogel KE (2009) Backcross breeding. In: Weeks DP, Yang B (eds) Transgenic maize: methods and protocols. Humana Press, pp 161–169
Wei WH, Zhao WP (2003) Genomic in situ hybridization analysis for identification of introgressed segments in alloplasmic lines from *Zea mays* × *Zea diploperennis*. Hereditas 138:21–26. https://doi.org/10.1534/hereditas.138.21
White SE, Doebley JF (1999) The molecular evolution of terminal ear1, a regulatory gene in the genus Zea. Genetics 153:1455–1462
Xin Li, Yunlong Li, Yan Sun, Sinan Li, Quan Cai, Shujun Li, Minghao Sun, Tao Yu, Xianglong Meng, Jianguo Zhang, Ying Zhao (2024) Harnessing Maize Genetic Diversity for Climate-Resilient Cultivars. Authorea. https://doi.org/10.22541/au.172505649.95280473/v1
Zaidi PH, Maniselvan P (2010) Genetic analysis of waterlogging tolerance in tropical maize. Maydica 55:17–26

Seeds of Success: The Legacy of Conventional Maize Breeding

4

Ramesh Kumar, Santosh Kumar, Chikkappa Gangadhar Karjagi, and Abhijit Kumar Das

Abstract

Maize is one of the world's most important cereals, contributing substantially to food, feed, and industrial sectors. Its exceptional genetic variability and cross-pollinated nature have enabled significant yield improvement and adaptation through conventional breeding approaches. The present work synthesizes the historical trajectory of maize improvement and elucidates the genetic principles and breeding methodologies that have shaped modern maize production. Early domestication from *teosinte* involved key genetic changes influencing ear morphology, kernel size, and plant architecture, forming the basis for subsequent selection. The discovery of heterosis and the development of inbred–hybrid systems triggered unprecedented yield gains, while recurrent, backcross, and pedigree selection strengthened disease resistance, abiotic stress tolerance, and grain quality. Major achievements include the development of high-yielding hybrids, quality protein maize, and region-specific climate-resilient cultivars. However, challenges persist, including genetic bottlenecks, slow breeding cycles, and increasing vulnerability to climate extremes and evolving pathogens. Integrating molecular tools such as marker-assisted selection, genomic selection, and targeted gene editing provides new opportunities to enhance selection efficiency while maintaining the genetic diversity essential for long-term sustainability. Overall, conventional

R. Kumar (✉) · A. K. Das
ICAR-Indian Institute of Maize Research, Ludhiana, India

S. Kumar
ICAR-Indian Agricultural Research Institute, Jharkhand, India

C. G. Karjagi
Regional Maize Research and Seed Production Center (ICAR-Indian Institute of Maize Research), Begusarai, India

M. Choudhary et al. (eds.), *Next-Generation Maize Research*,
https://doi.org/10.1007/978-981-92-0722-0_4

breeding continues to be central to global maize improvement and remains indispensable for meeting future food and nutritional security demands.

Keywords

Conventional maize breeding · Heterosis · Hybrid maize · Genetic diversity · Quality protein maize

4.1 Introduction

4.1.1 Global Importance of Maize

Maize, commonly known as corn, is one of the most important cereal crops worldwide, serving as a primary source of food, feed, and industrial raw material. Owing to its wide adaptability across diverse agro-climatic conditions, maize is cultivated in more than 170 countries over approximately 197 million hectares, with an annual global production exceeding 1148 million metric tons, making it the highest-producing cereal crop globally, surpassing both wheat and rice (FAOSTAT 2022). Its versatility and high productivity have positioned maize as a cornerstone of global food demand, economic development, and sustainable agricultural systems (Hossain et al. 2022; Shiferaw et al. 2011). In developing regions, maize contributes substantially to dietary energy intake and livelihood security. It accounts for nearly 61% of total caloric intake in Mesoamerica and about 45% in Eastern and Southern Africa, underscoring its role as a staple food crop for millions of people. In other regions such as the Andean countries, West and Central Africa, and South Asia, maize contributes between 4% and 22% of daily caloric intake, reflecting its broad but region-specific nutritional importance (Shiferaw et al. 2011). Beyond human consumption, maize is indispensable to global livestock production, contributing nearly 63% of total feed used in the poultry and piggery industries, and plays a critical role in industrial sectors including starch production, ethanol-based biofuels, biodegradable plastics, pharmaceuticals, and textiles (Yadav et al. 2015; Bao et al. 2012; Devi et al. 2017).

The demand for maize is projected to increase substantially due to population growth, rising income levels, expansion of livestock industries, and increasing reliance on biofuels. It is estimated that global maize demand may double by 2050, placing immense pressure on existing production systems, particularly under the constraints imposed by climate change, land degradation, and resource scarcity (Rosegrant et al. 2009). These challenges necessitate continuous genetic improvement of maize to ensure stable yields, resilience to biotic and abiotic stresses, and enhanced nutritional quality (Kumar et al. 2025a).

4.1.2 Rationale and Scope of Conventio+nal Maize Breeding

Maize is unique among cereal crops due to its high genetic plasticity, extensive natural outcrossing behavior, and remarkable responsiveness to selection, which together provide an exceptional foundation for genetic improvement through breeding. Conventional maize breeding, defined as the systematic selection and hybridization of maize genotypes using natural recombination without direct genetic modification, has been the principal driver of yield enhancement and trait improvement for more than a century (Dhillon and Malhi 2006; Hallauer et al. 2010). The exploitation of heterosis, first demonstrated in maize in the early twentieth century, revolutionized crop improvement and laid the foundation for modern hybrid breeding, resulting in unprecedented gains in productivity and yield stability (Shull 1909; Duvick 2005).

Historically, maize breeding has evolved from farmer-led selection of diverse landraces and open-pollinated varieties to scientifically managed hybrid breeding and population improvement programs. These efforts have led to the development of high-yielding hybrids, disease-resistant cultivars, drought-tolerant and climate-resilient varieties, and nutritionally enhanced maize such as QPM, significantly contributing to global food and nutritional security (Vasal 2002; Kumar et al. 2019, 2024a; Prasanna et al. 2020). In countries such as the United States, India, China, and several nations in Sub-Saharan Africa, conventional breeding has resulted in four- to five-fold increases in maize productivity since the early twentieth century, highlighting its transformative impact on agricultural systems (Hallauer et al. 2010; Yadav et al. 2015). Despite recent advances in molecular breeding, genomic selection, and gene editing technologies, conventional maize breeding remains the backbone of maize improvement worldwide. It provides the essential framework for exploiting genetic diversity, maintaining heterotic patterns, developing inbred lines, and translating genetic gains into farmer-preferred cultivars. The scope of conventional maize breeding extends beyond yield improvement to encompass resistance to evolving pests and diseases, tolerance to drought and heat stress, enhancement of grain quality and nutritional value, and conservation and utilization of maize genetic resources (Troyer 1999; Schnable 2015).

In this context, the present chapter synthesizes the historical evolution, principles, methodologies, and impacts of conventional maize breeding, with particular emphasis on its role in enhancing productivity, resilience, and food security. The chapter also examines the challenges confronting conventional breeding in the era of climate change and genetic erosion, and highlights how its integration with modern molecular tools can sustain its relevance and effectiveness in meeting future global maize demademand.

4.2 Global Significance and Utilization of Maize

Maize plays a central role in the global food system owing to its wide adaptability, high productivity, and diverse end uses. In many low- and middle-income countries, maize is a primary source of dietary energy. It contributes approximately 61% of total caloric intake in Mesoamerica and around 45% in Eastern and Southern Africa, where it is consumed daily in various traditional forms. In other regions, including the Andean countries, West and Central Africa, and South Asia, maize contributes between 4% and 22% of total caloric intake, underscoring its region-specific but indispensable role in human nutrition (Shiferaw et al. 2011). The ability of maize to produce relatively high yields under diverse environmental conditions, including marginal and rainfed systems, has made it a critical crop for ensuring food availability and livelihood security in vulnerable agro-ecosystems (Hossain et al. 2022).

Conventional maize breeding has been fundamental in strengthening maize production by enabling the development of high-yielding, disease-resistant, drought-tolerant, and nutritionally improved varieties. The widespread adoption of hybrid maize has significantly enhanced yield stability and productivity, thereby reducing the risk of food shortages and improving farmer incomes in major maize-growing regions (Duvick 2005; Hallauer et al. 2010).

4.2.1 Feed, Biofuel, and Industrial Uses

Beyond its importance as a food crop, maize is a cornerstone of the global feed and industrial economy. Approximately 63% of global maize production is utilized as animal feed, particularly in the poultry and piggery industries, making maize indispensable to the rapidly expanding livestock sector (Yadav et al. 2015). The high starch content and favorable energy profile of maize grains make them an ideal feed ingredient, supporting efficient meat, milk, and egg production. Maize also plays a major role in the bioenergy sector (Choudhary et al. 2020). In the United States, the world's largest maize producer, more than 30% of total maize production is used for ethanol manufacturing, contributing to renewable energy generation and reduced dependence on fossil fuels (Yadav et al. 2015). The expansion of maize-based ethanol production has further increased the strategic importance of maize in national and global energy policies.

In addition, maize serves as a vital raw material for a wide range of industrial applications. It is extensively used in the production of starch, corn syrup, and other processed food ingredients. Maize-derived products are also utilized in the manufacture of biodegradable plastics, adhesives, thickeners, and emulsifiers, with applications spanning the food processing, pharmaceutical, and textile industries (Bao et al. 2012; Devi et al. 2017). This multifaceted utilization underscores maize's unique position as both a food and an industrial crop, linking agricultural production with broader economic and industrial systems.

4.2.2 Future Demand and Emerging Challenges

Global demand for maize is projected to rise sharply in the coming decades, driven by population growth, increasing consumption of animal-based products, expansion of biofuel industries, and growing industrial uses. It has been estimated that maize demand may double by 2050, placing unprecedented pressure on existing production systems (Rosegrant et al. 2009). Meeting this rising demand will be particularly challenging under the constraints of climate change, including increased frequency of droughts, heat stress, and erratic rainfall patterns, which threaten maize productivity, especially in rainfed regions (Lobell et al. 2011; Cairns et al. 2013). In addition to climatic stresses, future maize production faces challenges related to land and water scarcity, declining soil fertility, and emerging pest and disease pressures. These factors highlight the need for sustained genetic improvement to enhance yield potential, resource-use efficiency, and resilience to biotic and abiotic stresses. Conventional maize breeding, with its proven capacity to deliver stable yield gains and adaptable cultivars, remains central to addressing these challenges and ensuring that maize continues to fulfill its critical role in global food, feed, and industrial systems (Hallauer et al. 2010; Prasanna et al. 2020).

4.3 Historical Evolution of Conventional Maize Breeding

4.3.1 Domestication and Early Selection

The history of conventional maize breeding begins with the domestication of maize from its wild progenitor, teosinte (*Zea mays* ssp. *parviglumis*), approximately 9000–10,000 years ago in Mesoamerica, particularly in present-day southern Mexico (Doebley and Stec 1991; Schnable 2015). Early indigenous farmers played a decisive role in this process through conscious selection for desirable traits such as larger ears, increased kernel size, reduced branching, and improved harvestability. These selection pressures resulted in profound morphological and genetic changes that transformed teosinte, characterized by small, hard, and shattering kernels, into cultivated maize with non-shattering cobs and exposed, soft kernels (Doebley 1992; Schnable 2015). Key genetic modifications accompanied domestication, including changes in plant architecture and ear morphology. The *tb1* (*teosinte branched1*) gene played a central role by suppressing lateral branching and promoting a single dominant stalk, thereby increasing apical dominance and ear size (Wang et al. 2017). Selection also led to the loss of natural seed shattering, enabling easier harvesting, and to extensive diversification in kernel composition, giving rise to flint, dent, floury, waxy, and sweet maize types adapted to diverse culinary, nutritional, and agro-ecological requirements (Doebley and Stec 1991; Schnable 2015).

Following domestication, maize rapidly spread across North and South America, adapting to a wide range of environments from temperate regions to tropical and high-altitude systems. By the time of European contact in the fifteenth century, maize had already become a staple crop throughout the Americas. The Columbian Exchange subsequently facilitated its introduction into Europe, Africa, and Asia, where further farmer-led selection enabled maize to adapt to new climatic and production systems, laying the foundation for its global significance (Doebley 1992; Wang et al. 2017).

4.3.2 OPVs Versus Hybrid Maize

Conventional maize breeding has historically progressed from the use of open-pollinated varieties (OPVs) to the widespread adoption of hybrid maize, with each system exhibiting distinct characteristics and implications for productivity. OPVs are genetically heterogeneous populations maintained through mass selection and natural cross-pollination. Their broad genetic base allows adaptation to variable environments and enables farmers to save seed for subsequent planting, making OPVs suitable for low-input and resource-constrained farming systems (Dhillon and Malhi 2006). However, the lack of genetic uniformity in OPVs results in variable yield performance and limits their maximum yield potential. In contrast, hybrid maize varieties are developed through controlled crosses between selected inbred lines, allowing breeders to deliberately exploit heterosis. Hybrid maize consistently outperforms OPVs in terms of grain yield, uniformity, and stability, particularly under favorable management conditions (Hallauer et al. 2010). The transition from OPVs to hybrids has been a major driver of global yield improvement, contributing to four- to fivefold increases in maize productivity since the early twentieth century (Duvick 2005). A key limitation of hybrid maize is that the heterotic advantage diminishes in subsequent generations due to genetic segregation, rendering farm-saved seed unsuitable for replanting. As a result, hybrid cultivation requires annual seed purchase, which can pose economic challenges for smallholder farmers. Despite this constraint, the superior agronomic performance and economic returns of hybrid maize have led to its dominance in global maize production systems (Duvick 2005; Hallauer et al. 2010).

4.3.3 From Open-Pollinated Varieties (OPVs) to Hybrid Maize

Until the early twentieth century, maize improvement relied primarily on mass selection within open-pollinated varieties. Farmers and early breeders selected seed from superior plants based on phenotypic performance, resulting in genetically diverse populations with broad adaptation but highly variable yields (Dhillon and Malhi 2006). While OPVs were well-suited to low-input systems and allowed seed saving, their limited yield potential and lack of uniformity constrained large-scale productivity gains. A major turning point in maize breeding occurred with the discovery of heterosis, or hybrid vigor, by George H. Shull in 1909. Shull demonstrated that crosses between genetically diverse inbred lines produced progeny with superior yield, vigor, and uniformity compared to their parents, establishing the scientific basis of hybrid maize breeding (Shull 1909). The subsequent development of inbred lines through repeated self-pollination enabled controlled hybridization and predictable exploitation of heterosis.

The practical application of heterosis was further advanced by Donald F. Jones, who introduced the double-cross hybrid system in 1918. By crossing two single-cross hybrids, Jones made hybrid seed production more feasible at a commercial scale, overcoming the low seed yield and poor agronomic performance of early inbred lines (Jones 1918). The release of the first commercial hybrid maize, "Copper Cross," in the 1920s marked the beginning of a rapid transition from OPVs to hybrids, particularly in the United States and Europe, resulting in substantial and sustained yield gains (Troyer 1999; Duvick 2005).

4.3.4 Green Revolution and Global Expansion of Hybrid Maize

The widespread adoption of hybrid maize accelerated during the mid-twentieth century, coinciding with the Green Revolution. By the 1940s and 1950s, hybrid maize had largely replaced OPVs in the United States, leading to dramatic improvements in yield stability, disease resistance, and agronomic uniformity (Duvick 2005). During the 1960s and 1970s, the benefits of hybrid maize expanded to developing regions, driven by coordinated international and national breeding efforts. Institutions such as the International Maize and Wheat Improvement Center (CIMMYT) and national agricultural research systems played a pivotal role in adapting hybrid maize to tropical and subtropical environments. The development of region-specific hybrids enabled maize cultivation under diverse stress conditions, including drought-prone and low-input systems (Prasanna et al. 2020). In India, the establishment of the All India Coordinated Maize Improvement Project (AICMIP) in 1957 marked a major milestone, leading to the development and release of high-yielding and disease-resistant hybrids tailored to different agro-climatic zones (Dhillon and Malhi 2006). As a result, maize productivity in India increased nearly fourfold, rising from about 547 kg ha^{-1} in 1950–51 to over 3.1 t ha^{-1} in recent decades (Singh et al. 2024a). By the late twentieth century, conventional maize breeding had firmly established hybrid breeding as the dominant approach worldwide. Continuous refinement through recurrent selection, backcross breeding, and pedigree selection further enhanced resistance to biotic and abiotic stresses and improved grain quality. These cumulative advances underscore how the transition from early domestication and OPVs to hybrid breeding during the Green Revolution fundamentally reshaped global maize production and laid the foundation for modern maize improvement systems (Hallauer et al. 2010; Yadav et al. 2015).

4.4 Foundations of Conventional Maize Breeding

4.4.1 Genetic Diversity and Heterosis

The foundation of conventional maize breeding lies in the crop's exceptional genetic diversity and its highly cross-pollinated nature, which together facilitate extensive recombination and the generation of novel trait combinations. This inherent variability has enabled breeders to exploit natural genetic differences for the improvement of grain yield, stress tolerance, and adaptability across diverse agro-climatic environments (Troyer 1999; Hallauer et al. 2010). Genetic diversity within maize populations provides a broad reservoir of alleles that can be selectively combined to enhance agronomic performance and sustain long-term genetic gain (Das et al. 2021). Among the most influential principles underpinning maize improvement is heterosis, or hybrid vigor, first demonstrated by George H. Shull in 1909. Heterosis refers to the phenomenon whereby hybrid offspring derived from crosses between genetically distinct inbred lines exhibit superior vigor, biomass accumulation, yield, and environmental adaptability compared to their parental lines (Shull 1909). The systematic exploitation of heterosis revolutionized maize breeding and remains the cornerstone of modern hybrid maize production. Subsequent refinement of heterotic groups and combining ability concepts has further enhanced the efficiency of hybrid development, enabling consistent yield gains and improved resilience to biotic and abiotic stresses (Duvick 2005; Hallauer et al. 2010).

4.4.2 Development of Inbred Lines and Hybrid Seed Production

The development of inbred lines is a critical prerequisite for hybrid maize breeding. Inbred lines are produced through repeated self-pollination over five to six generations, resulting in near-homozygous genotypes. Although inbreeding leads to inbreeding depression, characterized by reduced vigor and yield, it enables the fixation of desirable alleles and facilitates the assessment of combining ability among parental lines (Shull 1909). The subsequent crossing of complementary inbred lines restores vigor through heterosis, forming the basis of high-yielding hybrid maize varieties. Hybrid seed production involves a series of carefully managed steps, including the selection of parental inbred lines with strong general and specific combining ability, controlled pollination to prevent self-fertilization, and large-scale seed multiplication (Singh et al. 2019, 2021). Controlled pollination is typically achieved through manual detasseling or the use of male-sterile systems, ensuring genetic purity of hybrid seed (Hallauer et al. 2010). These practices have enabled the reliable and scalable production of single-cross, double-cross, and three-way hybrids for commercial cultivation. The integration of inbred line development with systematic hybrid seed production has been central to the success of conventional maize breeding. This framework has allowed breeders to translate genetic diversity and heterosis into stable, high-performing cultivars, underpinning sustained yield gains and reinforcing the foundational role of conventional breeding in global maize improvement (Hallauer et al. 2010; Duvick 2005).

4.5 Impact of Conventional Breeding on Maize Improvement

Conventional maize breeding has been the principal driver of sustained yield improvement over the past century, significantly strengthening global food security. The transition from open-pollinated varieties to hybrids, coupled with systematic selection and population improvement, has resulted in more than a fourfold increase in average maize yields since the early twentieth century (Hallauer et al. 2010; Duvick 2005). These genetic gains have been particularly evident in major maize-growing regions, including North America, China, Brazil, and India, where hybrid adoption has enhanced yield stability and farmer profitability (Yadav et al. 2015). In India, coordinated breeding programs led to a substantial rise in productivity, underscoring the role of conventional breeding in meeting food and livelihood security demands (Dhillon and Malhi 2006; Singh et al. 2024a, b).

4.5.1 Resistance to Diseases and Insect Pests

Maize is vulnerable to a wide range of fungal, bacterial, viral diseases, and insect pests, which can cause severe yield losses in susceptible cultivars. Conventional breeding approaches, particularly backcross breeding, recurrent selection, and hybridization, have been effective in developing resistant varieties and hybrids. Resistance to major diseases such as maize streak virus, downy mildew, gray leaf spot, and turcicum leaf blight has been successfully incorporated into elite germplasm, reducing crop losses and dependence on chemical control measures (Kumar et al. 2022; Duvick 2005; Hallauer et al. 2010; Singh et al. 2018). Similarly, breeding for tolerance and resistance to insect pests such as stem borers and fall armyworm has contributed to improved crop resilience and yield stability across diverse production environments (Yadav et al. 2015).

4.5.2 Drought and Abiotic Stress Tolerance

Abiotic stresses, particularly drought and heat, pose major constraints to maize production and can reduce yields by as much as 50% under severe conditions. Conventional breeding has addressed these challenges by exploiting genetic variation in landraces and related germplasm through recurrent selection, backcross breeding, and population improvement. Selection for traits such as deep rooting, improved water-use efficiency, and stress tolerance has led to the development of drought-resilient maize hybrids adapted to rainfed and marginal environments, especially in Sub-Saharan Africa, India, and Mexico (Cairns et al. 2013; Hallauer et al. 2010). These efforts have enhanced yield stability under climatic variability and remain central to climate-resilient maize production.

4.5.3 Nutritional Improvement: Quality Protein Maize (QPM)

One of the most notable achievements of conventional maize breeding is the development of QPM, which addresses protein malnutrition in maize-dependent populations. QPM was developed through the incorporation of the *opaque-2 (o2)* gene, which enhances lysine and tryptophan content, followed by selection of modifier genes to restore kernel hardness and agronomic performance (Vasal 2002). QPM varieties combine improved nutritional quality with competitive yield and stability and have been widely adopted in Africa, Asia, and Latin America. Their deployment has contributed significantly to improving dietary protein quality and nutritional security in regions where maize is a primary staple food (Vasal 2002; Prasanna et al. 2020).

4.6 Molecular Support to Conventional Breeding

While conventional breeding remains the foundation of maize improvement, its efficiency has been substantially enhanced through the integration of molecular tools. Marker-assisted selection (MAS), genomic selection (GS), and gene editing now complement phenotypic selection by improving precision, accelerating breeding cycles, and strengthening responses to complex stresses—without replacing conventional breeding frameworks (Xu and Crouch 2008; Prasanna et al. 2020).

4.6.1 Marker-Assisted Selection (MAS)

MAS uses DNA-based molecular markers linked to target genes or quantitative trait loci to enable indirect selection of desirable traits. By allowing selection at early growth stages and minimizing environmental effects on phenotyping, MAS improves selection accuracy and reduces breeding time (Collard and Mackill 2008). In maize, MAS has been widely applied for traits governed by major genes, particularly disease resistance and stress tolerance, and for gene pyramiding to enhance durability of resistance (Xu and Crouch 2008; Kumar et al. 2021; Yathish et al. 2022). However, its effectiveness is limited for highly polygenic traits, necessitating complementary approaches for complex trait improvement (Prasanna et al. 2020).

4.6.2 Genomic Selection (GS)

GS represents a major advancement for improving complex quantitative traits such as grain yield, drought tolerance, and nitrogen-use efficiency. Unlike MAS, GS uses genome-wide marker information to predict the breeding value of individuals, capturing both major and minor genetic effects simultaneously (Crossa et al. 2017). This approach enables rapid cycling and simultaneous selection for multiple traits, thereby shortening breeding cycles and increasing genetic gain per unit time. Large-scale maize breeding programs, including those at CIMMYT, have successfully implemented GS to accelerate hybrid development and improve performance under stress-prone environments (Crossa et al. 2017; Prasanna et al. 2020).

4.6.3 Gene Editing as a Complementary Tool

Gene editing technologies, particularly CRISPR-based systems, provide precise and targeted modification of specific genes and are emerging as powerful complements to conventional breeding. In maize, gene editing has been explored for enhancing drought tolerance (e.g., modification of stress-responsive genes such as *ZmDREB2A*), improving nutritional traits such as provitamin A content, and strengthening resistance to insect pests and herbicides (Qin et al. 2007; Zhu and Zhu 2022). These approaches offer the potential for rapid trait improvement while retaining elite genetic backgrounds developed through conventional breeding. Nevertheless, their deployment depends on regulatory frameworks and public acceptance, reinforcing the continued centrality of conventional breeding supported by molecular tools (Prasanna et al. 2020).

4.7 Case Studies of Successful Maize Breeding Programs

4.7.1 United States: Success of Hybrid Maize

The United States represents the most prominent example of successful conventional maize breeding through the large-scale adoption of hybrid maize. Following the discovery of heterosis by Shull (1909) and the development of double-cross hybrids by Jones (1918), hybrid maize was rapidly commercialized during the 1930s. By the 1950s, hybrid varieties had almost completely replaced open-pollinated varieties, leading to major gains in yield stability, uniformity, and disease resistance (Troyer 1999; Duvick 2005). Continuous refinement of inbred lines, heterotic grouping, and breeding methodologies resulted in sustained yield increases, making the United States the world's largest maize producer and establishing hybrid breeding as the global standard (Hallauer et al. 2010).

4.7.2 CIMMYT and IITA: Tropical Maize Improvement

In tropical and subtropical regions, international research organizations have played a pivotal role in adapting hybrid maize to challenging environments. The International Maize and Wheat Improvement Center (CIMMYT) has been involved in developing high-yielding, stress-resilient, and nutritionally enhanced maize varieties for Latin America, Africa, and Asia. Through systematic conventional breeding, CIMMYT has delivered drought-tolerant hybrids, disease-resistant germplasm, and Quality Protein Maize (QPM), significantly improving productivity and food security in maize-dependent regions (Vasal 2002; Prasanna et al. 2020).

Complementing CIMMYT's efforts, the International Institute of Tropical Agriculture (IITA) has focused on maize improvement in West and Central Africa. IITA's breeding programs have developed early-maturing, Striga-resistant, and stress-tolerant maize varieties adapted to humid and sub-humid tropics. These conventional breeding efforts have led to substantial yield gains and enhanced resilience of maize-based farming systems in resource-constrained environments (Yadav et al. 2015; Cairns et al. 2013).

4.7.3 Indian Council of Agricultural Research (ICAR) and Maize Improvement in India

In India, conventional maize breeding has been driven by coordinated national efforts under the ICAR. The establishment of the All India Coordinated Maize Improvement Project (AICMIP) in 1957 marked a major milestone in systematic maize improvement. Through mass selection, recurrent selection, and hybrid breeding, ICAR-led programs developed region-specific hybrids and composites with improved yield potential, disease resistance, and adaptability (Dhillon and Malhi 2006; Yadav et al. 2015). The adoption of hybrid maize in India resulted in a nearly fourfold increase in productivity, transforming maize from a subsistence crop into an important component of food, feed, and industrial systems. The development and dissemination of high-yielding hybrids and QPM varieties further strengthened nutritional security and farmer livelihoods, demonstrating the sustained impact of conventional breeding on maize improvement in the country (Vasal 2002; Singh et al. 2024a, b).

4.8 Challenges and Limitations

4.8.1 Long Breeding Cycles

A major limitation of conventional maize breeding is the long time required to develop improved varieties. Breeding approaches such as pedigree selection, recurrent selection, and backcross breeding typically require 8–12 years to release a new cultivar, owing to multiple generations of selection, extensive field evaluations, and multi-location testing to ensure yield stability and adaptability (Hallauer et al. 2010; Prasanna et al. 2020; Kumar et al. 2020, 2024a, b, 2025b; Singh et al. 2024a, b). This time-intensive process constrains the speed of genetic gain, particularly for complex traits such as drought tolerance and nutrient-use efficiency that are strongly influenced by genotype × environment interactions.

4.8.2 Genetic Bottlenecks and Narrowing of Diversity

The widespread adoption of hybrid maize has significantly improved productivity but has also contributed to a narrowing of the genetic base. Commercial hybrid development often relies on a limited number of elite inbred lines, increasing the risk of genetic bottlenecks and reducing overall diversity within cultivated maize (Troyer 1999). Such genetic erosion heightens vulnerability to emerging pests, diseases, and environmental stresses and limits the availability of novel alleles for future breeding. Although landraces and wild relatives harbor valuable genetic diversity, their underutilization in mainstream breeding programs remains a persistent challenge (Hallauer et al. 2010).

4.8.3 Climate Change Impacts

Climate change poses a growing threat to maize production, particularly in rainfed systems. Increased frequency of droughts, heat stress, and erratic rainfall patterns adversely affect maize growth, flowering, and grain filling, leading to substantial yield losses (Lobell et al. 2011; Cairns et al. 2013). Conventional breeding for climate resilience is challenging due to the polygenic nature of stress tolerance traits and the need for long-term, multi-environment testing. Limited availability of stress-tolerant genetic resources further slows progress in developing broadly adapted, climate-resilient cultivars (Prasanna et al. 2020).

4.8.4 Durability of Disease Resistance

Ensuring durable resistance to diseases remains a critical limitation in conventional maize breeding. Resistance genes that are effective against specific pathogens often lose efficacy over time as pathogen populations evolve new virulence mechanisms. The cultivation of genetically uniform hybrid varieties can accelerate the breakdown of resistance, increasing the risk of widespread disease outbreaks (Duvick 2005). Moreover, the scarcity of durable resistance sources in available germplasm and the slow pace of resistance breeding through conventional methods complicate efforts to maintain long-term protection against major maize diseases (Hallauer et al. 2010; Prasanna et al. 2020).

4.9 Future Prospects and Sustainability

4.9.1 Climate-Resilient and Resource-Efficient Hybrids

Future maize improvement must prioritize the development of climate-resilient and resource-efficient hybrids to address increasing drought, heat stress, and erratic rainfall associated with climate change. Conventional breeding will continue to play a central role by exploiting stress-adapted landraces and wild relatives to enhance drought tolerance, heat resilience, and water-use efficiency in elite germplasm. Breeding strategies focusing on deep-root systems, improved nitrogen-use efficiency, and stable performance under low-input conditions are essential for sustaining maize productivity while reducing environmental impacts (Cairns et al. 2013; Prasanna et al. 2020). These approaches align with the need for climate-smart maize cultivars capable of maintaining yield stability under increasingly variable agro-climatic conditions.

4.9.2 Germplasm Conservation and Utilization

The long-term sustainability of maize breeding depends on effective conservation and utilization of genetic diversity. Concerns over genetic erosion resulting from reliance on a narrow pool of elite inbred lines underscore the importance of conserving landraces and wild relatives as reservoirs of adaptive traits (Troyer 1999). Ex situ conservation through gene banks, such as those maintained by CIMMYT and national programs, and in situ conservation through on-farm maintenance of traditional landraces are critical strategies for safeguarding maize genetic resources (Prasanna et al. 2020). Conventional breeding approaches, including introgression and pre-breeding, enable the systematic incorporation of this conserved diversity into elite backgrounds, ensuring the availability of novel alleles for stress tolerance, disease resistance, and yield improvement.

4.9.3 Continued Role of Conventional Breeding in Food Security

Despite advances in molecular breeding technologies, conventional maize breeding remains the foundation of global maize improvement and food security. The development of high-yielding hybrids, nutritionally enhanced cultivars such as Quality Protein Maize, and region-specific stress-resilient varieties has substantially improved maize productivity and nutritional outcomes, particularly in developing countries (Vasal 2002; Yadav et al. 2015). Conventional breeding ensures the broad acceptance, adaptability, and accessibility of improved maize varieties, especially for smallholder farmers with limited resources. As global demand for maize continues to rise, the integration of conventional breeding with complementary molecular tools will enhance efficiency, but the core principles of selection, hybridization, and genetic diversity management will remain indispensable for sustainable maize-based food systems (Hallauer et al. 2010; Prasanna et al. 2020).

4.10 Conclusion

Conventional maize breeding has played a foundational and enduring role in transforming maize into a high-yielding, resilient, and globally important cereal crop. The systematic application of selection, hybridization, and heterosis exploitation has driven sustained improvements in yield, stability, stress tolerance, and nutritional quality, notably through the development of hybrid maize and Quality Protein Maize (Shull 1909; Duvick 2005; Vasal 2002). Despite challenges such as long breeding cycles, genetic narrowing, and climate change pressures, conventional breeding remains central to maize improvement and food security. When integrated with molecular tools, it continues to provide sustainable, widely accepted, and scalable solutions for meeting future food and nutritional demands (Hallauer et al. 2010; Prasanna et al. 2020).

References

Bao JD, Yao JQ, Zhu JQ, Hu WM, Cai DG, Li Y, Shu QY, Fan LJ (2012) Identification of glutinous maize landraces and inbred lines with altered transcription of waxy gene. Mol Breed 30:1707–1714

Cairns JE, Hellin J, Sonder K, Araus JL, MacRobert JF, Thierfelder C, Prasanna BM (2013) Adapting maize production to climate change in sub-Saharan Africa. Food Secur 5:345–360

Choudhary M, Singh A, Gupta M, Rakshit S (2020) Enabling technologies for utilization of maize as a bioenergy feedstock. Biofuels Bioprod Biorefin 14(2):402–416

Collard BC, Mackill DJ (2008) Marker-assisted selection: an approach for precision plant breeding in the twenty-first century. Philos Trans R Soc B Biol Sci 363(1491):557–572

Crossa J, Pérez-Rodríguez P, Cuevas J, Montesinos-López O, Jarquín D, De Los Campos G, Burgueño J, González-Camacho JM, Pérez-Elizalde S, Beyene Y, Dreisigacker S (2017) Genomic selection in plant breeding: methods, models, and perspectives. Trends Plant Sci 22(11):961–975

Das AK, Choudhary M, Kumar P, Karjagi CG, Yathish KR, Kumar R, Singh A, Kumar S, Rakshit S (2021) Heterosis in genomic era: advances in the molecular understanding and techniques for rapid exploitation. Crit Rev Plant Sci 40(3):218–242

Devi EL, Hossain F, Muthusamy V, Chhabra R, Zunjare RU, Baveja A, Jaiswal SK, Goswami R, Dosad S (2017) Microsatellite marker-based characterization of waxy maize inbreds for their utilization in hybrid breeding. 3 Biotech 7:1–9

Dhillon BS, Malhi HS (2006) Maize breeding in India—retrospective analysis and prospects. Indian J Plant Genet Resour 19(03):327–345

Doebley J (1992) Mapping the genes that made maize. Trends Genet 8(9):302–307

Doebley J, Stec A (1991) Genetic analysis of the morphological differences between maize and teosinte. Genetics 129(1):285–295

Duvick DN (2005) The contribution of breeding to yield advances in maize (*Zea mays* L.). Adv Agron 86:83–145

FAOSTAT (2022) The State of Food and Agriculture 2022: leveraging automation in agriculture for transforming agrifood systems. FAO, Rome

Hallauer AR, Carena MJ, Miranda Filho JD (2010) Quantitative genetics in maize breeding, vol 6. Springer Science & Business Media

Hossain F, Muthusamy V, Bhat JS, Zunjare RU, Kumar S, Prakash NR, Mehta BK (2022) Maize breeding. In: Fundamentals of field crop breeding. Springer Nature, Singapore, pp 221–258

Jones DF (1918) Segregation of susceptibility to parasitism in maize. Am J Bot 5(6):295–300

Kumar P, Choudhary M, Hossain F, Singh NK, Choudhary P, Gupta M, Singh V, Chikappa GK, Kumar R, Kumar B, Jat SL, Rakshit S (2019) Nutritional quality improvement in maize (Zea mays): progress and challenges. Indian J Agric Sci 89:8

Kumar P, Kumar S, Choudhary M, Chikappa GK, Kumar B, Singh NK, Sinha SK (2020) GGE biplot based stability analysis of experimental hybrids for baby corn purpose and green fodder. Range Manag Agrofor 41(1):60–66

Kumar P, Choudhary M, Jat BS, Dagla MC, Singh V, Das AK, Kumar S, Longmei N, Henry RJ, Wani SH (2021) Isolation of genes/quantitative trait loci for drought stress tolerance in maize. In: Molecular breeding in wheat, maize and sorghum: strategies for improving abiotic stress tolerance and yield. CABI, Wallingford, pp 267–281

Kumar B, Choudhary M, Kumar K, Kumar P, Kumar S, Bagaria PK, Sharma M, Lahkar C, Singh BK, Pradhan H, Rakshit S (2022) Maydis leaf blight of maize: update on status, sustainable management and genetic architecture of its resistance. Physiol Mol Plant Pathol 121:101889

Kumar R, Kumar S, Das AK, Dhonde S, Kaur Y, Kumar S, Shukla S, Rakshit S (2024a) Unveiling genotype× environment dynamics for grain yield in QPM hybrids through AMMI, GGE Biplot, and MTSI approach. Indian J Genet Plant Breed 84(03):449–460

Kumar P, Longmei N, Choudhary M, Gupta M, Kumar B, Jat BS, Bhushan B, Dagla MC, Aggarwal SK (2024b) Enhancement of nutritional quality in maize grain through QTL-based approach. Cereal Res Commun 52(1):39–55

Kumar B, Kumar S, Kumar P, Kumar A, Jat SL, Singh AK, Sharma P, Gami R, Choudhary M, Guleria SK, Singh SB, Hossain F, Kumar R, Jat HS, Rakshit S (2025a) Genotype and genotype by environment interaction (GGE) biplot analysis for assessing genotype stability in grain iron and zinc content of maize across diverse environments. J Food Compos Anal 142:107504

Kumar S, Kumar P, Dhar B, Kumar B, Dar ZA, Abhijith KP, Tirkey AE, Yathish KR, Singh P, Das AK, Nath V (2025b) Multi-environment evaluation of dual-purpose baby corn hybrids for yield stability and forage biomass in rainfed agro-ecological systems. Front Sustain Food Syst 9:1680866

Lobell DB, Bänziger M, Magorokosho C, Vivek B (2011) Nonlinear heat effects on African maize as evidenced by historical yield trials. Nat Clim Chang 1(1):42–45

Prasanna BM, Palacios-Rojas N, Hossain F, Muthusamy V, Menkir A, Dhliwayo T, Ndhlela T, San Vicente F, Nair SK, Vivek BS, Zhang X, Olsen M, Fan X (2020) Molecular breeding for nutritionally enriched maize: status and prospects. Front Genet 10:1392

Qin F, Kakimoto M, Sakuma Y, Maruyama K, Osakabe Y, Tran LSP, Shinozaki K, Yamaguchi-Shinozaki K (2007) Regulation and functional analysis of ZmDREB2A in response to drought and heat stresses in *Zea mays* L. Plant J 50(1):54–69

Rosegrant MW, Ringler C, Zhu T (2009) Water for agriculture: maintaining food security under growing scarcity. Annu Rev Environ Resour 34(1):205–222

Schnable JC (2015) Genome evolution in maize: from genomes back to genes. Annu Rev Plant Biol 66(1):329–343

Shiferaw B, Prasanna BM, Hellin J, Bänziger M (2011) Crops that feed the world 6: past successes and future challenges to the role played by maize in global food security. Food Secur 3:307–327

Shull GH (1909) Corn breeding. Bot Gaz 48(5):396–398

Singh SB, Karjagi CG, Hooda KS, Mallikarjuna N, Harlapur SI, Rajashekara H, Devlash R, Kumar S, Kasana RK, Kumar S, Gangoliya SS, Rakshit S (2018) Identification of resistant sources against turcicum leaf blight of maize (*Zea mays* L.). Maize J 7(2):64–71

Singh SB, Kumar S, Kasana RK, Singh SP (2019) Combining ability analysis and heterosis for yield and yield-attributing traits in late-maturing winter maize inbred lines (*Zea mays* L.). Front Crop Improv 7(1):42–51

Singh SB, Kumar P, Kasana RK, Choudhary M, Kumar S, Kumar R, Karjagi CG, Kumar B, Rakshit S (2021) Unveiling combining ability and heterotic grouping of newly developed winter maize (*Zea mays*) inbreds. Indian J Agric Sci 91:1586–1591

Singh SB, Kumar S, Kumar R, Kumar P, Yathish KR, Jat BS, Chikkappa GK, Kumar B, Jat SL, Dagla MC, Kumar S (2024a) Stability analysis of promising winter maize (*Zea mays* L.)

hybrids tested across Bihar using GGE biplot and AMMI model approach. Indian J Genet Plant Breed 84(1):73–80

Singh AP, Pandey PK, Joshi RP (2024b) Harnessing genetic diversity for climate-resilient maize: a comprehensive review. J Appl Biol Agric 1(1):21–28

Troyer AF (1999) Background of US hybrid corn. Crop Sci 39(3):601–626

Vasal SK (2002) Quality protein maize: overcoming the hurdles. J Crop Prod 6(1–2):193–227

Wang X, Ren Y, Zhang S, Chen Y, Wang N (2017) Applications of organic manure increased maize (*Zea mays* L.) yield and water productivity in a semi-arid region. Agric Water Manag 187:88–98

Xu Y, Crouch JH (2008) Marker-assisted selection in plant breeding: from publications to practice. Crop Sci 48(2):391–407

Yadav OP, Hossain F, Karjagi CG, Kumar BVKV, Zaidi PH, Jat SL, Chawla JS, Kaul J, Hooda KS, Kumar P, Dhillon BS (2015) Genetic improvement of maize in India: retrospect and prospects. Agric Res 4:325–338

Yathish KR, Karjagi CG, Gangoliya SS, Kumar A, Preeti J, Yadav HK, Srivastava S, Kumar S, Swamy HKM, Singh A, Gadag RN (2022) Introgression of the low phytic acid locus (lpa2) into elite maize (*Zea mays* L.) inbreds through marker-assisted backcross breeding (MABB). Euphytica 218(9):127

Zhu G, Zhu H (2022) Modified gene editing systems: diverse bioengineering tools and crop improvement. Front Plant Sci 13:847169

Molecular Markers in Maize Breeding: Past, Present, and Future

5

Abhijit Kumar Das, Rumesh Ranjan, K. R. Yathish, Ramesh Kumar, Bharat Bhushan, Sunandan Swain, Gulshan Gaur, Sudip Nandi, Sayantan Das, Anirban Roy, and Hanuman Sahay Jat

Abstract

The discovery of molecular markers and recent advances in sequencing technologies have revolutionized plant breeding. Molecular markers have led to the mapping of genes/QTLs controlling a trait on the genomic regions, and mapping of QTLs has led to the identification of linked molecular markers. The traditional QTL mapping often results in large genomic intervals spanning millions of base pairs and containing hundreds of genes, making functional validation and breeding applications challenging. QTL fine mapping is employed to refine broad chromosomal regions identified through initial QTL analysis, narrowing them down to precise genomic intervals that harbor causal genes. The genomic regions identified through QTL/fine mapping have been used for accelerated transfer of traits through marker-assisted backcross breeding. However, most of the complex traits are governed by polygenes with small effects, which makes it difficult to assemble a higher number of favorable alleles through MABC. Further, Genome-Wide Association Studies (GWAS) are an alternative mapping approach to QTL mapping and involve searching the entire genome of the organism with a large array of markers to identify genetic variations related to specific traits.

A. K. Das (✉) · R. Kumar · B. Bhushan · G. Gaur · S. Nandi · S. Das · H. S. Jat
ICAR-Indian Institute of Maize Research, Ludhiana, India
e-mail: Bharat.Bhushan@icar.org.in

R. Ranjan · S. Swain
Punjab Agricultural University, Ludhiana, India
e-mail: rumeshranjan@pau.edu; sunandan-24142003@pau.edu

K. R. Yathish
Winter Nursery Centre, IIMR Regional Centre, Hyderabad, India

A. Roy
Ramakrishna Mission Vivekananda Educational and Research Institute,
Kolkata, West Bengal, India

M. Choudhary et al. (eds.), *Next-Generation Maize Research*,
https://doi.org/10.1007/978-981-92-0722-0_5

GWAS usually has higher mapping resolution as it exploits historical recombination events. However, the effectiveness of GWAS depends on many factors, such as the level of linkage disequilibrium in the population, the size of the population, and the quality of phenotypic data. Genomic selection is one of the advanced breeding applications of modern days with multiple advantages, which include selection for traits that are complex and hard to phenotype for a large set of genotypes and need longer phenotypic trials. Advancement of sequencing technology and easy accessibility of SNP data for large sets of genotypes have emerged as one of the major reasons to adopt genomic selection.

Keywords

Markers · Genomic selection · QTLs · GWAS · MABC

5.1 Introduction

In plant breeding, conservation genetics, and evolutionary studies, the identification and characterization of genetic variation are fundamental. Traditional breeding relies largely on phenotypic selection, which is often influenced by environmental conditions, developmental stages, and growth factors, thereby limiting precision. This has led to the development of molecular marker systems that provide a more accurate, reliable, and high-throughput means of genetic analysis. Genetic markers are broadly classified into classical markers (i.e. morphological, cytological, and biochemical traits) and DNA (molecular) markers like restriction fragment length polymorphism (RFLP), simple sequence repeats (SSR), etc. Visually observable traits such as flower color, seed characteristics, and growth habits are considered as morphological markers, which can be assessed without specialized equipment and are widely used in conventional breeding, but their application is limited by low numbers, stage specificity, and strong environmental influence. Cytological markers involve variations in chromosome number and structure, including banding patterns such as G-, Q-, and R-bands, which aid in chromosome identification, mutation detection, physical mapping, and linkage analysis. However, their use in breeding and genetic mapping is restricted due to low polymorphism, labor-intensive procedures, high costs, and environmental sensitivity. Biochemical markers, or isozymes, represent allelic variants of enzymes encoded by different genes and are useful for estimating gene frequencies and studying genetic diversity, population structure, and gene flow. Despite being co-dominant, economical, and relatively simple to use, biochemical markers are limited by low polymorphism and sensitivity to tissue type, growth stage, and extraction methods.

To overcome the limitations of classical markers, molecular (DNA) markers have been widely adopted due to their high specificity, reproducibility, and environmental independence. Molecular markers represent variations in DNA nucleotide sequences that do not interfere with gene expression or trait performance. Depending on sequence variation, they may be monomorphic or polymorphic. Marker polymorphisms arise from point mutations, insertions, deletions, duplications, inversions, translocations, and replication errors in tandem repeats. An ideal DNA marker should be abundant across the genome, highly polymorphic, co-dominant, reproducible, cost-effective, amenable to automation and multiplexing, and suitable for inter-laboratory data sharing. Advances in next-generation sequencing (NGS) have enabled the use of molecular markers in genome-wide selection (GWS) and GWAS (Mir and Varshney 2013). The various molecular marker systems, along with their advantages and limitations, are summarized in Table 5.1.

Table 5.1 List of different molecular markers and their nature for use in molecular breeding

S. No.	Name of markers	Principles	Advantage	Disadvantage
1	Restriction Fragment Length Polymorphism (RFLP)	DNA digestion by restriction enzymes; polymorphism detected by differences in fragment sizes	Co-dominant, highly reproducible, locus-specific, no prior sequence info needed	Laborious, time-consuming, requires large DNA, costly, uses radioactive probes
2	Random Amplified Polymorphic DNA (RAPD)	PCR using short random primers binding at multiple loci; polymorphisms by the presence/absence of bands	Cheap, quick, no prior sequence info needed	Dominant marker, low reproducibility
3	Simple Sequence Repeat (SSR)	PCR amplification of microsatellite repeat regions; polymorphism results from variation in repeat numbers	Co-dominant, highly polymorphic, reproducible, and low DNA amount required	Needs flanking sequence information for primer design
4	Amplified Fragment Length Polymorphism (AFLP)	Restriction digestion followed by selective PCR amplification	High reproducibility, no prior sequence info needed	Dominant markers are possible, requiring high-quality DNA
5	Cleaved Amplified Polymorphic Sequences (CAPS)	PCR followed by restriction enzyme digestion; polymorphism via digest patterns	Codominant, sequence-specific, PCR amplification followed by restriction digestion	Needs sequence and enzyme info, codominant, reproducible
6	Sequence-Characterized Amplified Region (SCAR)	PCR amplification using specific primers derived from RAPD fragments	High reproducibility, locus-specific	Requires sequence information
7	Chloroplast Simple Sequence Repeat (CpSSR)	PCR amplification of microsatellites; repeat number variation—codominant	Highly polymorphic and reproducible	Requires sequence data, costly primers

(continued)

Table 5.1 (continued)

S. No.	Name of markers	Principles	Advantage	Disadvantage
8	Random Amplified Microsatellites Microsatellite Polymorphic (RAMP)	Combination of RAPD and SSR approaches	Polymorphic, useful for diversity	Complex, dominant
9	Selective Amplification of Microsatellite Polymorphic Loci (SAMPL)	PCR based on SSR sequences with selective primers	Polymorphic, reproducible	Requires sequence information
10	Inter Simple Sequence Repeats (ISSR)	PCR amplification using primers anchored in SSR regions, PCR amplification between SSR regions	Polymorphic, no sequence info needed, Polymorphic, no prior sequence data required	Dominant, sensitive to reaction conditions, less reproducible than SSR
11	Sequence-Related Amplified Polymorphism (SRAP)	Targets coding sequences by primer design	Reproducible, codominant	Requires specific primers
12	Single-Strand Conformation Polymorphism (SSCP)	Detects conformational changes in single-stranded DNA, denatured DNA single strands form unique conformations, migration differences detected by gel electrophoresis	High resolution, detects SNPs,	Technically demanding, Limited to short DNA fragments, labor-intensive
13	Direct Amplification of Length Polymorphisms (DALP)	Uses arbitrarily primed PCR with designed oligonucleotide primer pairs, produces multi-banded DNA fragment patterns, Fragments directly sequenced with universal M13 primers. Detects micro-deletions, insertions, and microsatellites	High resolution fingerprinting combined with polymorphism characterization, no prior sequence knowledge needed, applicable across many species, Useful in population genetics, genome mapping, QTL analysis.	Primer design and optimization can be complex, require high-quality DNA, and precise PCR conditions. Multi-locus nature complicates some genetic analyses

(continued)

Table 5.1 (continued)

S. No.	Name of markers	Principles	Advantage	Disadvantage
14	Single Nucleotide Polymorphism (SNP)	Variation of single base pairs detected by high-throughput genotyping	Very abundant, codominant, highly reproducible, Abundant, codominant, highly reproducible, automated	Requires advanced tech, costly, Needs expensive technology and reference genomes
15	Expressed Sequence Tag (EST)	DNA sequences derived from expressed regions (cDNA) to identify polymorphisms; polymorphisms detected in coding regions, Sequence-based markers derived from transcribed regions	Functional markers, codominant	Limited to expressed genes, sequence-dependent
16	Inter-Retrotransposon Amplified Polymorphism (IRAP)	PCR amplification between retrotransposons	Detects transposable elements	Dominant, pattern complexity
17	Retrotransposon-Microsatellite Amplified Polymorphism (REMAP)	Combines retrotransposon and SSR primers to amplify inter-retrotransposon regions	High polymorphism, Detects genetic diversity; targets mobile elements	Requires retrotransposon knowledge, complex band patterns, technically demanding
18	Diversity Arrays Technology (DaRT)	Hybridization-based markers detected on microarrays	High throughput, no sequence info needed	Requires specialized arrays and equipment
19	Retrotransposon-Based Insertion Polymorphism (RBIP)	Detects polymorphisms due to retrotransposon insertions	Co-dominant	Retrotransposon specific
20	Intron Polymorphism Based on SSR (IPBS)	Combines intron and SSR information	Polymorphic	Requires sequence data
21	Inter-SINE Amplified Polymorphism (ISAP)	PCR amplification between Short Interspersed Elements (SINEs)	Specific, reproducible in species	Primer design requires genomic info

(continued)

Table 5.1 (continued)

S. No.	Name of markers	Principles	Advantage	Disadvantage
22	Inter Small RNA Polymorphism (iSNAP)	PCR markers developed using primers around small non-coding RNAs	Sequence-based, reproducible	Limited application, needs RNA data
23	Start Codon Targeted Polymorphism (SCoT)	PCR markers based on conserved sequences flanking the ATG start codon	Reproducible, gene-targeted	Dominant, primer design critical
24	Conserved Region Amplification Polymorphism (CoRAP)	PCR with arbitrary and fixed primers targeting coding/ intron regions	Highly reproducible, gene-targeted	Requires conserved region sequence info
25	Target Region Amplification Polymorphism (TRAP)	PCR using fixed primers from EST sequences + arbitrary primers	Functional, gene-associated, useful for tagging traits	Requires known EST data

5.2 Use of Molecular Markers in QTL (Quantitative Trait Locus) Mapping Led to Important Discovery in Maize

The discovery of molecular markers has led to the mapping of genes/QTLs controlling a trait on the genomic regions, and the mapping of QTLs has led to the identification of trait-linked molecular markers, which have been used in different breeding programs. There are two types of traits: qualitative (controlled by few genes) and quantitative (controlled by many genes). The qualitative traits show discontinuous variation and are discrete traits for which the phenotype can be predicted. These classes can also help infer the genotypes of individuals. In contrast, quantitative traits exhibit continuous variation and are influenced by multiple genes and environmental factors, making their inheritance more complex. Molecular markers are ideal for studying QTLs and mapping them, which can be effectively used in marker-assisted selection (MAS) (Collard et al. 2005). This method is effective for detecting genes with large effects, but it struggles with small-effect genes (minor QTLs) and requires high heritability, robust phenotyping, and replications to ensure reliability. The key objectives of QTL mapping include pinpointing genomic regions affecting traits, quantifying their contribution to phenotypic variation, determining gene action (additive/dominant), and identifying favorable alleles. Essential prerequisites for QTL mapping include a contrasting parental mapping population, a dense marker-based linkage map, accurate phenotyping, and specialized statistical tools.

QTL mapping has been extensively utilized in maize for identifying QTLs for various traits, including yield-related traits, plant architecture, quality traits, resistance to biotic and abiotic stresses, etc. Wang et al. (2020) identified QTLs, namely qKW-1 and qKL-2, for kernel length (KL) on chromosome 9 through composite interval mapping (CIM) in F_2 and $F_{2:3}$ populations. Out of the two QTLs, qKL-2 was stably detected across three environments. Similarly, QTLs for cob weight (qCW2–1) and cob height (qCH1–2, qCH1–1) were reported on chromosomes 1 and 2, respectively (Sarkar et al. 2023). Yang et al. (2020) identified three main QTL hotspots using the doubled haploid (DH) lines derived from ZD958 on chromosome 2.02 (flanking markers umc1165-bnlg1017), 2.05–2.06 (umc1065-umc1637), and 6.05 (nc012-bnlg345) with various yield-related traits across multiple environments, which could be due to pleiotropic QTL or multiple tightly linked QTL. Drought stress is one of the most important factors affecting crop yield loss, and it is important to search for the QTLs/genes involved in drought tolerance. Zhao et al. (2018) identified 69 QTLs explaining 4.0%–17.2% of phenotypic variation for plant height, ear height, ear weight, ear length, ASI, and 100-kernel weight in two F2:3 populations under drought-stress and nonstress conditions. Out of the 69 QTLs, 52 QTLs were identified under water-stressed conditions, and 21 stable QTLs were validated under multiple water-stressed conditions. For stalk rot resistance, one major QTL, namely qRfg1, and one minor QTL, i.e. qRfg2, were mapped (Yang et al. 2010; Zhang et al. 2012), and later the ZmCCT gene was found to be the causal gene at qRfg1 QTL as reported by Wang et al. (2017). The detailed list of mapped QTLs for different traits is presented in Table 5.2.

Table 5.2 QTLs reported for different traits in maize

S. No.	Trait	QTL identified	Chromosome	Mapping method	Population	References
1	Kernel length (KL)	qKW-1, and qKL-2	Chr 9	Composite interval mapping (CIM)	F_2 and $F_{2:3}$	Wang et al. (2020)
2	Cob weight (CW)	qCW2–1	Chr 2	SNP Genotyping (GBS)	RIL	Sarkar et al. (2023)
3	Cob height (CH)	qCH1–2, qCH1–1	Chr 1	SNP Genotyping (GBS)	RIL	Sarkar et al. (2023)
4	Kernel width	Qkwid6	Chr 6	CIM	$F_{2:3}$	Jiang et al. (2023)
5	Grain weight	qKW 7	Chr 7	CIM	$F_{2:3}$	Zhao and Su (2019)
6	Fusarium ear rot resistance	*qmXX22–5* *qmXX22–6*	Chr 5 Chr 6	mSNP-based QTL mapping	DH	Meng et al. (2025)
7	Kernel starch content	qSC8–1	Chr 8	GWAS	Multiparent population	Yang et al. (2025)
8	*Striga* resistance	*qSC*7.1	Chr 1, 5, 9	QTL mapping (DArTseq markers)	F2:3	Badu-Apraku et al. (2023)

(continued)

Table 5.2 (continued)

S. No.	Trait	QTL identified	Chromosome	Mapping method	Population	References
9	Low-temperature germination rate	*qLTGR4–1*	Chr 4	–	RIL	Han et al. (2025)
10	Protein content	*qGPC6–30*	Chr 6	SNP Genotyping	F3, DH	Ndlovu et al. (2024)
11	Starch content	*qSC1–20* qSC5–55	Chr 1 Chr 5	SNP Genotyping	F3	Ndlovu et al. (2024)
12	Waterlogging (Root: brace root number)	*RBRNN*	Chr 4	GWAS	365 Inbred Lines	Guo et al. (2021)

5.3 Fine Mapping Using Molecular Markers Led Important Discovery in Maize

The traditional QTL mapping often results in large genomic intervals spanning millions of base pairs and containing hundreds of genes, making functional validation and breeding applications challenging (Wang et al. 2020). To overcome this limitation, QTL fine mapping is employed to refine broad chromosomal regions identified through initial QTL analysis, narrowing them down to precise genomic intervals that harbor causal genes or polymorphisms associated with complex traits. Fine mapping is essential for resolving genetic complexity, facilitating functional gene discovery, and addressing minor-effect QTLs. The genetic basis of quantitative traits can be more accurately understood through QTL mapping. In maize, numerous QTLs associated with kernel traits have been identified (Zhang et al. 2014; Xu et al. 2015; Liu et al. 2014) but a few have been positionally cloned due to certain challenges like the presence of and it is difficult to precisely locate specific loci due to the presence of quantitative genes that control the complex traits like kernel yield and kernel size with small effects (Peiffer et al. 2014). Notably, QTLs identified across different genetic backgrounds and multiple environments have a greater likelihood of being successfully positionally cloned, enhancing their utility in functional genomics and breeding programs.

Fine mapping in maize has been reported for numerous traits. Chen et al. (2016) fine-mapped the QTL, i.e. qGW4.05, to a 279.6-kb interval, and GRMZM2G039934 was identified as a candidate gene, significantly associated with kernel weight in maize. Similarly, Zhang et al. (2013) reported fine mapping for kernel number per row in maize, and the QTL (qKN) was fine-mapped to a 480 kb region on chromosome 10 between markers bnlg1360 and umc1645. Zhou et al. (2020) fine-mapped qHN7 on chromosome 7 to a 721.1 kb region, and four candidate genes were identified. Li et al. (2022) characterized the brittleness mutant brittle stalk-5 (bk-5) of the maize encoding a cellulose synthase catalytic subunit (CesA) involved with cellulose synthesis, and map-based cloning revealed that a single recessive nuclear gene controlled bk-5 and that it was located on chromosome 3 spanning a 90.2-kb region. Further, sequence analysis revealed a single non-synonymous missense mutation (T-to-A) in the last exon of Zm00001d043477 that was responsible for the change of leucine to histidine. A major QTL Tar spot complex (TSC) resistance (qRtsc8–1) was identified and fine-mapped to a 721 kb region on chromosome 8 at 81 Mb (Ren et al. 2022). Schmidt et al. (2024) fine-mapped the Barley yellow dwarf (BYDV) virus resistance genes in maize to ~0.3 Mbp, which comprised nine genes, and further analysis using association mapping and gene expression analysis reduced the number of candidate genes to two, viz., Zm00001eb428010 and Zm00001eb428020. Ducrocq et al. (2009) reported fine mapping of a major QTL located on maize chromosome 10 regulating flowering time through photoperiod sensitivity. The authors could map the QTL to an interval of 170 kb, and a homologous gene to the rice Ghd7 heading date regulator encoding a CCT domain protein was identified. Salvo et al. (2018) reported fine mapping of QTL associated with embryogenic tissue culture response and plant regeneration ability in maize and genetically fine-mapped in the region for controlling embryogenic and regenerable tissue culture response to a 23.9 Mb region on chromosome 3. Hou et al. (2022) characterized and fine-mapped two major QTLs for Rind Penetrometer Resistance governing stalk strength in maize, and the two QTLs, qRPR1–3 and qRPR3–1, were narrowed down to 4-Mb and 550-kb genomic intervals, respectively, using a sequential fine-mapping strategy based on the progeny test. The following table presents the progress made in QTL fine mapping of some important traits (Table 5.3).

Table 5.3 Genes identified through fine mapping experiments for different traits in maize

S. No.	Trait studied	QTL/Gene identified	Method	Key findings	References
1	Fusarium ear rot (FER)	qRfv2	QTL mapping using RIL	Fine-mapped to interval of 1.01 Mb (markers IDP8 & IDP10)	Li et al. (2025)
2	Kernel weight and size	qGW4.05	Association mapping	qGW4.05 was fine-mapped to a 279.6-kb interval. GRMZM2G039934 was identified as a candidate gene	Chen et al. (2016)
3	Kernel number per row	qKN	Fine mapping using NILs	qKN was fine-mapped to a 480 kb region on chromosome 10 between markers bnlg1360 and umc1645	Zhang et al. (2013)
4	Kernel size	qKL-2	Fine mapping with NILs and MAS; Transcriptome analysis	qKL-2 was fine-mapped to a 1.95 Mb region on chromosome 9. GRMZM2G006080, encoding a receptor-like protein kinase FERONIA	Wang et al. (2020)

(continued)

Table 5.3 (continued)

S. No.	Trait studied	QTL/Gene identified	Method	Key findings	References
5	Gray Leaf Spot (GLS)	qGLS8	Fine mapping using NIL (BC_3F_2)	qGLS8 was fine-mapped to a 124-kb region on chromosome 8, flanked by markers ym20 and ym51. MYB-related transcription factor 24 and a C3H transcription factor 347 were identified.	Qiu et al. (2021)
6	Plant height (PH) and ear height (EH)	*Qtl-chr1-EP*	GWAS	105 SNPs and 22 QTL identified for PH and EH *Qtl-chr1-EP* was confirmed via GWAS and linkage analysis	Yin et al. (2022)
7	Maize Rough Dwarf Disease (MRDD) Resistance	*qMrdd2*	RIL, Chromosome Segment Substitution Line	*qMrdd2* mapped to a 0.55 Mb region between markers MK807 and MK811-Fine-mapped to a 315-kb region between RD81 and RD87	Xu et al. (2022)
8	Kernel width (KW)	*qKW7a*, *qKW7b*	Fine mapping, NILs, genome collinearity	*qKW7b* was fine-mapped to 59 kb; *Zm00001d020460* (zinc finger homeodomain TF) was identified as a candidate gene	Tang et al. (2020)
9	Husk number (HN) and Husk length (HL)	*qHN7 on chromosome 7*	QTL mapping (RILs), fine mapping	*qHN7* was fine-mapped to a 721.1 kb region. Four candidate genes were identified	Zhou et al. (2020)

5.4 Transition from Classical Backcross Breeding to Marker-Assisted Backcross Breeding

5.4.1 Classical Backcross Breeding

The backcross breeding approach was proposed by Harlan and Pope in 1922 and has since been widely used across crop species. It is primarily employed to transfer one or a few desirable traits from a donor parent into an elite or adapted variety (recurrent parent) that lacks those traits. Backcrossing can be used to introgress both dominant and recessive genes, although transferring recessive traits requires more time due to the need for selfing to express the target allele. During backcrossing, individuals carrying the desired trait are repeatedly crossed with the recurrent parent, leading to progressive recovery of the elite genetic background. Backcrossing increases homozygosity for alleles of the recurrent parent, with the expected genome recovery calculated as $[1 - (1/2)^{\{m+1\}}]$, where m is the number of backcross generations. However, recovery can be affected by linkage drag, and the probability of eliminating an undesirable linked allele is given by $[1 - (1 - p)^{\{m+1\}}]$, where p represents the recombination frequency between loci. The proportion of completely homozygous plants in a backcross generation is expressed as $[(2^{\{m\}} - 1)/2^{\{m\}}]^{\{n\}}$, where n is the number of loci with contrasting alleles.

5.4.1.1 Transfer of a Dominant Trait

When a single dominant resistance gene is transferred from a donor (RR) to a recurrent parent (rr), resistant individuals can be selected in each backcross generation. The donor parent is preferably used as the male to avoid confusion with selfed resistant seed and to ensure recovery of cytoplasmic genes from the recurrent parent. The F_1 plants (Rr) are backcrossed repeatedly with the recurrent parent, and resistant plants are selected in each generation, typically up to six or seven backcrosses. Final selfing of BC_6F_1 or BC_7F_1 plants produces homozygous resistant lines that closely resemble the recurrent parent, which are confirmed through progeny testing.

5.4.1.2 Transfer of Recessive Trait

The transfer of characters governed by a recessive gene is carried out with a slight modification to provide for the identification of plants carrying the desirable recessive allele. Hence, the backcrosses can't be made one after another. After the BC1 and after every two subsequent backcrosses, the F2 generation must be grown to identify the resistant plant progenies. So, it takes around 12 seasons (6 years) to transfer a recessive trait.

5.4.1.3 Major Limitations of the Classical Backcross Method

As mentioned above, 5–6 years are usually required for gene introgression into a target variety through backcross breeding. Though the individuals in the backcross population have varying levels of background recovery, individuals with high background recovery cannot be identified. So, it takes a longer time to recover traits of recurrent parents as the breeder is dependent on the overall population mean. Moreover, as mentioned above, if the trait is recessive, additional self-pollination is required as heterozygotes cannot be identified through phenotypic selection. In addition to the method of backcrossing, the new hybrid needs to be tested in a station trial as well as in AICRP before it can be released, which will take another 3–4 years. The varieties in major crops have a life of 10–12 years. Thus, until the gene(s) have been introduced into an existing variety, it might already be outdated. So, accelerated transfer of the trait is required through marker-assisted backcross breeding.

5.4.2 Marker-Assisted Backcross Breeding (MABB)

MABB has three major steps, viz., (i) Foreground selection for the target gene(s), (ii) Background selection throughout the genome and (iii) Background selection near the target gene(s) to reduce linkage drag when introgressing wild or exotic germplasm.

5.4.2.1 Foreground Selection

Foreground selection is the process of selecting positive individuals in the BC populations using a molecular marker instead of phenotype. Foreground selection can differentiate between a heterozygote (Aa) from homozygote (AA) in the BC populations, which helps to avoid unnecessary selfing for the expression of a recessive trait. The suitable marker for use in foreground selection should possess the following properties. Linked with target gene(s) of interest. It should be a co-dominant marker to allow distinction between homozygotes and heterozygotes. The marker system should be reproducible so that the results can be reproduced across laboratories. Furthermore, the marker system should allow automation for high-throughput scale.

In Fig. 5.1, the transfer of recessive white kernel color (yy) has been demonstrated to transfer from the donor parent to the recurrent parent (YY). The F1 kernels will be yellow (Yy) and the BC_1 kernels will also be yellow, though the genotypes in the BC_1 population will be YY and Yy. So, in classical backcross breeding, the BC_1 individuals need to be selfed for the expression of white kernel.

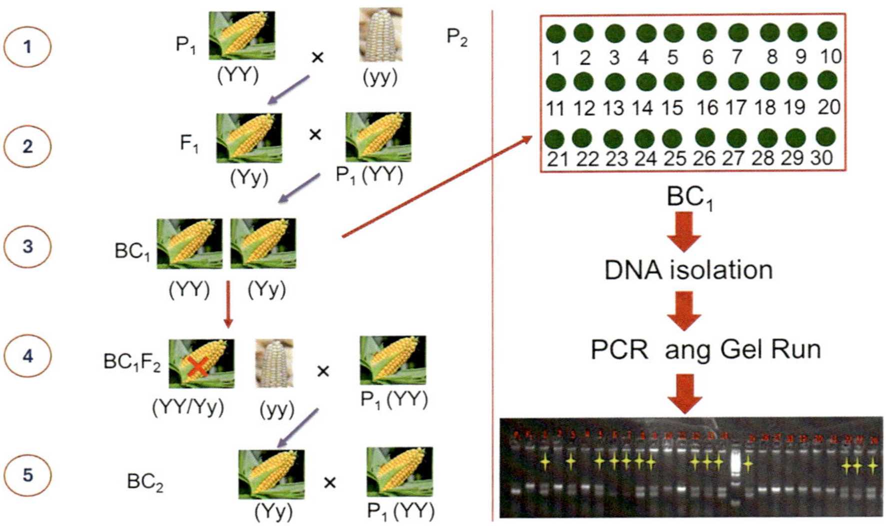

Fig. 5.1 Foreground selection reduces time by eliminating the need for selfing for expression of recessive trait (Though different colored cobs have been depicted above, but the kernel will appear in the same cob)

However, using a molecular marker linked to kernel color, the heterozygous (Yy) individuals can be identified in the BC_1F_1 population itself, and the selected individuals can be backcrossed with the recurrent parent to derive the BC_2F_1 population, instead of deriving BC_1F_2 generation. To identify the heterozygous (Yy) individuals in the backcross population, each individual in the backcross population must be tagged and leaf samples from each plant must be collected at the knee height stage before flowering to avoid contamination of DNA from pollen. The leaf samples should carry the same unique number as the tag of the plant from which it was collected. The leaf samples need to be stored at −80 °C until the DNA isolation. The DNA isolation has to be done separately for each genotype. The PCR has to be performed with the linked molecular marker for the target trait, and the PCR product has to be run on an agarose gel. From the gel image, it is clear that there are two types of bands, viz., double and single bands. The single bands represent YY, whereas the Yy represents the heterozygous individual. So, from the unique number, it is now possible to identify the heterozygous individuals in the BC populations. These individuals are now crossed with recurrent parent to develop the BC_2F_1 population. So, using molecular markers, the backcross generations can be grown continuously without selfing until we need to grow the BC_2F_2/BC_3F_2 generation for deriving homozygous genotypes (yy).

5.4.2.2 Background Selection

After carriers of the target trait were identified by foreground selection, the next issue is to efficiently recover the recurrent parent genome. The classical BC method requires 6–7 BC generations to recover 99% of the recurrent parent genome. This can be achieved by 2–3 backcrosses through background selection in MABB. The expected recovery of the background genome can be estimated using the following formula. $E_t = 1 - (1/2)^{t+1}$, where E_t = expected proportion of the recurrent parent genome and t = backcross generation. However, this recovery is an average recovery of the whole BC population. So, the proportion of background recovery in BC_1, BC_2, BC_3, BC_4 and BC_5 generation is 75.00%, 87.50%, 93.75%, 96.87% and 98.44%, respectively. So, in a BC_1 population, though the average recovery of the population is 75%, there will be individuals where recovery can vary, even from ~65% to 85% (hypothetical). This can be explained in detail with the following example (Fig. 5.2).

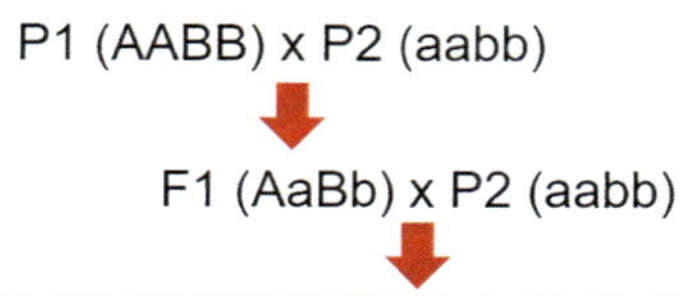

	ab	Recovery
AB	AaBb	2/4= 50%
Ab	Aabb	3/4= 75%
aB	aaBb	3/4= 75%
ab	aabb	4/4=100%

Fig. 5.2 Different genotypes and variation of background recovery in BC_1 individuals

In the above example, P1 is the donor parent and P2 is the recurrent parent, and the calculation of background recovery is based on 2 genes only. The crossing between P1 (AABB) and P2 (aabb) will lead to the development of F_1 with an AaBb configuration. Further, crossing the F1 with the recurrent parent (aabb) will lead to the development of BC_1 generation. In BC1 generation, four different types of genotypes will be there, viz., AaBb, Aabb, aaBb, and aabb. These individuals will vary in the background genome of the recurrent parent, which can be calculated by simply dividing the number of alleles from the recurrent parent by the total number of alleles. The background recovery of AaBb, Aabb, aaBb, and aabb genotypes become 50%, 75%, 75%, and 100%, respectively. However, the overall background recovery of this population is 75% [(50 + 75 + 75 + 100)/4]. So, using molecular markers, genotypes with high background recovery can be selected, that speed up the BC breeding program.

5.4.2.3 Selection of Polymorphic Markers

To perform the background selection, the DNA of both donor and recurrent parents must be isolated. Then, uniformly distributed SSR markers (SNP markers can also be used) must be selected from maize genome. The complete list of SSR markers can be accessed from the Maize Genetics and Genomics Database (maizeGDB). The markers should be selected carefully to represent every bin location of 10 maize chromosomes. The selected SSR markers will be used for PCR amplification of P1 and P2, and the polymorphic markers will be identified. The markers for which P1 and P2 amplify different bands are called polymorphic (Fig. 5.3c), whereas the

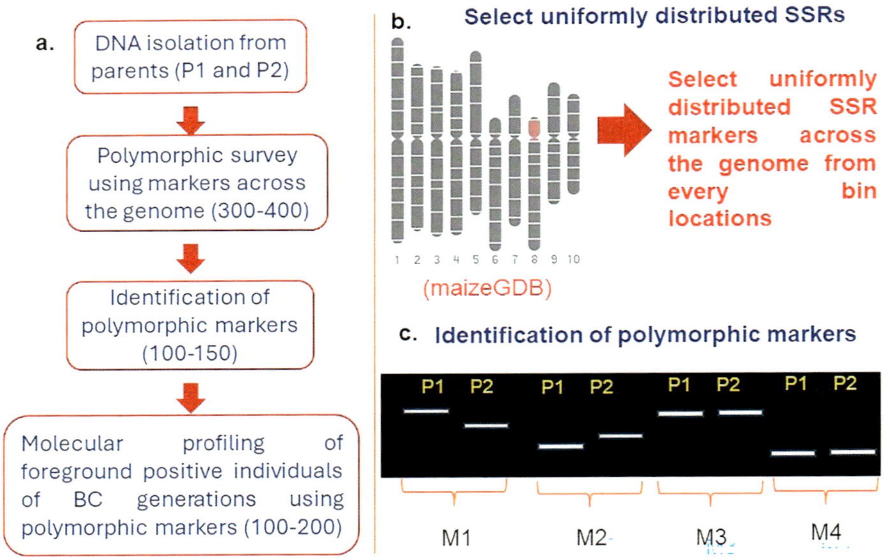

Fig. 5.3 (**a**) Flowchart of background selection. (**b**) Selection of background markers across the genome. (**c**) Identification of polymorphic markers

markers amplifying bands of the same size are termed monomorphic. In M1 and M2 are polymorphic markers, and M3 and M4 are monomorphic markers. The polymorphic markers are selected for background recovery. Then, molecular profiling of foreground positive individuals of BC generations is done using polymorphic markers (100–200).

5.4.2.4 Linkage Drag and Recombinant Selection

The donor DNA segment many a time remain even after many generations of backcross, particularly around the foreground marker, which may negatively affect the agronomic performance or may expose pathogen susceptibility of the newly derived genotype. Recombinant selection can be performed to reduce the effect of linkage drag in the MABB breeding program. Suppose we want to transfer Y from donor parent (P1) to recurrent parent (P2), but the susceptibility allele (S) is linked to Y, the recurrent parent is resistant (resistant allele R is linked to y). So, we must select flanking polymorphic SSR markers that cover these two genes and are located very closely, here A & a in the left and B & b in the right. So, in the backcross population, select 'A-Y-R-b' type of individuals, which will have only 'Y' from the donor parent but recovered the 'b' allele from the recurrent parent due to crossing over. However, to recover such an event, a larger backcross population is needed.

Significant advancements have been made in developing biofortified maize cultivars with quality protein maize (QPM), provitamin A, and kernel zinc using molecular breeding techniques. Further, there are other traits like kernel amylose, amylopectin, etc., where significant work has been done using marker-assisted selection around the world and in India. The summary of marker-assisted varietal/hybrid development is presented in Table 5.4.

5.4.3 Marker-Assisted Recurrent Selection (MARS) in Maize

MABC is an efficient method to transfer one or a few QTLs with a large genetic effect from the donor to the recurrent parent. However, most of the complex traits (grain yield, biotic, abiotic stresses, etc.) are governed by polygenes with small effects, which makes it difficult to assemble a higher number of favorable alleles through MABC. MARS, on the other hand, involves identification and crossing among genotypes selected using molecular markers for a number of generations, which can be useful to generate such an ideal genotype carrying favorable alleles for a number of genes using multi-trait selection indices (Peleman and Van der Voort 2003). Many QTLs, both for minor as well as major effects, can be accumulated in a population using markers that are significantly associated with target traits, which results in higher genetic gain as compared to the MABC (Bernardo 2008).

MARS has several advantages over phenotypic recurrent selection. Recurrent selection is one of the systematic breeding techniques that involves repeated cycles of selection and intermating to improve the genetic makeup of the population. It aims to increase the frequency of favorable alleles within the population. This technique is particularly useful for quantitative traits like yield, disease resistance,

Table 5.4 Summary of MABC-led varietal/hybrid development

S. No.	Trait	Gene	Level in original genotype	Level/increase in introgressed genotype	References
1.	QPM	*o2*	Lys-1.76%–2.08%, Trp-0.35%–52%	Lys-52%–95%, Trp-47%–118% Lys-48%–74%, Trp-55%–100%	Hossain et al. (2018)
2.	QPM	*o2*	Lys-0.18%, Trp-0.04%	Lys-0.–0.36%, Trp-0.09%	Sharma et al. (2024)
3.	QPM, Waxy	*o2, wx1*	Amylopectin-72.7%	Amylopectin-98.4%	Mishra et al. (2025)
4.	Waxy	*wx1*	Amylopectin-72.45%	Amylopectin-98.84%	Talukder et al. (2022b)
5.	β-carotene	LcyE, crtRB1	β-carotene: 0.75%–1.25%, β-cryptoxanthin: 0.22%–0.70%, provitamnin A-1.05–1.41.	β-carotene-4.95%–7.46%, β-cryptyoxabthin: 2.96%–4.47%, provitamnin A: 7.14%–9.28%	Singh et al. (2021)
6.	β-carotene, lycopene ε-cyclase, QPM	LcyE, crtRB1, o2	Lys (0.15%–0.25%), Trp (0.03%–0.04%), and provitamin-A (3–4 ppm)	Lys (0.390%), Trp (0.082%), proA (21.14 ppm)	Baveja et al. (2021)
7.	QPM	o16 into o2	Lys: 0.30%–0.35%, Trp: 0.06%–0.08%	Lys: 0.41%–0.56%, Trp: 0.09%–0.14%	Sarika et al. (2018)
8.	Provitamin A, QPM	o2 into crtRB1	Lys: 0.10% (average), Trpe: 0.01%, β-Carotene: About 1.0–1.3 μg/g	Lys: 0.314% (average), Trp: 0.077% (average), β-Carotene: About 6.80 μg/g	Chandran et al. (2019)
9.	Tocopherol, proA	*crtRB1, vte4*	proA: 3.33 ± 0.28 μg/g, α-tocopherol: (9.21 ± 0.33 μg/g)	proA: 19.52 ± 0.52 μg/g, α-tocopherol: 20.75 ± 0.44 μg/g	Chauhan et al. (2024)
10.	QPM	*o2*	Trpe: 0.37%–0.55%	Trpe: 0.72%–1.03%	Kaur et al. (2020)
11.	QPM	*o2*	Lys: 0.5%–2.0%; Trp: 0.3%–0.6%	Lys: 2.7%–4.5%; Trp: 0.7%–1.1%	Hossain et al. (2018)
12.	Oil content	*dgat1–2 and fatb*	Oil content: 4.80%, Oleic Acid: 26.7%–27.0%, Palmonic acid: 22%	Oil content: 6.73%, Oleic Acid: 40.49%, Palmonic acid: 14.23%	Katral et al. (2024)
13.	Provitamin-A	CrtRB1	–	7.056–9.232 μg/g	Natesan et al. (2020)

14.	Waxy	*wx1*	Amylopectin: range: 69.2%–71.6%	Amylopectin: range: 95.1%–98.7%	Talukder et al. (2022a)
15.	QPM, Waxy	*wx1, o2*	Lys: 0.31%, Amylopectin: 97.7%	Lys: 0.0.47%, Amylopectin: 98.6%	Zhou et al. (2016)
16.	QPM	*o2 & o16*	Lys: 0.24%–0.34%	Lys: 0.36%–0.54% (o2/wx1); 0.62 (o2/o16/wx1)	Zhang et al. (2013)
17.	β-carotene	crtRB1	β-carotene: 2.26 μg/g, β-cryptoxanthin: 2.26 μg/g, and proA: 3.38 μg/g	β-carotene: 9.22 μg/g, β-cryptoxanthin: 3.05 μg/g, and proA: 10.75 μg/g	Goswami et al. (2019)
18.	Provitamin A, QPM, α-tocopherol	crtRB1, o2, vte4	proA: 3.14 μg/g, Lys: 0.18%, Trp: 0.04%, α-Tocopherol: 11.16 μg/g	proA: 16.47 μg/g, Lys: 0.36% Trpe: 0.09%, α-tocopherol: 20.42 μg/g	Chauhan et al. (2022)
19.	β-carotene	crtRB1	β-carotene: 4.4 μg/g	β-carotene: 7.54 and 8.71 μg/g	Qutub et al. (2021)
20.	Provitamin A, QPM, α-tocopherol	crtRB1, o2, vte4	α-Tocopherol: 7.90–8.06,	α-Tocopherol: 16.13–16.83	Das et al. (2021)
21.	Phytic acid	lpa2	3.16–3.59 mg/g	2.37–2.40 mg/g	Yathish et al. (2022)
22.	Amylose	*ae1*	Amylose: 22.25%–26.39%	Amylose: 40.40%–58.10%	Arora et al. (2024)
23.	Amylose	*ae1*	Amylose: ~25%	Amylose: ~47.23%	Han et al. (2022)
24.	Provitamin A, QPM,	o2, crtRB1	Lys: 0.23%, Trp: 0.06%, pro-A: 3.12 μg/g	Lys: 0.39%, Trp: 0.10%, provitamin-A: 18.98 μg/g	Mehta et al. (2020)

drought tolerance, etc. Phenotypic recurrent selection is regarded as an effective strategy for the improvement of polygenic traits by means of increasing the frequency of desirable genes, and also helps in maintaining the high genetic variability in a heterozygous population. But its efficiency and effectiveness are not satisfactory in most of the cases because the phenotypic selection of genotypes is influenced by the environmental effects, and the genotypic selection may take a longer time, i.e. at least 2–3 crop seasons for one cycle of selection. On the other hand, molecular marker technology offers the possibility to speed up this process and makes the selection process more efficient, as this is not influenced by the environment. Use of molecular markers to enrich favorable alleles over multiple generations to improve complex traits is called Marker-assisted recurrent selection (MARS). It is an advanced molecular breeding strategy that combines traditional recurrent selection with molecular markers. MARS involves genotyping with molecular markers linked to quantitative trait loci (QTLs) and selecting superior individuals based on favorable alleles, rather than relying solely on phenotype. In MARS, molecular markers are used to identify the favorable alleles for complex traits like yield, disease resistance, or drought tolerance in the population and select multiple genomic regions involving the expression of these traits. These selected plants are then intercrossed, and the cycle is repeated, gradually increasing the frequency of desirable alleles in the population. Therefore, MARS could enhance the efficiency of recurrent selection and accelerate the process by integrating the favorable allele. This method also helps to integrate the multiple QTLs with small genetic effects governing the expression of complex traits in the population. This breeding technique could significantly improve the rate of genetic gain in a shorter time period.

5.4.4 Applications of MARS in Maize

MARS has been successfully applied in large-scale, realistic maize breeding programs in a number of studies (Ragot et al. 2000; Johnson 2004; Bankole et al. 2017; Beyene et al. 2016). Bankole et al. (2017) applied MARS to enhance drought resilience and grain yield (GY) in a biparental cross of two high-performing drought-resistant lines. S1 lines obtained from each selection type underwent genotyping using single-nucleotide polymorphism (SNP) markers. The molecular analysis indicated that the occurrence of favorable marker alleles rose from 0.510 at C0 to 0.515 at C2, whereas the effective number of alleles (Ne) per locus declined from C0 (1.93) to C2 (1.87). In ten biparental tropical maize (*Zea mays* L.) populations, Beyene et al. (2016) used MARS to evaluate genetic gains in grain yield (GY). The breeding value of individual plants is predicted using a subset of molecular markers that are highly associated with target attributes of interest. Rapid recombination and selfing come next. The best C0 families for recombination were chosen using a selection index based on phenotypic and marker data. 55–87 SNPs tagging particular quantitative trait loci (QTL) were used to genotype individual plants from chosen families. The best individuals from each cycle were either selfed (to generate C1S1 and C1S2) or intercrossed (to form C1). Using MARS, the total gain for GY in all ten populations was 51 kg ha^{-1} $year^{-1}$ under water stress and 105 kg ha^{-1} $year^{-1}$ under well-watered conditions. GY of C1S2-derived hybrids was 8.7%, 5.9%, and 16.2%, substantially higher than C0, founding parents, and commercial checks in all WW settings. The findings show that MARS has the ability to boost genetic growth in sub-Saharan Africa in both ideal and dry conditions.

5.5 Genome-Wide Association Studies (GWAS)

GWAS is an alternative and complementary approach to traditional QTL mapping. GWAS has become widely adopted in plant research. GWAS scans the entire genome using dense molecular markers to identify genetic variants associated with phenotypic traits. By exploiting historical recombination events accumulated over generations, GWAS provides higher mapping resolution than biparental QTL mapping. The success of GWAS depends on population size, marker density, phenotyping accuracy, and the extent of linkage disequilibrium (LD). LD refers to the non-random association of alleles at different loci and is commonly estimated using Lewontin's D′ and the squared correlation coefficient (r^2). The r^2 value indicates the strength of LD between SNPs; values greater than 0.2 suggest strong linkage or co-inheritance within the same QTL region (Alqudah et al. 2020). In maize, a highly cross-pollinated species, LD decays rapidly compared to self-pollinated crops, enabling fine mapping but requiring high marker density. A major advantage of association mapping is its ability to utilize all recombination events in a population's evolutionary history, resulting in improved resolution for complex traits. Additionally, GWAS allows simultaneous analysis of multiple traits within the same population using a single genotypic dataset, making it cost-effective compared with developing separate mapping populations for each trait. However, GWAS has limitations, including the need for extensive and precise phenotyping and constraints related to population structure. Rare alleles and QTLs with small effects may remain undetected due to low allele frequencies, unlike biparental mapping, where allele frequencies can be manipulated.

In maize, candidate gene-based association studies were first reported in the early twenty-first century (Thornsberry et al. 2001), followed by the first genome-wide association analysis for kernel fatty acid composition (Beló et al. 2008). The release of the B73 reference genome (Schnable et al. 2009) greatly accelerated GWAS applications by enabling high-resolution SNP discovery and functional annotation. GWAS can be applied to breeding as well as natural populations and is effective in identifying predictive or causative loci (Viana et al. 2016). Multivariate GWAS has further enabled the dissection of pleiotropic genes controlling complex traits, such as maize inflorescence architecture (Rice et al. 2020). GWAS has been widely applied in maize for agronomic, quality, biotic, and abiotic stress traits.

Significant SNPs associated with kernel iron and zinc concentrations were reported by Hindu et al. (2018). Wang et al. (2017) identified loci controlling flowering-related traits using a 55K SNP array. Li et al. (2016) detected loci associated with lignin, cellulose, and hemicellulose content in maize stalks, with candidate genes involved in cell wall metabolism, transcription regulation, and signaling pathways. GWAS has also elucidated resistance to maize lethal necrosis (Gowda et al. 2015), Gibberella ear rot (Han et al. 2018), and tar spot, where a major QTL (qRtsc8–1) explained 18%–43% phenotypic variation (Mahuku et al. 2016). Furthermore, GWAS has identified genes associated with tolerance to abiotic stresses, including cold and drought (Wang et al. 2016). Major GWAS-identified traits in maize are summarized in Table 5.5.

Table 5.5 List of SNPs and candidate genes identified through GWAS analysis

S. No.	Target trait	Population	SNPs used	Candidate genes/ Chromosomal regions for SNPs	References
1	Yield-related traits	362 inbred lines and RIL population	43,735 SNPs	127 annotated genes were predicted as candidate genes	Zhang et al. (2022)
2	Maize Stalk Rot	Testcross hybrids	2,00,681 high-quality SNPs	Eight SNPs span six genomic regions in Tuxpeño and four in non-Tuxpeño. Phenotypic variance explained 6.20%–9.09%	Song et al. (2024)
3	Drought Tolerance	240 elite inbred lines	29,619 SNPs	Two SNPs on chromosome 7 showed a drought-tolerant ERF transcription factor in their vicinity, which promotes ABA-dependent stomatal closure under WS.	Thirunavukkarasu et al. (2014)
4	Flowering time and Northern corn blight resistance	1487 inbred lines	8244 SNPs	SNP significantly associated with NCLB was traced to resistance genes *Ht1*, *HtP*, *GPC4*, *DBF1,* and SNPs NCLB resistance correlated with FT *was traced to GID1L2*, a gibberellin receptor.	Van Inghelandt et al. (2012)
5	Aflatoxin	300 diverse inbred lines.	261,184 SNPs	GRMZM2G052991, GRMZM2G158141, GRMZM2G158141	Warburton et al. (2015)

(continued)

Table 5.5 (continued)

S. No.	Target trait	Population	SNPs used	Candidate genes/ Chromosomal regions for SNPs	References
6	Thermo tolerance	261 inbred lines	259,973 SNPs	*GRMZM2G377194*, *GRMZM2G026892*, *GRMZM2G060349*, and *GRMZM2G122199*	Gao et al. (2019)
7	Fusarium Stalk Rot	165 maize inbred lines	4666 SNPs	GRMZM2G082709 (Chr 1), GRMZM2G080041(Chr 2), GRMZM5G841142(Chr 6) and GRMZM2G006745 (Chr 8).	Liu et al. (2021)
8	Cell Wall Components	368 diverse inbred lines	559,285 SNPs	GRMZM2G140817 (Chr 6), GRMZM2G031200 (Chr 6), GRMZM2G042627(Chr 7), GRMZM2G017186 (Chr 1).	Li et al. (2016)
9	Chilling-tolerant germination	190 inbred lines	4,886,919 SNPs	*Zm00001eb272370*, *Zm00001eb272380*, *Zm00001eb272390*, *Zm00001eb272400*.	Ma et al. (2023)
10	Starch content	263 maize inbred lines	52,370 SNPs	Total of 77 candidate genes were found in these associated regions.	Liu et al. (2016)
11	Ear-related Traits	292 maize inbred lines	25,331 SNPs	GRMZM2G173943 (Kernel length); GRMZM2G124502, RMZM2G092475, and GRMZM2G057441 (Kernel weight); GRMZM2G015287 (ED and CD)	Zhu et al. (2018)
12	Haploid male fertility	F1s were subjected to haploid induction	425,597 SNPs	GRMZM2G056236 (BIN 5.05) is annotated as being involved in sexual reproduction and may thus affect the restoration of haploid male fertility.	Ma et al. (2018)
13	Charcoal rot resistance	396 inbred lines	296,497 SNPs	GRMZM2G178767, RMZM2G170047, GRMZM2G168337, GRMZM2G465999, GRMZM2G050647	Rashid et al. (2021)

(continued)

Table 5.5 (continued)

S. No.	Target trait	Population	SNPs used	Candidate genes/ Chromosomal regions for SNPs	References
14	Carbon and nitrogen metabolism	NAM population composed of 5000 RILs	1.6 million SNPs from maize HapMap1	Three SNPs associated with chlorophyll a variation that tag carbonic anhydrases (CAs) on chromosome 3, and a significantly associated SNP for malate that is approximately 2 kb away from a malate transporter gene	Zhang et al. (2015)
15	Iron Homeostasis	302 maize inbred lines	437,650 SNPs	18 and 9 unique genes were linked to the significant (FDR = 0.05) associated SNPs under the Fe-deficient and Fe-sufficient regimes, respectively.	Benke et al. (2015)
16	Seedling root development	384 inbred lines	135,311 SNPs	GRMZM26322186 for seedling root trait SRL (Standard root length). GRMZM2G153722 for two traits, root diameter and surface area.	Pace et al. (2015)
17	Husk traits	508 diverse lines	543,641 SNPs	GRMZM2G029077 (chr 8) and GRMZM2G124239 (chr 9)	Cui et al. (2016)
18	Cadmium accumulation in maize leaf	269 maize accessions	43,737 SNPs	GRMZM2G175576, simultaneously encoding a cadmium/zinc-transporting ATPase, was significantly increased in roots, stems, and leaves in response to Cd stress.	Zhao et al. (2018)
19	Heat Tolerance	662 doubled haploid (DH) lines	187,000 SNPs	GRMZM2G887068, GRMZM2G31624, RMZM2G104620, GRMZM2G317287, GRMZM2G057557 and RMZM2G418432	Longmei et al. (2021)
20	Resistance to northern corn leaf blight	999 inbred lines	56,110 SNPs	GRMZM2G100107 (Chr 7), GRMZM2G158141 (Chr 10), GRMZM2G020254 (Chr 10)	Ding et al. (2015)

5.6 Genomic Selection

Technology development in plant breeding for new variety development is limited by genetic gain over a larger span of time. In recent advances in technology development, it focused on quick genetic gain, which includes the addition of a marker system in plant breeding. Since its inception, marker-assisted selection for crop improvement has been utilized with a unified focus for identifying large-effect QTLs through limited markers through mapping a biparental population. Association mapping in random populations later shifted the need for a biparental population and also reduced the time taken for marker trait association to quantify (Jannink et al. 2010). However, continuous selection over large-effect QTLs, minor QTLs with considerable potential, remained neglected due to over-reliance on conventional QTL mapping (Alemu et al. 2024). Genomic selection is one of the advanced breeding applications of modern days with multiple advantages, which include selection for traits that are complex and hard to phenotype for a large set of genotypes, and need longer phenotypic trials. Genomic selection gives additional advantage in selection efficiency with higher intensity, cost effectiveness over other methods, and identifies breeding values using high-density markers and utilizes limited time for breeding cycle (Alemu et al. 2024). Estimation of genetic values without applying costly phenotyping by prediction based on a specific set of genotypes with genomic data and tested in the actual environment is the function of genomic selection models, which enhances resource allocation, and allows continuous integration of new markers in subsequent generations to breeding program (Escamilla et al. 2025). Genomic selection currently emphasizes reproducibility of prediction models and modeling methodologies. However, there is no single specific limited model of GS. All the models increase genetic gain with variable predictive accuracy according to traits and the phenotyping environment concerned. Requirement for an update in GS models, including markers, genomic information, functional omics information, and phenotypic trait data shifts allele frequency (Jannink et al. 2010). Limited performance causes rejection of multiple cross products designed for the improvement of maize yield. In cross-pollinated crops, it takes a large quantity of space and intensive efforts to achieve genetic gain, which has been achieved at a pace after the introduction of MAS followed by GS models. GS differs in predicting breeding value based on training of the population based on whole genome markers following optimization, than any other method of breeding value estimation like BLUP-based mixed models. Prediction accuracy varies among crops, populations, and types of selected markers as limiting QTL-associated markers reduce accuracy (Kaler et al. 2022). Genomic prediction accuracy is highly sensitive in a variable environment that variable genetic populations (Rogers and Holland 2022).

The set of SNPs contributing to important characters was linked to GCA for successful identification of promising crosses and ranking of the crossing remains consistent, which indicates higher prediction accuracy (Zhang et al. 2025a, b). Sometimes finding high-density SNPs is costly and involves complex methods, which is eased by low-density SNPs and converting them into a model without sacrificing accuracy (Bernardeli et al. 2025). Evolving challenges of manual phenotyping and individual genotyping in prediction models have recently been smoothened by specific chip sequencing and support vector machine learning models integrating easy extraction of phenotyping data from low complexity tools, making GS easy (Wu et al. 2025). Challenges of GS cycles includes reduction in prediction accuracies in succeeding selection cycles, rectified by retraining the models incorporating early cycle data (Polzer et al. 2025). Achieving the target of climate-adapted variety development in maize has involved phenotypic plasticity, environmental stability-linked markers in the models integrating environmental response traits for GS models (He et al. 2025), and validation. Evolution of the genomic prediction model searches for the density of markers, robust models, integration of environmental response, etc. However, the selection of markers with biological information and gene ontology information adds robustness to model validation (Ali et al. 2025). Sparse Selection index has been utilized for selecting a needful training set for increasing prediction accuracy in maize (Lopez-Cruz et al. 2021). Microbiome-enabled genomic selection was also reported to enhance the prediction significantly for traits measured by machines (Yang et al. 2024). Parametric methods of genomic prediction for multiple quantitative traits in a large number of F1 crosses reached more than 55% prediction accuracy under maximum conditions (Xiang et al. 2024). An instance-based deep learning platform showed consistently higher accuracy than classical prediction methods (Lourenço et al. 2024). Lower prediction accuracy was increased to more than 80% incorporating environmental variation and response, including genomic information of candidate loci near to SNPs (Tolley et al. 2023). Promoting genetic gain by considering off-season grown.

Crop has gained higher prediction accuracy compared to direct one-season prediction for adaptive traits like flowering (Choquette et al. 2023). Hence, evolved genomic selection models with higher accuracy need to be selected accordingly. The major work on different traits of genomic selection has been mentioned in Table 5.6.

5.7 Conclusion

The molecular markers have been significantly used by the molecular breeders for the improvement of different traits in maize. It has helped in the identification of QTLs/genes, and identified alleles were also introgressed in target genotypes through marker-assisted breeding. Introgression of alleles governing single traits has been successfully done, and hybrids were made available to the public for traits like provitamin A, vitamin E, phytic acid, etc. However, work on introgression of multiple alleles governing different traits in a single background, like high-yielding genotypes with resilience to different biotic and abiotic stress, needs to be initiated on a large scale. Though marker-assisted breeding has been integrated in mainstream breeding in India, at least in large institutes, the application of SNP-based markers and advanced technologies like genomic selection still is in vogue. Though

Table 5.6 Genomic selection used for improvements in different traits in maize

S. No.	Trait	Population used	Marker used	Genetic gain/ important findings	References
1	Grain yield under drought tolerance	F2:3 for pedigree selection while single cross progenies of F2:3 with the tester are used for GS	SNPs using TaqMan assay (191–218 SNPs) and KASP assay (197–286 SNPs)	The efficiency of GS in maize, with an average gain per cycle of 0.086 mg ha^{-1} under drought stress. Higher gains in grain yield under drought stress were achieved using GS (Four-Fold) compared to conventional pedigree selection.	Beyene et al. (2016)
2	Nitrogen use efficiency	552 maize hybrid populations and their parental inbreed	26,769 SNPs were used	GS accuracy ranged from −0.12 to 0.78 with a larger Training population. Increased TP size (from 30 to 90 or 10 to 40 hybrids) improved GS accuracy by 5%, 19%, and 31%.	Mastrodomenico et al. (2019)
3	Shelling % and other traits	Test cross progenies	The maize 55K SNP array	The prediction accuracy for shelling percentage (33.7%) was higher than for grain yield (29.6%) and kernel number per ear (28.7%)	Sun et al. (2019)
4	Stalk lodging resistance and oil content	RIL population	Maize SNP 3K array for high oil/55 K SNP array for lodging	Prediction accuracies in the HO population using the UV model ranged from 0.06 to 0.52, and for the LR population, changed from 0.38 to 0.58 across environments	Liu et al. (2020)

(continued)

Table 5.6 (continued)

S. No.	Trait	Population used	Marker used	Genetic gain/ important findings	References
5	Husk traits	The maize association population, with 498 inbred lines belonging to four subpopulations	56,110 SNPs from the SNP array/ Affymetrix Axiom Maize 600K array SNPs	Husk thickness exhibited the highest prediction accuracies within and among most subpopulations. Thus, to improve husk traits by GS in maize breeding selection programs, it is recommended to begin with Husk thickness	Zhenhai et al. (2020)
6	Days to tasseling, pollen-shedding, and silking	Multi-parent DH population consisting of 379 DH lines	134,785 high-quality SNPs	Increased TP size (70%), the prediction accuracy for DTT (0.27–0.46), DTP (0.34–0.54), and DTS (0.34–0.54). When marker density reached 3000, the prediction accuracy reached a plateau for DTT (0.46), DTP (0.52), and DTS (0.53).	Fan et al. (2024)
7	Resistant to maize lethal necrosis	615 diverse tropical/ subtropical maize inbred lines	GBS SNP (a subset of 2000 SNPs)	GS results revealed higher selection gain per year for marker-based selection compared to phenotypic-based selection for MLND resistance.	Gowda et al. (2015)

(continued)

Table 5.6 (continued)

S. No.	Trait	Population used	Marker used	Genetic gain/ important findings	References
8	Resistance to *Stenocarpella maydis*	238 maize lines	23,154 DArT Markers	The PRK (percentage of rotten kernels) provided a higher predictive power, which is approximately 10.84% (rrBLUP) and 10.96% (BSSV) higher compared with the ERIS (ear rot incidence score).	dos Santos et al. (2016)
9	Combined drought and water-logged stress	Two multi-parent yellow synthetic (MYS) populations	1256 SNPs, a total of 342 and 312 polymorphic markers for MYS-1 and MYS-2, respectively	Under drought stress, MYS-1 showed an 8.4% gain in grain yield per cycle, whereas MYS-2 had a 9.7% gain per cycle. In the case of waterlogging stress, the genetic gain per GS cycle was low with MYS-1 (3.6%), whereas MYS-2 showed relatively high gains (11.3%).	Das et al. (2020)
10	Combine heat and draught stress	300 tropical and subtropical maize inbred lines	10,108 high-quality SNPs	GS accuracies obtained for marker-trait associated SNPs (0.28–0.75) were comparatively greater than those obtained for genome-wide SNPs (0.13–0.64)	Yuan et al. (2019)

(continued)

Table 5.6 (continued)

S. No.	Trait	Population used	Marker used	Genetic gain/ important findings	References
11	Northern Corn Leaf Blight Resistant	Two $F_{2:3}$ populations	768 public plex SNPs	Application of the rrBULP model achieved the prediction accuracies of 0.83 (Population 1) and 0.79 (Population 2) for NCBL resistance, respectively	Balasundara et al. (2021)
12	Resistant to Fall armyworm (FAW) and Weevils (MW)	A panel of 58 maize lines with diverse genetic backgrounds	28,919 DRSNPs	PAs achieved with BLUPs varied from 0.66 to 0.82 for MW-resistance traits, and for FAW resistance, 0.694–0.714 for Random Based Training Set (RBTS) of 37%, and 0.843–0.844 for RBTS of 85%, and these were at least two-fold those from BLUEs.	Badji et al. (2021)
13	Striga (weed) resistance	F2:3 of an early-maturing yellow bi-parental maize population	102 SNPs (1536 Illumina Golden Gate Array)	Gains in grain yield of 498 kg ha^{-1} $cycle^{-1}$ (16.9% $cycle^{-1}$) and 522 kg ha^{-1} $cycle^{-1}$ (12.6% $cycle^{-1}$) were obtained under *Striga*-infested and optimal environments, respectively.	Badu-Apraku et al. (2019)

the cost of sequencing technology has reduced over the period, the cost and availability of genomic tools still deter molecular breeding strategies, particularly in developing countries. So, accessible genomic platforms need to be developed for the developing nations with public and private support services. Though the progress in the area of genomics research has been tremendous, the importance of accurate field phenotyping should not be compromised, as inaccurate phenotyping will lead to erroneous results, regardless of the genomic tools being used. So, high-throughput phenotyping techniques like the use of light, sensors, cameras, computers, and highly modified devices for precision phenotyping are needed for achieving the goals successfully. So, constant innovations in molecular marker technology in the coming years will make it further productive, precise, and cost-effective for easy investigation of traits of interest.

References

Alemu A, Åstrand J, Montesinos-López OA, Isidro y Sánchez J, Fernández-González J, Tadesse W, Vetukuri RR, Carlsson AS, Ceplitis A, Crossa J, Ortiz R, Chawade A (2024) Genomic selection in plant breeding: key factors shaping two decades of progress. Mol Plant 17:552–578. https://doi.org/10.1016/j.molp.2024.03.007

Ali B, Mary-Huard T, Charcosset A, Moreau L, Rincent R (2025) Improvement in genomic prediction of maize with prior gene ontology information depends on traits and environmental conditions. Plant Genome 18:e20553. https://doi.org/10.1002/tpg2.20553

Alqudah AM, Sallam A, Baenziger PS, Börner A (2020) GWAS: fast-forwarding gene identification and characterization in temperate cereals: lessons from barley–a review. J Adv Res 22:119–135

Arora A, Bhamare D, Das AK, Dixit S, Venadan S, Yathish KR, Kumar R, Paul D, Sekhar JC, Neelam S, Nandi S, Rakshit S (2024) Development of high-amylose maize (*Zea mays* L.) genotypes adapted to Indian conditions through molecular breeding. Crop Pasture Sci 75. https://doi.org/10.1071/CP23343

Badji A, Machida L, Kwemoi DB, Kumi F, Okii D, Mwila N, Agbahoungba S, Ibanda A, Bararyenya A, Nghituwamhata SN (2021) Factors influencing genomic prediction accuracies of tropical maize resistance to fall armyworm and weevils. Plants 10:29. https://doi.org/10.3390/plants10010029

Badu-Apraku B, Talabi AO, Fakorede MAB (2019) Yield gains and associated changes in an early yellow bi-parental maize population following genomic selection for Striga resistance and drought tolerance. BMC Plant Biol 19:129. https://doi.org/10.1186/s12870-019-1740-z

Badu-Apraku B, Adewale S, Paterne A, Offornedo Q, Gedil M (2023) Mapping quantitative trait loci and predicting candidate genes for Striga resistance in maize using resistance donor line derived from Zea diploperennis. Front Genet 14:1012460. https://doi.org/10.3389/fgene.2023.1012460

Balasundara DC, Lohithaswa HC, Rahul M (2021) Genetic mapping and genomic prediction for northern corn leaf blight (*Exserohilum turcicum* (Pass.) Leonard and Suggs) resistance. Res Sq. Preprint, rs-618501. https://doi.org/10.21203/rs.3.Rs-618501/v1

Bankole F, Menkir A, Olaoye G, Crossa J, Hearne S, Unachukwu N, Gedil M (2017) Genetic gains in yield and yield related traits under drought stress and favorable environments in a maize population improved using marker assisted recurrent selection. Front Plant Sci 8:808

Baveja A, Muthusamy V, Panda KK, Zunjare RU, Das AK, Chhabra R, Hossain F (2021) Development of multinutrient-rich biofortified sweet corn hybrids through genomics-assisted selection of shrunken2, opaque2, lcyE and crtRB1 genes. J Appl Genet 62:419–429

Beló A, Zheng P, Luck S, Shen B, Meyer DJ, Li B, Tingey S, Rafalski A (2008) Whole genome scan detects an allelic variant of fad2 associated with increased oleic acid levels in maize. Mol Gen Genomics 279(1):1–10

Benke A, Urbany C, Stich B (2015) Genome-wide association mapping of iron homeostasis in maize. BMC Genet 16:1. https://doi.org/10.1186/s12863-014-0153-0

Bernardeli A, Guilhen JHS, Oliveira ICM, Guimarães LJM, Borém A, Jarquin D, Pastina MM (2025) Genomic selection of maize test-cross hybrids leveraged by marker sampling. Plant Genome 18(2):e70030

Bernardo R (2008) Molecular markers and selection for complex traits in plants: learning from the last 20 years. Crop Sci 48(5):1649–1664

Beyene Y, Semagn K, Crossa J, Mugo S, Atlin GN, Tarekegne A, Meisel B, Sehabiague P, Vivek BS, Oikeh S, Alvarado G (2016) Improving maize grain yield under drought stress and non-stress environments in sub-Saharan Africa using marker-assisted recurrent selection. Crop Sci 56(1):344–353

Chandran S, Pukalenthy B, Adhimoolam K, Manickam D, Sampathrajan V, Chocklingam V, Natesan S (2019) Marker-assisted selection to pyramid the opaque-2 (o2) and β-carotene (crtRB1) genes in maize. Front Genet 10:859

Chauhan HS, Muthusamy V, Rashmi T, Basu S, Anand A, Mehta BK, Hossain F (2022) Characterization of crtRB1- and vte4-based biofortified sweet corn inbreds for seed vigour and physico-biochemical traits. J Appl Genet 63:651–662

Chauhan HS, Zunjare RU, Rashmi T, Muthusamy V, Das AK, Mishra SJ, Gain N, Mehta BK, Singh AK, Gupta HS, Hossain F (2024) Enrichment of vitamin A and vitamin E in sweet corn kernels through genomics-assisted introgression of mutant versions of crtRB1 and vte4 genes. Appl Biochem Biotechnol 2024:1–7

Chen L, Li YX, Li C, Wu X, Qin W, Li X, Wang T (2016) Fine-mapping of qGW4.05, a major QTL for kernel weight and size in maize. BMC Plant Biol 16:1–13

Choquette NE, Weldekidan T, Brewer J, Davis SB, Wisser RJ, Holland JB (2023) Enhancing adaptation of tropical maize to temperate environments using genomic selection. G3: Genes Genomes Genet 13:jkad141. https://doi.org/10.1093/g3journal/jkad141

Collard BCY, Jahufer MZZ, Brouwer JB, Pang ECK (2005) An introduction to markers, quantitative trait loci (QTL) mapping and marker-assisted selection for crop improvement: the basic concepts. Euphytica 142:169–196

Cui Z, Luo J, Qi C, Ruan Y, Li J, Zhang A, Yang X, He Y (2016) Genome-wide association study reveals the genetic architecture of four husk traits in maize. BMC Genomics 17:946. https://doi.org/10.1186/s12864-016-3229-6

Das AK, Muthusamy V, Zunjare RU, Baveja A, Chauhan HS, Bhat JS, Guleria SK, Kumar B, Saha S, Hossain F (2020) Genetic variability for kernel tocopherols and haplotype analysis of γ-tocopherol methyl transferase (vte4) gene among exotic and indigenous maize inbreds. J Food Compos Anal 88:103446

Das AK, Gowda MM, Muthusamy V, Zunjare RU, Chauhan HS, Baveja A, Hossain F (2021) Development of maize hybrids with enhanced vitamin E, vitamin A, lysine and tryptophan through molecular breeding. Front Plant Sci 12:659381

Ding J, Ali F, Chen G, Li H, Mahuku G, Yang N, Narro L, Magorokosho C, Makumbi D, Yan J (2015) Genome-wide association mapping reveals novel sources of resistance to northern corn leaf blight in maize. BMC Plant Biol 15:206. https://doi.org/10.1186/s12870-015-0589-z

dos Santos JPR, Pires LPM, de Castro Vasconcellos RC, Guimarães CT, Parentoni SN, Guimarães PEO, Silva AR, Guimarães LJM (2016) Genomic selection for resistance to *Stenocarpella maydis* in maize lines using DArTseq markers. BMC Genet 17:86. https://doi.org/10.1186/s12863-016-0392-3

Ducrocq S, Giauffret C, Madur D, Combes V, Dumas F, Jouanne S, Coubriche D, Jamin P, Moreau L, Charcosset A (2009) Fine mapping and haplotype structure analysis of a major flowering time quantitative trait locus on maize chromosome 10. Genetics 183(4):1555–1563

Escamilla DM, Li D, Negus KL, Kappelmann KL, Kusmec A, Vanous AE, Schnable PS, Li X, Yu J (2025) Genomic selection: essence, applications, and prospects. Plant Genome 18(2):e70053. https://doi.org/10.1002/tpg2.70053

Fan Z, Lin S, Jiang J, Zeng Y, Meng Y, Ren J, Wu P (2024) Dual-model GWAS analysis and genomic selection of maize flowering time-related traits. Genes 15:740. https://doi.org/10.3390/genes15060740

Gao J, Wang S, Zhou Z, Wang S, Dong C, Mu C, Song Y, Ma P, Li C, Wang Z, He K (2019) Linkage mapping and genome-wide association reveal candidate genes conferring thermotolerance of seed-set in maize. J Exp Bot 70(18):4849–4864

Goswami R, Zunjare RU, Khan S, Baveja A, Muthusamy V, Hossain F (2019) Marker-assisted introgression of rare allele of β-carotene hydroxylase (crtRB1) gene into elite quality protein maize inbred for combining high lysine, tryptophan and provitamin A in maize. Plant Breed 138(2):174–183

Gowda M, Das B, Makumbi D, Babu R, Semagn K, Mahuku G, Olsen MS, Bright JM, Beyene Y, Prasanna BM (2015) Genome-wide association and genomic prediction of resistance to maize lethal necrosis disease in tropical maize germplasm. Theor Appl Genet 128(10):1957–1968. https://doi.org/10.1007/s00122-015-2559-0

Guo Z, Zhou S, Wang S, Li WX, Du H, Xu Y (2021) Identification of major QTL for waterlogging tolerance in maize using genome-wide association study and bulked sample analysis. J Appl Genet 62(3):405–418. https://doi.org/10.1007/s13353-021-00629-0

Han S, Miedaner T, Utz HF, Schipprack W, Schrag TA, Melchinger AE (2018) Genomic prediction and GWAS of Gibberella ear rot resistance traits in dent and flint lines of a public maize breeding program. Euphytica 214(1):6

Han J, Guo Z, Wang M, Liu S, Hao Z, Zhang D, Yong H, Weng J, Zhou Z, Li M, Li X (2022) Using the dominant mutation gene Ae1-5180 (amylose extender) to develop high-amylose maize. Mol Breed 42(10):57

Han Q, Ren Z, Zhu Q, Zhou Y, Zhu M, He J, Zhao G (2025) Maize SERRATE 1B positively regulates seed germinability under low-temperature. Plant Sci 355:112458

He K, Yu T, Gao S, Chen S, Li L, Zhang X, Huang C, Xu Y, Wang J, Prasanna BM, Hearne S, Li X, Li H (2025) Leveraging automated machine learning for environmental data-driven genetic analysis and genomic prediction in maize hybrids. Adv Sci 12(17):e2412423. https://doi.org/10.1002/advs.20241242

Hindu V, Palacios-Rojas N, Babu R, Suwarno WB, Rashid Z, Usha R, Saykhedkar GR, Nair SK (2018) Identification and validation of genomic regions influencing kernel zinc and iron in maize. Theor Appl Genet 131(7):1443–1457

Hossain F, Muthusamy V, Pandey N, Vishwakarma AK, Baveja A, Zunjare RU, Gupta HS (2018) Marker-assisted introgression of opaque2 allele for rapid conversion of elite hybrids into quality protein maize. J Genet 97(1):287–298

Hou X, Cheng S, Wang S, Yu T, Wang Y, Xu P, Xu X, Zhou Q, Hou X, Zhang G, Chen C (2022) Characterization and fine mapping of qRPR1-3 and qRPR3-1, two major QTLs for rind penetrometer resistance in maize. Front Plant Sci 13:944539

Jannink J-L, Lorenz AJ, Iwata H (2010) Genomic selection in plant breeding: from theory to practice. Brief Funct Genomics 9(2):166–177. https://doi.org/10.1093/bfgp/elq001

Jiang T, Zhang C, Zhang Z, Wen M, Qiu H (2023) QTL mapping of maize (*Zea mays* L.) kernel traits under low-phosphorus stress. Physiol Mol Biol Plants 29(3):435–445

Johnson R (2004) Marker-assisted selection. Plant Breed Rev 24(1):293–310

Kaler AS, Purcell LC, Beissinger T, Gillman JD (2022) Genomic prediction models for traits differing in heritability for soybean, rice and maize. BMC Plant Biol 22:87. https://doi.org/10.1186/s12870-022-03479-y

Katral A, Hossain F, Zunjare RU, Mishra SJ, Ragi S, Kasana RK, Muthusamy V (2024) Enhancing kernel oil and tailoring fatty acid composition by genomics-assisted selection for dgat1-2 and fatb genes in multi-nutrient-rich maize: a new avenue for food, feed and bioenergy. Plant J 119(5):2402–2422

Kaur R, Kaur G, Vikal Y, Gill GK, Sharma S, Singh J, Chawla JS (2020) Genetic enhancement of essential amino acids for nutritional enrichment of maize protein quality through marker-assisted selection. Physiol Mol Biol Plants 26(11):2243–2254

Li K, Wang H, Hu X, Liu Z, Wu Y, Huang C (2016) Genome-wide association study reveals the genetic basis of stalk cell wall components in maize. PLoS One 11(8):e0158906. https://doi.org/10.1371/journal.pone.0158906

Li Q, Nie S, Li G, Du J, Ren R, Yang X, Liu B, Gao X, Liu T, Zhang Z, Zhao X (2022) Identification and fine mapping of the recessive gene BK-5, which affects cell wall biosynthesis and plant brittleness in maize. Int J Mol Sci 23(2):814

Li Y, Tao X, Yao L, Tang S, Zhang X, Tong L, Xu M (2025) qRfv2, a quantitative resistance locus against Fusarium ear rot in maize. Crop J 13(1):41–50

Liu Y, Wang L, Sun C, Zhang Z, Zheng Y, Qiu F (2014) Genetic analysis and major QTL detection for maize kernel size and weight in multi-environments. Theor Appl Genet 127(5):1019–1037

Liu N, Xue Y, Guo Z, Li W, Tang J (2016) Genome-wide association study identifies candidate genes for starch content regulation in maize kernels. Front Plant Sci 7:1046. https://doi.org/10.3389/fpls.2016.01046

Liu X, Hu X, Li K, Wang H, Yang C, Zhang Z, Chen J (2020) Genetic mapping and genomic selection for maize stalk strength. BMC Plant Biol 20:196. https://doi.org/10.1186/s12870-020-2270-4

Liu S, Fu J, Shang Z, Song X, Zhao M (2021) Combination of genome-wide association study and QTL mapping reveals the genetic architecture of Fusarium stalk rot in maize. Front Agron 2:590374. https://doi.org/10.3389/fagro.2020.590374

Longmei N, Gill GK, Zaidi PH, Kumar R, Nair SK, Hindu V, Vinayan MT, Vikal Y (2021) Genome-wide association mapping for heat tolerance in sub-tropical maize. BMC Genomics 22:154. https://doi.org/10.1186/s12864-021-07463-y

Lopez-Cruz M, Beyene Y, Gowda M, Crossa J, Pérez-Rodríguez P, de Los Campos G (2021) Multi-generation genomic prediction of maize yield using parametric and non-parametric sparse selection indices. Heredity 127(5):423–432. https://doi.org/10.1038/s41437-021-00474-1

Lourenço VM, Ogutu JO, Rodrigues RAP, Posekany A, Piepho HP (2024) Genomic prediction using machine learning: a comparison of the performance of regularized regression, ensemble, instance-based and deep learning methods on synthetic and empirical data. BMC Genomics 25:152. https://doi.org/10.1186/s12864-023-09933-x

Ma H, Li G, Würschum T, Zhang Y, Zheng D, Yang X, Li J, Liu W, Yan J, Chen S (2018) Genome-wide association study of haploid male fertility in maize (*Zea mays* L.). Front Plant Sci 9:353588. https://doi.org/10.3389/fpls.2018.00974

Ma Y, Yao L, Zhang L, Su A, Wang R, Song W, Li Z, Zhao J (2023) Genome-wide association analysis of chilling-tolerant germination in a new maize association mapping panel. Food Energy Secur 12(3):e445

Mahuku G, Chen J, Shrestha R, Narro LA, Guerrero KVO, Arcos AL, Xu Y (2016) Combined linkage and association mapping identifies a major QTL (qRtsc8-1),conferring tar spot complex resistance in maize. Theor Appl Genet 129:1217–1229

Mastrodomenico AT, Bohn MO, Lipka AE, Below FE (2019) Genomic selection using maize ex-plant variety protection germplasm for the prediction of nitrogen-use traits. Crop Sci 59(1):278–291. https://doi.org/10.2135/cropsci2018.06.0398

Mehta BK, Muthusamy V, Zunjare RU, Baveja A, Chauhan HS, Chhabra R, Hossain F (2020) Biofortification of sweet corn hybrids for provitamin-A, lysine and tryptophan using molecular breeding. J Cereal Sci 96:103093

Meng B, Wang S, Li WX, Guo Z, Tang J (2025) QTL mapping of Fusarium ear rot resistance using genotyping by target sequencing (GBTS) in maize. J Appl Genet 66:787–796

Mir RR, Varshney RK (2013) Future prospects of molecular markers in plants. In: Molecular markers in plants. Blackwell Publishing Ltd, pp 169–190

Mishra SJ, Hossain F, Zunjare RU, Chhabra R, Katral A, Gopinath I, Muthusamy V (2025) Genomics-assisted stacking of waxy1, opaque2, and crtRB1 genes for enhancing amylopectin in biofortified maize for industrial utilization and nutritional security. Funct Integr Genomics 25(1):18

Natesan S, Duraisamy T, Pukalenthy B, Chandran S, Nallathambi J, Adhimoolam K, Rajasekaran R (2020) Enhancing β-carotene concentration in parental lines of CO6 maize hybrid through marker-assisted backcross breeding (MABB). Front Nutr 7:134

Ndlovu N, Kachapur RM, Beyene Y, Das B, Ogugo V, Makumbi D, Gowda M (2024) Linkage mapping and genomic prediction of grain quality traits in tropical maize (*Zea mays* L.). Front Genet 15:1353289

Pace J, Gardner C, Romay C, Ganapathysubramanian B, Lübberstedt T (2015) Genome-wide association analysis of seedling root development in maize (*Zea mays* L.). BMC Genomics 16(1):47. https://doi.org/10.1186/s12864-015-1226-9

Peiffer JA, Romay MC, Gore MA, Flint-Garcia SA, Zhang Z, Millard MJ, Gardner CAC, McMullen MD, Holland JB, Bradbury PJ, Buckler ES (2014) The genetic architecture of maize height. Genetics 196(4):1337–1356

Peleman JD, Van der Voort JR (2003) Breeding by design. Trends Plant Sci 8(7):330–334

Polzer C, Auinger HJ, Terán-Pineda M, Hölker AC, Mayer M, Presterl T, Rivera-Poulsen C, da Silva S, Ouzunova M, Melchinger AE, Schön CC (2025) Rapid cycling genomic selection in maize landraces. Theor Appl Genet 138(4):75. https://doi.org/10.1007/s00122-025-04855-6

Qiu H, Li C, Yang W, Tan K, Yi Q, Yang M, Bai G (2021) Fine mapping of a new major QTL-qGLS8 for gray leaf spot resistance in maize. Front Plant Sci 12:743869

Qutub M, Chandran S, Rathinavel K, Sampathrajan V, Rajasekaran R, Manickam S, Natesan S (2021) Improvement of a yairipok chujak maize landrace from north eastern Himalayan region for β-carotene content through molecular marker-assisted backcross breeding. Genes (Basel) 12(5):762

Ragot M, Gay G, Muller JP, Durovray J (2000) Efficient selection for the adaptation to the environment through QTL mapping and manipulation in maize. In: Molecular approaches for the genetic improvement of cereals for stable production in water-limited environments. International Maize and Wheat Improvement Center (CIMMYT), pp 128–130

Rashid Z, Kaur H, Babu V, Singh PK, Harlapur SI, Nair SK (2021) Identification and validation of genomic regions associated with charcoal rot resistance in tropical maize by genome-wide association and linkage mapping. Front Plant Sci 12:726767. https://doi.org/10.3389/fpls.2021.726767

Ren J, Wu P, Huestis GM, Zhang A, Qu J, Liu Y, Zheng H, Alakonya AE, Dhliwayo T, Olsen M, San Vicente F (2022) Identification and fine mapping of a major QTL (qRtsc8-1) conferring resistance to maize tar spot complex and validation of production markers in breeding lines. Theor Appl Genet 135(5):1551–1563

Rice BR, Fernandes SB, Lipka AE (2020) Multi-trait genome-wide association studies reveal loci associated with maize inflorescence and leaf architecture. Plant Cell Physiol 61(8):1427–1437

Rogers AR, Holland JB (2022) Environment-specific genomic prediction ability in maize using environmental covariates depends on environmental similarity to training data. G3 (Bethesda, Md.) 12(2):jkab440. https://doi.org/10.1093/g3journal/jkab440

Salvo S, Cook J, Carlson AR, Hirsch CN, Kaeppler SM, Kaeppler HF (2018) Genetic fine-mapping of a quantitative trait locus (QTL) associated with embryogenic tissue culture response and plant regeneration ability in Maize (*Zea mays* L.). Plant Genome 11(2):170111

Sarika K, Hossain F, Muthusamy V, Zunjare RU, Baveja A, Goswami R, Gupta HS (2018) Marker-assisted pyramiding of opaque2 and novel opaque16 genes for further enrichment of lysine and tryptophan in sub-tropical maize. Plant Sci 272:142–152

Sarkar B, Varalaxmi Y, Vanaja M, RaviKumar N, Prabhakar M, Yadav SK, Singh VK (2023) Mapping of QTLs for morphophysiological and yield traits under water-deficit stress and well-watered conditions in maize. Front Plant Sci 14:1124619

Schmidt M, Guerreiro R, Baig N, Habekuß A, Will T, Ruckwied B, Stich B (2024) Fine mapping a QTL for BYDV-PAV resistance in maize. Theor Appl Genet 137(7):163

Schnable PS, Ware D, Fulton RS, Stein JC, Wei F, Pasternak S, Liang C, Zhang J, Fulton L, Graves TA, Minx P (2009) The B73 maize genome: complexity, diversity, and dynamics. Science 326(5956):1112–1115

Sharma P, Sharma V, Dhaliwal HS, Kumar R (2024) Introgression of opaque2 allele into sweetcorn composite through marker-assisted selection. Cereal Res Commun 52(4):1769–1778

Singh J, Sharma S, Kaur A, Vikal Y, Cheema AK, Bains BK, Hossain F (2021) Marker-assisted pyramiding of lycopene-ε-cyclase, β-carotene hydroxylase1 and opaque2 genes for development of biofortified maize hybrids. Sci Rep 11(1):12642

Song J, Pacheco A, Alakonya A, Cruz-Morales AS, Muñoz-Zavala C, Qu J, Wang C, Zhang X, San Vicente F, Dhliwayo T (2024) Genome-wide association mapping and genomic prediction of stalk rot in two mid-altitude tropical maize populations. Crop J 12(2):558–568

Sun Q, Wang P, Li W, Li W, Lu S, Yu Y, Zhao M, Meng Z (2019) Genomic selection on shelling percentage and other traits for maize. Breed Sci 69(2):266–271. https://doi.org/10.1270/jsbbs.18141

Talukder ZA, Muthusamy V, Chhabra R, Bhatt V, Reddappa SB, Mishra SJ, Hossain F (2022a) Combining higher accumulation of amylopectin, lysine and tryptophan in maize hybrids through genomics-assisted stacking of waxy1 and opaque2 genes. Sci Rep 12(1):706

Talukder ZA, Muthusamy V, Chhabra R, Bhatt V, Reddappa SB, Mishra SJ, Hossain F (2022b) Enrichment of amylopectin in sub-tropically adapted maize hybrids through genomics-assisted introgression of waxy1 gene encoding granule-bound starch synthase (GBSS). J Cereal Sci 105:103443

Tang B, Li Y, Mu Z, Chen L, Guo H, Chen Z, Wang T (2020) Fine mapping and candidate gene analysis of qKW7b, a major QTL for kernel width in maize. Mol Breed 40(7). https://doi.org/10.1007/s11032-020-01146-4

Thirunavukkarasu N, Hossain F, Arora K, Sharma R, Shiriga K, Mittal S, Mohan S, Namratha PM, Dogga S, Rani TS, Katragadda S (2014) Functional mechanisms of drought tolerance in subtropical maize (*Zea mays* L.) identified using genome-wide association mapping. BMC Genomics 15(1):1182

Thornsberry JM, Goodman MM, Doebley J, Kresovich S, Nielsen D, Buckler ES (2001) Dwarf8 polymorphisms associate with variation in flowering time. Nat Genet 28(3):286–289

Tolley SA, Brito LF, Wang DR, Tuinstra MR (2023) Genomic prediction and association mapping of maize grain yield in multi-environment trials based on reaction norm models. Front Genet 14:1221751. https://doi.org/10.3389/fgene.2023.1221751

Van Inghelandt D, Melchinger AE, Martinant JP, Stich B (2012) Genome-wide association mapping of flowering time and northern corn leaf blight (*Setosphaeria turcica*) resistance in a vast commercial maize germplasm set. BMC Plant Biol 12(1):56

Viana JMS, Mundim GB, Silva FFE, Garcia AAF (2016) Efficiency of genome-wide association study in open-pollinated populations. bioRxiv:050955

Wang X, Wang H, Liu S, Ferjani A, Li J, Yan J, Yang X, Qin F (2016) Genetic variation in ZmVPP1 contributes to drought tolerance in maize seedlings. Nat Genet 48(10):1233–1241

Wang C, Yang Q, Wang W, Li Y, Guo Y, Zhang D, Ma X, Song W, Zhao J, Xu M (2017) A transposon-directed epigenetic change in ZmCCT underlies quantitative resistance to Gibberella stalk rot in maize. New Phytol 215(4):1503–1515

Wang G, Zhao Y, Mao W, Ma X, Su C (2020) QTL analysis and fine mapping of a major QTL conferring kernel size in maize (*Zea mays*). Front Genet 11:603920

Warburton ML, Tang JD, Windham GL, Hawkins LK, Murray SC, Xu W, Boykin D, Perkins A, Williams WP (2015) Genome-wide association mapping of aspergillus flavus and aflatoxin accumulation resistance in maize. Crop Sci 55(5):1857–1867

Wu H, Han R, Zhao L, Liu M, Chen H, Li W, Li L (2025) AutoGP: an intelligent breeding platform for enhancing maize genomic selection. Plant Commun 6(4):101240. https://doi.org/10.1016/j.xplc.2025.101240

Xiang Y, Xia C, Li L, Wei R, Rong T, Liu H, Lan H (2024) Genomic prediction of yield-related traits and genome-based establishment of heterotic pattern in maize hybrid breeding of Southwest China. Front Plant Sci 15:1441555. https://doi.org/10.3389/fpls.2024.1441555

Xu M, Jiang L, Ge M, Zhao H, Zhang T (2015) Analysis of heterosis and quantitative trait loci for kernel shape related traits using triple testcross population in maize. PLoS One 10(4):e0124779

Xu Z, Wang F, Zhou Z, Meng Q, Chen Y, Han X, Tie S, Liu C, Hao Z, Li M, Zhang D, Han J, Wang Z, Li X, Weng J (2022) Identification and fine mapping of a major QTL conferring resistance to maize rough dwarf disease. Plant Dis 106(1):65–72. https://doi.org/10.1094/PDIS-03-20-0495-RE

Yang Q, Yin G, Guo Y, Zhang D, Chen S, Xu M (2010) A major QTL for resistance to Gibberella stalk rot in maize. Theor Appl Genet 121(4):673–687

Yang J, Liu Z, Chen Q, Qu Y, Tang J, Lübberstedt T, Li H (2020) Mapping of QTL for grain yield components based on a DH population in maize. Sci Rep 10(1):7086

Yang Z, Zhao T, Cheng H, Yang J (2024) Microbiome-enabled genomic selection improves prediction accuracy for nitrogen-related traits in maize. G3 (Bethesda, Md.) 14(3):jkad286. https://doi.org/10.1093/g3journal/jkad286

Yang X, Shaw RK, Li L, Jiang F, Fan X (2025) Novel candidate genes and genetic basis analysis of kernel starch content in tropical maize. BMC Plant Biol 25(1):105

Yathish KR, Karjagi CG, Gangoliya SS, Kumar A, Preeti J, Yadav HK, Srivastava S, Kumar S, Swamy HKM, Singh A, Phagna RK (2022) Introgression of the low phytic acid locus (lpa2) into elite maize (*Zea mays* L.) inbreds through marker-assisted backcross breeding (MABB). Euphytica 218:127

Yin X, Bi Y, Jiang F, Guo R, Zhang Y, Fan J, Fan X (2022) Fine mapping of candidate quantitative trait loci for plant and ear height in a maize nested-association mapping population. Front Plant Sci 13:963985

Yuan Y, Cairns JE, Babu R, Gowda M, Makumbi D, Magorokosho C, Zhang A, Liu Y, Wang N, Hao Z, San Vicente F, Olsen MS, Prasanna BM, Lu Y, Zhang X (2019) Genome-wide association mapping and genomic prediction analyses reveal the genetic architecture of grain yield and flowering time under drought and heat stress conditions in maize. Front Plant Sci 9:1919. https://doi.org/10.3389/fpls.2018.01919

Zhang D, Liu Y, Guo Y, Yang Q, Ye J, Chen S, Xu M (2012) Fine-mapping of qRfg2, a QTL for resistance to Gibberella stalk rot in maize. Theor Appl Genet 124(3):585–596

Zhang G, Wang X, Wang B, Tian Y, Li M, Nie Y, Peng Q, Wang Z (2013) Fine mapping a major QTL for kernel number per row under different phosphorus regimes in maize (*Zea mays* L.). Theor Appl Genet 126:1545–1553

Zhang Z, Liu Z, Hu Y, Li W, Fu Z, Ding D, Li H, Qiao M, Tang J (2014) QTL analysis of kernel-related traits in maize using an immortalized F2 population. PLoS One 9(2):e89645

Zhang N, Gibon Y, Wallace JG, Lepak N, Li P, Dedow L, Chen C, So Y, Kremling K, Bradbury PJ, Brutnell T, Stitt M, Buckler ES (2015) Genome-wide association of carbon and nitrogen metabolism in the maize nested association mapping population. Plant Physiol 168(2):575–583. https://doi.org/10.1104/pp15.00025

Zhang X, Ren Z, Luo B, Zhong H, Ma P, Zhang H, Hu H, Wang Y, Zhang H, Liu D, Wu L (2022) Genetic architecture of maize yield traits dissected by QTL mapping and GWAS in maize. Crop J 10(2):436–446

Zhang Z, Wang X, Zhang Y, Zhou K, Yu G, Yang W, Li F, Guan X, Zhang X, Yang Z, Xu C, Xu Y (2025a) SPDC-HG: an accelerator of genomic hybrid breeding in maize. Plant Biotechnol J 23(5):1847–1861. https://doi.org/10.1111/pbi.70011

Zhang D, Yang F, Li J, Liu Z, Han Y, Zhang Q, Pan S, Zhao X, Wang K (2025b) Progress and perspectives on genomic selection models for crop breeding. Technol Agron 5:e006. https://doi.org/10.48130/tia-0025-0002

Zhao Y, Su C (2019) Mapping quantitative trait loci for yield-related traits and predicting candidate genes for grain weight in maize. Sci Rep 9(1):16112. https://doi.org/10.1038/s41598-019-52222-5

Zhao X, Luo L, Cao Y, Liu Y, Li Y, Wu W, Lan Y, Jiang Y, Gao S, Zhang Z, Shen Y, Pan G, Lin H (2018) Genome-wide association analysis and QTL mapping reveal the genetic control of cadmium accumulation in maize leaf. BMC Genomics 19(1):91. https://doi.org/10.1186/s12864-017-4395-x

Zhenhai C, Haixiao D, Ao Z, Yanye R, Yan H, Zhiwu Z (2020) Assessment of the potential for genomic selection to improve husk traits in maize. G3 (Genes|Genomes|Genetics) 10(10):3741–3749. https://doi.org/10.1534/g3.120.401600

Zhou Z, Song L, Zhang X, Li X, Yan N, Xia R, Zhu H, Weng J, Hao Z, Zhang D, Yong H (2016) Introgression of opaque2 into waxy maize causes extensive biochemical and proteomic changes in endosperm. PLoS One 11(7):e0158971

Zhou G, Mao Y, Xue L, Chen G, Lu H, Shi M, Zhang Z, Huang X, Song X, Hao D (2020) Genetic dissection of husk number and length across multiple environments and fine-mapping of a major-effect QTL for husk number in maize (*Zea mays* L.). Crop J 8(6):1071–1080. https://doi.org/10.1016/j.cj.2020.03.009

Zhu M, Shao Y, Pei H, Guo M, Li J, Song Y, Zhao A (2018) Genetic diversity and genome-wide association study of major ear quantitative traits using high-density SNPs in maize. Front Plant Sci 9:966. https://doi.org/10.3389/fpls.2018.00966

Next-Generation Maize: Tailoring Traits with Accelerated Breeding

6

R. K. Bhavyasree, Deepak, Arpit Chaudhary, Bhupender Kumar, Bahadur Singh Jat, Pravin Kumar Bagaria, and Mukesh Choudhary

Abstract

Maize (*Zea mays* L.) is one of the most important crops for global food security and industrial value chains, yet its production is increasingly threatened by a changing climate. Conventional breeding pipelines in maize delivered significant genetic gains, which, despite historic successes, are too slow to address the rising demands of maize. This chapter synthesizes recent advances in next-generation maize improvement, emphasizing how accelerated breeding technologies and strategic trait prioritization converge to deliver climate-resilient, resource-efficient, and nutritionally enriched cultivars. Trait priorities now extend beyond maximizing yield potential to enhancing yield stability under variable ecologies, climate resilience, and developing semi-dwarf and smart-canopy architectures suited for high-density and low-input systems. Greater nitrogen, water, and radiation use efficiency, alongside enhanced kernel nutritional profiles such as provitamin A biofortification and Quality Protein Maize, illustrate the shift toward sustainability and human and poultry nutrition. Durable resistance to fall armyworm, maize lethal necrosis, and other evolving biotic stresses remains central. Enabling traits, including high haploid induction rate and increased doubled-haploid responsiveness, are highlighted as critical accelerators of genetic gain. Accelerated breeding approaches such as speed breeding, doubled haploidy, genomic selection, optimized cross selection, next-generation bulked segregant analysis, and precision genome editing compress breeding cycles from years to months while improving prediction accuracy and maintaining genetic diversity.

R. K. Bhavyasree
Punjab Agricultural University, Regional Research Station, Faridkot, Punjab, India

Deepak · A. Chaudhary · B. Kumar · B. S. Jat · P. K. Bagaria · M. Choudhary (✉)
ICAR-Indian Institute of Maize Research, Ludhiana, Punjab, India
e-mail: pravin.bagaria@icar.org.in; mukesh.choudhary1@icar.org.in

M. Choudhary et al. (eds.), *Next-Generation Maize Research*,
https://doi.org/10.1007/978-981-92-0722-0_6

The integrated deployment of key traits and advanced tools should ultimately focus on maximizing per-day productivity. Collectively, these innovations position maize improvement at the forefront of modern crop improvement, enabling a transition toward predictive, automated, and globally scalable design-based breeding systems to accelerate genetic gains.

Keywords

Maize · Resilience · Breeding · Genetic gain · Improvement

6.1 Introduction

Maize (*Zea mays L.*) is a very important global crop with wide adaptation and enormous diversity. It is an important crop with multi-dimensional roles in food security, livestock feed, and industrial supply chains. Its versatility is unmatched: maize kernels are consumed directly, processed into starches and oils, converted into ethanol, and used in bioplastics. Stover and cobs fuel livestock and bioenergy systems. Maize is cultivated in more than 200 million hectares globally due to its multifaceted use, making it the most demanding crop in both developed and developing countries (Liu et al. 2025). However, climate extremes are intensifying, with heat waves during flowering, prolonged droughts, and erratic rainfall patterns becoming more frequent. These stresses most occur together, making compound events-such as heat plus drought or flooding plus pathogen outbreaks increasingly common. Climate change-led pest and disease pressures are common, exemplified by the spread of fall armyworm (FAW) across Africa and Asia. Meanwhile, population growth and urbanization compel researchers to develop high-yielding hybrids for farmers to produce more in less land.

Maize is a highly diverse crop with genetic diversity, spanning tropical, subtropical, and temperate germplasm pools. Its heterotic patterns are well characterized, enabling systematic exploitation of general and specific combining ability. Its genome has vast structural variations, offering ample opportunities for allele mining. Its physiology-C4 photosynthesis, planting density responsiveness, and architectural plasticity make it a key crop for ideotype redesign. Semi-dwarf, nitrogen-efficient and high-density friendly hybrids are now envisioned as the maize equivalent or superior of the Green Revolution's wheat and rice breakthroughs (Xiao et al. 2025).

Conventional breeding has delivered remarkable gains over the past century, placing maize from a subsistence crop into a global commodity. Hybrid breeding, heterosis exploitation, and doubled haploid (DH) technology are credited for yield improvement (Gupta et al. 2022). However, the returns from the conventional breeding are low and slow. Breeders have to deliver faster genetic gains for trait targets, including climate resilience, resource use efficiency, and nutritional quality (Choudhary et al. 2019). These challenges and advances in genomics, transcriptomics, metabolomics, and phenomics catalysed the rise of next-generation breeding (Choudhary et al. 2019). Sequencing costs have dropped, enabling whole-genome resequencing of diverse germplasm (Kumar et al. 2021). Pangenomics has revealed extensive presence-absence variation and structural variants that shape phenotype (Aggarwal et al. 2022). High-throughput phenotyping (HTP) platforms, such as UAVs, ground rovers, and greenhouse imaging systems, made it possible to generate terabytes of trait data per season (Singh et al. 2019). The availability of artificial intelligence (AI) and machine learning (ML) models helps to integrate the huge datasets, predicting performance across environments and making selection decisions. With advances in genome editing tools such as CRISPR-Cas9, it is possible to precisely tweak alleles, while speed breeding methods shortened generation intervals significantly (Kumar et al. 2020).

The important feature of next-generation maize breeding is integration. Speed breeding × DH × genomic prediction pipelines reduce cycle time from 6–7 years to a year or a couple. AI-augmented phenotyping improves prediction accuracy for complex polygenic traits. Genome editing introduces beneficial alleles such as *ARGOS8* variants for drought tolerance, *Brachytic2* for semi-dwarfism, and teosinte ligule alleles for canopy optimization (Shi et al. 2017; Kumar et al. 2020). Pangenome-driven allele mining uncovers rare variants linked to climate resilience (Aggarwal et al. 2022). Digital breeding platforms model genotype × environment × management (G × E × M) interactions, improving the design for mating parents and trial allocation. This integration helps in decision-making by switching from empirical crossing to optimized cross selection (OCS), where genomic estimated breeding values (GEBVs), co-ancestry/genetic relatedness, and predicted genetic gain guide the parental choice (Cowling et al. 2019; Zhu et al. 2024). Decadal vision of Maize2035 envisions autonomous phenotyping, hologenomics, single-cell omics, and synthetic biology converging into programmable breeding ecosystems (Liu et al. 2025). In short, maize, a crop that historically symbolized heterosis, now stands at the forefront of next-generation smart breeding for the future. The following sections cover trait prioritization, accelerated breeding approaches, digital tools, and future visions to redefine maize improvement for the next decade.

6.2 Trait Prioritization in Modern Maize Breeding

Trait prioritization is the fundamental step of modern breeding programs. It defines which traits breeders should emphasize when selecting donors, designing crosses, and deploying genomic selection/prediction models. In the past, yield stood as the key trait, but yield now must be considered as part of a multidimensional framework that integrates resilience, sustainability, nutritional quality, and market demand (Zhang et al. 2025).

6.2.1 Yield Potential and Biomass Productivity

Yield remains the main framework of breeding objectives, but its definition has evolved over time. Breeders now target hybrids that can provide sustained yields under high plant density, low nitrogen input, and climate stress conditions. Kernel number, a component trait for grain yield, is critical as heat or drought during flowering can drastically reduce grain set. Biomass accumulation and canopy efficiency are also important, with radiation use efficiency (RUE) and crop growth rate during grain filling as key determinants for the accumulation of carbohydrate reserves. In the context of a changing climate, the focus is shifting from increasing single plant yield to the 'smart canopy' traits for high-density planting. High-throughput phenomics platforms, including UAV-based multispectral imaging and LiDAR, now allow breeders to quantify canopy photosynthesis and biomass distribution with precision (Zhang et al. 2025).

6.2.2 Stress Resilience and Climate Adaptation

Climate change-induced abiotic and biotic stresses threaten maize production. Deep root systems, efficient carbohydrate remobilization, and reproductive resilience, such as silk elongation under stress, are key traits for drought tolerance, whereas good pollen viability, silk receptivity, and canopy cooling mechanisms and low anthesis silking interval are important traits for heat tolerance in maize. Waterlogging tolerance involves aerenchyma formation and adventitious roots. Next-generation bulked segregant analysis (NG-BSA) fast-tracks the identification of QTLs and candidate genes for these climate-resilient traits by integrating phenotyping with genomic data (Wang et al. 2023a). Resistance to FAW, blights (maydis leaf blight and turcicum leaf blight), bacterial leaf and sheath blight, maize streak virus, maize lethal necrosis, and charcoal rot remains on high priority for biotic stress resistance. NG-BSA, in combination with, can be integrated to accelerate identification of resistance alleles and their deployment in elite germplasm (Singh et al. 2019).

6.2.3 Semi-Dwarfism, Smart Canopy Design, and Plant Architecture

Plant architecture holds immense importance in producing more in less area. Semi-dwarf ideotypes, engineered through GA and BR pathway edits, reduce lodging and improve nitrogen use efficiency (NUE). Upright or erect leaves enhance light penetration, robust brace roots, and stronger stalks in semi-dwarf plants that resist lodging, enabling ultra-high-density systems, holding importance for land-scarce regions in Asia and Africa (Xiao et al. 2025). Moreover, uniformity/consistency in plant height, maturity, and ear placement promotes mechanical harvesting in maize.

6.2.4 Resource-Use Efficiency

Resource-use efficiency is an important trait for sustainable maize production, exploring the full potential of the genotype. NUE can be improved by better N uptake, assimilation, and remobilization of N. Similarly, water use efficiency depends upon stomatal regulation and transpiration efficiency. RUE benefits from a smart-canopy design like erect leaves. Furthermore, the genome editing tools are being implemented for the trait-specific improvements. For example, the editing of *ARGOS8* and nitrate transporters can accelerate progress toward low-input, efficient maize (Shi et al. 2017).

6.2.5 Market Driven Traits

Market-driven traits such as Quality Protein Maize (with enhanced lysine and tryptophan), high oil, high starch, high carotenoid content, high methionine, low phytate, and higher ethanol recovery are of immense importance for industry-driven breeding (Kumar et al. 2019). Early flowering, strong anther exertion, and DH responsiveness are key traits for hybrid breeding and seed industry competitiveness.

Future maize breeding should prioritize per-day productivity as the fundamental selection criterion, as it captures true genetic potential by integrating yield, duration, and environmental response, allowing meaningful comparison across locations, climates, and germplasm types. Traits should be evaluated by their contribution to daily biomass and grain accumulation rather than absolute yield alone. Aligning all breeding targets toward per-day productivity will accelerate genetic gain and climate resilience.

6.3 Accelerated Breeding Approaches

The rapid evolution of breeding tools and technologies has transformed modern maize breeding into a dynamic, accelerated pipeline. The breeder's equation-genetic gain (ΔG) = (selection intensity × accuracy × genetic variance)/cycle length, depicts that shortening cycle length is the most powerful option for enhancing genetic gains. Hence, modern maize breeding aims to reduce the generation interval, raise accuracy, and preserving diversity.

6.3.1 Speed Breeding: Enabling Rapid Generation Turnover

Speed breeding is based on the use of desirable modifications in photoperiod, temperature, CO_2 concentration, and light spectra to shorten generation intervals. Extended photoperiods (20–22 h of light), optimized temperature regimes (28–30 °C day, 22–24 °C night), and full-spectrum LEDs accelerate vegetative and reproductive development. Early seed harvest and embryo rescue are also rapid generation advancement approaches for reducing the generation interval time.

For maize, achieving rapid flowering was historically challenging due to photoperiod sensitivity and complex vegetative reproductive transitions. However, targeted manipulation of flowering genes such as *ZmCOL3* and *ZCN8*, combined with circadian regulators, now enables 4–5 generations per year (Jin et al. 2018; Guo et al. 2018). Speed breeding can be integrated with genomic selection, allowing rapid generation advancement before phenotyping, enhancing annual genetic gains.

6.3.2 Doubled Haploid (DH) Technology: Instant Homozygosity

DH technology is extensively used in global maize breeding programs. Haploid induction, typically using maternal inducers carrying *qhir1* and *qhir8*, produces haploids. *R1-nj*, oil content (NIR-based), or anthocyanin markers help in the haploid identification. Followed by it, colchicine-based chromosome doubling fixes homozygosity in a single generation. Furthermore, loss-of-function mutations of *ZmPLA1*, *DUF679* domain membrane protein (*ZmDMP*) and Gamete Expression protein 1 (*ZmGEX1*) are also used for haploid breeding, as they help to induce haploidy in maize (Zhong et al. 2019; Kumar et al. 2020; Sun et al. 2025). Recent genomic studies have also identified regulatory factors controlling haploid induction rate (HIR), improving efficiency and applicability across germplasm pools. Use of in vivo haploid induction with speed breeding allows breeders to generate DH populations within a single year, accelerating mapping population development, allele mining, and hybrid development (Gupta et al. 2022; Chen et al. 2024).

6.3.3 Genomic Selection (GS): Predictive Breeding at Scale

Genomic selection uses genome-wide markers to predict breeding values without needing phenotypic data for each cycle. In maize, GS is mainly useful due to rapid linkage disequilibrium decay and wide marker coverage. Use of GS in maize provides several advantages, such as reduced cycle time, increased accuracy for low-heritability traits, early-generation selection, and multi-trait and multi-environment modelling. Deep learning models such as Cropformer improved predictive accuracy by capturing non-linear interactions and epistasis, providing interpretable understandings of functional loci (Wang et al. 2024). The genetic selection models like Bayes B, have been found to be very accurate for the complex traits like drought tolerance (Crossa et al. 2017; Amadu et al. 2025).

Integration of genomics data with high-throughput phenotyping has revolutionized performance predictions in maize breeding. UAV-based accurate phenotyping of diverse maize germplasm for canopy temperature, NDVI, plant height, and flowering time increases GS's predictive performance for yield stability under stress (Araus and Cairns 2018). The integrative approaches like Environmental Genomic Selection (EGS), an extension of genome environment association, use environmental covariates to eliminate the gaps of genomic prediction models and facilitate climate-smart breeding (Halpin-McCormick et al. 2025; Kim et al. 2026). Novel databases like MaizeGEP, which include a large set of information on genotypic and phenotypic data over environments can be used to increase the precision of genomic selection (Zhang et al. 2026).

6.3.4 Optimized Cross Selection (OCS): Intelligent Crossing Design

Traditionally, breeders relied on intuition to plan crosses. OCS replaces intuition with quantitative scientific frameworks, which makes a valid sense. It evaluates parental combinations based on GEBVs, predicted within-family variance, coancestry, and inbreeding penalties. This approach balances short-term gain with long-term diversity, crucial in accelerated breeding pipelines where strong selection pressure can reduce genetic variance. OCS can be used to design crosses for high-density adapted architecture, combining drought and heat resilience alleles, and maintaining heterotic pools during rapid cycles (Allier et al. 2019; Cowling et al. 2023). For example, Cowling et al. (2023) applied OCS in a global spring oilseed rape (canola) breeding program across three $S_{0,1}$ family selection cycles (2016, 2018, and 2020) with multi-environment field trials in Australia and Canada, using an index that emphasized improving grain yield, seed oil, meal protein, and blackleg resistance while maintaining key agronomic and quality traits and reducing glucosinolate content. Hence, a balanced genetic gain for traits can be achieved through simultaneous improvement of multiple traits.

6.3.5 Next-Generation Bulked Segregant Analysis (NG-BSA): Rapid Gene Discovery

Bulked segregant analysis (BSA) is a good alternative to linkage-based QTL mapping. NG-BSA expands traditional BSA by integrating pangenome variants, RNA-seq, epigenomics, and machine learning-based candidate gene ranking. NG-BSA enhances mapping resolution, reduces population size requirements, and enables simultaneous mapping of multiple traits. By coupling ML-driven candidate prioritization with transcriptomics, NG-BSA accelerates use of natural diversity in targeted allele introgression breeding pipelines. It has been successfully used for the improvement of drought and heat tolerance, plant architecture, male sterility, and early vigour traits (Wang et al. 2023a).

6.3.6 Genome Editing: Precise Trait Engineering

CRISPR-Cas systems have revolutionized maize breeding by enabling targeted modification of genes and regulatory sequences (Kumar et al. 2020). Some of the examples include: (i) Semi-dwarf ideotype creation through edits to GA and BR pathways (Xiao et al. 2025); Abiotic stress tolerance via modifications in *ARGOS8*, *ZmNAC111*, and stomatal regulatory genes-*ZmSTOMAGEN1/2* (Shi et al. 2017; Liu et al. 2022; Xu et al. 2025); Enhanced kernel nutrition, including carotenoid and amino acid pathway edits (Kumar et al. 2020); Male sterility systems-Seed Production Technology for hybrid seed production (Yang et al. 2025); Yield contributing characters through editing of *1-aminocyclopropane-1-carboxylate oxidase2* (*ZmACO2*) gene and promoter editing of *ZmCLE7/ZmFCP1* (Liu et al. 2021; Ning et al. 2021). Furthermore, base and prime editing expand the scope by enabling single-nucleotide changes without double-strand breaks, reducing off-target effects.

6.3.7 Pangenomics and Structural Variant Mining

The maize pangenome has revealed extensive presence-absence variation, copy number variation, and structural variants (SVs) that significantly influence phenotype. Pangenomics enables the discovery of rare beneficial alleles, novel alleles, SV-based marker development, and improved genomic prediction accuracy. This is fundamental for stress adaptation, particularly for traits governed by SVs such as root architecture, kernel composition, and flowering time (Aggarwal et al. 2022; Zhang et al. 2025).

6.3.8 Multi-Environment Testing and G × E Modelling

Genotype × environment (G × E) interactions play an important role in understanding the genotypic response to a changing climate. Modern breeding pipelines integrate AI-driven environmental clustering, reaction norm modelling, and digital twins to predict genotype × environment × management outcomes. These models help to reduce trial costs while improving prediction accuracy for performance stability, resulting in a truly intelligent, automated, and predictive breeding system (Zhang et al. 2025). Modern maize breeding increasingly uses the integrated genomic, phenomic, and computational tools to achieve faster and more consistent genetic gains for yield and yield-associated complex traits. Pangenomics and NG-BSA enable the discovery of candidate genes, structural variants, and adaptive haplotypes underlying economically important traits (Aggarwal et al. 2022). Discovered genetic loci are translated into breeding decisions through genomic selection frameworks that incorporate optimized cross selection and multi-environment testing with G × E modelling, allowing accurate prediction of performance, stability, and environment-specific adaptation. The coordinated deployment of GS with G × E Modelling creates a data-driven, predictive, and accelerated breeding pipeline to achieve sustained multi-trait genetic gains in maize under diverse environments (Fig. 6.1).

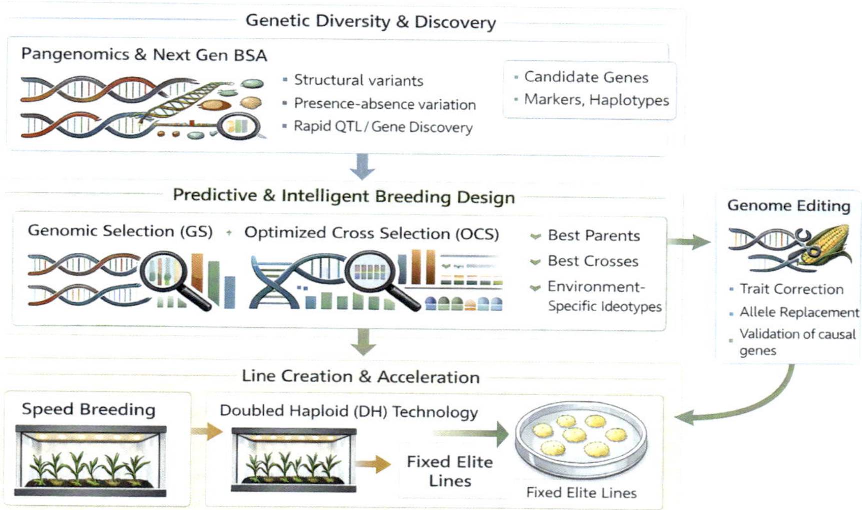

Fig. 6.1 The schematic illustrates an integrated breeding pipeline in which genetic diversity and trait discovery through pangenomics and NG-BSA feed into predictive and intelligent breeding design using GS, OCS, and multi-environment testing with G × E modelling. Accelerated line development can be achieved through speed breeding and DH technology, which generate fixed elite lines, while genome editing enables targeted gene tweaking and gene validation

6.4 Digital and AI-Driven Breeding Tools

Digital and AI-driven technologies are becoming an essential component of maize breeding. Modern breeding programmes have started transitions from empirical decision-making to predictive and automated systems, making the use of digital tools a must for managing high-dimensional data. This helps to optimize breeding decisions and hasten the path from gene discovery to cultivar release (Zhang et al. 2025). In the current era, phenotyping is no longer a bottleneck to limit genetic gain. High-throughput phenotyping (HTP) platforms have made possible the implementation of automated, sensor-rich systems that collect large-scale, high-precision, multi-temporal data.

6.4.1 High Throughput Phenomics

6.4.1.1 UAV and Satellite-Based Phenomics

The use of unmanned aerial vehicles (UAVs) equipped with multispectral, hyperspectral, RGB, thermal, or LiDAR sensors has increased in maize breeding to estimate plant height, biomass, canopy temperature depression, chlorophyll indices, and leaf senescence. The use of UAV-derived traits has significantly improved prediction accuracy for yield, flowering time, stay-green, and stress resilience under field conditions (Araus and Cairns 2018).

6.4.1.2 Ground-Based HTP Systems

Ground vehicles deploy LiDAR, thermal cameras, and high-definition imaging for recording traits like stem diameter, brace roots architecture, leaf area, and stem hardiness, providing a platform for centimetre-level resolution. Such platforms enable the early-generation selection and plant architectural traits.

6.4.1.3 Indoor and Controlled-Environment Phenomics

Indoor or controlled-environment phenomics is based on the integrated use of automated imaging, conveyor systems, and environmental sensors to observe and record traits under controlled conditions. Measurement of seedling vigour, leaf growth, root traits, transpiration efficiency, and stress acclimation can be easily done using such controlled facilities. Such facilities complement the data of field HTP platforms, providing accurate and robust data for multi-environment genomic prediction models.

6.4.2 Machine Learning (ML) and Deep Learning (DL) in Breeding

Genomic prediction, trait classification, stress detection, and decision support in large breeding programs are driven by ML and DL.

6.4.2.1 ML Models for Genomic Prediction

ML helps the GS programmes to shift from the use of GBLUP, Bayesian ridge regression, and RKHS models to random forests, gradient boosting machines, support vector machines, and elastic net variants. ML has been proven to be effective when dealing with the presence of epistasis and nonlinear interactions in complex traits (Crossa et al. 2017).

6.4.2.2 DL for Phenotype Extraction

Deep learning, particularly convolutional neural networks (CNNs) and transformer-based vision models, automatically extract trait values from raw images or hyperspectral data. Applications of DL in maize breeding include ear counting, disease lesion identification, kernel set prediction, leaf segmentation, tassel detection, and tassel architecture quantification. The use of DL models eliminates manual annotation, increases throughput, and improves trait accuracy under field conditions (Singh et al. 2019).

6.4.2.3 Multimodal AI for Integrating Genomics, Phenomics, and Enviromics

AI multi-factor models integrating genotype, phenotype, environmental covariates, and management variables significantly outperform traditional single-factor models. Graph neural networks, multimodal transformers, and variational autoencoders help to model G × E × M responses, enabling the development of hybrids for specific environments (Zhu et al. 2024).

6.4.3 Digital Twins and Robotics

Digital twins simulate crop systems virtually, integrating genotype-specific physiology, weather conditions, soil sensors-based information, HTP data, and crop growth models. The application of digital twins includes prediction of yield stability, stress resilience, optimal density, fertilizer responsiveness, and disease and pest risk. Digital twins are also helpful in parental selection, nursery management, trial allocation, and regional deployment by running "what if" scenarios across seasons and testing sites (Zhang et al. 2025).

The use of robotics and automation helps to increase consistency and reduce labour bottlenecks in the labour-intensive maize crop. Automated detasseling, pollination benches, seed sorting systems, root phenotyping stations, and greenhouse robots standardize operations, minimizing human error and improving phenotyping quality. Robotics can also be integrated with speed breeding to enhance high-throughput generation turnover (Jeon et al. 2023).

6.5 Global Success Stories and Future Ideotypes

The impact of the next generation tools and technologies on genetic gain and farmer impact can be assessed from the real breeding programs. For example, in Sub-Saharan Africa, the Drought-Tolerant Maize for Africa pipeline combined multi-environment testing, marker-assisted introgressions, and genomic selection to develop and deploy over 200 drought-tolerant hybrids and OPVs, with on-farm yield increases of ~20–30% under moisture stress (Cairns et al. 2013). Similarly, nutritional breeding pipelines for provitamin A biofortification, built on *crtRB1* and *lcyE* variants with DH fixation and market-aligned product profiling, have delivered vast adoption of biofortified maize in Zambia and Nigeria, illustrating a classic example of integrating trait prioritization, seed system alignment, and end-user preferences for significant impact in maize (Pixley et al. 2012). DH programs have optimized haploid inducer genetics (*qhir1*, *qhir8*; *mtl/zmpla1/nld*; *zmdmp*) and refined the haploid identification (*R1-nj*, *NIR-oil*), whose integration with speed breeding has made possible the line development in a single year. Moreover, NG-BSA can be used in those DH populations for rapid allelic variant identification and introgression (Wang et al. 2023a; Chen et al. 2024).

Global maize breeding models have started working on "designed pipelines" that harness speed breeding × DH × genomic selection × OCS, with high-throughput phenotyping and deep learning models closing the loop between data flow and decision making. Programs deploying OCS achieved better short-term gain without losing genetic diversity because cross usefulness, within-family variance, and coancestry penalties are modelled or taken care of (Allier et al. 2019; Cowling et al. 2023). Integration of digital twins in such pipelines simulates G × E × M scenarios before execution of large field trials, guiding trial placement, parental choice, and risk management across diverse climate gradients (Zhang et al. 2025). This "predict-then-deploy" approach is being considered in both public and private sectors.

Future ideotypes tailored to climate volatility and resource constraints are outcomes of such breeding pipelines. A high-density, climate-resilient semi-dwarf ideotype is a combination of traits such as compact internodes (GA/BR pathway edits), upright leaves for deep-canopy light use, strong rind and lignin for lodging resistance, and brace/nodal root robustness and good kernel set under stress (Xiao et al. 2025). Additionally, traits such as reproductive resilience (pollen viability, silk receptivity), canopy cooling (CTD), and plastic root hydraulics for rapid post-flood recovery are also equally important. NG-BSA has been reported to identify favourable alleles, helping in quantitative stacking without excessive linkage drag (Wang et al. 2023a). A nutrient-efficient ideotype has better nitrate uptake and assimilation (including edits for *ARGOS8* for regulatory tuning) with semi-dwarf biomass partitioning to sustain yield at low N (Shi et al. 2017). Finally, a mechanization-friendly smart-canopy ideotype is based on the combination of traits such as uniform plant and ear height, reduced tassels, synchronized maturity, and kernel integrity for clean mechanical harvest.

6.6 Future Prospects and Global Innovation Frontiers

Maize breeding is going to witness a significant transformation in the coming decade. The convergence of genomics, phenomics, AI, and synthetic biology will lead to intelligent and programmable breeding decisions. Decadal vision frameworks such as *Maize2035* foresee breeding ecosystems will be autonomous, predictive, and climate-resilient, making product development at a faster pace (Liu et al. 2025; Zhang et al. 2025).

6.6.1 Autonomous Phenotyping and Edge-AI Systems

Unmanned phenotyping with automation features is increasing in maize research. UAVs can fly as per scheduled missions, field robots can monitor plant architecture and stress signatures, and greenhouse sensors can likely adjust microclimates in real time. Edge-AI systems can process data directly at the source, enabling immediate feedback to the researchers. Breeders do not need to wait for end-of-season measurements; instead, continuous daily measurements for dynamic trait profiles can be obtained, guiding selection decisions in real time. For example, analysis of 679 maize inbred lines with 2.8M SNPs and 1M+ RGB images (image-based traits referred to as i-traits) quantified 67 i-traits, revealed dynamic growth patterns. GWAS identified 857 QTLs for dynamic growth variation, mostly period-specific, with cryptic pleiotropy and key genes shaping architecture, supporting predictive breeding strategies (Wu et al. 2026).

6.6.2 Hologenomics and Plant-Microbiome Breeding

Future breeding is likely to extend beyond the plant genome to investigate its associated beneficial microbiomes. Hologenomics integrates host and microbial genomes to harness the symbiotic interactions for enhanced climate resilience. Favourable alleles recruiting nitrogen-fixing bacteria, stress-mitigating endophytes, or disease-suppressive rhizosphere communities can be harnessed to achieve sustainable breeding targets. Selecting for microbiome-mediated resilience can reduce fertilizer dependence and enhance stress tolerance, developing cultivars for marginal soils (Mehta et al. 2021).

6.6.3 Multi-Scale Genotype-to-Phenotype (G2P) Prediction

Multi-scale G2P models that integrate gene networks, organ physiology, whole-plant dynamics, and field-scale enviromics can act as a guiding force for implementing better breeding decisions. For example, Wang et al. (2023b) made a G2P container that integrates 16 GS models and 13 metrics for maize genotype-to-phenotype prediction. It enables parallel evaluation, auto-ensemble selection, and optimized training sets, enhancing precision while reducing phenotyping costs, supporting efficient genomic selection in breeding programs. Such G2P models can help to envision the expression of alleles under diverse climates, better management practices, and soil conditions (Wang et al. 2023b). Reaction norm modelling integrated with digital twins enables breeders to develop hybrids tailored to future climatic scenarios rather than past environmental conditions. Real-time IoT data can be used to recalibrate these models at continuous time intervals, making predictions precise as environments change.

6.6.4 Synthetic Biology and Programmable Genetic Circuits

The advances in synthetic biology have made it possible to move beyond single-gene edits to modular genetic circuits. The advantage of such programmable stress switches lies in minimizing growth penalties, as it will activate resilience pathways only under severe conditions. Flowering-related genetic circuits can help to fine-tune the phenology in response to day length or heat. Shoot system architecture circuits can help to fine-tune the internode elongation and leaf angle under high planting density. Metabolic rewiring can also be employed to optimize nitrogen and carbon use in maize. Such programmable gene circuits have the potential to develop climate "smart" maize cultivars (Xiao et al. 2025).

6.6.5 Semi-Dwarf Ideotype Integration

Semi-dwarf ideotypes are best suited to high plant density and low-input systems. Such ideotypes have compact stems, upright leaves, and robust roots, making them a perfect fit for better performance under ultra-high planting densities. Semi-dwarf genotypes have less lodging risk and improved NUE, making them ideal for climate-smart production systems, particularly in Asia and Africa.

6.7 Conclusion

Maize breeding is witnessing a new era defined by the integration of smart digital tools and AI. Modern breeding has refined the prioritization of traits from yield alone to integration of climate resilience, resource use efficiency, and nutritional enrichment. Digital and AI-driven technologies have become the core of contemporary maize breeding, transforming the discipline from empirical selection to predictive, data-intensive, and increasingly automated pipelines. High-throughput phenomics has addressed the phenotyping bottlenecks by enabling precise, scalable, and multi-temporal measurement of complex traits across field and controlled environments. Furthermore, integration of machine learning and deep learning can help to convert the image data into biologically meaningful phenotypes, strengthening the genomic prediction accuracy for yield, stability, and stress resilience. The integration of genomic selection with optimized cross selection and multi-environment testing allows breeders to predict not only performance and adaptability across diverse genotype × environment × management scenarios, but also to sustain the genetic diversity. Speed breeding, doubled haploid technology, and robotics help to reduce the duration of breeding cycles, resulting in accelerated gains. Further, genome editing adds an additional layer, enabling targeted allele modification and validation of causal alleles within elite backgrounds. Together, these tools form an integrated, closed-loop breeding systems that connect gene discovery to cultivar development. Furthermore, integrated use of digital twins, edge AI, and multi-scale genotype-to-phenotype modelling will further advance the predictive breeding for evaluation of genotypes across future climate and management regimes before large-scale field testing. Emerging frontiers such as hologenomics and programmable genetic circuits allow the investigation of dynamic responses to environmental signals. In a nutshell, accelerated but sustained genetic gain in maize breeding will depend on making breeding pipelines as integrated systems where prediction, acceleration, and precision engineering help to deliver high-yielding, resource-efficient, and climate-ready maize cultivars.

References

Aggarwal SK, Singh A, Choudhary M, Kumar A, Rakshit S, Kumar P, Bohra A, Varshney RK (2022) Pangenomics in microbial and crop research: progress, applications, and perspectives. Genes 13(4):598

Allier A, Lehermeier C, Charcosset A, Moreau L, Teyssèdre S (2019) Genomic prediction of within-family variance can help maintain long-term genetic gain in low-input breeding programs. Genetics 211(3):1233–1248

Amadu MK, Beyene Y, Chaikam V, Tongoona PB, Danquah EY, Ifie BE, Burgueno J, Prasanna BM, Gowda M (2025) Genome-wide association mapping and genomic prediction analyses reveal the genetic architecture of grain yield and agronomic traits under drought and optimum conditions in maize. BMC Plant Biol 25(1):135

Araus JL, Cairns JE (2018) Field high-throughput phenotyping: the new crop breeding frontier. Trends Plant Sci 19(1):52–61

Cairns JE, Hellin J, Sonder K, Araus JL, MacRobert JF, Thierfelder C, Prasanna BM (2013) Adapting maize production to climate change in sub-Saharan Africa. Food Secur 5(3):345–360

Chen YR, Lübberstedt T, Frei UK (2024) Development of doubled haploid inducer lines facilitates selection of superior haploid inducers in maize. Front Plant Sci 14:1320660

Choudhary M, Wani SH, Kumar P, Bagaria PK, Rakshit S, Roorkiwal M, Varshney RK (2019) QTLian breeding for climate resilience in cereals: progress and prospects. Funct Integr Genomics 19(5):685–701

Cowling WA, Li L, Siddique KH, Banks RG, Kinghorn BP (2019) Modeling crop breeding for global food security during climate change. Food Energy Secur 8(2):e00157

Cowling WA, Castro-Urrea FA, Stefanova KT, Li L, Banks RG, Saradadevi R, Sass O, Kinghorn BP, Siddique KH (2023) Optimal contribution selection improves the rate of genetic gain in grain yield and yield stability in spring canola in Australia and Canada. Plants 12(2):383

Crossa J, Pérez-Rodríguez P, Cuevas J, Montesinos-López O, Jarquín D, De Los Campos G et al (2017) Genomic selection in plant breeding: methods, models, and perspectives. Trends Plant Sci 22(11):961–975

Guo L, Wang X, Zhao M, Huang C, Li C, Li D, Yang CJ, York AM, Xue W, Xu G, Tian F (2018) Stepwise cis-regulatory changes in ZCN8 contribute to maize flowering-time adaptation. Curr Biol 28(18):3005–3015

Gupta M, Choudhary M, Kumar H, Kaswan V, Kaur Y, Choudhary JR, Yadav S (2022) Doubled haploid technology in maize (*Zea mays*): status and applications. Indian J Agric Sci 92(3):283–291

Halpin-McCormick A, Campbell Q, Negrão S, Morrell PL, Hübner S, Neyhart JL, Kantar MB (2025) Environmental genomic selection to leverage polygenic local adaptation in barley landraces. Commun Biol 8(1):618

Jeon J, Kim Y, Lee J (2023) Digital breeding: integrating automation, AI, and phenotyping in crop improvement pipelines. Plant Phenomics 5:100123

Jin M, Liu X, Jia W, Liu H, Li W, Peng Y, Du Y, Wang Y, Yin Y, Zhang X, Yan J (2018) *ZmCOL3*, a CCT gene represses flowering in maize by interfering with the circadian clock and activating expression of ZmCCT. J Integr Plant Biol 60(6):465–480

Kim KH, Park D, Lee BM (2026) Biotechnological strategies to enhance maize resilience under climate change. Biology 15(2):161

Kumar P, Choudhary M, Hossain F, Singh NK, Choudhary P, Gupta M, Singh V, Chikappa GK, Kumar R, Kumar B, Jat SL, Rakshit S (2019) Nutritional quality improvement in maize (*Zea mays*): progress and challenges. Indian J Agric Sci 89(6):895–911

Kumar K, Gambhir G, Dass A, Tripathi AK, Singh A, Jha AK, Yadava P, Choudhary M, Rakshit S (2020) Genetically modified crops: current status and future prospects. Planta 251(4):91

Kumar P, Choudhary M, Jat BS, Kumar B, Singh V, Kumar V, Singla D, Rakshit S (2021) Skim sequencing: an advanced NGS technology for crop improvement. J Genet 100(2):38

Liu L, Gallagher J, Arevalo ED, Chen R, Skopelitis T, Wu Q, Bartlett M, Jackson D (2021) Enhancing grain-yield-related traits by CRISPR–Cas9 promoter editing of maize CLE genes. Nat Plants 7(3):287–294

Liu S, Liu X, Zhang X, Chang S, Ma C, Qin F (2022) Co-expression of *ZmVPP1* with *ZmNAC111* confers robust drought resistance in maize. Genes 14(1):8

Liu H, Wang X, Chen S (2025) Integration of rapid generation technologies with genomic prediction for accelerated breeding. Front Plant Sci 16:122–145

Mehta S, Singh B, Patra A, Tripathi A, Easwaran M, Choudhary JR, Choudhary M, Aggarwal SK (2021) Maize microbiome: current insights for the sustainable agriculture. In: Microbiomes and plant health. Academic Press, pp 267–297

Ning Q, Jian Y, Du Y, Li Y, Shen X, Jia H, Zhao R, Zhan J, Yang F, Jackson D, Liu L (2021) An ethylene biosynthesis enzyme controls quantitative variation in maize ear length and kernel yield. Nat Commun 12(1):5832

Pixley K, Rojas NP, Babu R, Mutale R, Surles R, Simpungwe E (2012) Biofortification of maize with provitamin A carotenoids. In: Carotenoids and human health. Humana Press, Totowa, pp 271–292

Shi J, Gao H, Wang H, Lafitte HR, Archibald RL, Yang M, Hakimi SM, Mo H, Habben JE (2017) ARGOS8 variants generated by CRISPR-Cas9 improve maize grain yield under field drought conditions. Plant Biotechnol J 15(2):207–216

Singh AK, Ganapathysubramanian B, Singh A, Sarkar S (2019) Machine learning for high-throughput stress phenotyping in plants. Trends Plant Sci 24(2):97–109

Sun Y, Hou L, Li C, Zhou Z, Wu J, Yu H (2025) Maternal haploid induction in maize via mutation of Gamete Expression protein 1. Plant Biotechnol J 23(7):2595–2607

Wang X, Han L, Li J, Shang X, Liu Q, Li L, Zhang H (2023a) Next-generation bulked segregant analysis for Breeding 4.0. Cell Rep 42(9):113039

Wang Q, Jiang S, Li T, Qiu Z, Yan J, Fu R, Ma C, Wang X, Jiang S, Cheng Q (2023b) G2p provides an integrative environment for multi-model genomic selection analysis to improve genotype-to-phenotype prediction. Front Plant Sci 14:1207139

Wang Y, Li Z, Zhang H (2024) Cropformer: interpretable deep learning for genomic prediction and variant prioritization. Bioinformatics 40(7):201

Wu C, Geng Z, Li W, Ye J, Hao X, Xu J, Jin M, Wu X, Du Y, Chen Y, Ma C (2026) Genetic dynamics drive maize growth and breeding. Genome Biol. https://doi.org/10.1186/s13059-026-03957-8

Xiao S, Song W, Wang R, Zhang S, Liu M, Zhang R, Xu T, Liu H, Hu J, Xing J, Wang X (2025) Engineering semi-dwarf and climate-adapted maize ideotypes through targeted hormonal pathway modulation and structural optimization. Nat Plants 11:110–128

Xu JJ, Chen QY, Qin LF, Liu Y, Li YZ, Fan XW (2025) The regulatory role of ZmSTOMAGEN1/2 in maize stomatal development is elucidated via gene editing and metabolic profiling. PLoS One 20(7):0328433

Yang P, Sun Y, Zheng T, Xu C, Lin C, Shen Z (2025) A male sterility system with a novel bifunctional selection strategy for hybrid seed production in maize. New Phytol 247(6):2726–2732

Zhang Y, Huang G, Zhao Y, Lu X, Wang Y, Wang C, Guo X, Zhao C (2025) Revolutionizing crop breeding: next-generation artificial intelligence and big data-driven intelligent design. Engineering 44:245–255

Zhang D, Han Y, Pan S, Liu Z, Zhao X, Zhang Q, Wang X, Sun J, Wang K (2026) MaizeGEP: a maize hybrids dataset with genotype, phenotype, and envirotype to develop genomic selection models. Genomics Proteomics Bioinformatics:qzaf 140. https://doi.org/10.1093/gpbjnl/qzaf140

Zhong Y, Liu C, Qi X, Jiao Y, Wang D, Wang Y, Liu Z, Chen C, Chen B, Tian X, Li J (2019) Mutation of *ZmDMP* enhances haploid induction in maize. Nat Plants 5:575–580

Zhu W, Li W, Zhang H, Li L (2024) Big data and artificial intelligence-aided crop breeding: Progress and prospects. J Integr Plant Biol 67:722–739

Leveraging Artificial Intelligence, Machine Learning, and Big Data in Maize Breeding

7

R. K. Bhavyasree, Vishal Singh, Arpit Chaudhary, Pardeep Kumar, Mamta Gupta, and Mukesh Choudhary

Abstract

Maize (*Zea mays* L.) is a global staple central to food, feed, and industrial systems, valued for its nutritional composition and adaptability across diverse agroecologies. However, its production is increasingly challenged by climate variability, emerging pests and diseases, soil degradation, and the rising global demand for grain, which continues to outpace annual yield gains. Modern genomics, phenomics, and environmental monitoring technologies now generate vast and complex datasets, yet their value is fully realized only when coupled with advanced analytical frameworks. Consequently, artificial intelligence (AI) and machine learning (ML) have become essential tools for next-generation maize breeding. Big data sources-including whole-genome sequencing, multi-omics profiling, high-throughput phenotyping, drone and satellite imagery, and weather and soil datasets enable a comprehensive understanding of genotype–phenotype–environment dynamics. AI and ML facilitate parent selection, genomic prediction, trait forecasting, G × E modelling, image-based phenotyping, genomic mapping, disease detection, and breeding strategy optimization. Approaches such as deep learning, convolutional neural networks, random forests, federated learning, and automated prediction tools like AutoGP improve accuracy, accelerate decision-making, and support climate-resilient ideotype development. Despite this promise, challenges persist, including heterogeneous data formats, limited standardization, high computational demands, and the interpretability issues inherent in "black-box" models. Addressing these limita-

R. K. Bhavyasree
Punjab Agricultural University, Regional Research Station, Faridkot, Punjab, India

V. Singh · A. Chaudhary · P. Kumar · M. Gupta · M. Choudhary (✉)
ICAR-Indian Institute of Maize Research, Ludhiana, Punjab, India
e-mail: pardeep.kumar1@icar.org.in; mukesh.choudhary1@icar.org.in

M. Choudhary et al. (eds.), *Next-Generation Maize Research*,
https://doi.org/10.1007/978-981-92-0722-0_7

tions requires improved data harmonization, interdisciplinary expertise, and accessible computational infrastructure. Furthermore, AI-driven multi-omics integration, intelligent genome editing, robotics, and autonomous phenotyping will boost maize breeding from Breeding 4.0 to Breeding 5.0, an era defined by hyper-intelligent, predictive, and sustainable breeding pipelines capable of meeting future food and nutritional security needs.

Keywords

Maize breeding · Artificial intelligence · Machine learning · Big data analytics · Genomic prediction · High-throughput phenotyping

7.1 Introduction

Maize (*Zea mays* L., commonly known as corn) is one of the most important staples in the world. It plays a crucial role in maintaining global food security due to its adaptation to diverse climatic conditions and widespread consumption. It is a cornerstone crop of agriculture, not only due to its human food value but also due to its importance in animal nutrition as well as its industrial value. Maize grains naturally contain ample amounts of carbohydrates, proteins, and micronutrients. With the help of biofortification, it is further modified by adding vitamins, which makes the crop contribute to food as well as nutritional security, thus aligning with the sustainable development goals (Wang and Zhang 2024).

Maize was domesticated in Southern Mexico, 1000 years after the major domestication event (wheat) in the Fertile Crescent. Despite its isolated domestication event, it disseminated across the globe very quickly to rise into the status of a major global staple along with rice and wheat (Erenstein et al. 2022). The demand for this crop is increasing day by day due to its use as animal and poultry feed, industrial input, and as an energy crop. Despite the demand for 2.4% increased maize production to meet the projected demand of 2050, the growth is only 1.6% annually (Vieira et al. 2025). There have been several advancements in production technologies in the past decades. But still, the maize production faces several challenges affecting yield and sustainability due to factors like climate change, land degradation, biotic and abiotic stresses (Wang and Zhang 2024). A major portion of maize yield variability can be explained by the influence of several climatic variables, thus challenging maize production even more in the future (Zhao et al. 2025).

Over the years, plant breeding has progressed through distinct evolutionary phases, each driven by advances in scientific knowledge and technology, resulting in progressively higher genetic gains and improved efficiency (Khan et al. 2022). The earliest phase, *Breeding 1.0*, began with the domestication, selection, and cultivation of food crops around 10–12 thousand years ago, where farmers relied entirely on phenotypic selection for visible traits such as yield, adaptability, and quality (Fig. 7.1). This laid the foundation for systematic crop improvement. The second phase, *Breeding 2.0*, emerged in the late nineteenth century following the

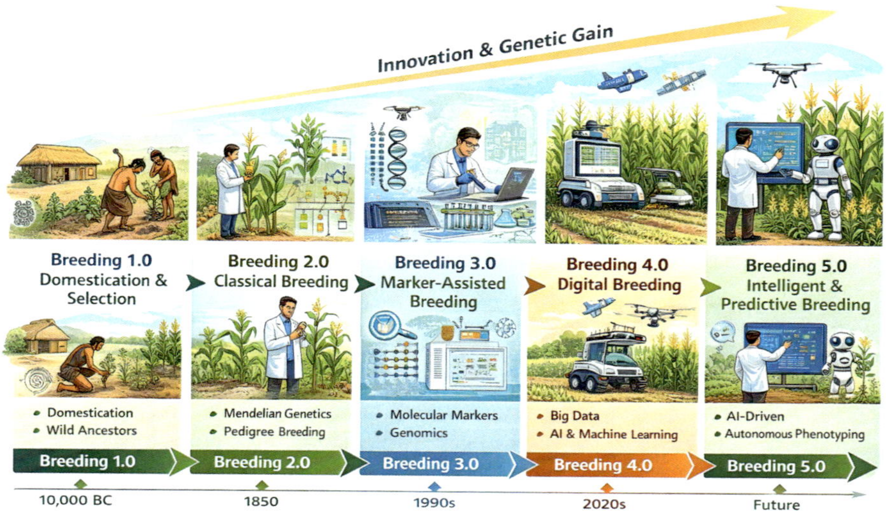

Fig. 7.1 Evolution of plant breeding from early domestication to intelligent, AI-driven Breeding 5.0, highlighting the transition from phenotypic selection to genomics, digital tools, and predictive analytics that together accelerate genetic gain and enable development of climate-resilient, high-yielding, and resource-efficient crop varieties

rediscovery of Mendelian genetics and marked a major breakthrough in breeding science. It introduced controlled hybridization, pedigree-based selection, recurrent selection, and properly designed and replicated field trials, enabling structured improvement of complex traits such as yield and adaptation in crops like maize. The advent of molecular biology and the use of molecular markers approximately three decades ago ushered in *Breeding 3.0*, characterized by marker-assisted selection, QTL mapping, and later genomic selection, which significantly accelerated breeding cycles and improved selection accuracy, particularly for complex and low-heritability traits (Wallace et al. 2018). Rapid advances in next-generation sequencing, high-throughput phenotyping, and information technology have led to an explosion of genomics and phenomics data, driving the transition to *Breeding 4.0*, where digital tools, advanced statistical methods, and machine learning enable data-driven breeding and better understanding of genotype × environment interactions (Zhang et al. 2025). Building on this foundation, the emerging *Breeding 5.0* phase integrates artificial intelligence, big data analytics, automation, and robotics to support predictive and optimized breeding strategies, addressing challenges such as climate change, limited land resources, and increasing biotic and abiotic stresses, while ensuring global food and nutritional security (Fig. 7.1).

Despite the enormous volume of data produced by modern technologies, their utility is limited unless they are effectively processed and analysed. So proper algorithms or data analysis platforms have to be developed and incorporated into the plant sciences to make them applicable for crop improvement. So, incorporation of modern technologies like artificial intelligence (AI) and machine learning (ML) into plant breeding has become essential, as they can help in handling and analysing data, predicting the outcome and development of superior varieties.

7.2 Sources of Big Data in Maize

Modern maize breeding mainly aims in developing climate resilient crop varieties with high yield, disease resistance, and good nutritional value. To achieve this goal in this era of growing demands for food and feed, the use of technologies involving big data handling and analysis is crucial to enhance the precision, speed, and decision-making accuracy in crop improvement. Recent advances in conveniently generating dense SNP panels through either genotyping-by-sequencing (GBS), or whole-genome data, along with the co-existence of high-throughput phenotyping (HTP), drone and satellite imagery, and streaming weather layers, have opened avenues of implementing a multi-modal approach in predictive breeding. The breadth of modalities challenges traditional statistical workflows, but AI's comparative advantage in integrating signals from various types and scales opens avenues for such endeavors (Araus and Cairns 2014; Chen et al. 2023; Singh et al. 2016).

The genomic information generated through the advanced sequencing methods contributes a major portion of big data in maize. The whole genome sequencing generates a huge amount of data, which is further used for the generation of transcriptomes, methylomes, or annotation studies. The structural variations are studied using this data to uncover the diversity, population genomics, evolution, and functionality of genes in maize (Chen et al. 2023; Hu et al. 2021; Hufford et al. 2021). Bulk data on markers like SNPs/Indels are available from the genomic data generated through whole genome sequencing or resequencing, which are further utilized for association studies, genomic selection, diversity analysis, and for the detailed studies of adaptive variations of the maize crop worldwide (Chen et al. 2022; Grzybowski et al. 2023; Liu et al. 2015). Also, NGS-based technologies assisted association mapping, progeny selection, gene mapping, and bulk segregant analysis (BSA), increasing the pace of yield and resistance breeding in maize (Best and McSteen 2022; Heuermann et al. 2019; Nowak et al. 2023; Peng et al. 2021).

High-throughput next-generation plant phenomics tools are now replacing the labor-intensive traditional phenotyping methods. These techniques allow the precise phenotyping of large plant populations in a short time with a higher level of precision at the plant level. Different imaging techniques like 3-D imaging, fluorescence imaging, infrared and visible light scanning, RGB imaging, X-ray imaging, magnetic resonance imaging, and near infrared reflectance spectroscopy (NIRS) are used for this purpose (DeSalvio et al. 2024; Pasala and Pandey 2020; Zhang et al. 2015). There are different platforms for using unmanned aerial vehicle (UAV) for the field phenotyping to those methods that are specific to the phenotyping of the whole plant, shoots, roots, or cobs, and even a low-cost portable pocket phenomics platform which can generate a vast amount of real-time data (Angidi et al. 2025; Kienbaum et al. 2021; Shu et al. 2022). The drone and satellite imaging, complemented with automatic sensor systems using robotics, also contribute to phenomics data. Besides the genomics and phenomics data, the environmental and crop management data, including pest and disease records, are very important in analyzing the trends and implementing need-based crop breeding in maize.

7.3 Application of AI and Machine Learning in Maize Breeding

With the enormous amount of data generated, the application of AI and machine learning has found a wide scope in maize breeding. The integration of these techniques has been done in different areas, supplementing breeding such as parent selection, trait prediction, analyzing Genotype × environmental (G × E) interactions, phenotyping and image analysis, genomic mapping, and marker/gene discovery and breeding strategy optimization.

7.3.1 Parent Selection

Selecting superior parental lines to develop inbred lines is crucial for the success of a hybrid breeding program. Capturing co-ancestry and diversity to balance short-term gains with long-term genetic variance is important and has been implemented in simulation studies (Gaynor et al. 2021). Apart from the simulation methods developed in AlphaSimR, an AI-based black-box approach has been investigated to select mating pairs in plant breeding (Hamazaki and Iwata 2024).

7.3.2 Trait Prediction and Genomic Selection

The prediction of complex traits like yield, biotic and abiotic stress resistance, and nutrient use efficiency is very much relevant as well as difficult in this era of climate change. So, the critical application of AI and machine learning made revolutionary changes in genomic prediction by enabling the analysis of complex and nonlinear marker trait relationships (Fig. 7.2). The huge quantum of multi-omics data can be processed for the accurate prediction of complex trait like yield in maize using different types of prediction models assisted with the machine learning methods like partial least squares (PLS), Random Forest (RF), and Gaussian process with radial basis function kernel (GaussprRadial) (Wu et al. 2024). Machine learning methods can help capture non-linear relationships and hidden patterns in the data and provide an opportunity to add non-genetic data into the model. Exploring Multi-Layer Perceptron and Convolutional Neural Networks can help when training data are large and structured (Montesinos-López et al. 2018).

Several algorithms or machine learning methods, like Deep Learning for genomic selection, are complex. These can be replaced by automated user-friendly interfaces like AutoGP, which can enhance the accuracy and a broad model pool (Wu and Zhang 2025). The AI models like Random Forest ranking and regression are well

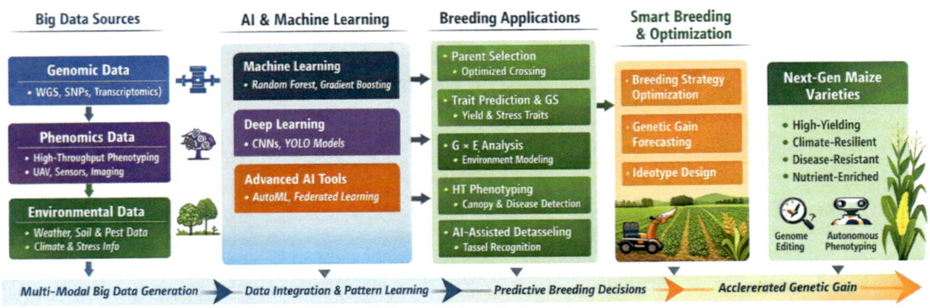

Fig. 7.2 Integrated framework integration of genomic, phenomic, and environmental big data with AI and machine learning to enable predictive breeding, optimize breeding strategies, and accelerate genetic gain in maize

studied for the agricultural applications like breeding value and yield predictions, even from the vegetative characters or genome-wide SNPs (Ramos et al. 2020; Sarkar et al. 2015). Centralized data can be used for the development of a base model to increase prediction accuracy by the joined efforts of different scientific groups, leading to a federated paradigm, and such approaches as federated random forest techniques are found to be very suitable for maize yield predictions (Zhang et al. 2023).

7.3.3 G × E Interactions

Stability of the genotype is very important in the crop improvement programs, which is totally dependent on G × E interactions. There are several practical difficulties in evaluating the G × E interactions from the multi-environmental trials (MET) with high precision, which can be simplified by using the integration of AI and machine learning techniques. Using machine learning, you can even simulate the performance of maize hybrids in different environments, and their yield can be predicted with accuracy rather than going for MET (Barreto et al. 2024). Likewise, the prediction of the performance of the genotypes in the specific stress environments, like drought, is also possible using the integration of genomic selection and machine learning (Wu et al. 2025). The integration of environmental predictors considering weather and soil variables for the forecast of yield performance using the non-linear gradient boosting methods like XGBoost and LightGBM can increase the prediction accuracy (Westhues et al. 2021). Thus, machine learning methods allow us to incorporate environmental data into genomic prediction models and can improve prediction accuracy for unknown environments. Under the Genomes-To-Field initiative, different models used machine learning methods for genomic prediction in maize (Fernandes et al. 2024).

7.3.4 High-Throughput Phenotyping and Image Analysis

The utilization of conventional phenotyping in this modern era is cumbersome because it is labor-intensive and error-prone. So, due to the introduction of High-throughput phenotyping (HTP), massive datasets are generated that demand advanced computational analysis. Multispectral and hyperspectral images are being collected by using drones or ground-based sensors, which can be used for the detection of plant traits using AI-powered image analysis tools (Fig. 7.2). The real-time assessment of stress responses, plant growth monitoring, and disease assessment in the early stages is now possible through machine learning models (Farooq et al. 2024). These automated systems remove human errors as well as increase the speed of data collection, facilitating the breeder to screen a greater number of plants. HTP provides an opportunity for non-destructive imaging of plants on a scale. Implementing HTP has evolved from hand-curated feature extraction to end-to-end learning with deep neural networks. HTP allows recording canopy-level traits such as leaf area, tassel density, plant height, and disease severity, to mention a few. Recent advances in HTP platforms paired with AI have made it central to quantitative, scalable phenotyping, aiding breeding decisions (Murphy et al. 2024).

7.3.5 AI Assisted Detasseling

The potential areas of deployment in maize breeding might be tasseling detection in seed production, tassel counting, hybrid seed-purity check, etc. An AI-based model called CA-YOLO, which is based on deep learning, provides a way to detect tassels for detasseling (Jia et al. 2024). It uses a lightweight Convolutional Neural Network (CNN) for the classification of tassel presence/absence from drone-based images. Another study conducted a detasseling audit to detect any missed tassels after detasseling drive for seed production (Qi et al. 2025). The study utilizes a YOLO (You Only Look Once) model framework for real-time object detection using UAV imagery and pays special attention to early tasseling stages, which are crucial for seed purity. The proposed MT-YOLO model achieves a precision of 93.1%, outperforming other deep learning models.

7.3.6 Genomic Mapping and Marker/Gene Discovery

Machine learning techniques can analyze large-scale phenotypic and genotypic data available to unfold quantitative trait loci (QTLs) associated with yield and stress tolerance (Govaichelvan et al. 2024). There are several algorithms, like CNNs and random forests, which can be utilized for the prediction of gene trait associations in maize. These AI tools not only process the data efficiently to accelerate the gene discovery and functional validation of genes, but also help in analyzing the changes in gene expression, co-expression in different regulatory mechanisms in complex metabolic networks, as in stress responses like drought (Xiong et al. 2025). The disease detection and prediction platforms assisted with AI and Internet of Things (IoT) VGNet (for disease detection) can also supplement to increase the pace of resistance breeding (Al-Otaibi et al. 2024; Fan and Guan 2023; Kundu et al. 2022). Several deep learning tools and big models like multi-modal models are being used for phenomics and genomics prediction in plant diseases, which can lead to drastic progress in resistance breeding programs (Ma et al. 2025).

7.3.7 Breeding Strategy Optimization and Smart Breeding

The integration of advanced technologies to accelerate the genetic gain and development of need-based designer maize varieties transforms conventional breeding into "smart breeding." This strategy mainly relies upon high-throughput phenotyping and genotyping associated with predictive modelling (Liu et al. 2025). The algorithms can also model genetic gains over generations and can help to design breeding strategies for maximal improvement of the targeted traits by reducing field trials and resources (Xu et al. 2022; Zhang et al. 2025).

7.4 Challenges in Application of AI and Machine Learning in Plant Breeding

Integration of AI and machine learning is leading to the next-generation plant breeding era in crop plants. Despite the transformative potential, this field faces several limitations in the complex domain of plant breeding. Plant breeding data is highly diverse, consisting of a bulk amount of phenotypic data from the field trials, enormous genomic data, and environmental data (Atapattu et al. 2024). AI faces difficulty in standardization of such data without pre-processing and domain-specific expertise (Cooper et al. 2014). Another problem is the unavailability of a universally accepted data standard in plant breeding. So, for every new data set, the AI models must be custom-trained. The AI models are highly sensitive to data quality. And with the missing values, inconsistencies, and environmental variability, the plant breeding data is difficult to handle.

So, there is a need for multidisciplinary collaborations as well as intensive scientific training for the plant breeders for the proper utilization of the AI tools and machine learning models in plant breeding. Without the proper understanding of the model interpretations, there will be a negative impact on the breeding decisions because some of the models, like deep learning techniques, are often referred to as "black-box" systems, making the data difficult to interpret (Montesinos-López et al. 2018). Another limitation is the availability of computational infrastructure and expertise needed for the application of AI and machine learning. Though there are several limitations, the advancement of technologies is so rapid that it simplifies the complex algorithms and several use-friendly and simplified algorithms like AutoGP (Wu and Zhang 2025). Further, multiorganizational collaborations and federated paradigms will help the AI to revolutionize the science of plant breeding (Zhang et al. 2023).

7.5 Future Prospects

The revolutionary science of AI and machine learning is offering great opportunities in plant breeding to address the global challenges like climate change, food security, nutritional security and sustainable agriculture. By integrating these technologies with advanced plant breeding methods and predictive modelling, breeders can develop climate-resilient, high-yielding, tailored crop plants for different environments (Farooq et al. 2024; Sheikh et al. 2024). AI can significantly shorten the breeding cycles by using the AI algorithms to predict the plant traits and outcomes in an early stage, which can even save so many resources and multi-environment trials. Also, they can be used for the analysis of complex genetic and phenotypic data for target-based breeding and precise selection for yield, quality, as well as stress tolerance (Khan and Khare 2025). A multi-omics-based study by integrating genomics, transcriptomics, proteomics, and metabolomics will provide a holistic understanding of plant growth and development. This can be made possible only using AI (Fu et al. 2025; Yadav and Kumaraswamy 2024).

Advanced AI-powered tools like drones and robotics can be utilized for the automation, high-throughput phenotyping, and real-time monitoring of plant breeding trials (Li et al. 2021). The designer crops which are customized for a specific region, and the crops that are resilient in this changing climate scenario can be made possible through AI and machine learning-assisted data handling and prediction (Fu et al. 2025; Zhu and Li 2024). AI can also complement the modern biotechnological interventions like genome editing, which can lead to "intelligent gene editing" with great accuracy and precision. For example, in the CRISPR/Cas system, AI can be used, starting from the optimization of CRISPR design to the prediction of the outcomes. The scope of editing can be broadened even more using AI-assisted automation in genome editing processes (Jiang et al. 2025; Kim et al. 2025).

The biotechnology-assisted breeding, along with integration and handling big data, led to the Breeding 4.0 for precision breeding, which was only possible due to the introduction of AI. But now it is the era of Breeding 5.0, in which AI will lead the breeding processes by hyper intelligence and cross-disciplinary convergence of genomics, phenomics, and environmental data. The deep learning model-based processing of this complex biological information will convert the resourceful gene banks to dynamic and analysable 'intelligent data sources'. Along with these, the technological advances like simulated breeding design, fast breeding, and automation using robotics will also increase the precision and efficiency in crop improvement (Fu et al. 2025).

Conclusively, the AI and machine learning-assisted maize breeding is a need of the hour, as the global food and feed demand is increasing at a very rapid rate despite the challenges, including climate change. The rapid advancement of technologies is simplifying the complex models, and interdisciplinary approaches are required for the implementation of these models to advance breeding science for precise and target-based breeding.

References

Al-Otaibi S, Khan R, Ali J, Ahmed A (2024) Artificial intelligence and Internet of Things-enabled decision support system for the prediction of bacterial stalk root disease in maize crop. Comput Intell 40(1):e12632

Angidi S, Madankar K, Tehseen MM, Bhatla A (2025) Advanced high-throughput phenotyping techniques for managing abiotic stress in agricultural crops—a comprehensive review. Crops 5(2):8

Araus JL, Cairns JE (2014) Field high-throughput phenotyping: the new crop breeding frontier. Trends Plant Sci 19(1):52–61

Atapattu AJ, Perera LK, Nuwarapaksha TD, Udumann SS, Dissanayaka NS (2024) Challenges in achieving artificial intelligence in agriculture. In: Artificial intelligence techniques in smart agriculture. Springer, pp 7–34

Barreto CAV, das Graças Dias KO, de Sousa IC, Azevedo CF, Nascimento ACC, Guimarães LJM, Guimarães CT, Pastina MM, Nascimento M (2024) Genomic prediction in multi-environment trials in maize using statistical and machine learning methods. Sci Rep 14(1):1062

Best NB, McSteen P (2022) Mapping maize mutants using bulked segregant analysis and next generation sequencing. Curr Protoc 2(11):e591

Chen L, Luo J, Jin M, Yang N, Liu X, Peng Y, Li W, Phillips A, Cameron B, Bernal JS (2022) Genome sequencing reveals evidence of adaptive variation in the genus Zea. Nat Genet 54(11):1736–1745

Chen J, Wang Z, Tan K, Huang W, Shi J, Li T, Hu J, Wang K, Wang C, Xin B (2023) A complete telomere-to-telomere assembly of the maize genome. Nat Genet 55(7):1221–1231

Cooper M, Messina CD, Podlich D, Totir LR, Baumgarten A, Hausmann NJ, Wright D, Graham G (2014) Predicting the future of plant breeding: complementing empirical evaluation with genetic prediction. Crop Pasture Sci 65(4):311–336

DeSalvio AJ, Adak A, Murray SC, Jarquín D, Winans ND, Crozier D, Rooney WL (2024) Near-infrared reflectance spectroscopy phenomic prediction can perform similarly to genomic prediction of maize agronomic traits across environments. Plant Genome 17(2):e20454

Erenstein O, Jaleta M, Sonder K, Mottaleb K, Prasanna BM (2022) Global maize production, consumption and trade: trends and R&D implications. Food Secur 14(5):1295–1319

Fan X, Guan Z (2023) Vgnet: a lightweight intelligent learning method for corn diseases recognition. Agriculture 13(8):1606

Farooq M, Hussain M, Wahid A (2024) Artificial intelligence in plant breeding: applications and prospects. Trends Plant Sci 29(2):123–135

Fernandes IK, Vieira CC, Dias KOG, Fernandes SB (2024) Using machine learning to combine genetic and environmental data for maize grain yield predictions across multi-environment trials. Theor Appl Genet 137(8):189

Fu J, Zheng S, Fan L, Zheng X, Qian Q (2025) Breeding 5.0: artificial intelligence (AI) decoded germplasm for accelerated crop innovation. J Integr Plant Biol. https://doi.org/10.1111/jipb.70008

Gaynor RC, Gorjanc G, Hickey JM (2021) AlphaSimR: an R package for breeding program simulations. Genet Sel Evol 53:69

Govaichelvan KN, Pathmanathan D, Zainal Abidin R, Abu A (2024) Machine learning for major food crops breeding: applications, challenges, and ways forward. Agron J 116(3):1112–1125

Grzybowski MW, Mural RV, Xu G, Turkus J, Yang J, Schnable JC (2023) A common resequencing based genetic marker data set for global maize diversity. Plant J 113(6):1109–1121

Hamazaki K, Iwata H (2024) AI-assisted selection of mating pairs through simulation-based optimized progeny allocation strategies in plant breeding. Front Plant Sci 15:1361894

Heuermann MC, Rosso MG, Mascher M, Brandt R, Tschiersch H, Altschmied L, Altmann T (2019) Combining next generation sequencing and progeny testing for rapid identification of induced recessive and dominant mutations in maize M2 individuals. Plant J 100(4):851–862
Hu Y, Colantonio V, Müller BSF, Leach KA, Nanni A, Finegan C, Wang B, Baseggio M, Newton CJ, Juhl EM (2021) Genome assembly and population genomic analysis provide insights into the evolution of modern sweet corn. Nat Commun 12(1):1227
Hufford MB, Seetharam AS, Woodhouse MR, Chougule KM, Ou S, Liu J, Ricci WA, Guo T, Olson A, Qiu Y (2021) De novo assembly, annotation, and comparative analysis of 26 diverse maize genomes. Science 373(6555):655–662
Jia Y, Fu K, Lan H, Wang X, Su Z (2024) Maize tassel detection with CA-YOLO for UAV images in complex field environments. Comput Electron Agric 217:108562
Jiang B, An Z, Niu L, Qin D (2025) Precise genome editing process and its applications in plants driven by AI. Funct Integr Genomics 25(1):109
Khan I, Khare BK (2025) Integrating AI in plant science: a review of applications and future prospects. Plant Gene 44:100542
Khan M, Wang S, Wang J, Ahmar S, Saeed S, Khan SU, Xu X, Chen H, Bhat JA, Feng X (2022) Applications of artificial intelligence in climate-resilient smart-crop breeding. Int J Mol Sci 23(19):11156
Kienbaum L, Correa Abondano M, Blas R, Schmid K (2021) DeepCob: precise and high-throughput analysis of maize cob geometry using deep learning with an application in genebank phenomics. Plant Methods 17(1):91
Kim M, Go M, Kang S-H, Jeong S, Lim K (2025) Revolutionizing CRISPR technology with artificial intelligence. Exp Mol Med 57(7):1–13
Kundu N, Rani G, Dhaka VS, Gupta K, Nayaka SC, Vocaturo E, Zumpano E (2022) Disease detection, severity prediction, and crop loss estimation in MaizeCrop using deep learning. Artif Intell Agric 6:276–291
Li D, Quan C, Song Z, Li X, Yu G, Li C, Muhammad A (2021) High-throughput plant phenotyping platform (HT3P) as a novel tool for estimating agronomic traits from the lab to the field. Front Bioeng Biotechnol 8:623705
Liu J, Qu J, Yang C, Tang D, Li J, Lan H, Rong T (2015) Development of genome-wide insertion and deletion markers for maize, based on next-generation sequencing data. BMC Genomics 16(1):601
Liu H-J, Liu J, Zhai Z, Dai M, Tian F, Wu Y, Tang J, Lu Y, Wang H, Jackson D (2025) Maize2035: a decadal vision for intelligent maize breeding. Mol Plant 18(2):313–332
Ma J, Cheng Z, Cao Y (2025) Artificial intelligence-assisted breeding for plant disease resistance. Int J Mol Sci 26(11):5324
Montesinos-López OA, Montesinos-López A, Crossa J, Gianola D, Hernández-Suárez CM, Martín-Vallejo J (2018) Multi-trait, multi-environment deep learning modeling for genomic-enabled prediction of plant traits. G3 8(12):3829–3840
Murphy KM, Ludwig E, Gutierrez J, Gehan MA (2024) Deep learning in image-based plant phenotyping. Annu Rev Plant Biol 75(1):771–795
Nowak B, Tomkowiak A, Sobiech A, Bocianowski J, Kowalczewski PŁ, Spychała J, Jamruszka T (2023) Identification and analysis of candidate genes associated with yield structure traits and maize yield using next-generation sequencing technology. Genes 15(1):56
Pasala R, Pandey BB (2020) Plant phenomics: high-throughput technology for accelerating genomics. J Biosci 45(1):111
Peng C, Mei Y, Ding L, Wang X, Chen X, Wang J, Xu J (2021) Using combined methods of genetic mapping and nanopore-based sequencing technology to analyze the insertion positions of G10evo-EPSPS and Cry1Ab/Cry2Aj transgenes in maize. Front Plant Sci 12:690951
Qi J, Ding C, Zhang R, Xie Y, Li L, Zhang W, Chen L (2025) UAS-based MT-YOLO model for detecting missed tassels in hybrid maize detasseling. Plant Methods 21(1):21
Ramos APM, Osco LP, Furuya DEG, Gonçalves WN, Santana DC, Teodoro LPR, da Silva CA Jr, Capristo-Silva GF, Li J, Baio FHR (2020) A random forest ranking approach to predict yield in maize with UAV-based vegetation spectral indices. Comput Electron Agric 178:105791

Sarkar RK, Rao AR, Meher PK, Nepolean T, Mohapatra T (2015) Evaluation of random forest regression for prediction of breeding value from genomewide SNPs. J Genet 94(2):187–192

Sheikh M, Iqra F, Ambreen H, Pravin KA, Ikra M, Chung YS (2024) Integrating artificial intelligence and high-throughput phenotyping for crop improvement. J Integr Agric 23(6):1787–1802

Shu M, Fei S, Zhang B, Yang X, Guo Y, Li B, Ma Y (2022) Application of UAV multisensor data and ensemble approach for high-throughput estimation of maize phenotyping traits. Plant Phenomics 2022:9802585. https://doi.org/10.34133/2022/9802585

Singh A, Ganapathysubramanian K, Singh AK, Sarkar ST (2016) Machine learning for high-throughput stress phenotyping in plants. Trends Plant Sci 21(2):110–124

Vieira RA, Uhdre R, Scapim CA, van Becelaere G (2025) Global corn area from 1960 to 2030: patterns, trends and implications. J Agric Sci 163(3):1–12

Wallace JG, Rodgers-Melnick E, Buckler ES (2018) On the road to breeding 4.0: unraveling the good, the bad, and the boring of crop quantitative genomics. Annu Rev Genet 52(1):421–444

Wang G, Zhang X (2024) Maize in global food security: role and challenges. Field Crop 7(3):124–133

Westhues CC, Mahone GS, da Silva S, Thorwarth P, Schmidt M, Richter J-C, Simianer H, Beissinger TM (2021) Prediction of maize phenotypic traits with genomic and environmental predictors using gradient boosting frameworks. Front Plant Sci 12:699589

Wu H, Zhang Y (2025) AutoGP: an intelligent breeding platform for enhancing maize genomic selection. Nat Biotechnol 43(3):345–356

Wu C, Luo J, Xiao Y (2024) Multi-omics assists genomic prediction of maize yield with machine learning approaches. Mol Breed 44(2):14

Wu J, Xu H, Li Q (2025) Integrating genomic selection and machine learning for predicting maize yield under drought. Maize Genom Genet 16(3):139–148

Xiong Y, Wang R, Yang Q (2025) Artificial intelligence empowers maize drought resistance breeding: current status, technological breakthroughs, and future prospects. Adv Resour Res 5(2):817–833

Xu Y, Zhang J, Zhang Z (2022) Smart breeding driven by big data, artificial intelligence, and multi-omics. Mol Plant 15(9):1291–1303

Yadav KK, Kumaraswamy B (2024) The integration of artificial intelligence (AI) and high-throughput phenotyping (HTP) to estimate agricultural traits in crop development. Agric Biotechnol J 16(4):349–362

Zhang Y, Guo X, Du J, Zhao C (2015) Review on characterization of maize phenotypic diversity: from genome and genotyping to phenomics and high-throughput phenotyping. Res Crops 16(2):351

Zhang Q, Zhao X, Han Y, Yang F, Pan S, Liu Z, Wang K, Zhao C (2023) Maize yield prediction using federated random forest. Comput Electron Agric 210:107930

Zhang Y, Huang G, Zhao Y, Lu X, Wang Y, Wang C, Guo X, Zhao C (2025) Revolutionizing crop breeding: next-generation artificial intelligence and big data-driven intelligent design. Engineering 44:245–255

Zhao Z-W, Leng P, Han X-J, Shang G-F (2025) Global maize yield responses to essential climate variables: assessment using atmospheric reanalysis and future climate scenarios. Comput Electron Agric 232:110140

Zhu W, Li X (2024) Big data and artificial intelligence-aided crop breeding. J Integr Plant Biol 66(1):1–12

Specialty Corns Shaping the Future of Peri-Urban Agriculture

8

Chikkappa Gangadhar Karjagi, Kondajji Rangappa Yathish, and Sain Dass

Abstract

Specialty corns are cultivated for specialized end uses. Historically, the most widely known specialty corn includes sweet corn and popcorn. As scientific understanding advanced, differentiation of maize at the morphological, genetic, biochemical, and molecular levels led to the identification and inclusion of several specialized maize types under specialty corn. These include hard endosperm maize with elevated lysine and tryptophan content, popularly known as quality protein maize (QPM); waxy corn, containing nearly 100% amylopectin in the starch; amylomaize, with starch containing as high as 55–60% amylose; and high-oil corn, possessing 6–13% oil content. Furthermore, the recognition of the nutritional value of immature maize ears (specifically the cob), harvested 1–3 days after silk emergence and before pollination, when the ears are approximately 3–4 cm long and unfertilized, led to the coining of the term "baby corn." However, unlike other specialty corns, baby corn cannot be distinguished genetically. Thus, kernel phenotype or the presence of specific biochemical constituents has remained the primary basis for classification and nomenclature of different specialty corns. The awareness about specialty corns has not only increased in the most populous metropolitan cities in India but also even penetrated into "Tier 2" and "Tier 3" cities. The increasing importance of specialty corn has opened up innumerable opportunities to develop specialty corn cultivars suitable for commercial cultivation in different agro-climatic conditions. In India and across the world, a large number of specialty corn cultivars with enhanced nutritional value and improved quality traits have been developed and released. The specialty corn market worldwide, including India, is rapidly expanding with several post-harvest processing and value addition opportunities.

C. G. Karjagi (✉) · K. R. Yathish
ICAR-Indian Institute of Maize Research (ICAR-IIMR), Ludhiana, Punjab, India

S. Dass
Directorate of Maize Research, New Delhi (Presently ICAR-IIMR, Ludhiana), Ludhiana, Punjab, India

M. Choudhary et al. (eds.), *Next-Generation Maize Research*,
https://doi.org/10.1007/978-981-92-0722-0_8

Keywords

Sweet corn · Popcorn · Baby corn · Specialty corn · Waxy corn · QPM

8.1 Introduction

Maize (*Zea mays* ssp. *mays* L.) is one of the most important cereal crops and possesses tremendous genetic diversity. It is classified into numerous types based on variation observed in morphological, biochemical, or other features. The simplest classification of maize is based on kernel color, such as yellow maize, the most prominent and widely cultivated type globally; white maize, extensively used for food purposes in several countries, particularly in Africa, Latin America, and parts of Asia. Some of the landraces of maize are mixed kernel color with varying proportions and combinations of different colors like yellow, white, blue, and purple. However, such mixed kernels are often not traded commercially in large scale but cultivated in small areas for household consumption often among tribal communities. Maize is also classified into dent, semi-dent, and flint based on the kernel texture. However, commercially, maize grain is traded as yellow dent, yellow flint, white dent, white flint, and mixtures of flint and dent types in both the yellow and white color categories; waxy maize is also traded occasionally. At international market, trade in maize grain, often termed "commodity corn," is regulated by uniform grades and standards without any mention of its suitability for specific end uses. On the contrary, specialty corns are meant for specific end uses. Specialty corn is defined as any corn that is improved for specific end use compared to commodity corn (Hallauer 2000).

Historically, the most widely known specialty corn like sweet corn and popcorn are identified based on the distinctive morphological and genetic characteristics. However, baby corn cannot be differentiated genetically, but it is considered a specialty corn. Baby corn is the immature ear of maize and gained acceptance as a nutritious vegetable during the mid-twentieth century owing to its rich nutrient composition. For example, baby corn contains approximately 86 mg of phosphorus (P) on dry matter basis, which is considerably higher than that of some leafy vegetables, such as spinach (Yodpet 1979; Das et al. 2008). In addition, baby corn is also considered a good source of vitamin A, vitamin C, and folate (B_9). Other specialty corns that can differentiate genetically are hard endosperm maize kernel with high-lysine and high-tryptophan (popularly called as quality protein maize (QPM)), waxy corn (contains nearly 100% amylopectin in the starch), amylomaize (amylose content as high as 55–60% in the starch), and high-oil corn (6–13% oil) (Kumar et al. 2019). Unlike commodity corn, the commercial cultivation of specialty corn requires special considerations for both production and marketing. For example, the commercial cultivation of sweet corn and baby corn requires specific logistical facilities and proper storage to minimize post-harvest losses due to their highly perishable nature. Therefore, these specialty corns require careful handling throughout the production chain, similar to other vegetable crops, from cultivation to final disposal at retail outlets. The cultivation of sweet corn and baby corn is often recommended in and around urban settlements (peri-urban areas) because of their high market demand and perishability. Hence, they are considered ideal crops for peri-urban agriculture. In countries like India, these specialty corns are increasingly promoted to enhance profitability and net returns from agriculture. Moreover, sweet corn and baby corn have attained the status of high-value commodities because of their superior nutritive value. Apart from sweet corn and baby corn, popcorn is another specialty corn that has gained popularity worldwide. It is considered a nutritious snack and is rich in antioxidants such as polyphenols. On the other hand, popcorn, another specialty corn, demand in India has increased substantially in recent years, mostly after mid-2000s. Presently, almost 80–90% of total popcorn consumed in India is being imported from countries like the United States of America (USA), Brazil, and Argentina. The awareness and demand for specialty corns such as sweet corn, baby corn, and popcorn have increased significantly not only in metropolitan cities of India but also in Tier 2 and Tier 3 cities. Consequently, investment in research and development related to specialty corns has expanded considerably in India and other countries during the last one and a half decades. Sweet corn and baby corn are consumed both in fresh form and through a wide range of recipes and value-added processed products. Besides these, other specialty corns, such as quality protein maize (QPM), waxy corn, amylomaize, and high-oil maize, possess diverse applications in food processing, animal feed, and various industries. QPM serves as a highly nutritious food for humans as well as quality feed for animals. Waxy corn is extensively utilized in processed foods to improve stability, uniformity, and texture, particularly in dairy products, and is also used in quick-energy drinks for athletes. Amylomaize is used as a thickening agent in puddings and processed foods, especially those intended for diabetic patients, and also finds application in corrugating and adhesive industries. Similarly, corn oil is regarded as one of the finest edible oils because of its high content of polyunsaturated fatty acids, particularly linoleic acid (54–60%), while high-oil maize is an important component of poultry and pig feed. It is increasingly evident that most specialty corns—including sweet corn, popcorn, waxy corn, amylomaize, QPM, and high-oil maize—are produced and marketed for specific end uses owing to their unique kernel biochemical composition. These characteristics primarily arise from allelic variations or recessive mutant alleles involving one or a few genes.

8.2 Mutants Imparting Special Utility in Maize

Kernel color in maize plays an important role in determining its end use and market preference. Although there is no strict classification based on kernel color, yellow maize is predominantly preferred for animal feed and industrial applications, whereas white maize is generally used for human consumption. The color of maize kernels is mainly governed by two groups of pigments: carotenoids and anthocyanins, which are produced through different metabolic pathways. Carotenoids are synthesized in the starchy endosperm and impart a yellow color to the kernels. In contrast, anthocyanins accumulate in the aleurone layer, the outermost layer surrounding the endosperm, and are responsible for purple, red, or blue pigmentation. Therefore, the final kernel color depends on the interaction and expression of genes controlling these pigments. Yellow kernels are visible only when the aleurone layer is colorless, while the absence of both carotenoids and anthocyanins results in white or colorless kernels. One of the earliest genes identified for controlling yellow and white kernel color in maize was *Y1*, a structural gene encoding the enzyme phytoene synthase (Coe et al. 1988). Similarly, genes such as *Pr1*, *C1*, and *R1* regulate pigmentation in the aleurone layer and produce purple or red kernel colors depending on their allelic forms and genetic interactions. In addition to kernel color, the biochemical composition of maize kernels—including starch, amino acids, lipids, and micronutrients such as provitamin A, vitamin E, iron, and zinc—is also influenced by allelic variation or mutant alleles at one or a few genes. Certain genes also affect the accumulation of antinutritional compounds such as phytate. The major mutant alleles and allelic variants influencing these biochemical traits, along with their effects on kernel morphology, are discussed below.

8.2.1 Mutants Affecting Kernel Starch and Its Components

Starch is the major storage compound in maize kernels, accounting for nearly 66–75% of the kernel weight. In normal commodity maize, starch typically consists of about 25–30% amylose and 70–75% amylopectin (Kumar et al. 2019). The biosynthesis of starch is a complex biochemical process involving multiple intermediate steps catalyzed by several enzymes encoded by numerous genes. Variations or mutations in these genes can alter the starch biosynthetic pathway, thereby affecting the total starch content as well as the relative proportions of amylose and amylopectin in the kernel. Extensive research on starch metabolism in plants has led to the identification of several mutant alleles and allelic variants associated with these changes in starch composition and functionality. One of the earliest mutant alleles identified based on the opaque kernel phenotypes was the *waxy1* (*wx1*) mutant. Subsequently, numerous allelic variants of *wx1* and also several other mutant alleles, namely, *shrunken1* (*sh1*), *shrunken2* (*sh2*), *shrunken4* (*sh4*), *amylose extender1* (*ae1*), *brittle1* (*bt1*), *brittle2* (*bt2*), *dull1* (*du1*), *sugary1* (*su1*), and *sugary2* (*su2*) have been identified. The mutant alleles or allelic variants identified later showed overlapping phenotypes with the previously reported alleles. For example, shrunken, opaque to translucent phenotype is common for *brittle1* (*bt1*), *brittle2* (*bt2*), *dull1* (*du1*), *shrunken1* (*sh1*), *shrunken2* (*sh2*), and *shrunken4* (*sh4*). Allelic tests, also known as genetic complementation or cis–trans tests, are commonly used to determine whether newly identified mutants are allelic or non-allelic to previously characterized mutants. Genes, allelic variants, and mutant alleles involved in starch metabolism are broadly classified into two groups—class I and class II mutants—based on their effects on kernel dry weight and starch accumulation. Class I mutants, such as brittle (bt) and shrunken (sh), cause a drastic reduction in kernel dry weight and produce severely shrunken kernels with very low starch content. In contrast, class II mutants, including sugary (su), amylose extender (ae), and waxy (wx), also affect kernel weight and starch composition, but their effects are comparatively less severe than those of class I mutants. Detailed genetic studies have shown that class II mutants can mask or modify the phenotypic effects of class I mutants through epistatic interactions. Furthermore, interactions among multiple class II mutants may also produce extreme kernel phenotypes resembling those caused by individual class I mutants. Thus, mutants affecting starch metabolic steps show inter-allelic or intra-genic as well as non-allelic or inter-genic epistatic interactions.

8.2.1.1 Mutants Affecting Kernel Protein

Amino acids are the building blocks of proteins. In general, protein content in the maize kernels varies between 8% and 10% on a dry-weight basis but is deficient in two important essential amino acids (lysine and tryptophan). Thus, the biological value (BV) of maize protein (after milling) is low (~32%) as compared to other cereals like rice (~75%) and wheat (~60%). BV is a measure of the presence of essential amino acids in proportion to needs, based on the proportion of protein retained in the body for growth and/or maintenance, and is expressed in percent of nitrogen absorbed. The discovery of the mutant gene *opaque2*, followed by biochemical analysis of opaque maize kernel composition, revealed significant improvement in the amino acid profile of maize protein (Mertz et al. 1964). The levels of lysine and tryptophan were found to increase substantially, almost doubling compared to commodity/normal maize, thereby providing an opportunity to enhance the BV of maize protein. The enhanced nutritional value of opaque maize has led to its consideration for specific end use in food and feed industries.

The characterization of *opaque2* mutant allele revealed its regulatory role; this allele encodes a DNA-binding protein that binds to the DNA sequences encoding 22 kD subunit of α-zein fraction of maize protein. In general, maize proteins are composed of four different classes of proteins, namely prolamins, referred to as zeins (52%), glutelins (25%), albumins (7%), and globulins (5%). Since a substantial proportion of maize proteins consists of the zein fraction, which is primarily composed of prolamines, the overall nutritional quality of maize protein is poor. This is because prolamines are deficient in lysine and tryptophan, two essential amino acids. The DNA-binding protein coded by *o2* mutant fails to bind effectively to the DNA sequences encoding the 22 kD subunit of zeins, which leads to reduced levels of prolamin with enhanced levels of non-zein fractions, leading to enhanced levels of lysine and tryptophan.

8.2.1.2 Mutants Affecting Kernel Lipid Profile

Maize kernels contain around 3–4% oil, the third most abundant biochemical component in the kernel (Bergquist et al. 1998). Maize is an important ingredient of animal/poultry feed which make up 40–60% of the feed proportion to provide the energy. However, in order to increase the feed efficiency and also enhance the energy density, attention was paid long back to increase the oil percentage of the maize kernel through directed, focused, long-term breeding efforts. As a result, oil content in the maize kernel has been increased to >20% in some of the maize genotypes, Illinois high-oil (IHO) lines specifically bred at the University of Illinois. The IHO lines are the result of long-term artificial selection breeding experiment, which started in 1896 by C. G. Hopkins (Laurie et al. 2004). The directional selection over 120 years for high oil and low oil has resulted in development of IHO and Illinois low-oil (ILO) lines with oil content of 1.1–19.3% as compared to base population containing 4.7% oil. The corn oil contains both saturated and unsaturated fatty acids (triacylglycerides) to provide balanced composition that combines nutritional value with desirable cooking and storage properties. Based on the long-term studies on IHO lines, key gene(s) have been identified which enhance the oil content in maize kernel by enhancing specific fatty acids like linoleic acid; the *ln1* (*linoleic acid*1) gene mediates the desaturation of oleic fatty acid at the Δ 12–13 position to linoleic; and *fad2* (*fatty acid desaturase2*) gene enhances the oleic acid content.

8.3 Sweet Corn

Sweet corn is one of the most popular and highly valued vegetable crops worldwide because of its high sugar content and pleasant taste. It is commonly consumed either raw or cooked in various culinary preparations. The characteristic sweetness of sweet corn is primarily attributed to the accumulation of phytoglycogen, a highly branched and water-soluble α-glucan composed of α-(1→4)-linked glucan chains with α-(1→6) branch points. Owing to its appealing flavor and nutritional value, sweet corn is widely enjoyed by people of all age groups across the world. Although sweet corn cultivation, consumption, and marketing as fresh, canned, or frozen were extended across continents from the Americas through Europe to Africa and East Asia. Its commercial prominence originated in the USA, where it is the second most important canned vegetable after tomato, the second frozen vegetable after potato, and the fourth largest fresh vegetable. The cultivation and use of sweet corn in the Americas dates back to the pre-Columbian period, and several ancient landraces remain under cultivation. Notably, Chullpi, grown by Quechua-speaking communities in Peru, is believed to have dispersed from the southern Sierra region, where sweet corn is cultivated at elevations of 2400–3400 m. Other pre-Columbian sweet corn landraces, such as Maíz Dulce of Mexico and Ducillo del Noroeste, were grown by indigenous groups including the Hopi, Zuni, Pima, Papagos, and Pueblo peoples.

8.3.1 Modern Sweet Corn

It is quite intriguing to know the origin of present-day sweet corn. Different hypotheses have been proposed based on historical documented materials as well as archeological evidence. Based on the multiple proposed theories and hypotheses, the conclusive and unambiguous evidence for definite origin of sweet corn is not available. However, circumstantial evidence suggests that the pre-Columbian sweet corn landraces described above likely served as the primary progenitors of modern sweet corn. Thereafter, continuous human-mediated selection and systematic breeding efforts, particularly through hybridization with commodity corn, resulted in remarkable improvements in several quality and agronomic traits, including sweetness, tenderness, crispness, texture, mouthfeel, shelf life, yield potential, adaptability, and desirable ear and plant phenotypes. A major proportion of these advancements in sweet corn breeding was achieved in the United States of America. Subsequently, owing to trade, migration, and intercontinental human movement, sweet corn gradually disseminated to several other regions of the world.

8.3.1.1 Historical Account of Sweet Corn Improvement—USA

Garden Book, written by Thomas Jefferson in 1810, contains the earliest known written evidence of proper sweet corn with shriveled kernels. The first known sweet corn cultivar, 'Darling Early', was released for commercial cultivation in the USA in 1844. The cultivar is a cross between early yellow flint and white sweet corn. As per the above hypothesis regarding the origin of modern sweet corn, Darling Early is probably the first evidence. Subsequently, other varieties like Stowell's Evergreen (Menomony soft corn × northern sweet corn) and Crosby were released for cultivation. By 1899, more than 60 sweet corn varieties were being cultivated in the USA, most of which possessed white kernels. This preference for white-kernel sweet corn parallels the traditional preference for white maize for human consumption as food. Golden Bantam (Darlings Early × Early Yellow Flint), the yellow sweet corn variety released by the Burpee company in 1902, has changed the perspective of the color preference. The excellent quality and wide adaptability of 'Golden Bantam' across diverse agro-climatic regions helped dispel the prevailing preference for white sweet corn, leading to its recognition as the standard for best-quality sweet corn. As a result, Golden Bantam variety was used in crossing to convert a large number of white sweet corn varieties into yellow. Therefore, most of the present-day sweet corn inbred lines and hybrids have Golden Bantam in their pedigree.

8.3.1.2 Genetic Basis of Sweetness in Sweet Corn

Endosperm mutants imparting sweetness in maize kernels are associated with starch biosynthesis, and 14 genes (*Ae1*, *Bt1*, *Bt2*, *Du1*, *Isa2*, *Sh1*, *Sh2*, *SbeIa*, *SbeIIa*, *Su1*, *Su2*, *Se1*, *Wx1*, and *Zpu1*) have been identified in this pathway. Among these, eight mutant alleles (ae1, bt1, bt2, du1, sh2, su1, se1, and wx1) are commercially exploited in specialty corn, including sweet corn. These mutants reduce starch accumulation in the endosperm, thereby increasing kernel sweetness. The *su1* mutant, one of the earliest identified, enhances sweetness by elevating water-soluble polysaccharides (WSP), particularly phytoglycogen (~17-fold increase, ~35% WSP), along with sugars. Based on their effects on starch metabolism, these genes are classified into class I and class II mutants. Class I mutants (*bt1*, *bt2*, and *sh2*) markedly reduce starch synthesis while increasing sugar levels by 4–8 times compared to non-mutant corn, resulting in kernels with negligible total carbohydrates at maturity. Sweet corn hybrids carrying class I mutants are therefore termed supersweet or extra-sweet, with *sh2* being the most widely used endosperm mutant after *su1* and currently surpassing *su1* in fresh sweet corn markets. Class II mutants (*ae1*, *du1*, *su1*, and *wx1*) do not substantially reduce total starch content at maturity but alter the composition and structure of endosperm polysaccharides. Individually, these mutants do not elevate sugar levels at 21 days after pollination (DAP) to the extent observed in class I mutants; however, combinations of recessive class II alleles exhibit complementary gene action, resulting in sugar levels comparable to class I mutants. Accordingly, genotypes carrying multiple recessive alleles, such as *ae1 du1 wx1*, have been developed and commercially released. Elevated phytoglycogen levels in *su1* genotypes are maintained when combined with *wx1*, *du1*, or *bt1*, but are suppressed when combined with *ae1*, *bt2*, or *sh2*. The *se1* gene, which does not fall into either class, enhances sugar levels close to those of *sh2* when present with *su1* by maintaining WSP levels similar to unmodified *su1*. Owing to its dosage effect and favorable influence on sweetness, *se1* is frequently incorporated into sweet corn breeding programs.

WSP contributes to smooth texture and kernel creaminess in *su1*-based genotypes. However, combining sweetness-imparting mutants presents several challenges. The *se1* allele is often associated with pale yellow kernels and is difficult to identify in certain genetic backgrounds, complicating the development of high-quality *su1 se1* hybrids. Moreover, *su1* genotypes have a narrow harvest window due to rapid sugar decline before and after harvest, whereas class I mutants maintain higher sucrose levels up to 28 DAP and nearly double the sugar content of *su1* at 21 DAP, thereby extending the harvest window. Nevertheless, other quality traits such as pericarp tenderness continue to decline, necessitating harvest within 18–24 DAP depending on environmental conditions. Despite higher initial sucrose levels, *su1 se1* genotypes exhibit sugar loss rates similar to *su1*, with ≥50% sugar loss within 24 h at room temperature. Triple-mutant *ae1 du1 wx1* genotypes show slower sugar loss than *su1* or *su1 se1*, though still faster than class I mutants. Therefore, optimal combinations of mutant alleles, along with germination behavior and post-harvest quality, must be carefully considered in sweet corn cultivar development.

8.3.1.3 Sweet Corn Breeding

Genetically diverse germplasm as base material is essential for the development of high-yielding and superior-quality sweet corn cultivars. Prominent contributors include Golden Bantam for the yellow fresh market and Evergreen and Country Gentleman for processed corn. Further, prominent sweet corn breeders such as D. F. Jones and W. R. Singleton (Connecticut), W. A. Huelson (Illinois), E. S. Haber (Iowa State), and G. M. Smith (Purdue) played key roles in early improvement efforts. Sweet corn inbred lines released during 1930–1960, along with their *sh2* versions, continue to be used in commercial hybrid breeding. Sweet corn initially spread slowly from the USA to other parts of the world, followed by the development of varieties adapted to new environments and systematic germplasm enhancement. The wide availability of base germplasm in the USA has enabled the use of diverse breeding methods for developing high-quality hybrids. Selection of appropriate base germplasm is critical to achieve breeding objectives, with pedigree breeding often initiated using backcross, single, three-way, or double crosses. Subsequently, population improvement approaches, such as recurrent selection (RS), reciprocal recurrent selection (RRS), heterotic grouping, and identification of heterotic patterns, are applied. The identification of molecular markers linked to sweetness-determining mutant alleles has further facilitated marker-assisted selection (MAS) and marker-assisted backcross breeding (MABB) for rapid hybrid development. In India, sweet corn varieties were initially developed using limited exotic germplasm imported during the 1980s–2000s. Renewed emphasis since the late 2010s has led to the development of several single-cross hybrids with de-husked fresh ear yields reaching 15 MT ha^{-1}. Currently, both public and private sectors are actively engaged in sweet corn improvement. Public sector releases include Central Maize VL Sweet Corn-1, Central Maize VL Sweet Corn-2, ASKH-1, ASKH-4, and ISCH-601, while some hybrids have been biofortified for essential micronutrients and vitamins through MABB and MAS in India. Private sector hybrids, such as Candy (KSCH-333), VMH-4106, Hy-brix-53, Hi-brix-39, and NUZI-260, have also been released. The demand for sweet corn is increasing globally and in India, which has transitioned from a net importer to an emerging exporter due to strengthened national maize improvement programs. Consequently, several hybrids have been released in recent years. Commercial cultivation of sweet corn in India has expanded mainly in peri-urban areas around metropolitan cities such as New Delhi, Bengaluru, and Pune. Large-scale promotion of sweet corn through peri-urban agriculture could significantly enhance farmers' profitability and ensure sustained income.

8.4 Popcorn

Popcorn has emerged globally as one of the most popular and nutritious snack foods. North America, particularly the United States of America, has traditionally been the leading producer, consumer, and exporter of popcorn. However, its consumption and demand have expanded considerably in other countries, including Brazil, Argentina, China, Japan, and several European nations. In India as well, the demand for popcorn has increased steadily over the years. Its popularity, once confined mainly to metropolitan cities, has now spread to smaller cities and towns. Although India has traditionally been a major importer of popcorn, commercial cultivation has recently expanded in states such as Andhra Pradesh, Telangana, Karnataka, and Maharashtra to meet the growing domestic demand. India offers immense market potential for popcorn due to rapid urbanization, a growing consumer base, the presence of numerous tourist destinations, and expanding entertainment sectors such as cinema and multiplex theaters. In addition, street vendors constitute a significant segment of the popcorn market in the country. The Indian popcorn market has been growing rapidly, with an estimated compound annual growth rate (CAGR) of around 15% during 2018–2022 (India Popcorn Market 2018 Industry – Key Players, Size, Trends, Opportunities, Growth-Analysis To 2022 | ABNewswire).

8.4.1 What Makes Popcorn to Pop?

Starch packing in popcorn kernels is the primary factor responsible for popping, producing the characteristic "pop" sound when heated to about 180 °C. The popcorn endosperm consists of two starch types: soft starch at the core and hard starch surrounding it. Popcorn kernels are typically small, round, or flint-type, with a lower soft-to-hard starch ratio than field corn. Additionally, popcorn pericarp is thicker than that of other maize types due to structural differences. Studies have shown that popcorn pericarp cellulose exhibits a more crystalline arrangement and a higher degree of fibrillar packing, resulting in thermal diffusivity and conductivity that are 2.0–2.9 times higher than those of field corn pericarp. When heated using microwave ovens, pressure cookers, earthen pots, kettles, or stoves, heat diffuses rapidly to the kernel core due to tight starch packing. As a result, internal pressure of about 135 psi builds up in the endosperm, causing kernel swelling and eventual bursting, accompanied by the "pop" sound. During popping, starch granules gelatinize and release flavor, adding value to popcorn. Efficient heat transfer to the kernel core is important to increase the internal temperature without burning the pericarp. In contrast, field corn pericarp, owing to lower thermal diffusivity, tends to char and calcine under high temperatures. Increased core temperature leads to vaporization of moisture, and the gas generated escapes through kernel rupture. Greater pericarp thickness allows the kernel to sustain higher pressure, resulting in increased expansion or popping volume. The final popping volume, a key quality parameter, is determined by the combined effects of soft starch proportion, kernel moisture content, and pericarp thickness. As pericarp thickness and soft starch proportion are genotype-dependent, maintaining optimal kernel moisture is critical. Maximum popping volume is generally achieved at 11–13% moisture, with some studies reporting an optimum of 11.6%. Popcorn kernels are sometimes coated with oil or butter to retain moisture and enhance flavor and mouthfeel. Pericarp thickness has shown a strong positive correlation with popping or expansion volume (Silva et al. 1993). Kernel size also affects popping volume, and recent preferences have shifted toward relatively larger kernels due to higher expansion and improved profitability.

8.4.2 Classification of Popcorn

Popcorn is classified into two major types (butterfly or snowflake and mushroom) based on the characteristic shape of flake (a popped kernel of corn, more specifically popcorn). Butterfly popcorn produces flakes with irregular wing-like shapes and a crispy texture, although the flakes are relatively fragile and break easily. In contrast, mushroom-type popcorn produces flakes that are round and compact, resembling a mushroom cap with an uneven and rough surface (Hallauer 2000). The flake shape of popcorn also possesses unique utility for consumer. The butterfly popcorn, being delicate, easily breaks and is used for direct consumption with minimal value addition, such as adding salt/butter/flavor. On the contrary, the mushroom popcorn is mainly used for further value addition/processing and packing. The shape of mushroom is unique, which facilitates coating with different flavors, creams, etc. The rough surface of mushroom-type popcorn holds cream and coating firmly and allows for packing, storing, and long-distance marketing. In fact, several types of coating due to wider availability of flavors in the market like caramel, chocolate, or range of savory. Even though the market share of butterfly popcorn is higher, the demand for both types of popcorn is more or less stable over the years in the market because of their unique utility, consumer preference, and market segment. The advances in packaging and value addition have increased the ready-to-eat (RTE) popcorn marketing, which is mainly driving the market, and it is expected to increase continuously in the near future.

8.4.3 Nutritional Value of Popcorn

Popcorn contains several micronutrients like minerals (Ca, Fe, Mg, P, K, Na, Zn, Cu, and Mn) and vitamins [B1 (Thiamin), B2 (Riboflavin), B3 (Niacin) and B6 (Pyrodoxine), B9 (Folate), A, E, and K]. It is rich in fiber, low in fat with no cholesterol (higher unsaturated fatty acids), sugar, and other carbohydrates and considered low-calorie nutritious snack food. The levels of the above micronutrients vary depending on type of cultivars, geographical regions, and environmental conditions. In addition to its nutritional importance, popcorn provides multiple health benefits. Being low in calories and highly satiating, it may assist in weight regulation. The insoluble fiber present in popcorn improves digestion, supports gut health, and helps maintain regular bowel movement. Moreover, the presence of polyphenols and antioxidant compounds contributes to protection against cardiovascular diseases, vascular inflammation, hypertension, and certain forms of cancer.

8.4.4 Popcorn Breeding

In India, research on popcorn hybrid development began in the early 2000s in response to rising demand. The first single-cross popcorn hybrid, BPCH 6, was released in 2015 by Professor Jayashankar Telangana Agricultural University (PJTSAU), erstwhile ANGRAU, Hyderabad. Subsequently, several open-pollinated popcorn cultivars such as KDPC-2, VL Popcorn-1, and Bajaura Popcorn-2 were released. However, these early released hybrids did not meet international quality standards, despite increasing market demand for premium popcorn. Currently, several improved popcorn hybrids have been released in India, including DMRHP 1402, LPCH 2, LPCH 3, IPCH 501, APCH 2, and APCH 3 by ICAR—Indian Institute of Maize Research. Among these, LPCH 3 has entered into an MoU with Gourmet Popcornica Ltd. for commercial seed production and marketing. The yield levels of recently released hybrids range from 3 to 4 MT ha^{-1} and are adapted to diverse agro-climatic regions. These hybrids exhibit superior popcorn quality, particularly in popping percentage and popping volume, high yield and adaptability due to their tropical genetic background. There is significant scope for promoting public sector popcorn hybrids to reduce dependence on imported seed and popcorn, thereby conserving foreign exchange. Given India's favorable climate, popcorn can be cultivated year-round in different regions, and the availability of multiple seed production sites enables large-scale hybrid seed production at affordable costs. The availability of locally adapted, high-quality hybrids is expected to substantially reduce imports of both seed and commercial popcorn in the coming years. Although popcorn yields are relatively low (about half of field corn), its market price is 3–4 times higher, making popcorn cultivation economically viable. Key considerations include cultivar choice, yield potential, popcorn quality parameters, and maintenance of spatial or temporal isolation to ensure quality. Contract farming further ensures assured returns, adoption of best management practices, and large-scale cultivation, benefiting both farmers and procurers. Hybrid popcorn seed is costlier than field corn seed mainly due to the lower seed yield of popcorn parental lines. At present, mushroom-type popcorn hybrids have not yet been developed by public sector institutions in India. Efforts are underway to diversify popcorn germplasm through exploration and introduction of elite and exotic materials. Future breeding programs aim to enhance popcorn quality traits such as popping percentage and expansion ratio, along with the development of mushroom-type popcorn hybrids.

8.5 Baby Corn

Immature ear (precisely cob) of maize, harvested before pollination on 1–3 days after silk (3–4 cm long) emergence (unfertilized) is called baby corn. The discovery of nutritional value of baby corn has found its special end use as vegetable. As a result, baby corn is considered one of the specialty corns. The main quality parameters for baby corn include baby corn length, baby corn diameter, baby corn color, straight ovule arrangement, and its mouth feel (although it is highly subjective). The additional quality parameters include the nutritional composition, often not considered for defining the baby corn. The desirable size of baby corn includes 10–12 cm long and 1–1.5 cm diameter. The additional features include regular ovule (also called row) arrangement and creamy/light yellow color along with crunchy texture upon bite.

8.5.1 Baby Corn Significance and Breeding

Baby corn is rich source of phosphorous as compared to other vegetables. Apart from Phosphorous, it is also an important source of vitamins (thiamine, ascorbic acid), and minerals (Calcium) including dietary fiber (Yodpet 1979). The greater significance of baby corn is reflected in the livelihood of the baby corn growing farmers, especially small and marginal farmers. In India, >85% of the farmers are small and marginal. In this context, baby corn cultivation by such small and marginal farmers, especially in peri-urban areas, has tremendous potential to improve the livelihood of such farmers. Baby corn has the potential to generate regular income and also the duration to generate income is relatively short period, provided proper planning by the small and marginal farmer. Its cultivation also supports dairy industry through the by-products like tassel part, baby corn husk, and green stalk, generated after baby corn picking. It also provides an opportunity for post-harvest processing and value addition for baby corn and green stalk as silage. For example, it can be used as main ingredient in preparation of several cuisines like sweet candy, jam, halwa, murabba, pickeles, kheer, raita, pakoda, soup, and mixed vegetable subji. Indian farmers have tremendous opportunities to export baby corn around the world due to its demand. It can be processed into canned or frozen product for long-term storage and export.

Due to the growing demand for baby corn, maize breeders started baby corn breeding in countries like India and Thailand. The presence of more than one ear per plant, called "prolificacy," is an important breeding objective (Singh et al. 2019). Initially, the field corn hybrids, which possess desirable features of baby corn, were directly used for baby corn cultivation. However, over a period of time, sophistications have been introduced in baby corn breeding, especially through development of male sterile baby corn hybrids. Baby corn, unfertilized ear/cob harvested young (immature) before pollination (unfertilized), needs to be harvested before pollination to maintain highest quality. Detasseling, emasculation process in maize (similar to the one undertaken during hybrid seed production in maize operation), is a mandatory and an important operation which is advocated while undertaking baby corn cultivation. However, the removal of the tassel part before anthesis (detasseling) to prevent pollination of baby corn adds extra production cost to farmers. Therefore, maize breeders have started investing resources to develop male sterile baby corn hybrids. In India, HM 4 hybrid was released as the first baby corn hybrid. Subsequently, several other baby corn hybrids including male sterile baby corn hybrids have been developed and released for commercial cultivation in India. The male sterile baby corn hybrids released in India in recent years includes Pusa Male Sterile Baby Corn Hybrid 3 (ABHS-27), Pusa HM 4 (Male Sterile Baby Corn 2) (ABSH4-2), IBCH 402 (IBH 11-227), IBCH 403 (IBH 11-223), and Punjab Baby Corn 1 (JH 32434). Others are fertile which require detasseling like IBCH 401 (IMHSB 19KB-2), Rajendra baby corn 1 (MBC 11-15), Ladhowal Baby Corn Hybrid 3 (LBCH 3 or DMRHB 1305), Baby corn GAYMH 1 (GAYMH 1), AH 7043, IMHB1532, and IMHB1539.

8.6 Waxy Maize

Corn starch is composed of two major polysaccharides, namely, amylose and amylopectin. Amylose has a linear structure with α-1,4 glycosidic bonds, whereas amylopectin possesses a highly branched structure containing both α-1,4 and α-1,6 glycosidic bonds. Variations in the proportion of these two starch components are reflected in changes in maize kernel phenotype. Waxy maize, characterized by dull, opaque kernels with a waxy or marble-like appearance, was first identified in China during the early 1900s while selecting maize populations. Hence, waxy maize is often considered to have originated in China. The waxy phenotype can be visually distinguished across a wide range of genetic backgrounds; however, the characteristic dull and waxy appearance becomes clearly visible only when kernel moisture content falls below approximately 16%. Biochemical analyses have shown that starch in waxy maize consists almost entirely of amylopectin and contains little to no amylose. Nevertheless, varying levels of amylose and amylopectin have been reported among genotypes carrying different allelic variants of the *wx* gene. For instance, genotypes possessing the Argentine waxy allele (*wx*-a) contain more than 95% amylopectin along with small amounts (<5%) of amylose. Other kernel characteristics, such as hardness, are generally similar to those of normal non-waxy maize. An important feature of the *wx* allele is its epistatic effect over all known endosperm mutants in maize. Therefore, genotypes carrying the recessive *wx* allele typically produce starch composed almost entirely of amylopectin, with only a few reported exceptions. Genetic studies have established that the waxy phenotype is controlled by a single recessive gene, waxy (*wx*). The original allele identified in China was designated as *wx*-c. The normal dominant *Wx* gene encodes the enzyme Granule Bound Starch Synthase I (GBSSI), which catalyzes an important step in amylose synthesis during starch biosynthesis. Mutation in the wx allele inactivates GBSSI, thereby blocking amylose synthesis and resulting in the accumulation of nearly pure amylopectin starch (Klosgen et al. 1986).

Waxy corn is popular as a fresh corn market type, particularly in several Asian countries, owing to its highly digestible nature. It is also extensively utilized in the preparation of value-added processed food products, including instant energy drinks. Recently, waxy maize is also known to have shown high bioethanol recovery as compared to commodity corn (30% amylose and 70% amylopectin). Due to the recessive nature of the waxy trait, planned and well-managed commercial cultivation of waxy corn is preferred to maintain trait purity while achieving yield levels comparable to those of commodity corn. Recently, efforts are being made in different countries to develop waxy maize by employing different breeding approaches or methods. The most efficient and fast-breeding approach would be rapid conversion of high-yielding, widely adopted, popular commodity corn hybrid into waxy hybrid through introgression of *wx* allele into the parental lines of the hybrid through marker-assisted selection. Recently, CRISPR (clustered regularly interspaced short palindromic repeats) along with CRISPR-associated (Cas) proteins like Cas9 have been employed for rapid conversion of commodity corn into waxy corn through precise editing of key nucleotide positions in the *Wx* gene to develop waxy maize.

8.6.1 Breeding for Waxy Maize

The waxy maize was of only academic interest, especially for undertaking genetic and biochemical studies until the 1930s. The major reason could be lack of information regarding the advantages associated with waxy maize, absence of appropriate and suitable germplasm to take up any waxy corn improvement programs, etc. The biochemical investigation of waxy maize at the Iowa Agricultural Experiment Station, USA, revealed that the starch quality of waxy maize matches with that of tapioca (starch extracted from the roots of cassava plant (*Manihot utilissima*)). During World War II, the import of tapioca from the Far East countries into the USA has faced severe difficulties. As result, commercial cultivation of waxy corn has started in the USA. Substantial breeding efforts during the initial period, particularly in the 1950s and 1960s, were undertaken at the Iowa Agricultural Experiment Station, USA. Subsequently, even some private seed companies which are involved in hybrid maize seed production and marketing in the USA also undertook waxy corn breeding program in their maize research and development (R&D) centers. The major factors driving focused breeding efforts in waxy corn were its simple inheritance pattern and the masking effect of the wx allele on other endosperm mutant alleles, which made waxy corn breeding relatively simple, straightforward, and less labor-intensive. At the same time, there was substantial market demand for waxy corn in the USA as well as in countries such as Canada and several European nations. Visual identification of the waxy kernel phenotype is often difficult when kernels contain high moisture levels. However, the development of a simple pollen-staining technique greatly facilitated the selection of the waxy trait during breeding. Under diluted potassium iodide staining, pollen grains carrying the recessive *wx* gene appear reddish-brown, whereas pollen grains carrying the dominant *Wx* (non-waxy) allele stain dark blue to black.

The waxy corn breeding program got a substantial boost due to the abovementioned advantages, especially while undertaking the backcross breeding program to selectively transfer the *wx* allele into different genetic background for rapid and large-scale conversion of commodity corn into waxy corn. In general, the breeding methods that are used for development of commodity corn are being used for waxy corn breeding as well. However, the extensive population improvement and dedicated pedigree breeding program are not common unlike in commodity corn breeding. Nonetheless, backcross breeding has played a major role in boosting waxy corn breeding.

In India, the dedicated waxy corn breeding is being undertaken at different research institutes under the aegis of the Indian Council of Agricultural research (ICAR). For example, Indian Agricultural Research Institute (IARI) has developed and released the first waxy corn, Pusa Waxy Corn Hybrid 1 (APWH 4), by rapid conversion of parental lines, HKI1105 and HKI323 of HM 4 hybrid through marker-assisted backcross breeding (MABB). Similarly, Indian Institute of Maize Research has also converted two female parental lines, UMI1200, UMI1201, and a common male parent UMI1230 into their waxy versions through MABB. The parental lines are the parents of maize hybrids CMH-08-282 (CO-6) and CMH-08-292 (CoH(M)-8), which were developed at Tamil Nadu Agricultural University (TNAU), Coimbatore. Presently, large-scale conversion of elite maize genotypes into their waxy version is underway in India as part of germplasm diversification and rapid improvement of maize germplasm for waxy trait.

8.7 Amylomaize

The amylose and amylopectin in maize kernel starch are present in 1:3 ratios (20–25:75–80% of starch) by weight. The amylose content in the starch is known to enhance the nutritional quality of starch, determined by glycemic index (GI). The GI is a measure of digestibility rate of starch and absorption of glucose, released after starch digestion into the blood. Based on the digestibility, assessed in vitro, three components can be identified, namely, rapidly digestible (RDS), slowly digestible (SDS), and resistant starch (RS). The amylose content of corn starch represents the part of SDS and/or RS. Therefore, high-amylose corn starches or amylomaize starch can be used as a source of SDS or RS (ungelatinized resistant granules).

Detailed surveys of maize germplasm have shown that amylose content in normal maize generally ranges from 22% to 28%, although some exotic genotypes may contain up to 36% amylose. However, the discovery of novel endosperm mutants through visual identification of distinct kernel morphology during maize breeding provided new opportunities for altering amylose content in corn starch. Kernels exhibiting tarnished endosperm and varying degrees of pericarp wrinkling were first identified visually during selection. Subsequent biochemical analyses revealed that these kernels contained significantly higher amylose levels compared to normal commodity maize. This discovery led to the classification of such mutants as high-amylose corn, and the term "Amylomaize" was later introduced to describe maize genotypes possessing elevated amylose content in kernel starch. Another major breakthrough was the successful isolation of amylose and amylopectin in pure forms, which further stimulated interest in the breeding and development of high-amylose maize. Owing to the unique functional properties of amylose in food processing and various industrial applications, high-amylose corn has gained importance as a specialty corn type.

Similar to waxy corn, amylomaize or high amylose content is determined by another endosperm mutant gene, amylose extender (*ae*) involved in starch biosynthesis. The wild-type allele *Ae* codes for starch branching enzymes. In contrast to waxy corn, amylomaize or high-amylose corn does not completely replace amylopectin with amylose in the kernel starch. Instead, the amylose content increases substantially from the normal level of 25–30% in commodity maize to about 55–70% in amylomaize. In certain exceptional cases, the amylose content may reach as high as 80–90% due to the combined effects of double or triple endosperm mutants; however, such genotypes generally possess substantially lower total starch content. Detailed information on the mutant allele and its effect on the starch component, followed by development of analytical method for isolating pure form of amylose, together led to increased interest and efforts toward the development of amylomaize. Thus, breeding efforts have started to develop amylomaize with desirable agronomic performance as compared to commodity maize in both the private and public sectors.

8.7.1 Breeding Maize for High Amylose Content

The discovery of the *ae* mutant allele greatly expanded the possibilities for developing high-amylose maize. Since a single recessive ae allele is sufficient to substantially increase amylose content, transferring this allele through backcross breeding became a simple, rapid, and effective approach for developing high-amylose genotypes. However, high-amylose maize kernels often exhibit undesirable morphological features such as tarnished endosperm and varying degrees of pericarp wrinkling. Therefore, careful selection during breeding is essential to maintain the desirable kernel appearance of commodity maize while simultaneously increasing amylose content. Biochemical analyses of kernels carrying the *ae* gene have shown considerable variation in amylose content, generally ranging from 55% to 65%. This variation is largely attributed to the genetic background of the genotype carrying the ae allele. The influence of genetic background is likely due to the action of modifier genes that interact, either epistatically or otherwise, with the major recessive *ae* gene. Accurate estimation of amylose content is also critical for effective selection in high-amylose maize breeding programs. However, analytical estimation of amylose often involves substantial experimental error depending on the method used, laboratory equipment, technical expertise, and sampling procedures. Such errors can significantly affect the efficiency of selection. In addition, several physiological and environmental factors influence amylose accumulation, including genotype, environmental conditions, kernel developmental stage, kernel position on the ear, genotype × environment interactions (GEI), and variation in starch granule size. Collectively, these factors make breeding for high-amylose maize considerably more challenging than breeding for waxy maize.

The primary objective of maize breeding is to develop high-yielding and superior genotypes possessing desirable agronomic traits along with wide adaptability across diverse agro-climatic conditions. Since high amylose content is primarily governed by a single recessive *ae* gene, transferring this gene into elite genetic backgrounds with proven agronomic performance is considered the most effective breeding strategy. Most successful examples of high-amylose maize breeding have been reported from the USA. This is largely because breeding for high-amylose maize requires sophisticated infrastructure, particularly for the chemical analysis of starch composition and amylose content. In addition, the recessive nature of the *ae* allele necessitates strict maintenance of genetic purity through isolated seed production and segregated marketing systems to prevent contamination from normal maize. Large-scale isolated cultivation areas are therefore essential for commercial amylomaize production. Organized breeding efforts for the development of amylomaize began in the early 1950s following the discovery of the *ae* mutant in 1952.

The Bear Hybrid Corn Company initiated large-scale introgression of the *ae* allele into agronomically superior maize genotypes with support from the National Starch and Chemical Company. Additional support was provided by the Northern Utilization Research and Development Division Laboratory of the U.S. Department of Agriculture at Peoria, Illinois, which conducted chemical analyses of starch to estimate amylose content in the breeding populations used for amylomaize improvement. As breeding efforts progressed, the utilization of amylomaize in wet milling and starch industries increased significantly. Breeding programs adopted systematic backcross breeding, the use of diverse germplasm, and intensive selection in segregating populations to identify desirable combinations of high amylose content and acceptable kernel phenotype. Special emphasis was placed on selecting recurrent parents with favorable modifier genes influencing amylose accumulation and kernel characteristics. With the development of genetically stable germplasm, breeding strategies such as pedigree breeding, recurrent selection, and population improvement were also incorporated into amylomaize improvement programs. These efforts led to the development of different classes of amylomaize, including Class V and Class VI, containing approximately 50–60% and 60–70% amylose, respectively. Current breeding efforts are focused on developing higher classes (Class VII–IX) through the application of advanced breeding tools and techniques. However, the development of such ultra-high-amylose maize requires long-term breeding programs involving large population sizes and intensive selection. In India, systematic marker-assisted backcross breeding (MABB) programs for amylomaize development are currently underway, and several promising lines have already reached advanced stages of multi-location evaluation under the All India Coordinated Research Project (AICRP) on Maize (Talukder et al. 2022). The amylomaize version of HQPM 5 is presently in the final year of testing under AICRP during kharif 2025. In this program, the parental lines HKI163 and HKI161 were introgressed with the *ae1* gene through MABB. Particularly, the amylomaize versions of these lines already possessed three important genes: opaque2 for quality protein maize (QPM), and crtRB1 and lcyE for enhanced provitamin A content. Additionally, three other inbred lines—HKI1344, HKI1348-6-2, and HKI1378—have also been converted into amylomaize versions. These lines are the parental components of the previously released commercial hybrids HM-5 and HM-12 (HKH-313). The commercial cultivation of amylomaize requires special care, as a yield penalty of about 20–30% is generally reported. Therefore, contract agreement before undertaking commercial cultivation is crucial, which ensures purchase of amylomaize by paying a premium price to compensate the yield penalty. In the USA, the contract farming of amylomaize is common, but the countries like India need to develop mechanisms to make commercial cultivation of amylomaize practically successful.

8.8 High-Oil Corn

More than 60% of the total maize produced worldwide is utilized as feed for poultry, livestock, swine, and other animals. To prepare nutritionally balanced feed, vegetable oils and animal fats are commonly added as ingredients. In 1908, Smith suggested that increasing the oil content of maize kernels could substantially reduce the need for added animal fats in feed formulations. Although efforts to improve kernel oil content through conventional breeding began in the USA during the 1890s, the commercial cultivation of high-oil maize hybrids became practical only in the late 1980s. The primary objective of developing high-oil maize was to improve the nutritional quality and energy value of animal feed. As a result, the quantity of feed required can be reduced by approximately 5–6%, thereby lowering overall production costs.

8.8.1 Significance of Corn Oil

Oil is considered a high-energy-density biochemical macromolecule (Lambert 2001). Corn oil is regarded as a high-quality edible oil because it contains a high proportion of unsaturated fatty acids, particularly polyunsaturated fatty acids (PUFAs) such as linoleic acid (~50%) and linolenic acid (~1%), along with monounsaturated fatty acids (MUFAs) such as oleic acid (~40%) (Olmos et al. 2018). In addition, corn oil also contains saturated fatty acids, mainly palmitic acid (~10–12%) and stearic acid (~2%). The quality of corn oil is determined not only by the ratio of saturated to unsaturated fatty acids but also by the triglyceride composition associated with these fatty acids. Due to its desirable nutritional and functional properties, corn oil is widely used in cooking and in the preparation of various processed food products. India is currently a major importer of edible oils, and therefore increasing the oil content in maize could contribute to domestic oil production, particularly through starch-processing industries. Studies have demonstrated that replacing normal field corn (3.5–4.5% oil) with high-oil corn (6–13% oil) in poultry feed significantly improves feed efficiency and weight gain. Experimental results have shown superior feed conversion efficiency (FCE) with high-oil corn (2.58) compared to normal field corn (2.75), as lower FCE is more acceptable.

8.8.1.1 Constraints Associated with High-Oil Corn

A strong xenia effect has been reported to influence kernel oil content in maize. Therefore, commercial cultivation of high-oil corn requires proper isolation from low-oil maize genotypes to maintain genetic purity and desired oil levels. Another major limitation of high-oil maize is its comparatively lower grain yield than normal maize genotypes. High-oil corn also requires relatively higher temperatures during the reproductive and harvesting stages to promote rapid kernel drying, as these genotypes tend to dry more slowly after attaining physiological maturity. In addition, since kernel oil content is an invisible trait, efficient marketing and quality assessment require non-destructive and high-throughput methods for oil estimation. Despite these challenges, substantial efforts have been made worldwide, including in India, to develop high-oil maize genotypes through conventional breeding approaches.

8.8.1.2 Breeding Efforts to Enhance Corn Oil Content

Evaluation of a large number of maize germplasm lines has revealed substantial variation in kernel oil content. Increasing oil concentration in maize kernels was recognized as an important value-added trait because of its nutritional benefits in poultry, swine, and livestock feed. Consequently, breeding efforts were initiated to enhance kernel oil content through systematic selection.

Early breeding programs primarily adopted divergent selection in both directions, aiming to develop genotypes with either high or low oil content. Several public and private sector breeding programs focused on the development of high-oil maize germplasm. During the initial phases (1940s–1970s), backcross breeding was the most widely used breeding approach. As discussed earlier, the Illinois breeding program developed several Illinois High Oil (IHO) lines through long-term divergent selection. These IHO lines were later used as donor parents in other breeding programs to improve the oil content of elite hybrid parental lines through backcross breeding, using elite inbred lines as recurrent parents. Initially, parental lines of well-known double-cross hybrids, such as U.S. 13, were commonly used as recurrent parents for developing high-oil hybrids.

One of the major drawbacks of high-oil maize was its significant yield penalty compared to normal maize hybrids. To overcome this limitation, breeding strategies gradually shifted from merely converting existing elite inbred lines into high-oil versions toward broadening the genetic base through the development of high-oil synthetic populations. As a result, several high-oil synthetics such as R Alexho Elite (M), RDO (S1), R Iowa 2-ear (M), and RHORYD (M) were developed. The objective behind this strategy was to derive new high-oil inbred lines from these synthetics through inbreeding and subsequently use them in the development of improved high-oil hybrids, rather than simply modifying existing hybrids.

However, hybrids developed from high-oil synthetic-derived inbred lines still failed to compete with normal maize hybrids in grain yield. On average, high-oil hybrids exhibited approximately 7% lower grain yield and 4–5% higher kernel moisture content at harvest than normal hybrids.

To overcome the yield limitation, a novel breeding strategy known as the TopCross® method was developed, which significantly improved the commercial viability of high-oil maize hybrids. In this approach, seeds of cytoplasmic male sterile (CMS) single-cross hybrids with normal oil content were mixed with fertile high-oil pollinator seeds. During commercial cultivation, the fertile high-oil plants served as pollen donors, resulting in normal kernel development while simultaneously increasing oil content through the xenia effect. Thus, the TopCross® method effectively exploited the xenia effect, which had previously been considered a major constraint in high-oil maize production.

At present, the TopCross® method is widely used in the USA for the commercial cultivation of high-oil maize hybrids. In India, as well, breeding efforts for the development of high-oil maize hybrids have recently begun, although the programs are still at an early stage.

8.9 Quality Protein Maize (QPM)

In India, nearly 47–50% of the total maize production is utilized as poultry feed, while 13–15% is used as animal feed and about 10–12% is consumed as human food. Therefore, biofortification in maize is considered a vital intervention for improving the nutritional quality of maize kernels. In general, field corn kernels lack essential amino acids such as lysine and tryptophan, resulting in lower biological value. The discovery of opaque mutants in maize, followed by biochemical profiling of their kernels, revealed substantial alterations in amino acid composition. Subsequently, the development of hard endosperm opaque mutant genotypes led to the development of quality protein maize (QPM). The lysine and tryptophan content in QPM is nearly double that of conventional field corn. As many countries across the world continue to face malnutrition-related problems, QPM is considered a viable option for enhancing nutritional security compared to normal field corn. The biological value of QPM is comparable to that of milk. Moreover, the use of QPM in the poultry industry can significantly reduce production costs, as it improves the feed-to-biomass conversion ratio compared to conventional maize. Consequently, the overall feed requirement is reduced.

8.9.1 QPM Breeding

The growing significance of quality protein maize (QPM) due to its enhanced nutritional value has drawn the attention of maize breeders toward its improvement. Several countries across the world, including India, initiated focused research programs on the development of QPM hybrids several decades ago (Hossain et al. 2022). In India, sustained efforts in this direction began as early as the 1970s. The most notable contribution came from the International Maize and Wheat Improvement Center (CIMMYT), Mexico, which shared large-scale QPM germplasm with several Asian, Latin American, and African countries. As a result, several QPM hybrids were developed and released across different countries and continents. Furthermore, the development of QPM hybrids was accelerated through the application of marker-assisted selection (MAS). Among the various MAS approaches, marker-assisted backcross breeding (MABB) became the most widely adopted method in India. The application of MABB in India led to the development of the first MAS-derived QPM hybrid, Vivek QPM Hybrid 9 (Gupta et al. 2013). At present, several QPM hybrids have been developed and released in India for diverse agro-climatic conditions. These include HQPM 1, HQPM 5, HQPM 7, HQPM 29, Shaktiman 1, Shaktiman 2, Shaktiman 3, Shaktiman 4, Shaktiman 5, LQMH 1, IQMH 202, IQMH 203, VLQPM Hybrid 45 (FQH 165), VLQPM Hybrid 61, VLQPM Hybrid 63, and VLQPM Hybrid 59 (FQH 106). In addition, several improved versions of HQPM 1, HQPM 5, and Vivek QPM Hybrid 9, enriched with additional nutritional traits such as provitamin A, vitamin E, and low phytate content through gene pyramiding, have also been developed and released.

8.10 Conclusion

Specialty corns, characterized by their enhanced functional, nutritional, and industrial value compared to conventional maize, have significantly increased the importance of the maize crop. Their unique traits contribute to improved product quality and expanded end-use applications. However, the development, cultivation, production, and distribution of specialty corn varieties continue to face several challenges. Participatory breeding approaches provide an effective strategy to address these constraints by actively involving farmers and industry stakeholders in the breeding and dissemination process. Such approaches enhance awareness regarding the availability, nutritional benefits, and specific production requirements of specialty corns, thereby promoting their adoption. Furthermore, the continuous development of value-added corn products and varieties with improved and diverse functional properties will support the sustained growth of the specialty corn market, ensuring its continued and important role in agricultural development.

References

Bergquist RR, Nubel DS, Thompson DL (1998) U.S. Patent No. 5,704,160. U.S. Patent and Trademark Office, Washington, DC

Coe EH, Neuffer G, Hoisington DA (1988) The genetics of corn. In: Sprague GF, Dudley JW (eds) Corn and corn improvement. American Society of Agronomy, Madison, pp 81–258

Das S, Ghosh G, Kaleem MD, Bahadur V (2008) Effect of different levels of nitrogen and crop geometry on the growth, yield and quality of baby corn (*Zea mays* L.) cv. "golden baby". In: International symposium on the socio-economic impact of modern vegetable production technology in tropical Asia, vol 809. ISHS, pp 161–166

Gupta HS, Raman B, Agrawal PK, Mahajan V, Hossain F, Thirunavukkarasu N (2013) Accelerated development of quality protein maize hybrid through marker-assisted introgression of opaque-2 allele. Plant Breed 132(1):77–82

Hallauer AR (2000) Specialty corns. CRC Press

Hossain F, Zunjare RU, Muthusamy V, Bhat JS, Mehta BK, Sharma D, Talukder ZA, Chhabra R, Katral A, Dutta S, Chand G (2022) Biofortification of maize for nutritional security. In: Biofortification of staple crops. Springer Singapore, Singapore, pp 147–174

Klosgen RB, Gierl A, Schwarz-Sommer Z, Saedler H (1986) Molecular analysis of the waxy locus of *Zea mays*. Mol Gen Genet 203(2):237–244

Kumar P, Choudhary M, Hossain F, Singh NK, Choudhary P, Gupta M, Singh V, Chikappa GK, Kumar R, Kumar B, Jat SL, Rakshit S (2019) Nutritional quality improvement in maize (*Zea mays*): progress and challenges. Indian J Agric Sci 89:895–911. https://doi.org/10.56093/ijas.v89i6.90756

Lambert RJ (2001) High-oil corn hybrids. In: Hallau AR (ed) Special corn. CRC Press, Boca Raton, pp 131–153

Laurie CC, Chasalow SD, LeDeaux JR, McCarroll R, Bush D, Hauge B, Lai C, Clark D, Rocheford TR, Dudley JW (2004) The genetic architecture of response to long-term artificial selection for oil concentration in the maize kernel. Genetics 168(4):2141–2155

Mertz ET, Bates LS, Nelson OE (1964) Mutant gene that changes protein composition and increases lysine content of maize endosperm. Science 145(3629):279–280

Olmos SE, Trejo JM, Percibaldi NM, Eyhérabide GH (2018) Genetic diversity near the DGAT1-2 gene for high oleic acid content and kernel trait variation in a maize breeding collection. Mol Breed 38(8):98

Silva WD, Vidal BC, Martins MEQ, Vargas H, Pereira AC, Zerbetto M, Miranda LC (1993) What makes popcorn pop. Nature 362(6419):417–417

Singh SP, Neupane MP, Sravan US, Kumar S, Yadav T, Choudhary SK (2019) Nitrogen management in baby corn: a review. Curr Appl Sci Technol 34(5):1–11

Talukder ZA, Muthusamy V, Chhabra R, Gain N, Reddappa SB, Mishra SJ, Kasana R, Bhatt V, Chand G, Katral A, Hossain F (2022) Combining higher accumulation of amylopectin, lysine and tryptophan in maize hybrids through genomics-assisted stacking of waxy1 and opaque2 genes. Sci Rep 12(1):706

Yodpet C (1979) Studies on sweet corn as potential young cob corn (*Zea mays* L.). PhD thesis, University of the Philippines, Philippines

Maize Nutritional Enhancement Through Biochemical Interventions

9

Bharat Bhushan, Jomika Devi, Veena Devi, Pardeep Kumar, Arpit Chaudhary, and Abhijit Kumar Das

Abstract

Maize is a vital crop in global agriculture, providing sustenance to billions as a primary food, feed, and industrial resource. However, its inherent nutritional limitations, including deficiencies in essential amino acids like lysine and tryptophan and micronutrients such as iron, zinc, and vitamin A, contribute to widespread malnutrition. In regions where maize constitutes the primary dietary component, such deficiencies exacerbate issues like protein-energy malnutrition and "hidden hunger," impacting health, productivity, and economic stability. This chapter examines the trans-formative role of biochemical interventions in enhancing the nutritional value of maize. Conventional breeding approaches, such as the development of Quality Protein Maize (QPM), have successfully improved amino acid profiles. Advanced metabolic engineering strategies are also explored, with examples including β-carotene-enriched "Golden Maize" for combating vitamin A deficiency and biofortified varieties addressing iron and zinc bioavailability. Agronomical management practices, such as fertigation or soil treatment and food fortification, are discussed as complementary methods to boost maize's nutritional and yield capabilities. Omics technologies, encompassing genomics, proteomics, and metabolomics, have revolutionized precision interventions by identifying key genetic controls and metabolic pathways for nutrient enhancement. The chapter also addresses challenges such as regulatory hurdles, public skepticism, environmental sustainability, and socioeconomic accessibility, which impede the widespread adoption of biofortified maize. Future directions highlight the integration of multi-omics technologies, climate-resilient biofortification, and precision agriculture to create nutrient-dense maize varieties tailored to diverse global needs. By leveraging science-driven innova-

B. Bhushan (✉) · J. Devi · V. Devi · P. Kumar · A. Chaudhary · A. K. Das
ICAR-Indian Institute of Maize Research, Ladhowal, Ludhiana, Punjab, India
e-mail: bharat.bhushan@icar.org.in

M. Choudhary et al. (eds.), *Next-Generation Maize Research*,
https://doi.org/10.1007/978-981-92-0722-0_9

tions, maize can be elevated from a staple crop to a nutritional solution, playing a pivotal role in reducing malnutrition and advancing global food security.

Keywords

Maize nutrition · Biofortification · Protein-energy malnutrition · Metabolic engineering · Omics technologies · Global food security.

9.1 Introduction

Maize (*Zea mays* L.) is one of the most important crops in global agriculture, serving as a critical source of food, feed, and industrial raw material (Revilla et al. 2022). It is a staple food for billions of people, particularly in regions like sub-Saharan Africa, Latin America, and parts of Asia, where it provides a significant portion of the daily caloric intake (Erenstein et al. 2022). In these regions, maize is consumed in various forms such as porridge, tortillas, and fermented dishes, making it an essential component of the diet. In sub-Saharan Africa, for example, maize constitutes up to 30–50% of daily calories in many countries. The global significance of maize is further underscored by its role as a crucial ingredient in the livestock industry. It is one of the most widely used feed grains, providing a high-energy source for poultry, swine, cattle, and other livestock. In the United States, a large portion of the maize harvest is used for animal feed, supporting the country's meat, dairy, and poultry industries. Similarly, in China, maize is primarily used to feed pigs and poultry, driving the nation's vast meat production sector. Its high starch content makes it an ideal feed grain, and its versatility ensures it can be incorporated into a variety of animal feed formulations, contributing to the global supply of affordable animal-based protein (Loy and Lundy 2019).

Economically, maize is a lifeline for millions of farmers worldwide, particularly in developing countries. It serves as both a subsistence crop and a cash crop, enabling smallholder farmers to generate income and improve their livelihoods. In industrialized nations, large-scale maize farming contributes significantly to national economies. The United States is the largest producer of maize globally and a major exporter, with maize trade ensuring food security in regions that rely on imports for both human consumption and animal feed. Maize's adaptability to diverse agro-climatic conditions is one of the key reasons for its widespread cultivation, which is impossible for any other crop. It is cultivated across diverse environments, ranging from 58° N to 40° S, below sea level to altitudes above 3000 masl, with annual rainfall between 250 mm and 5000 mm, and a growing cycle of 3–13 months (Murdia et al. 2016). Moreover, the development of climate-resilient maize varieties has made it standalone to the impacts of climate variability, which is crucial for smallholder farmers in regions affected by adverse abiotic environments (Martey et al. 2020). The crop's ability to grow in a variety of soil types, from sandy to clay-rich, further enhances its suitability for global cultivation. In addition to its role as a food and feed crop, maize is an important raw material for various industrial applications. Maize is a significant source of starch, which is used in food processing, textiles, and pharmaceuticals, demonstrating its versatile economic importance. Maize starch-derived ethanol is a key component of the biofuel industry in the United States, providing a renewable energy source that helps reduce reliance on fossil fuels (Choudhary et al. 2020). Even the by-products of maize, like silk and shank, are getting their awards in the field of health and environment, respectively (Colbert 2023).

Despite its global prevalence and significant role in food security and industrial applications, maize often falls short nutritionally, creating a paradox where its ubiquity does not equate to its ability to sustain adequate health (Kumar et al. 2019, 2024). Maize is rich in carbohydrates but deficient in several essential nutrients, including some essential amino acids, vitamins, and minerals (Palacios-Rojas et al. 2020). This nutritional gap poses a significant challenge, particularly in regions where maize is the primary or sole dietary staple. The pressing need to improve the nutritional value of maize has led to increased focus on biochemical and genetic interventions. Biochemical interventions have solved the majority of mysteries associated with the seed deficit of a particular nutrient. Advances in genetics and biotechnology have opened new methodologies for enhancing the crop's nutritional profile. Biofortification strategies, such as the development of QPM and genetically modified (GM) maize varieties with higher levels of essential nutrients like provitamin A, iron, and zinc, are already being implemented. Additionally, the agronomic innovations like the application of trace elements hold promise in boosting nutrient availability and enhancing maize's resilience. The supplementation of nutrients to the wholesome maize meal is the easiest method, and it also serves its purpose to tackle the deficiency. This chapter delves into these multifaceted strategies, exploring the challenges, recent advancements, and the transformative potential of biochemical tools to bridge the gap between food security and nutritional adequacy and offering a comprehensive, modern road map for developing nutrient-rich, climate-resilient maize.

9.2 The Necessity for Nutritional Improvement

Despite its ubiquity and versatility, maize exhibits inherent nutritional limitations that pose significant challenges to the health and well-being of maize-dependent populations. This results in a paradox where maize, despite being a staple energy crop, contributes to various forms of malnutrition in some regions (Choudhary et al. 2021).

9.2.1 Current Nutritional Profile of Maize and Its Global Role: A Regional Perspective

Maize is an energy crop, rich in starch and proteins. However, its overall nutritional profile is marked by key deficiencies. Maize is notably low in essential amino acids, such as lysine and tryptophan, which are vital for the growth, development, and overall health of consumers. According to the WHO (2007), the daily requirements of lysine and tryptophan are 35 and 4.8 mg/kg body weight for children. These deficiencies were putting populations at risk of protein-energy malnutrition (PEM), particularly in children and pregnant women (Hossain et al. 2022). Additionally, maize lacks critical micronutrients like iron, zinc, and vitamin A (Goredema-Matongera et al. 2021). These deficiencies, collectively known as "hidden hunger," cause widespread health issues, including anemia due to iron deficiency, impaired immune function and growth from zinc deficiency, and impaired retinal functions caused by vitamin A deficiency (Mudadirwa et al. 2023). Table 9.1 depicts the comparative nutritional profile of traditional and biofortified maize in terms of biochemical constituents.

In Latin America, particularly in Mexico and Central America, maize holds cultural and nutritional significance, forming the base of traditional diets (Tanumihardjo et al. 2020). While *milpa* farming systems help to offset some nutritional gaps, populations in these areas remain at risk of iron, zinc, and vitamin A deficiencies (Mutungi et al. 2022). In parts of Asia, especially China and India, maize consumption is growing, though it is not traditionally a staple (García-Lara and Serna-Saldivar 2019). Over-reliance on maize as a staple exacerbates issues of protein-energy malnutrition (PEM) and micronutrient deficiencies.

9.2.2 Impact of Nutritional Deficiencies

The impact of nutritional deficiencies in maize-based diets can extend from immediate to chronic and on the individual to community health and economies. Malnutrition reduces labor productivity (Choge 2020). This perpetuates cycles of poverty, as malnourished individuals are less able to contribute to the economic development of their communities. Furthermore, women and children are disproportionately affected by nutritional deficiencies. In many societies, women bear the primary responsibility for food preparation and childcare, making them particularly vulnerable to the effects of poor nutrition. Children who already suffer from birth defects and developmental delays have lifelong challenges for their earning potential and societal equality.

Table 9.1 This table highlights the nutritional profile of traditional vs. biofortified maize

Nutritional components	Traditional maize	Quality protein maize (QPM)	Golden maize	Biofortified maize
Lysine (%)	Low	4.0	–	–
Tryptophan (%)	Low	0.8	–	–
Methionine (%)	Low	2.0		
Vitamin A (μg/g)	None	–	15 ppm	–
α/γ-tocopherol combined	Low			45 ppm
Iron (mg/kg)	Low	–	–	22 ppm
Zinc (mg/kg)	Low	–	–	37 ppm

9.3 Biofortification Strategies for Improving Nutritional Value

Biofortification has emerged as a sustainable solution to address malnutrition and "hidden hunger" in resource-limited settings. The most identified biofortification strategies are broadly categorized into (1) conventional and molecular breeding, (2) metabolic engineering, and (3) external agronomical treatments (Fig. 9.1).

9.3.1 Biofortification Through Conventional Breeding

It is most practical, acceptable, and endogenous fortification where existing nutrient content can be enhanced through a biological process without disturbing other biochemical metabolites or without introducing foreign genes (Garg et al. 2018). This process relies on the natural genetic variations within maize populations to develop nutrient-enriched varieties. By selecting inbreds and crossbreeding parent lines with desirable traits, breeders have enhanced the levels of key micronutrients such as provitamin A, iron, and zinc, as well as essential amino acids like lysine and tryptophan (Sheoran et al. 2022). One of the most notable successes in maize biofortification is the development of QPM. The discovery of opaque mutants in maize was the first lead in the process. The double-recessive *opaque-2* (*o2o2*) gene of mutants decreases zein-like protein bodies and enhances key amino acids. These were equivalent to *floury-2* mutants, but the problem of grain chalkiness disappointed the engaged scientists, Dr. SK Vasal and Dr. Evangelina Villegas. Their gradual efforts in making a favorable combination of modifier alleles for the QPM development addressed both the deficiency of amino acids as well as kernel hardness (Jilo 2022). Importantly, this was achieved without compromising yield or the agronomic performance of the crop. Over the years, QPM has been successfully cultivated in heavily maize-dependent countries like Ethiopia, Zambia, and Guatemala, where its introduction has significantly enhanced child nutrition (Hossain et al. 2019; Galani et al. 2022).

Traditional maize varieties contain phytic acid, a sugar-based compound present in the germ that binds to iron and zinc, rendering them poorly bioavailable (Kumar

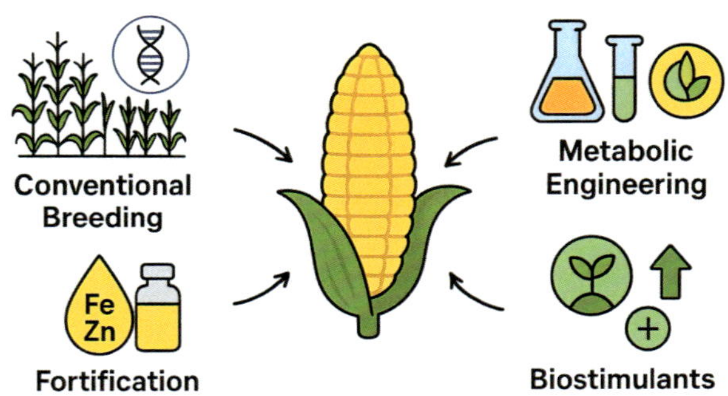

Fig. 9.1 Broad categorization of strategies for enhancing maize nutritional quality

et al. 2021). The physical removal of germ can increase the availability of iron and zinc from endosperm and aleurone significantly. Through breeding, low phytate maize hybrids have been developed with the help of low phytic acid (*lpa*) mutants. There were certain perturbations in the central metabolism of plants, but these were within the natural variations of parent plants. The *lpa1* mutants belong to the phytic acid transporter family, where some mutants like *lpa1-241* have seed viability and reduced yield issues, but with the hard work of breeders, 1 out of 10 high-yielding *lpa* lines can be achieved. The *lpa1-1* dent corn hybrids have 66% reduced phytic acid, which ultimately helps in increasing the bioavailability of calcium in humans who preferentially go for their tortillas. Similarly, the iron bioavailability was also enhanced by 49% (Mendoza et al. 1998; Raboy et al. 2000), and the fractional zinc absorption in *lpa* grain has been doubled as compared to normal grain (Hambidge et al. 2004). The inclusion of *lpa* seeds in the diet helped the pigs gain lean body mass and a healthy lipid profile (Stilborn et al. 2002). Similarly, various QPM, Vit A, and Vit E-rich hybrids were developed using molecular breeding.

One of the primary advantages of conventional and molecular breeding for biofortification is its compatibility with existing diverse agricultural landscapes and systems. Farmers can easily adopt nutritionally enhanced maize varieties without significant changes in cultivation practices. Furthermore, partnerships among research institutions, governments, and non-governmental organizations have facilitated the dissemination of biofortified maize to niche or catchment regions. Additionally, it circumvents much of the public resistance and regulatory hurdles associated with transgenic crops (Marques et al. 2021). Despite its many advantages, the success of conventional breeding for biofortification relies heavily on the availability and preservation of a broad genetic base or diversity. The loss of genetic diversity due to monocropping and habitat destruction poses a significant threat to the long-term success of breeding programs. Maize landraces, wild relatives, and introductions are invaluable genetic reservoirs. The physical parameters like size and hardness are impaired under geo-spatial introductions and must be addressed through classical genetics. The exotic germplasms must have both harboring traits and environmental adaptability. Additionally, awareness campaigns among farming communities are essential to promote the cultivation of biofortified maize and ensure its acceptance by the consumers.

9.3.2 Metabolic Engineering

Traditional breeding methods have limitations in introducing novel traits or significantly boosting certain nutrients. Metabolic engineering leverages advanced molecular tools for modifying precise parts of complex metabolic networks to enhance the ongoing production of metabolites, e.g. bio-fortification or introduce novel traits, e.g. enrichment (Singh et al. 2021; Das 2024). For example, transgenic maize with transgenes *Crbkt* and *BrCrtZ* yielded 16.77 mg/kg dry weight astaxanthin inside its endosperm after being transformed using a particle delivery technique (Farré et al. 2016). The key regulatory enzyme in oleic acid production is diacyl glycerol acyl transferase (DGAT). A phenylalanine insertion in DAGT1–2 at position 469 is responsible for the increased oil and oleic acid contents. Ectopic expression of the high-oil DAGT1–2 allele also increases oil and oleic acid contents by up to 41% and 107%, respectively (Zheng et al. 2008). This is the first report of direct metabolic engineering for fatty acid content in maize, a strategy that has been employed in model plants and oil crops extensively (Truksa et al. 2009). In addition, the *FAD2* mutants have been successful in oleic acid (MUFA) production (Beló et al. 2008). The breeding target of >25% oleic acid in total maize oil has been set internationally. It has also been anticipated that overexpression of the tonoplast sugar transporter may enhance sugar transport from leaves to the kernel and finally increase the oil content. Plants usually do not produce VLC-fatty acids like EPA and DHA. Hence, transgenes of elongases and desaturases are required for the production of essential fatty acids. The multigene transformation strategy under the control of seed-specific promoters in maize can be used for enriching these fatty acids in the maize germ. Fatty acid export from plastids to the endoplasmic reticulum can be a good strategy for increasing the density of germ oil. Stability of triacyl glycerol (TAG) is also an issue in plant fatty acid metabolism, so increasing the stability of lipid droplets by olesins may improve oil yield.

Another most celebrated example of metabolic engineering is the development of "Golden Maize, biofortified with β-carotene." This maize is similar to golden rice in terms of biofortification strategy. The phytoene synthase, *CrtB,* and its isoform *psy1* form the ground for enhancing the level of carotenoids in maize endosperm. Further, transposon-based cloning, which identified the gene product, *y1* helped molecular biologists for enhanced production of carotenoids in maize (Aluru et al. 2008). The multiple hydroxylases, e.g. *crtRB* are also playing their role in determining the ratio of provitamin activity to total carotenoid levels. The *CrtB*/*crtRB* functions well only in the presence of *y1,* which is necessary for the development of provitamins (Yan et al. 2010). These bioengineered maize varieties can produce 10–15μg/g of β-carotene, meeting or exceeding the recommended daily allowance for children consuming maize meal as a staple. The lycopene cyclase β and ε control the flow of metabolites for α and β-carotenes. Specifically, LcyE alleles favor the formation of non-provitamin A carotenoids like lutein and zeaxanthin.

The other prominent vitamins present in maize germ are tocopherols and tocotrienols, collectively known as tocols or vitamin E. The landmark achievement for this trait was the identification of the key enzyme *hggt* (homogentisic acid geranylgeranyl transferase) for vitamin E biosynthesis. Its induced expression from barley to maize helped molecular breeders to increase the tocopherol and tocotrienol contents (Cahoon et al. 2003). This shows how redirecting the metabolic flux of isoprenoids can enhance the antioxidant potential of grain.

Another significant focus of metabolic engineering is the biofortification of maize with iron and zinc. The major intervention has been done at the level of transport. The transport of iron in maize plants requires mugineic acid (MA). In the synthesis of MA, the enzymes NAS (nicotianamine synthase) and NAAT (nicotianamine aminotransferase) are the key enzymes whose overexpression can lead to increased grain iron content in cereals (Masuda et al. 2013). Concurrently, the overexpression of genes encoding iron and zinc transporter proteins, e.g. natural resistance and macrophage proteins (NRAMP), ZIP/ZIF (metal-regulated transporters or transporter-like proteins), MTP (metal tolerance proteins), and HMA (heavy metal ATPases) further increases their concentration in the kernels.

Through molecular interventions, scientists have developed crop varieties with minimal antinutrient factors, broadly known as altered products, e.g. maize varieties with reduced phytic acid content. Since phytic acid plays a role in seed germination and antioxidative response, hydrolyzing the excess phytic acid using the controlled phytase activity seems more useful. Overexpression of the phytase gene from *Aspergillus* (*Phya*) using the endosperm-specific promoter of rice *glutelin-1* (*Gt-1*) in combination with increased levels of Fe-binding ferritin protein from soy origin, increased the iron levels up to 40 ppm in the kernel, and three times increased bioavailability of Fe in maize (Drakakaki et al. 2005). A transgenic company in China, namely Origin Agritech, has released the BVLA4 30101 biofortified variety known for its phytate degradation. On the other hand, without disturbing the phytic acid, scientists have exploited various transporter proteins and developed the zinc biofortified maize whose kernel has 34 μg Zn per gram seed as compared to a normal kernel with 21 μg. Despite high phytic acid, the bioavailability of Zn has been doubled. One might wonder about the better effects in combining both, decreasing phytic acid along with enhancing Zn, in the near future. It is noteworthy that there is a positive correlation between Zn and protein content. With this assumption, QPM may acquire higher Zn and lower phytic acid in multi-nutrient breeding programs.

Beyond micronutrient enhancement, metabolic engineering also addresses the amino acid deficiencies inherent in maize. In maize, dihydrodipicolinate synthase (DHDPS), the first rate-limiting enzyme in lysine biosynthesis, has been the target of metabolic engineering. There are two strategies: one is making the enzyme insensitive to feedback inhibition through protein engineering, and the second is the introduction of the exogenous dapA gene coding for this enzyme, but feedback-insensitive (Ufaz and Galili 2008). But expression should be seed-specific, and hence, a seed-specific promoter has been used by the researchers. The regulation of lysine catabolism also plays a key role in overall lysine status. The key enzyme LKR/SDH (Lysine-Ketoglutarate Reductase and Saccharopine Dehydrogenase) is bifunctional and targeted to increase the lysine content in seeds. On this working line, Yang et al. (2020) increased lysine content in corn by modulating both lysine biosynthesis and catabolic pathways (Yang et al. 2020). The third amino acid, which is limiting in maize, is methionine. The biosynthesis of threonine is competitive with that of methionine. Hence, the downregulation of threonine synthase would increase the resource flow towards methionine and increase the overall methionine content. The methionine content has been increased in maize by modifying the cis-acting site for *Dzs10* (Lai and Messing 2002).

By engineering the expression of synthetic storage proteins, such as zeins with altered composition, scientists have created maize varieties with a more balanced amino acid profile (Hasan and Rima 2021). Mondal et al. (2020) exploited the expression of potato sb401, which encodes lysine-rich proteins and enhances lysine levels in transgenic QPM maize seeds (Mondal et al. 2020). These engineered proteins mimic the nutritional composition of animal-based proteins rich in essential amino acids, providing a plant-based alternative. Huang et al. (2006) used antisense dsRNA technology and improved tryptophan along with lysine contents in transgenic maize by reducing both multigene family proteins, e.g. 19- and 22-kD α-zeins (Huang et al. 2006). Khulbe et al. (2020) explored a similar way to modulate the stability of mRNA to enhance methionine content in maize seeds. Innovations like metabolic engineering enable the creation of "designer crops" tailored to specific nutritional needs. The reports on the level and profile of phenolics, terpenoids, and alkaloids are still scarce because their pathways are controlled by multigene families with conserved motifs. However, these innovations also face challenges, including regulatory hurdles, public perception issues, environmental concerns, and equitable access to these technologies in low-resource settings.

9.3.3 Fortification

External interventions like supplementation of minerals to soil and foliar parts serve as a complement to other bio-fortification strategies, offering practical and immediate solutions. The mineral or hormonal treatments work in synergy with the plant's physiological processes to enhance nutrient uptake, assimilation, and overall crop production quality. These interventions can be tailored to specific regional soil and climatic conditions, enabling targeted improvements in maize nutrition and productivity. Food supplements also help in alleviating nutritional disorders or prevention of diseases. For example, Zinc supplements for children and adolescents have helped them to prevent pneumonia and hormonal imbalances, respectively. Although some food fortification has been implemented in the long run to curb the nutritional disorders like goitre, it has some disadvantages like acceptability, price, and overdose issues. The application of plant growth hormones and rhizosphere-promoting cultures of microorganisms also helps in mineral acquisition and photo-assimilate efficiency.

Calcium deficiency in maize often manifests as poor kernel formation, reduced yield, and diminished nutritional quality (Gaj et al. 2018). Calcium fortification is just an example where it has shown remarkable results in terms of kernel development (He et al. 2024). Calcium-enriched soils improve the plant's structural integrity and enhance kernel density indirectly, while foliar sprays ensure a direct and efficient supply of calcium to developing tissues. Application of zinc fertilizers in Zn-rich soil can be helpful in increasing the content of Zn in the kernel, but this strategy may not work well in Zn-deficient soils (Zhang et al. 2013b). Since it is a costly fertilizer, small-scale farmers are reliant on the usage of NPK. The other important mineral application is iron. Although foliar application of iron seems to be a suitable application, application of N, P, and/or K has also been shown to increase the absorption of Fe in the soil. Once the soil has been enriched, then foliar application may enhance the sink strength of this mineral in the kernel. The opposite effect may be observed when the application is done in excess. Selenium (Se) is a very mobile element and enriching Malawi fertilizer with Se, along with the selection of lines with high selenium, will reduce the prevalence of inadequate Se supplies from 82 to 57%, increase the antioxidant potential of grain and the cost per alleviated case of dietary Se deficiency will be meagre, just approximately USD 0.36 per year. An application of 5 g Se/ha to maize crops in Malawi would increase dietary Se intake by 26–37μg Se/person/day (Chilimba et al. 2012).

The benefits of mineral fortification extend beyond improved nutritional profiles. Many of the farm-applied minerals either are part of enzymes or they activate the enzymes related to economically important metabolites like phenolics, flavonoids, isoprenoids, and alkaloids. The mild stresses also have a positive impact on the biosynthesis of flavonoids like anthocyanins. The exploitation of landraces, regulatory metabolic network mechanisms of secondary metabolites, and application of mineral fertilizers altogether can enhance phenolic compounds (ferulic acid) beyond the set targets.

9.4 Advances in Omics Technologies

High-throughput omics technologies, which are the hallmarks of modern biological research, provide an unprecedented level of insight into the genetic, proteomic, and metabolic foundations of bio-sciences. These tools have proven instrumental in the pursuit of nutritional enhancement in maize. By leveraging these holistic approaches, scientists can design precise interventions that are efficient and scalable, driving advancements in maize biofortification.

9.4.1 Genomics

It lies at the core of omics technologies, offering a comprehensive understanding of the maize genome and its applications. Genomics, encompassing sequencing, lays the groundwork by uncovering complete arrays of genes and genetic variations, and providing insights into the plant's inherent potential for trait expression. Transcriptomics builds on genomics by examining gene expression and regulation patterns, offering clues into the active metabolic pathways under specific environmental conditions.

The marker-assisted recurrent selection (MARS) has been used to accelerate the introgression of temperate traits like carotenoids or anthocyanins into locally adapted tropical germplasm. This strategy is helpful in population improvement programs with a goal of a wider genetic base. While this was a classical physical map approach, advanced genome map-based studies have identified the genes alongside the regions that are involved in quality trait description, e.g. modifying protein fractions as well as constituent components like amino acids. Additionally, genomics provided the trait linkage of provitamin A marker with lysine and tryptophan. It helped the breeders through marker-assisted backcrossing (MABC) to convert normal QPM to PVA-rich QPM and vice versa. Now, SNPs have made this process faster and cheaper (Shutu et al. 2012). Early generation screening by using

Table 9.2 Key traits, applied breeding strategy, and quality improvement of maize

Quality parameter	Gene(s)	Approach	References
Protein	*Opaque 2*	MABC (marker-assisted backcrossing)	Babu et al. (2005)
Lysine/ tryptophan	*Opaque 2, Opaque 16*	MABC	Zhang et al. (2013a)
Provitamin A	*Psy1, lcyE, crtRB1*	Development and evaluation of a functional marker	Azmach et al. (2013), Babu et al. (2013)
	crtRB1, lcyE	MABC	Liu et al. (2015), Singh et al. (2021)
Vitamin E	*vte4* (α-tocopherol methyl transferase)	MABC	Li et al. (2012), Das et al. (2021)
Oil content	*DGAT1-2*	MABC	Hao et al. (2014)
		Development and evaluation of a functional marker	Chai et al. (2012)
Carotenoid levels	Multiple genes	Genomic prediction	Owens et al. (2014)
Starch content	Multiple genes	GWAS study	Leng et al. (2019), Hu et al. (2021)
Phytic acid	*lpa* 2-2	Marker-assisted backcrossing	Sureshkumar et al. (2014)
Zinc and iron	*nas 5*	GWAS study	Wu et al. (2021)

the most predictive SNP markers have proven to be an alternative to genotyping. KASP assay is a good example for screening of PVA-rich germplasm. Some major quality traits of maize, along with their improvement strategy, have been presented in Table 9.2.

High-throughput sequencing platforms have made it possible to generate detailed genetic maps, revealing loci associated with desirable traits such as balanced amino acid profiles and reduced anti-nutritional factors like phytic acid. Previously, the clues of metabolic pathways and expression analysis of various genes in maize-related cereals led to the development of transgenic maize. Mutation breeding is another approach that can produce desirable traits, but it also leads to many undesired traits. One of the most transformative tools in genomics is the CRISPR-Cas9 system, a precise genome-editing technology. CRISPR allows researchers to edit specific genes directly for a desired output, say for nutrient enrichment, without introducing foreign DNA (Zhang et al. 2021). For example, targeted edits to genes controlling β-carotene biosynthesis have led to maize varieties capable of addressing vitamin A deficiency (Gautam et al. 2021). The same strategy can be used to restrict the activity of Δ^6-desaturase and divert the flow of maize linolenic acid towards the production of essential fatty acids like EPA/DHA. This approach not only accelerates the development of nutritionally superior maize but also circumvents some of the regulatory and public acceptance challenges associated with transgenic crops (Kumar et al. 2020).

9.4.2 Proteomics

Proteomics allows scientists to identify and quantify the expression of a multitude of proteins associated with nutrient biosynthesis, storage, transport, and degradation under normal or stress conditions. Proteomics studies have identified key regulatory proteins that govern metabolism in crops. For instance, a study by Chen et al. (2015) utilized MALDI-TOF/MS and 2-DE to uncover novel regulators of iron homeostasis in rice. Similarly, maize crops under phosphate-deficient conditions change their protein profile in roots, which are severely affected. Under stress, the prevalence of purine element binding proteins, some cytoskeletal proteins and small GTPases like Ran B1 might be involved in initiating events of signaling to root morphology and physiological movements.

The classical example of QPM maize is also based on proteome studies where replacement of α-zeins, especially 22 kDa, with albumins or globulins, significantly balances the amino acids and retains the kernel hardness (Zhao et al. 2023). Proteomic insights ensure that genetic interventions do not inadvertently disrupt essential metabolic pathways based on enzymes. The distinction of GM maize from non-GM maize can also be done by untargeted proteome analyses. For example, chitinase A was detected in non-GM and ribitol dehydrogenase in GM-maize exclusively. The speculative absence of chitinase A by the insertional event and presence of ribitol dehydrogenase homolog in the wild type may add disturbances to the pool of proteins in single transgene events. The authors further concluded that variability is variety-specific and cannot be attributed to a single event of transgenics. By characterizing the proteomic landscape of maize, researchers can fine-tune biochemical processes to maximize the crop's nutritional potential while maintaining its agronomic performance.

9.4.3 Metabolomics

The holistic study of the entire set of small molecules and metabolites within a biological system in a real-time manner adds another critical dimension to nutritional enhancement efforts. The performance of inbred populations can be assessed through metabolomics of primary carbon and nitrogen metabolism (Mashabela et al. 2022). Basically, metabolomics has integrated the genomic and metabolic network by which one can understand the impact of genetic alteration on phloem sap metabolites (yield), nutritional quality, stress response, and natural metabolic diversity. The robustness of C_4 photosynthesis under different environmental cues can also be understood by metabolic characterization. Several other researchable issues, like the effect of pollutants on metabolome and metabolic mechanisms in specific tissue under its influence, must be studied.

Scientists have used metabolomics tools both in supportive roles or in core roles. The various mutants of glutamine synthetase and their enzymatic non-functionality have elaborated the supportive role of enzymes in ear development and grain filling (Broyart et al. 2010). The core role of metabolomics has shown that different growth zones of leaves have different metabolons and are linked to protein synthesis. The metabolic response in growth zones during carbon depletion was less pronounced and delayed as compared to mature zones of leaves (Czedik-Eysenberg et al. 2016). Through metabolomic profiling, scientists can identify bottlenecks in nutrient biosynthesis pathways, enabling metabolic engineering to enhance the efficiency of these processes (van Der Hooft et al. 2020). Additionally, metabolomics plays a crucial role in assessing the bioavailability of nutrients. During the development of high oil maize lines, Cocuron et al. (2019) applied metabolomics and metabolic flux analysis to optimize fatty acid synthesis in maize kernels, focusing on boosting oil content without disturbances in other pathways or reducing crop yield. The production of TAG in the non-edible part of maize can be possible through an integrated approach of genomics and metabolic engineering. Recently, the oil content in maize leaves has been enhanced to two times by enhanced TAG assembly, efficient packaging of neutral lipids in cytoplasmic droplets, and Wri1, transcription factor usage (Alameldin et al. 2017).

The integration of these omics technologies, i.e., genomics, proteomics, and metabolomics, creates a powerful framework for understanding and improving maize nutrition. Integration of metabolomic data with proteomic data in a time series manner can help us to identify the basis of hybrid vigor. Combining transcriptomic data with metabolic data can help us to distinguish the photosynthetic mechanisms of C_3 and C_4 plants and also the evolutionary history. Chen et al. (2016) demonstrated the power of metabolomics in understanding the genetic regulation of metabolism in rice grain through phenotypic genome-wide association studies (pGWAS), which link genetic variation with observable traits, and metabolic genome-wide association studies (mGWAS), which associate genetic markers with specific metabolite profiles. Moreover, the ability to monitor the effects of interventions across these interconnected layers ensures that improvements are both effective and sustainable.

9.5 Challenges and Future Directions

Despite substantial advancements in maize biofortification, several challenges hinder the widespread adoption and impact of these innovations.

9.5.1 Regulatory and Public Perception Issues

Many countries impose stringent regulatory frameworks on genetically modified organisms (GMOs), slowing the approval process for transgenics even when such varieties demonstrate clear nutritional and agronomic benefits (Caradus 2023). These hurdles are compounded by widespread consumer skepticism, fueled by misinformation regarding health issues and environmentalists' resistance to GMOs (Broad 2023). To overcome these challenges, it is essential to implement robust public awareness campaigns that emphasize the scientific basis, safety, and tangible benefits of biofortified maize. Collaboration with local stakeholders, including farmers, community leaders, and policymakers, can help dispel misconceptions and build trust.

9.5.2 Environmental Sustainability

While biochemical interventions through genetic modifications offer powerful tools for enhancing nutrition, they must align with sustainable agricultural practices to minimize ecological disruption. For example, large-scale deployment of biofortified maize varieties should avoid reliance on environmentally harmful inputs such as excessive fertilizers or pesticides for yield enhancement. Additionally, the integration of biofortified crops into diverse farming systems like *milpa* can mitigate risks associated with monoculture practices. Future efforts in maize biofortification must also address concerns about the potential unintended consequences, such as genetic modifications on non-target organisms or ecosystems. Rigorous environmental impact assessments, coupled with long-term monitoring programs, are essential to ensure that biofortified maize varieties contribute to agricultural sustainability and economic viability, too.

9.5.3 Socioeconomic Accessibility

High costs associated with research, development, and popularization of biofortified seeds often limit their reach to poor farmers and consumers, leaving vulnerable populations devoid of their benefits. Bridging this gap requires strategic partnerships between public and private sector stakeholders to subsidize costs and streamline unbiased distribution channels. Additionally, fostering technology adoption through training programs and participatory front-line demonstrations for marginal farmers ensures that bio-fortified maize varieties meet both local agronomic conditions and nutritional requirements worldwide (Abdul Aziz et al. 2022).

9.6 Future Directions

The future of maize biofortification lies in leveraging cutting-edge technologies and innovative approaches to overcome existing limitations. This combination of omics approaches can uncover novel gene targets and metabolic conversions, offering precise need-based improvements to address futuristic nutritional demands across diverse populations. Data-driven precision farming techniques, such as remote sensing, soil health monitoring, and nutrient mapping, are essential for optimizing resource utilization and sustainable crop potential of bio-fortified maize in various agroecological settings. With climate change posing significant threats to global food security, the development of resilient maize varieties has become imperative.

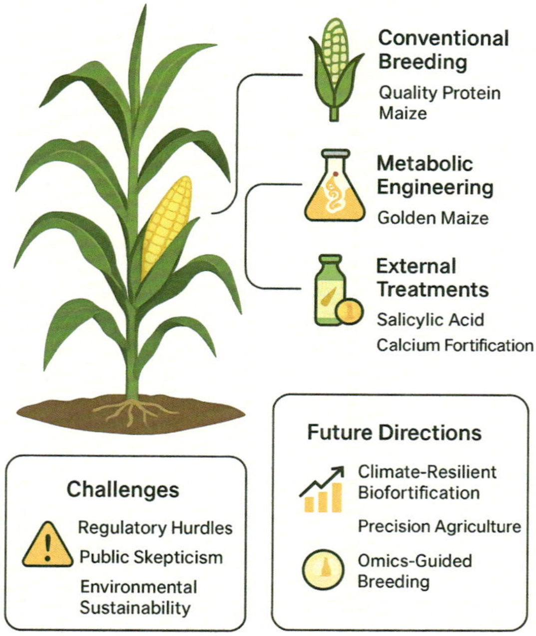

Fig. 9.2 Strategies, current challenges, and future directions for nutritional improvement in maize

These climate-resilient strategies will safeguard the crop's role as a cornerstone of global food systems and a critical tool in combating malnutrition on a global scale (Fig. 9.2).

9.7 Conclusion

The nutritional enhancement of maize through biochemical interventions marks a pivotal step in combating global malnutrition. Advances in genetics, biotechnology, and agronomy have transformed maize into a powerful tool for addressing calorie deficiencies and "hidden hunger," improving health and nutrition on a global scale. Biofortification, metabolic engineering, and multi-omics technologies have significantly elevated maize's nutrient profile, aligning with the Sustainable Development Goals, particularly SDG 2: Zero Hunger. However, regulatory barriers, public skepticism, and socioeconomic disparities hinder widespread adoption. Addressing these issues requires public education, sustainable farming practices, and equitable access facilitated through partnerships among governments, NGOs, and private stakeholders. The benefits of biofortified maize extend beyond nutrition, reducing healthcare costs, boosting productivity, and fostering social progress, particularly for women and children who are most vulnerable to deficiencies. By incorporating climate-resilient traits and farmer-centric approaches, maize can meet the challenges of a changing environment while promoting sustainable development. In summary, nutritionally enhanced maize through biochemical interventions offers a transformative solution to global malnutrition, symbolizing resilience and progress. With continued innovation, collaboration, and investment, maize can become a cornerstone of food security and a pathway to a healthier, more equitable world.

References

Abdul Aziz M, Brini F, Rouached H, Masmoudi K (2022) Genetically engineered crops for sustainably enhanced food production systems. Front Plant Sci 13:1027828

Alameldin H, Izadi-Darbandi A, Smith SA, Balan V, Jones AD, Sticklen M (2017) Production of seed-like storage lipids and increase in oil bodies in corn (Maize; *Zea mays* L.) vegetative biomass. Ind Crop Prod 108:526–534

Aluru M, Xu Y, Guo R, Wang Z, Li S, White W, Wang K, Rodermel S (2008) Generation of transgenic maize with enhanced provitamin A content. J Exp Bot 59(13):3551–3562

Azmach G, Gedil M, Menkir A, Spillane C (2013) Marker-trait association analysis of functional gene markers for provitamin A levels across diverse tropical yellow maize inbred lines. BMC Plant Biol 13:1–16

Babu R, Nair SK, Kumar A, Venkatesh S, Sekhar JC, Singh NN, Srinivasan G, Gupta HS (2005) Two-generation marker-aided backcrossing for rapid conversion of normal maize lines to quality protein maize (QPM). Theor Appl Genet 111:888–897

Babu R, Rojas NP, Gao S, Yan J, Pixley K (2013) Validation of the effects of molecular marker polymorphisms in *LcyE* and *CrtRB1* on provitamin A concentrations for 26 tropical maize populations. Theor Appl Genet 126:389–399

Beló A, Zheng P, Luck S, Shen B, Meyer DJ, Li B, Tingey S, Rafalski A (2008) Whole genome scan detects an allelic variant of *fad2* associated with increased oleic acid levels in maize. Mol Gen Genomics 279:1–10

Broad GM (2023) Improving the agri-food biotechnology conversation: bridging science communication with science and technology studies. Agric Hum Values 40(3):929–938

Broyart C, Fontaine JX, Molinié R, Cailleu D, Tercé-Laforgue T, Dubois F, Hirel B, Mesnard F (2010) Metabolic profiling of maize mutants deficient for two glutamine synthetase isoenzymes using 1H-NMR-based metabolomics. Phytochem Anal 21(2):102–109

Cahoon EB, Hall SE, Ripp KG, Ganzke TS, Hitz WD, Coughlan SJ (2003) Metabolic redesign of vitamin E biosynthesis in plants for tocotrienol production and increased antioxidant content. Nat Biotechnol 21(9):1082–1087

Caradus JR (2023) Processes for regulating genetically modified and gene edited plants. GM Crops Food 14(1):1–41

Chai Y, Hao X, Yang X, Allen WB, Li J, Yan J, Shen B, Li J (2012) Validation of *DGAT1-2* polymorphisms associated with oil content and development of functional markers for molecular breeding of high-oil maize. Mol Breed 29:939–949

Chen L, Ding C, Zhao X, Xu J, Mohammad AA, Wang S, Ding Y (2015) Differential regulation of proteins in rice (*Oryza sativa* L.) under iron deficiency. Plant Cell Rep 34(1):83–96

Chen W, Wang W, Peng M, Gong L, Gao Y, Wan J, Luo J (2016) Comparative and parallel genome-wide association studies for metabolic and agronomic traits in cereals. Nat Commun 7(1):12767

Chilimba AD, Young SD, Black CR, Meacham MC, Lammel J, Broadley MR (2012) Agronomic biofortification of maize with selenium (Se) in Malawi. Field Crop Res 125:118–128

Choge JK (2020) Malnutrition: current challenges and future perspectives. In: Malnutrition, pp 39–63, IntechOpen, 10.5772/intechopen.77861

Choudhary M, Singh A, Gupta M, Rakshit S (2020) Enabling technologies for utilization of maize as a bioenergy feedstock. Biofuels Bioprod Biorefin 14(2):402–416

Choudhary M, Grover K, Singh M (2021) Maize significance in Indian food situation to mitigate malnutrition. Cereal Chem 98(2):212–221

Cocuron JC, Koubaa M, Kimmelfield R, Ross Z, Alonso AP (2019) A combined metabolomics and fluxomics analysis identifies steps limiting oil synthesis in maize embryos. Plant Physiol 181(3):961–975
Colbert TA (2023) Acceptance, attitudes and nutritional content of novel nixtamalized maize products developed for the South African consumer (Doctoral dissertation, University of the Free State)
Czedik-Eysenberg A, Arrivault S, Lohse MA, Feil R, Krohn N, Encke B, Stitt M (2016) The interplay between carbon availability and growth in different zones of the growing maize leaf. Plant Physiol 172(2):943–967
Das A (2024) Metabolic engineering in plants: enhancing crop yield and quality. East Europ J Eng Technol 3(1):32–39
Das AK, Gowda MM, Muthusamy V, Zunjare RU, Chauhan HS, Baveja A, Hossain F (2021) Development of maize hybrids with enhanced vitamin-E, vitamin-A, lysine, and tryptophan through molecular breeding. Front Plant Sci 12:659381
Drakakaki G, Marcel S, Glahn RP, Lund EK, Pariagh S, Fischer R, Stoger E (2005) Endosperm-specific co-expression of recombinant soybean ferritin and *Aspergillus phytase* in maize results in significant increases in the levels of bioavailable iron. Plant Mol Biol 59(6):869–880
Erenstein O, Jaleta M, Sonder K, Mottaleb K, Prasanna BM (2022) Global maize production, consumption and trade: trends and R&D implications. Food Secur 14(5):1295–1319
Farré G, Perez-Fons L, Decourcelle M, Breitenbach J, Hem S, Zhu C, Sandmann G (2016) Metabolic engineering of astaxanthin biosynthesis in maize endosperm and characterization of a prototype high oil hybrid. Transgenic Res 25(4):477–489
Gaj R, Budka A, Górski D, Borowiak K, Wolna-Maruwka A, Bak K (2018) Magnesium and calcium distribution in maize under differentiated doses of mineral fertilization with phosphorus and potassium. J Elem 23(1):137
Galani YJH, Orfila C, Gong YY (2022) A review of micronutrient deficiencies and analysis of maize contribution to nutrient requirements of women and children in Eastern and Southern Africa. Crit Rev Food Sci Nutr 62(6):1568–1591
García-Lara S, Serna-Saldivar SO (2019) Corn history and culture. In: Corn, pp 1–18,Woodhead Publishing and AACC International Press, 10.1016/C2016-0-01986-1
Garg M, Sharma N, Sharma S, Kapoor P, Kumar A, Chunduri V, Arora P (2018) Biofortified crops generated by breeding, agronomy, and transgenic approaches are improving the lives of millions of people around the world. Front Nutr 5:12
Gautam V, Rana GS, Kumar P, Watpade S (2021) Development of carotenoids rich grains by genome engineering. In: Genome engineering for crop improvement, pp 83–96, Wiley Online Library.
Goredema-Matongera N, Ndhlela T, Magorokosho C, Kamutando CN, van Biljon A, Labuschagne M (2021) Multinutrient biofortification of maize (*Zea mays* L.) in Africa: current status, opportunities and limitations. Nutrients 13(3):1039
Hambidge KM, Huffer JW, Raboy V, Grunwald GK, Westcott JL, Sian L, Krebs NF (2004) Zinc absorption from low-phytate hybrids of maize and their wild-type isohybrids. Am J Clin Nutr 79(6):1053–1059
Hao X, Li X, Yang X, Li J (2014) Transferring a major QTL for oil content using marker-assisted backcrossing into an elite hybrid to increase the oil content in maize. Mol Breed 34:739–748
Hasan MM, Rima R (2021) Genetic engineering to improve essential and conditionally essential amino acids in maize: transporter engineering as a reference. Transgenic Res 30(2):207–220
He Z, Shang X, Zhang T, Yun J (2024) Effect of calcium and magnesium on starch synthesis in maize kernels and its physiological driving mechanism. Front Plant Sci 14:1332517
Hossain F, Muthusamy V, Zunjare RU, Gupta HS (2019) Biofortification of maize for protein quality and provitamin-A content. In: Nutritional quality improvement in plants, pp 115–136, Springer, Cham.
Hossain F, Zunjare RU, Muthusamy V, Bhat JS, Mehta BK, Sharma D, Talukder ZA, Chhabra R, Katral A, Dutta S, Chand G (2022) Biofortification of maize for nutritional security. In: Biofortification of staple crops. Springer Singapore, Singapore, pp 147–174

Hu S, Wang M, Zhang X, Chen W, Song X, Fu X, Fang H, Xu J, Xiao Y, Li Y, Bai G (2021) Genetic basis of kernel starch content decoded in a maize multi-parent population. Plant Biotechnol J 19(11):2192–2205

Huang S, Frizzi A, Florida CA, Kruger DE, Luethy MH (2006) High lysine and high tryptophan transgenic maize resulting from the reduction of both 19- and 22-kD alpha-zeins. Plant Mol Biol 61(3):525–535

Jilo T (2022) Nutritional benefit and development of quality protein maize (QPM) in Ethiopia. Cereal Res Commun 50(4):559–572

Khulbe RK, Pattanayak A, Sharma D (2020) Biofortification of maize using accelerated breeding tools. In: Accelerated plant breeding, volume 1: cereal crops. Springer International Publishing, Cham, pp 293–308

Kumar P, Choudhary M, Hossain F, Singh NK, Choudhary P, Gupta M, Singh V, Chikappa GK, Kumar R, Kumar B, Jat SL, Rakshit S (2019) Nutritional quality improvement in maize (*Zea mays*): Progress and challenges. Indian J Agric Sci 89:895–911. https://doi.org/10.56093/ijas.v89i6.90756

Kumar K, Gambhir G, Dass A, Tripathi AK, Singh A, Jha AK, Yadava P, Choudhary M, Rakshit S (2020) Genetically modified crops: current status and future prospects. Planta 251(4):91

Kumar A, Singh B, Raigond P, Sahu C, Mishra UN, Sharma S, Lal MK (2021) Phytic acid: blessing in disguise, a prime compound required for both plant and human nutrition. Food Res Int 142:110193

Kumar P, Longmei N, Choudhary M, Gupta M, Kumar B, Jat BS, Bhushan B, Dagla MC, Aggarwal SK (2024) Enhancement of nutritional quality in maize grain through QTL-based approach. Cereal Res Commun 52(1):39–55. https://doi.org/10.1007/s42976-023-00378-2

Lai JS, Messing J (2002) Increasing maize seed methionine by mRNA stability. Plant J 30:395–402

Leng P, Ouzunova M, Landbeck M, Wenzel G, Lübberstedt T, Darnhofer B, Eder J (2019) QTL mapping of improving forage maize starch degradability in European elite maize germplasm. Plant Breed 138(5):524–533

Li Q, Yang X, Xu S, Cai Y, Zhang D, Han Y, Yan J (2012) Genome-wide association studies identified three independent polymorphisms associated with α-tocopherol content in maize kernels. PLoS One 7(5):e36807

Liu L, Jeffers D, Zhang Y, Ding M, Chen W, Kang MS, Fan X (2015) Introgression of the *crtRB1* gene into quality protein maize inbred lines using molecular markers. Mol Breed 35(8):154

Loy DD, Lundy EL (2019) Nutritional properties and feeding value of corn and its coproducts. In: Corn. Woodhead and AACC International Press, pp 633–659

Marques E, Darby HM, Kraft J (2021) Benefits and limitations of non-transgenic micronutrient biofortification approaches. Agronomy 11(3):464

Martey E, Etwire PM, Kuwornu JK (2020) Economic impacts of smallholder farmers' adoption of drought-tolerant maize varieties. Land Use Policy 94:104524

Mashabela MD, Masamba P, Kappo AP (2022) Metabolomics and chemo-informatics in agricultural biotechnology research: complementary probes in unraveling new metabolites for crop improvement. Biology 11(8):1156

Masuda H, Kobayashi T, Ishimaru Y, Takahashi M, Aung MS, Nakanishi H, Nishizawa NK (2013) Iron-biofortification in rice by the introduction of three barley genes participated in mugineic acid biosynthesis with soybean ferritin gene. Front Plant Sci 4:132

Mendoza C, Viteri FE, Lönnerdal B, Young KA, Raboy V, Brown KH (1998) Effect of genetically modified, low–phytic acid maize on absorption of iron from tortillas. Am J Clin Nutr 68(5):1123–1127

Mondal S, Gayen D, Karmakar S (2020) Improvement of nutritional quality of rice seed through classical breeding and advance genetic engineering. In: Rice research for quality improvement: genomics and genetic engineering: volume 2: nutrient biofortification and herbicide and biotic stress resistance in rice. Springer, Singapore, pp 541–562

Mudadirwa D, Ndhlela T, Olakojo SA, Gonhi T (2023) The potential of provitamin A maize (Zea mays L.) biofortification in sub-Saharan Africa: current status, opportunities, challenges and perspectives: a review. CABI Rev 18:1

Murdia LK, Wadhwani R, Wadhawan N, Bajpai P, Shekhawat S (2016) Maize utilization in India: an overview. Am J Food Nutr 4(6):169–176

Mutungi C, Tungu J, Amri J, Gaspar A, Abass A (2022) Nutritional benefits of improved post-harvest handling practices for maize and common beans in Northern Tanzania: a quantitative farm-level assessment. J Stored Prod Res 95:101918

Owens BF, Lipka AE, Magallanes-Lundback M, Tiede T, Diepenbrock CH, Kandianis CB, Rocheford T (2014) A foundation for provitamin A biofortification of maize: genome-wide association and genomic prediction models of carotenoid levels. Genetics 198(4):1699–1716

Palacios-Rojas N, McCulley L, Kaeppler M, Titcomb TJ, Gunaratna NS, Lopez-Ridaura S, Tanumihardjo SA (2020) Mining maize diversity and improving its nutritional aspects within agro-food systems. Compr Rev Food Sci Food Saf 19(4):1809–1834

Raboy V, Gerbasi PF, Young KA, Stoneberg SD, Pickett SG, Bauman AT, Murthy PP, Sheridan WF, Ertl DS (2000) Origin and seed phenotype of maize low phytic acid 1-1 and low phytic acid 2-1. Plant Physiol 124(1):355–368

Revilla P, Alves ML, Andelković V, Balconi C, Dinis I, Mendes-Moreira P, Redaelli R, Ruiz de Galarreta JI, Vaz Patto MC, Žilić S, Malvar RA (2022) Traditional foods from maize (*Zea mays* L.) in Europe. Front Nutr 8:683399

Sheoran S, Kumar S, Ramtekey V, Kar P, Meena RS, Jangir CK (2022) Current status and potential of biofortification to enhance crop nutritional quality: an overview. Sustainability 14(6):3301

Shutu X, Dalong Z, Ye C, Yi Z, Shah T, Ali F, Qing L, Zhigang L, Weidong W, Jiansheng L, Xiaohong Y (2012) Dissecting tocopherols content in maize (*Zea mays* L.), using two segregating populations and high-density single nucleotide polymorphism markers. BMC Plant Biol 12:1–14

Singh J, Sharma S, Kaur A, Vikal Y, Cheema AK, Bains BK, Hossain F (2021) Marker-assisted pyramiding of lycopene-ε-cyclase, β-carotene hydroxylase1 and opaque2 genes for development of biofortified maize hybrids. Sci Rep 11(1):12642

Stilborn HL, Crum RC, Rice DW, Hinds MA, Ertl DS, Beach LR, Hu WE, Kleese RA (2002) Method of reducing cholesterol in eggs. U.S. Patent 6,391,348, 21 May 2002. Pioneer Hi-Bred International Inc., US Department of Agriculture

Sureshkumar S, Tamilkumar P, Senthil N, Nagarajan P, Thangavelu AU, Raveendran M, Vellaikumar S, Ganesan KN, Balagopal R, Vijayalakshmi G, Shobana V (2014) Marker assisted selection of low phytic acid traits in maize (*Zea mays* L.). Hereditas 151(1):20–27

Tanumihardjo SA, McCulley L, Roh R, Lopez-Ridaura S, Palacios-Rojas N, Gunaratna NS (2020) Maize agro-food systems to ensure food and nutrition security in reference to the Sustainable Development Goals. Glob Food Sec 25:100327

Truksa M, Vrinten P, Qiu X (2009) Metabolic engineering of plants for polyunsaturated fatty acid production. Mol Breed 23:1–11

Ufaz S, Galili G (2008) Improving the content of essential amino acids in crop plants: goals and opportunities. Plant Physiol 147(3):954–961

van Der Hooft JJ, Mohimani H, Bauermeister A, Dorrestein PC, Duncan KR, Medema MH (2020) Linking genomics and metabolomics to chart specialized metabolic diversity. Chem Soc Rev 49(11):3297–3314

World Health Organization, & United Nations University (2007) Protein and amino acid requirements in human nutrition, vol 935. World Health Organization, Geneva, Switzerland.

Wu D, Tanaka R, Li X, Ramstein GP, Cu S, Hamilton JP, Gore MA (2021) High-resolution genome-wide association study pinpoints metal transporter and chelator genes involved in the genetic control of element levels in maize grain. G3 11(4):jkab059

Yan J, Kandianis CB, Harjes CE, Bai L, Kim EH, Yang X, Skinner DJ, Fu Z, Mitchell S, Li Q, Fernandez MGS (2010) Rare genetic variation at *Zea mays crtRB1* increases β-carotene in maize grain. Nat Genet 42(4):322–327

Yang Q, Zhao D, Liu Q (2020) Connections between amino acid metabolisms in plants: lysine as an example. Front Plant Sci 11:928

Zhang W, Yang W, Wang M, Wang W, Zeng G, Chen Z, Cai Y (2013a) Increasing lysine content of waxy maize through introgression of opaque-2 and opaque-16 genes using molecular assisted and biochemical development. PLoS One 8(2):e56227

Zhang YQ, Pang LL, Yan P, Liu DY, Zhang W, Yost R, Zou CQ (2013b) Zinc fertilizer placement affects zinc content in maize plant. Plant Soil 372(1):81–92

Zhang D, Zhang Z, Unver T, Zhang B (2021) CRISPR/Cas: a powerful tool for gene function study and crop improvement. J Adv Res 29:207–221

Zhao HL, Yao QIN, Xiao ZY, Qin SUN, Gong DM, Qiu FZ (2023) Revealing the process of storage protein rebalancing in high quality protein maize by proteomic and transcriptomic. J Integr Agric 22(5):1308–1323

Zheng P, Allen WB, Roesler K, Williams ME, Zhang S, Li J, Glassman K, Ranch J, Nubel D, Solawetz W, Bhattramakki D (2008) A phenylalanine in DGAT is a key determinant of oil content and composition in maize. Nat Genet 40(3):367–372

Designing Future Maize Through Biotechnological Innovations

10

Shreya Mehindiratta, Pankaj Pandey, Shourat Fatma, Komal Rani, Nikhil Mehta, Aayushi Singla, Mamta Gupta, Alla Singh, Bhupender Kumar, and Krishan Kumar

Abstract

Maize is cultivated in more than 170 countries worldwide, making it one of the most versatile crops with a broad range of adaptability. While classical plant breeding plays a crucial role in improving maize yield, its effectiveness is often limited by the scarcity of usable germplasm diversity. Consequently, achieving certain desirable traits through conventional breeding alone is not always feasible. In such cases, integrating modern biotechnology approaches, such as genetic engineering (transgenesis, RNAi), genome editing (CRISPR-based tools), and related technologies, is useful for achieving desired traits quickly. These biotechnological innovations enable the design and development of maize genotypes with higher productivity, greater resilience to climate change, and improved nutritional quality, thereby meeting future challenges. To date, more than 300 transgenic or GM maize events have become available for cultivation or feed/food use. Despite the significant contributions of transgenic technology to maize improvement, public concerns about potential environmental risks and human health issues hinder its widespread adoption. To address these concerns and overcome low consumer acceptance of transgenic crops, alternative technologies like genome editing have been developed. Genome-edited (GEd) lines, particularly SDN-I and SDN-II types, can be readily generated as transgene-free plants. As a result, such transgene-free genome-edited plants are genetically similar, or even indistinguishable, from conventionally bred plants. Because these GEd crops are not regulated as transgenic and are not classified as GMOs in many

Shreya Mehindiratta and Pankaj Pandey contributed equally with all other contributors.

S. Mehindiratta · P. Pandey · S. Fatma · K. Rani · N. Mehta · A. Singla · M. Gupta · A. Singh · B. Kumar · K. Kumar (✉)
ICAR-Indian Institute of Maize Research, Ludhiana, Punjab, India
e-mail: krishan.kumar6@icar.org.in

M. Choudhary et al. (eds.), *Next-Generation Maize Research*,
https://doi.org/10.1007/978-981-92-0722-0_10

countries, their commercialization is generally easier and faster than that of transgenic crops. For instance, GEd maize such as low phytate maize and waxy corn is exempt from stringent regulatory processes and permitted for commercialization in the United States. Consequently, GEd maize is expected to achieve significantly higher consumer acceptance worldwide than transgenic maize. In summary, genome editing is, in most cases but not always, a more efficient, accurate, flexible, reliable, precise, and convenient alternative to transgenic technology.

Keywords

Maize improvement · Genetic engineering · Transgenic maize · Transgene-free crops · Genome editing

10.1 Introduction

Maize (*Zea mays* L.) is one of the world's most important cereal crops due to its wide adaptability and diverse end uses. It ranks third after wheat and rice and is extensively cultivated for food, feed, fodder, biofuel, and industrial applications. Owing to its high genetic yield potential, it is also known as "Queen of cereals" (Scott et al. 2019; Kumar et al. 2024). Rich in dietary fiber, vitamins, minerals, and antioxidants such as lutein and zeaxanthin, maize provides substantial nutritional benefits, including improved digestion, vision, and cardiovascular health (Ai and Jane 2016; Kazerooni et al. 2019). Despite its versatility and high yield potential, maize productivity often faces a variety of challenges, including a lack of improved/best-quality seeds, unfavorable climate, poor soil fertility, and post-harvest losses. Developing high-yielding, stress-resilient (disease-resistant and abiotic stress-tolerant), and improved single-cross maize hybrids is a promising strategy to overcome these challenges and enhance farmer profitability (Mdoda et al. 2025). Conventional plant breeding has long been used for crop improvement; however, it often requires several generations to develop new varieties because of its time-consuming selection cycles and reliance on natural genetic recombination (Sattar et al. 2025). Sometimes, conventional breeding may fail to achieve desired traits due to a scarcity of genetic diversity. Under such circumstances, biotechnological tools like genetic engineering and genome editing may offer promising solutions. In recent years, genetically engineered crops, particularly genetically modified and genome-edited varieties, have significantly contributed to improved yields, pest and disease resistance, herbicide tolerance, enhanced nutritional qualities and increased resilience to abiotic stresses in various crops, including maize (Kumar et al. 2020; Gupta et al. 2024). Unlike conventional breeding, these approaches offer fast, precise, and targeted modification of genetic traits (Fig. 10.1). Furthermore, integrating biotechnological innovations such as genetic engineering, genome editing, and genomics with advanced breeding tools like the double haploid system into conventional breeding programs can accelerate varietal development and contribute to

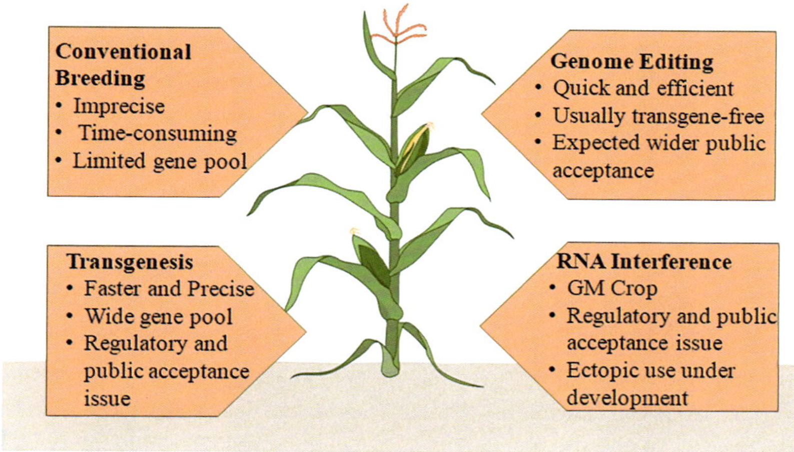

Fig. 10.1 Common tools for crop improvement: their advantages and limitations

improved productivity and sustainability (Pathirana and Carimi 2024). These innovations aim to enhance tolerance to abiotic and biotic stresses, improve resource-use efficiency, and ensure sustainable maize production under changing environmental conditions.

10.2 Biotechnological Innovations: A Modern Approach to Conventional Breeding

The production of maize is often compromised due to a variety of factors, such as diminishing agricultural land, rising populations, and biotic and abiotic stresses. To address these constraints, the development of improved maize genotypes with higher yield potential and enhanced stress tolerance is essential (Mdoda et al. 2025). Achieving such improvements requires the combined application of conventional breeding and modern biotechnological approaches (Kumar et al. 2020). Traditional plant breeding aims to improve crop traits through crossing of plants with desirable characteristics and selecting progeny with appropriate trait combinations. It has been a powerful tool for crop improvement that has contributed significantly to enhancing food security and agricultural productivity (Gedil and Menkir 2019). However, conventional breeding, being tedious and less precise, requires a much longer time for crop improvement with specific traits (Sattar et al. 2025). Moreover, a major limitation of conventional breeding is that the desired gene(s) can be chosen only from a sexually compatible gene pool. This poses a major limitation to the development of improved varieties if the desired gene governing the trait does not exist within its gene pool. In contrast, in a biotechnological approach, any desired and useful gene, even from sexually incompatible gene pools (i.e. other species), can be utilized and improved, or novel traits can be achieved in crop plants.

Consequently, biotechnological approaches offer powerful alternatives by enabling faster, more precise, and reliable crop improvement. A wide range of biotechnological tools—including transgenesis, RNA interference (RNAi), marker-assisted breeding, and genome editing—are available for crop improvement. Some of these approaches are discussed in detail in the following sections.

10.2.1 Transgenesis

Transgenesis refers to the stable insertion and expression of a transgene or foreign gene, from a sexually incompatible gene pool (species) into a recipient species in which desirable traits (governed by the transgene) do not occur naturally, and are thereby unattainable through conventional breeding (Kumar et al. 2018, 2020; Gupta et al. 2024). It is a highly effective approach that enables the transfer of desired genes from one species into the genome of another or within the same species to confer useful traits, like enhanced yields, herbicide tolerance, pest resistance, and improved nutritional quality (Blanco and Blanco 2017; Tiwari et al. 2024). For instance, *Bt* crops are improved varieties that are genetically modified to express a specific insecticidal crystal protein (cry) derived from the bacterium *Bacillus thuringiensis*, thus making them resistant to certain insect pests (Castagnola and Jurat-Fuentes 2012). The introduction of a desired gene is accomplished through plant transformation methods, viz., *Agrobacterium*-mediated gene transfer or direct gene transfer (Griffiths et al. 2005; Subburayalu et al. 2025). *Agrobacterium tumefaciens* is often referred to as a natural genetic engineer due to its natural ability to bring about genetic alteration in plant cells and horizontal transfer of bacterial genes (Gelvin 2003; Lacroix and Citovsky 2019). It has the innate ability to insert transfer DNA (T-DNA) from its tumor-inducing (Ti) plasmid into host plant cell genomes, which was first reported in 1977 (Chilton et al. 1977). The discovery of this mechanism led to the development of Ti plasmids as vectors for foreign gene delivery and the subsequent generation of transgenic plants. *Agrobacterium*-mediated gene transfer offers efficient integration of single or low-copy-number genes in dicots, as it enables stable gene integration and substantial DNA fragment transfer. Nevertheless, this method is associated with several limitations, like restricted host range, low efficiency in monocots, low transformation efficiency rates, and issues with transgene integration and off-target effects. To overcome such restraining challenges, direct gene transfer could be a rescuer; it can be efficiently employed for genetic transformation in monocots and *Agrobacterium*-resistant plant species. Direct gene transfer is a method of introducing foreign genes into the host cell with the help of physical or chemical methods that include microinjection, electroporation, and biolistic particle delivery. Nevertheless, the occurrence of random and multiple gene insertions remains a major limitation of direct gene transfer approaches (Gelvin 2010). The first transgenic plants, antibiotic-resistant tobacco and petunia, were produced in 1983 using *Agrobacterium*-mediated transformation (Fraley et al. 1983; Herrera-Estrella et al. 1983). This milestone was followed by the commercialization of the delayed-ripening *Flavr Savr* tomato in 1994. Subsequently, several transgenic crops, including *Bt* maize, *Bt* potato, herbicide-resistant cotton, canola, and soybean, received regulatory approval. In 1996, the first transgenic *Bt* maize cultivar was approved for commercial cultivation in the United States (James 1997; Gupta et al. 2024).

Maize currently has the highest number of approved transgenic events among all crops, with nearly 300 events authorized for cultivation or food and feed use across 46 countries (ISAAA 2025, https://www.isaaa.org/gmapprovaldatabase/). These events target key traits such as insect resistance, herbicide tolerance, abiotic stress tolerance, disease resistance, yield enhancement, and product quality improvement (ISAAA 2025). In 2019, transgenic maize occupied about 60.9 million hectares, accounting for nearly one-third of the global maize cultivation area (Gupta et al. 2024). To date, many genes have been well-characterized, functionally validated, and demonstrated to confer useful traits through the transgenic approaches, some of which are listed in Table 10.1. In addition to enabling the development of improved crop varieties, transgenesis has also contributed substantially to addressing global food and nutritional security challenges. In many studies, transgenic crops, including maize, have been proven to increase productivity, reduce pesticide and insecticide use, reduce greenhouse gas emissions, and increase economic advantages for farmers worldwide (Noack et al. 2024; Tiwari et al. 2024). Transgenic technology has immense potential to address challenges in maize cultivation that cannot be overcome through a conventional breeding approach. For example, it can be used to develop maize cultivars with higher yield, improved water-use efficiency, and enhanced nutrient-use efficiency, thereby supporting sustainable increases in maize production. This technology can also be employed to develop aflatoxin-resistant cultivars without any yield penalty, significantly enhancing overall productivity. Additionally, gene stacking or pyramiding approaches can be used to develop cultivars with resistance to multiple pests and diseases or to combine several desirable traits in a single variety. However, transgenic crops have been criticized for concerns related to biosafety (toxicity and allergenicity), environmental and ecological impacts (effects on non-target organisms and biodiversity), and socioeconomic issues, such as technology monopolization or the potential exploitation of farmers by companies charging high prices for seeds (Kumar et al. 2020). Therefore, robust regulatory frameworks are essential to ensure the safe, ethical, and responsible deployment of transgenic technologies (Table 10.1).

10.2.2 RNA Interference

RNA interference (RNAi) is a conserved biological process that inhibits gene expression by degrading specific mRNA molecules or translational repression (Pal 2021). This process involves small RNA molecules, such as microRNA (miRNA) and small interfering RNA (siRNA), which trigger the degradation or translation repression of homologous RNA targets (Aliu et al. 2013). RNAi functions as a defence mechanism against transposons and viruses, and it has also been widely utilized to silence genes involved in plant-pathogen interactions, thereby increasing

Table 10.1 A few proof-of-concept studies in maize for key trait improvement using biotechnological approaches

S. no.	Target gene	Encoded protein	Target trait	References
Representative examples of transgenic-based approach				
1.	*cspB*	Cold shock protein B	Drought tolerance	Wang et al. (2015)
2.	*zmm28*	MADS-box transcription factor	Higher yield	Wu et al. (2019)
3.	*ZmSCE1e*	SUMO conjugating enzyme	Salt and drought tolerance	Wang et al. (2019)
4.	*ZmVPS29*	Vacuolar protein sorting 29	Enhanced grain yield	Chen et al. (2020)
5.	*ZM-BG1H1*	Auxin transport	Increased grain yield	Simmons et al. (2020)
6.	*ZmPHYC2*	Phytochrome C	Plant height	Li et al. (2020)
7.	*ZmIPT2*	Isopentenyl transferase (IPT)	Enhances yield	Song et al. (2022)
8.	*ZmEREBP60*	AP2/ERF transcription factor	Drought tolerance	Zhu et al. (2022)
9.	*ZmKL9*	*Kernel length 9*	Yield	Gong et al. (2022)
10.	*ZmSUS1*	Sucrose synthase	Improving maize starch content	Li et al. (2023)
11.	*ZmPMP3g*	Plasma membrane proteolipid 3	Drought tolerance	Lei et al. (2023)
12.	*ZmPRX1*	Peroxidase	Drought tolerance	Zhai et al. (2024)
Representative examples of RNAi-based approach				
13.	*V-ATPase A*	V-type proton ATPase catalytic subunit A protein	Increased resistance to Western corn rootworm	Baum et al. (2007)
14.	*MDMV protease gene (P1)*	P1 protein (protease)	Increased resistance to maize dwarf mosaic virus	Zhang et al. (2010)
15.	*V-ATPase B*	V-type proton ATPase catalytic subunit B protein	Increased resistance to corn plant hopper (*Peregrinus maidis*)	Yao et al. (2013)
16.	*V-ATPase D*	V-type proton ATPase catalytic subunit D protein	Increased resistance to corn plant hopper (*Peregrinus maidis*)	Yao et al. (2013)
17.	*Aspergillus flavus amy1*	α-Amylase enzyme	Gene lowers aflatoxin buildup and fungal colonization in maize kernels	Gilbert et al. (2018)
18.	*Sec 23*	Sec23	Escalated resistance to Western corn rootworm	Vélez et al. (2020)

(continued)

Table 10.1 (continued)

S. no.	Target gene	Encoded protein	Target trait	References
19.	*Aspergillus flavus omtA*	*O*-methyl transferase gene (*omtA*)	Increased maize tolerance to aflatoxin contamination	Omolehin et al. (2024)
20.	Locust mAChR-C	C-type muscarinic acetylcholine receptor (mAChR-C)	Conferring control in locust	Zheng et al. (2025)
Representative examples of the genome editing-based approach				
21.	*ZmIPK1*	Inositol-1,3,4,5,6-pentakisphosphate 2-kinase	Insertional disruption of this gene results in altered inositol phosphate profile and herbicide tolerance	Shukla et al. (2009)
22.	*ZmPDS*	Phytoene desaturase	Albino phenotype	Liang et al. (2014)
23.	*Hkt1*	High-affinity K^+ transporter	Tolerance to salt stress	Xing et al. (2014)
24.	*ZmMRP4*	Phytoene desaturase	Albino phenotype	Liang et al. (2014)
25.	*ZmMs26*	Cytochrome P450-like protein	Male sterility	Svitashev et al. (2015)
26.	*gl2*	Glossy2 protein (acyltransferase)	Reduced epicuticular wax	Char et al. (2015)
27.	*lg1*	LG1 protein	Effect on maize leaf development	Svitashev et al. (2015)
28.	*PSY1*	Phytoene synthase	Targeted mutagenesis	Zhu et al. (2016)
29.	*MTL*	A pollen-specific phospholipase	Male sterility	Kelliher et al. (2017)
30.	*MS33*	Glycerol-3-phosphate acyltransferase	Male sterility	Xie et al. (2018)
31.	*ZmCCT9*	A CCT transcription factor	Promotes flowering during the long days of higher latitudes	Huang et al. (2018)
32.	*SH2*	Large subunit of endosperm ADP glucose pyrophosphorylase	Supersweet and waxy maize	Dong et al. (2019)
33.	*WX*	GRANULE-BOUND STARCH SYNTHASE I	Supersweet and waxy maize	Dong et al. (2019)
34.	*ZmRAVL*	*B3-domain transcription factor*	Improved maize yield under high-density planting	Tian et al. (2019)
35.	*ZmSRL5*	CASPL protein	Reduced drought tolerance	Pan et al. (2020)
36.	*ZmVLHP, ZmGW2*	*VRS1-LIKE HOMEOBOX PROTEIN, GRAIN WEIGHT2*	Increased total kernel number and/or kernel weight	Metje-Sprink et al. (2020)

(continued)

Table 10.1 (continued)

S. no.	Target gene	Encoded protein	Target trait	References
37.	*ZmALS1 and ZmALS2*	*Acetolactate syntheses 1 and 2*	Herbicide tolerance	Jiang et al. (2020)
38.	*ZmLOX3*	9-Lipoxygenase	Smut resistance	Pathi et al. (2020)
39.	*ZmMs25*	Fatty acyl reductases	Defective anther and pollen, male fertility	Zhang et al. (2021)
40.	*ZmCLCg*	Chloride transport	Reduced salt tolerance	Luo et al. (2021)
41.	*ZmACO2*	Ethylene biosynthesis	Enhanced ear length, kernel number, and grain yield	Ning et al. (2021)
42.	*CLE genes: ZmFCP1, ZmCLE7 & ZmCLEIE5*	CLE peptide	Escalated grain yield	Liu et al. (2021)
43.	*ZmBADH2a/2b*	Betaine aldehyde dehydrogenase 2	Aromatic maize	Wang et al. (2021)
44.	*ZmCOIa*	Receptor protein coronatine insensitive 1	Immunity to gibberella stalk rot	Ma et al. (2024)
45.	*ZmThx20*	Trihelix transcription factor	Improved grain yield	Li et al. (2021)
46.	*ZmKRN2*	Kernel rows number 2	Increase number of kernel rows in maize	Chen et al. (2022)
47.	*ZmPAT7*	Protein S-acyltransferases	Increased tassel branch number	Guan et al. (2022)
48.	*ZmCOI2a/b*	Receptor of jasmonate signal	Defective anther, male sterility	Qi et al. (2023)
49.	*ZmGA20ox3*	Gibberellin 20-oxidase	Shorter plant stature	Liu et al. (2024)
50.	*ZmASR1*	ABA/WDS domain protein	Enhanced drought tolerance	Yang et al. (2024)
51.	*ZmTKPR1-1 and ZmTKPR1-2*	Tetraketide α-pyrone reductases	Male sterility	An et al. (2024)
52.	*ZmPOR1/POR2*	Protochlorophyllide reductases	Tocopherols biosynthesis	Herr et al. (2025)
53.	*ZmEREB167*	APETALA2/ethylene-responsive family (AP2/ERF)	Grain yield	Qing et al. (2025)
54.	*ZmPL1*	T-Phylloplanin-like protein	Drought tolerance	Wang et al. (2025)

plant tolerance to biotic stressors. Targeting virulence factors or host susceptibility genes through RNAi can disrupt the infection process and improve plant resistance (Elbana and Saad-Allah 2024). The RNAi mechanism is conserved across various eukaryotic species and associated with key processes such as transposon silencing and gene regulation (Lax et al. 2020). RNAi can be induced both in vitro and in vivo using siRNAs or by expressing short hairpin RNA (shRNA) through vectors (Aliu et al. 2013). The discovery and development of RNAi have significantly advanced functional genomics research in plants, animals, and humans.

A major advantage of RNAi technology is its high specificity, which enables targeted suppression of pests and pathogens while minimizing effects on non-target organisms, making it an attractive tool for crop protection and improvement (Munkácsy et al. 2007). One of the successful applications of the RNAi approach is in the management of the western corn rootworm (WCR), which is considered one of the most economically important pests of maize in the USA (Darlington et al. 2022). The US Environmental Protection Agency has approved the RNAi-based product for controlling agricultural insect pests, including genetically modified corn that produces hairpin RNAs complementary to the *Snf7* gene of the WCR, in combination with *Bacillus thuringiensis* toxins, to control western corn rootworm (*Diabrotica virgifera virgifera*) (Liu et al. 2020). The potential of RNAi extends to the management of insect pests, plant pathogens, and nematodes, as well as addressing pesticide resistance. While RNAi-based plants are regulated under existing GMO frameworks, there is a need for science-based risk assessment procedures for topical RNAi applications, such as spray-based delivery systems (Silver et al. 2021; Arango et al. 2024). To overcome current limitations in RNAi deployment, researchers are exploring diverse delivery mechanisms, including bacterium-mediated RNA interference (bmRNAi), which has shown promising results for agricultural applications (Qi et al. 2024).

10.2.2.1 General Mechanism of RNAi

The first step in the RNAi machinery, known as the RNAi initiating step, involves the binding of RNA nucleases to a large double-stranded RNA (dsRNA) molecule. This binding leads to the cleavage of the dsRNA into distinct fragments that are approximately 21–25 nucleotides long, known as small interfering RNAs (siRNAs) (Agrawal et al. 2003). In the second step, these siRNAs associate with a multinuclease complex called the RNA-induced silencing complex (RISC), which is responsible for degrading the corresponding single-stranded messenger RNAs (mRNAs). Currently, there is limited knowledge about the intermediates, RNA-protein complexes, and the mechanisms involved in forming different complexes during the RNAi process. Furthermore, several aspects of RNAi remain unclear, particularly the molecular basis for its systemic spread within an organism (Mezzetti et al. 2020).

10.2.2.2 RNAi Mechanism in Plants

RNAi in plants involves sequence-specific identification of mRNAs followed by repression of their expression via either translation inhibition or ribonuclease-directed destruction (Goodfellow et al. 2019). It mostly involves three important phases: (i) initiation, where the dsRNA is cleaved into smaller fragments of siRNA or miRNA, (ii) effector phase, where either of them is complexed with proteins to form the RISC complex, and (iii) the degradation of homologous RNA (Agrawal et al. 2003). Two key players in plants are miRNA and siRNA, single-stranded

RNAs (21–24 nucleotides) involved in gene silencing through RNAi. In the miRNA-mediated pathway, transcription begins in the nucleus, where the primary miRNA (pri-miRNA) is processed by Dicer-like protein 1 (DCL1) to form the precursor miRNA (pre-miRNA) and eventually the miRNA duplex (around 22 nucleotides) (Goodfellow et al. 2019). The complex is transported to the cytoplasm for the formation of the RISC (RNA-induced silencing complex) complex with the help of Argonaute (Ago) proteins (Agrawal et al. 2003). The guide strand of the miRNA duplex is kept while the passenger strand is discarded. The target RNA is cleaved or has its translation stopped when the RISC-bound guide RNA attaches to it (via Watson–Crick base-pairing). AGO proteins identify the appropriate target mRNAs and suppress gene expression by using the retained sRNAs as guidance (Hung and Slotkin 2021). Whereas, in the siRNA-mediated pathway, initiated in the cytoplasm, mostly double-stranded RNA is cleaved by DCLs (DCL1, DCL2, DCL3, or DCL4) with ribonuclease activity to form siRNA (Mezzetti et al. 2020).

10.2.2.3 RNAi Applications in Crop Improvement

RNAi has emerged as an effective tool for crop improvement by enabling precise suppression of endogenous gene expression across diverse plant species. Plant viruses are a major cause of yield losses and are difficult to manage due to their multiple modes of transmission. RNAi provides an efficient strategy for broad-spectrum viral resistance by targeting different regions of viral genomes, often through silencing of coat protein (CP) genes (Rajput et al. 2021). For example, RNAi constructs containing CP and disease-specific (SP) genes successfully suppressed rice stripe disease caused by Rice stripe virus (RSV) in susceptible Japonica cultivars (Rajput et al. 2021). Similarly, expression of hairpin RNA (hpRNA) in tomato (*Solanum lycopersicum*) conferred resistance against Potato spindle tuber viroid (PSTVd) (Saurabh et al. 2014).

In maize, RNAi-mediated genetic modifications have led to improvements in pest resistance, nutritional quality, and stress tolerance (Zhang et al. 2017). For example, maize dwarf mosaic virus (MDMV), a widely prevalent pathogen causing severe yield losses, has been effectively targeted using an hpRNA construct designed against the viral P1 protease gene, resulting in MDMV-resistant maize lines (Zhang et al. 2010). RNAi has also demonstrated strong potential in insect and fungal pest management. For instance, knockdown of the C-type muscarinic acetylcholine receptor (mAChR-C), a gene essential for chitin metabolism, in the migratory locust (*Locusta migratoria*) via dsRNA delivery caused developmental abnormalities and impaired molting and metamorphosis (Zheng et al. 2025). Additionally, RNAi targeting the *Aspergillus flavus* alpha-amylase (amy1) gene in maize kernels significantly inhibited fungal growth and reduced aflatoxin accumulation during infection (Gilbert et al. 2018) (Table 10.1). Furthermore, Host-Induced Gene Silencing (HIGS) of the *O*-methyltransferase gene (*omtA*), a key gene in aflatoxin biosynthesis, has been successfully employed to enhance maize resistance to aflatoxin contamination. Transformation of maize inbred line B104 with an RNAi construct targeting *omtA*, followed by crossing with elite inbred lines, resulted in hybrids with significantly reduced aflatoxin levels, demonstrating a stable and heritable strategy for mitigating mycotoxin contamination in maize (Omolehin et al. 2024). These findings show that HIGS-induced *omtA* suppression can increase maize's resistance to aflatoxin contamination, which can then be passed on to elite backgrounds, offering an effective and heritable strategy for reducing aflatoxin contamination in maize (Omolehin et al. 2024). The above-mentioned examples demonstrate that RNA interference offers an effective strategy for improving maize resistance to biotic stresses and reducing mycotoxin contamination, thereby supporting the development of resilient and sustainable genotypes.

10.2.3 Genome Editing

Genome editing is a transformative technology that enables precise modification of plant genomes to improve existing traits or introduce novel, beneficial characteristics. This approach allows the creation of stable mutations, gene knockouts through disruption of endogenous gene function, or precise gene insertions and substitutions to correct or replace specific DNA sequences. Genome-editing strategies are based on engineered sequence-specific nucleases (SSNs) that recognize and cleave DNA at defined genomic locations. Commonly used SSNs include meganucleases, zinc-finger nucleases (ZFNs), transcription activator-like effector nucleases (TALENs), and the clustered regularly interspaced short palindromic repeats–associated (CRISPR/Cas) systems. These nucleases introduce double-stranded breaks (DSBs) at target sites, which are repaired by the cell through either non-homologous end joining (NHEJ) or homology-directed repair (HDR) (Voytas and Gao 2014; Kumar et al. 2020; Gao 2021). While NHEJ is an error-prone repair pathway that often results in insertions or deletions, HDR enables accurate repair using a homologous donor template.

Gene knockouts caused by small insertions or deletions, frequently leading to frameshift mutations, are predominantly generated via NHEJ, whereas HDR facilitates precise point mutations, targeted gene replacement, or site-specific gene insertion (Kim et al. 2014; Kumar et al. 2020). Meganucleases, ZFNs, and TALENs recognize target sequences through protein–DNA interactions, whereas CRISPR/Cas systems achieve target specificity through Watson–Crick base pairing between the guide RNA (gRNA) and the target DNA (Fig. 10.2). Owing to its simplicity, low cost, high efficiency, and ability to edit multiple targets simultaneously, CRISPR/Cas has become the most widely adopted genome-editing tool in plant research and crop improvement (Yin et al. 2017; Gao 2021).

The CRISPR/Cas9 system comprises two core components: the Cas9 endonuclease and a programmable gRNA. The gRNA guides Cas9 to the target genomic sequence through sequence complementarity, enabling precise DNA cleavage (Fig. 10.2). Following cleavage, the induced DSB is repaired by endogenous cellular pathways via NHEJ or HDR, depending on the desired editing outcome (Chen et al. 2018; Gupta et al. 2024).

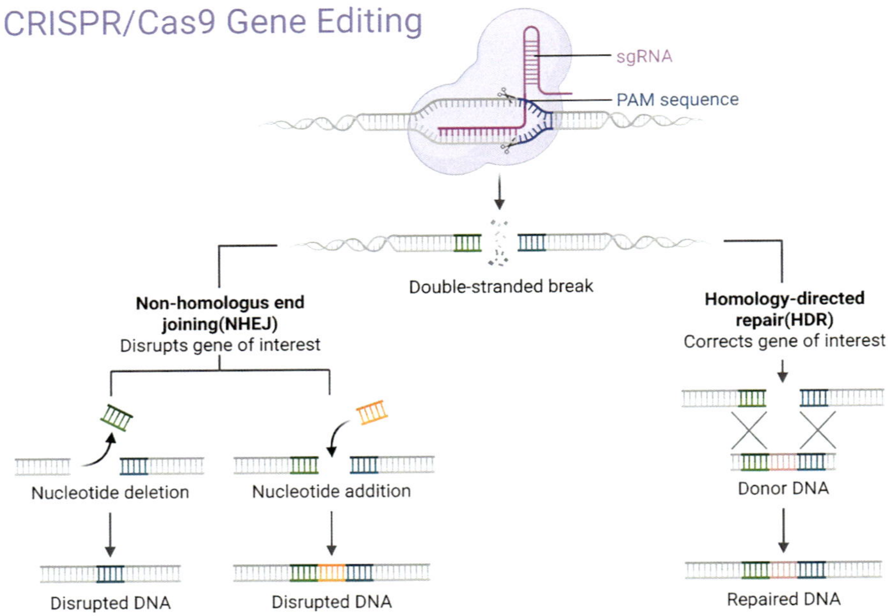

Fig. 10.2 Diagram illustrating the CRISPR/Cas genome-editing mechanism. (Created with BioRender.com)

10.2.3.1 Applications of Genome-Editing Tools in Maize Improvement, with Special Reference to CRISPR

Specifically, CRISPR/Cas, one of the most powerful genome-editing tools, has found wide-ranging applications across various fields, including agriculture. It enables precise and predictable genome modification in crops, thereby facilitating the development of improved maize varieties with traits such as enhanced yield, improved nutritional quality, tolerance to abiotic and biotic stresses, herbicide resistance, and optimized plant architecture, as well as specialized applications in male sterility and haploid induction (Table 10.1). Recently, several genome-edited maize events have been deregulated in certain parts of the world, as the final products are null segregants (lacking foreign DNA) and, therefore, are not expected to pose a plant pest risk under existing regulatory frameworks. These events are either already commercialized or are expected to be commercialized in the near future (Fig. 10.3).

10.2.3.1.1 Escalating Yield Potential via Genome-Editing Tools

Enhancing kernel yield has always been a primary objective of maize breeding. Several yield-related traits, including ear diameter, ear length, ear row number, ear weight, and kernel number per row, collectively determine yield potential (Lu et al. 2020; Wang et al. 2022). Consequently, genome editing aimed at improving these traits offers a promising strategy for yield enhancement. For instance, knockout of *ZmCEP1*, a gene involved in sugar and nitrate transport to developing kernels, has

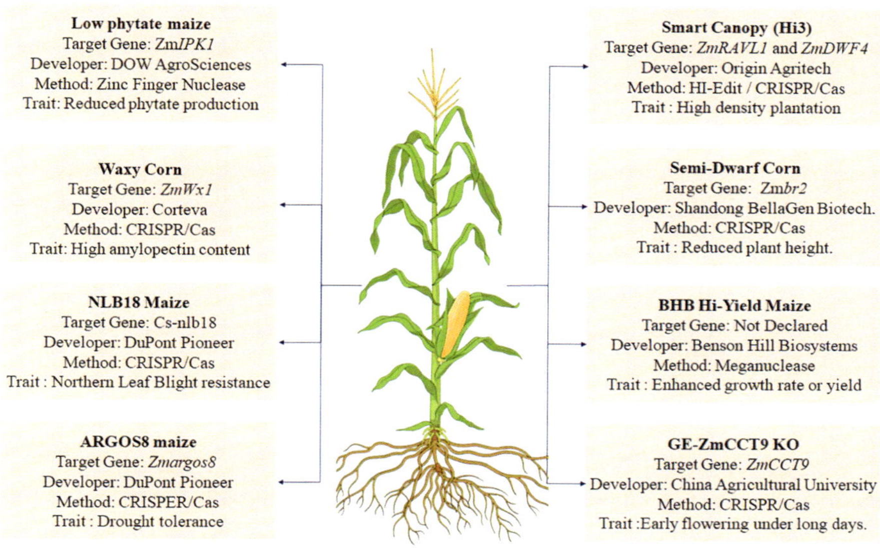

Fig. 10.3 Genome-edited maize events at the near-commercialization stage or already commercialized in the USA and other countries. (Information retrieved from https://euginius.eu/euginius/pages/gmo_searchresults.jsf?rpp=101&page=1&col=GMO&dir=false)

been reported to increase maize kernel yield (Xu et al. 2021; Wang et al. 2022). Likewise, editing of *ZmACO2* has been shown to improve some of these yield-associated traits, and ultimately overall grain yield in maize (Ning et al. 2021). In addition, disruption of *ZmKRN2*, a WD40 protein-encoding gene and major negative regulator underlying a key QTL for kernel row number, has been shown to increase kernel number per plant and results in approximately a 10% yield improvement under field conditions, without apparent trade-offs (Chen et al. 2022) (Table 10.1). In addition, optimizing plant architecture for high-density planting can further enhance yield by accommodating more plants per unit area. For example, *ZmRAVL1* acts as a positive regulator of leaf angle, and its editing results in a more upright leaf architecture, thereby facilitating high-density planting and increasing yield per unit area (Tian et al. 2019) (Fig. 10.3).

10.2.3.1.2 Enhancing Stress Tolerance Through Genome-Editing Tools

Biotic and abiotic stresses are major causes of yield loss worldwide. Genome editing has emerged as an efficient approach for developing new germplasm resources and resistant maize varieties. For example, knockout of *lox3* enhances resistance to *Ustilago maydis* (DC.) Corda by promoting reactive oxygen species bursts (Pathi et al. 2020). Similarly, CRISPR-Cas9–mediated editing of *ZmGDIα* provides increased resistance to maize rough dwarf virus without any agronomic penalty (Liu et al. 2022). Additionally, *ZmFBL41*, a recessive gene, confers resistance against Banded Leaf and Sheath Blight (BLSB); targeted base substitution via

Prime editing or knockout via CRISPR/Cas9 could serve as an effective strategy to develop BLSB-resistant maize in the future (Kumar et al. 2023). Moreover, precision editing to upregulate the promoter of *ARGOS8* has been shown to enhance maize kernel yield under drought stress (Shi et al. 2017) (Fig. 10.3).

10.2.3.1.3 Enhancing Nutritional Quality Through Genome-Editing Tools

Other than yield enhancement, improving the nutritional traits of maize has gained significant importance. Genome editing represents a powerful tool for biofortification, enabling enhancement of key nutrients, flavor, fragrance, texture, shelf life, digestibility, composition, and oil content. For example, precise editing of *ZmBADH2a* and *ZmBADH2b* increases levels of 2-acetyl-1-pyrroline, resulting in aromatic corn (Wang et al. 2021). Similarly, CRISPR-Cas9–mediated editing of *SH2* and *WX* genes can be used to develop super sweet and waxy maize varieties (Dong et al. 2019). Furthermore, CRISPR-Cas9 editing of *starch branching enzyme IIb* (*SBEIIb*) has been shown to enhance amylose and resistant starch content (Ma et al. 2024).

10.2.3.1.4 Additional Applications of Genome Editing in Maize

Development of novel germplasm for maize breeding programs represents a key specialized application of genome editing. Major examples include the following.

Male Sterility

The development of male-sterile lines plays a key role in the efficient production of hybrid seeds. Genome editing, particularly CRISPR/Cas-based approaches, has enabled the generation of stable male-sterile lines through targeted editing of male fertility–related genes such as *TMS5*, *Ms7*, *MS8*, *Ms26*, and *Ms45* (Svitashev et al. 2015; Li et al. 2017; Chen et al. 2018).

Haploid Induction

Double-haploid technology and in vivo haploid induction are widely employed to enhance the efficiency of maize breeding. These approaches facilitate rapid fixation of recombinant haploid genomes, enabling the development of completely homozygous lines and accelerating phenotypic evaluation. The integration of CRISPR/Cas technology with doubled haploid systems offers a precise and efficient strategy for crop improvement (Jacquier et al. 2021). For example, CRISPR/Cas-mediated editing of *ZmPOD65*, *ZmPLD3*, *ZmPLA1*, and *ZmDMP* has successfully resulted in haploid induction in maize (Liu et al. 2017; Zhong et al. 2019; Jiang et al. 2022).

10.3 Conclusion

The development of climate-resilient, high-starch, and high-yielding maize genotypes is imperative in the context of rising demand, climate change, and shrinking arable land. To achieve this, effective strategies must be adopted. Although conventional plant breeding has been effective, it is often time-consuming, imprecise, and limited by the available gene pool. These limitations can be overcome by integrating modern approaches—such as marker-assisted breeding, RNA interference, transgenesis, double haploids, speed breeding, genomic selection, and genome editing—with traditional breeding. Together, these approaches enable more efficient, precise, and accelerated crop improvement. Despite the proven benefits of transgenesis and RNAi, their adoption remains limited in many countries due to concerns about human health and environmental safety. In this context, genome-editing technologies, including CRISPR/Cas, base editing, and prime editing, have emerged as promising alternatives. Their high efficiency, precise and targeted modification, ease of use, flexibility, and ability to generate transgene-free plants (especially SDN-I and SDN-II outcomes) have led to wider acceptance. In recent years, targeted editing of endogenous genes has enabled the development of genome-edited maize with improved and novel traits, paving the way for next-generation maize cultivars. An effective strategy for maize improvement is the integration of conventional breeding with genomics, genome editing, artificial intelligence-assisted trait prediction, high-throughput phenotyping, and speed breeding. Advanced editing tools will enable precise improvement of complex traits, thereby supporting sustainable maize production and global food, feed, and fuel security.

References

Agrawal N, Dasaradhi PVN, Mohmmed A, Malhotra P, Bhatnagar RK, Mukherjee SK (2003) RNA interference: biology, mechanism, and applications. Microbiol Mol Biol Rev 67(4):657–685

Ai Y, Jane JL (2016) Macronutrients in corn and human nutrition. Compr Rev Food Sci Food Saf 15(3):581–598

Aliu H, Beadini N, Beadini S, Iseni G (2013) RNA interference and cancer. J Nat Sci Res 3:120–124

An X, Zhang S, Jiang Y, Liu X, Fang C, Wang J, Wan X (2024) CRISPR/Cas9-based genome editing of 14 lipid metabolic genes reveals a sporopollenin metabolon ZmPKSB-ZmTKPR1-1/-2 required for pollen exine formation in maize. Plant Biotechnol J 22(1):216–232

Arango AV, Darlington M, Jurat-Fuentes JL, Kogel K, Rathore K, Smagghe G (2024) RNA interference in agriculture: methods, applications, and governance. Issue paper 72. Council for Agricultural Science and Technology, Ames

Baum JA, Bogaert T, Clinton W, Heck GR, Feldmann P, Ilagan O, Johnson S, Plaetinck G, Munyikwa T, Pleau M, Vaughn T (2007) Control of coleopteran insect pests through RNA interference. Nat Biotechnol 25(11):1322–1326

Blanco A, Blanco G (2017) The genetic information (II). In: Medical biochemistry. Academic Press, pp 493–524

Castagnola AS, Jurat-Fuentes JL (2012) Bt crops: past and future. In: *Bacillus thuringiensis* biotechnology. Springer, pp 283–304

Char SN, Unger-Wallace E, Frame B, Briggs SA, Main M, Spalding MH (2015) Heritable site-specific mutagenesis using TALENs in maize. Plant Biotechnol J 13(7):1002–1010

Chen R, Xu Q, Liu Y, Zhang J, Ren D, Wang G, Liu Y (2018) Generation of transgene-free maize male sterile lines using the CRISPR/Cas9 system. Front Plant Sci 9:1180

Chen L, Li YX, Li C, Shi Y, Song Y, Zhang D, Wang H (2020) The retromer protein ZmVPS29 regulates maize kernel morphology likely through an auxin-dependent process(es). Plant Biotechnol J 18(5):1004–1014

Chen W, Chen L, Zhang X, Yang N, Guo J, Wang M, Yang X (2022) Convergent selection of a WD40 protein that enhances grain yield in maize and rice. Science 375(6587):eabg7985

Chilton MD, Drummond MH, Merlo DJ, Sciaky D, Montoya AL, Gordon MP, Nester EW (1977) Stable incorporation of plasmid DNA into higher plant cells: the molecular basis of crown gall tumorigenesis. Cell 11(2):263–271

Darlington M, Reinders JD, Sethi A, Lu AL, Ramaseshadri P, Fischer JR, Boeckman CJ, Petrick JS, Roper JM, Narva KE, Vélez AM (2022) RNAi for western corn rootworm management: lessons learned, challenges, and future directions. Insects 13(1):57

Dong L, Qi X, Zhu J, Liu C, Zhang X, Cheng B, Mao L, Xie C (2019) Supersweet and waxy: meeting the diverse demands for specialty maize by genome editing. Plant Biotechnol J 17(10):1853

Elbana NM, Saad-Allah KM (2024) Implementing RNA interference to boost plant resistance and enhance crop characteristics: a review. Plant Mol Biol Report 43:1–18

Fraley RT, Rogers SB, Horsch RB (1983) Use of a chimeric gene to confer antibiotic resistance to plant cells. In: Advances in gene technology: molecular genetics of plants and animals, vol 20. Academic Press, New York, pp 211–221

Gao C (2021) Genome engineering for crop improvement and future agriculture. Cell 184(6):1621–1635

Gedil M, Menkir A (2019) An integrated molecular and conventional breeding scheme for enhancing genetic gain in maize in Africa. Front Plant Sci 10:1430

Gelvin SB (2003) Agrobacterium-mediated plant transformation: the biology behind the "gene-jockeying" tool. Microbiol Mol Biol Rev 67(1):16–37

Gelvin SB (2010) Plant proteins involved in agrobacterium-mediated genetic transformation. Annu Rev Phytopathol 48(1):45–68

Gilbert MK, Majumdar R, Rajasekaran K, Chen ZY, Wei Q, Sickler CM, Wang K (2018) RNA interference-based silencing of the alpha-amylase (amy1) gene in *Aspergillus flavus* decreases fungal growth and aflatoxin production in maize kernels. Planta 247:1465–1473

Gong D, Wang Y, Zhang H, Liang K, Sun Q, Qiu F (2022) Overexpression of ZmKL9 increases maize hybrid hundred kernel weight. Plant Biotechnol J 21(3):451

Goodfellow S, Zhang D, Wang MB, Zhang R (2019) Bacterium-mediated RNA interference: potential application in plant protection. Plants 8(12):572

Griffiths AJF, Wessler SR, Lewontin RC, Gelbart WM, Suzuki DT, Miller JH (2005) Introduction to genetic analysis, 8th edn. W.H. Freeman, New York

Guan H, Chen X, Wang K, Liu X, Zhang D, Li Y, Li Y (2022) Genetic variation in ZmPAT7 contributes to tassel branch number in maize. Int J Mol Sci 23(5):2586

Gupta M, Kumar K, Singh A, Singh B, Jha AK, Kumar B (2024) Transgenic and genome-edited maize: status and prospect. In: Genetic engineering of crop plants for food and health security, vol 2. Springer Nature Singapore, pp 373–403

Herr S, Li X, Wu D, Hunter CT, Magallanes-Lundback M, Wood JC, Gore MA (2025) Total tocopherol levels in maize grain depend on chlorophyll biosynthesis within the embryo. BMC Plant Biol 25(1):328

Herrera-Estrella L, De Block M, Messens EHJP, Hernalsteens JP, Van Montagu M, Schell J (1983) Chimeric genes as dominant selectable markers in plant cells. EMBO J 2(6):987–995

Huang C, Sun H, Xu D, Chen Q, Liang Y (2018) ZmCCT9 enhances maize adaptation to higher latitudes. Proc Natl Acad Sci USA 115(2):E334–E341

Hung YH, Slotkin RK (2021) The initiation of RNA interference (RNAi) in plants. Curr Opin Plant Biol 61:102014

ISAAA (2025) GM approval database. https://www.isaaa.org/gmapprovaldatabase/. Accessed 2025

Jacquier NM, Gilles LM, Martinant JP, Rogowsky PM, Widiez T (2021) Maize in planta haploid inducer lines: a cornerstone for doubled haploid technology. In: Doubled haploid technology, vol 2. Humana, pp 25–48

James C (1997) Global status of transgenic crops in 1997. ISAAA Briefs 5:1–31

Jiang YY, Chai YP, Lu MH, Han XL, Lin Q, Zhang Y, Chen QJ (2020) Prime editing efficiently generates W542L and S621I double mutations in two ALS genes in maize. Genome Biol 21:1–10

Jiang C, Sun J, Li R, Yan S, Chen W, Guo L, Qin G, Wang P, Luo C, Huang W, Zhang Q (2022) A reactive oxygen species burst causes haploid induction in maize. Mol Plant 15(6):943–955

Kazerooni EG, Atia Sharif HN, Rehman R, Nisar S (2019) Maize (corn)-a useful source of human nutrition and health: a critical review. Int J Chem Biochem Sci 15:35–41

Kelliher T, Starr D, Richbourg L, Chintamanani S, Delzer B, Nuccio ML (2017) MATRILINEAL, a sperm-specific phospholipase, triggers maize haploid induction. Nature 542(7639):105–109

Kim JM, Kim D, Kim S, Kim JS (2014) Genotyping with CRISPR-Cas-derived RNA-guided endonucleases. Nat Commun 5(1):3157

Kumar K, Aggarwal C, Sapna B, Singh I, Yadava P (2018) Microbial genes in crop improvement. In: Crop improvement through microbial biotechnology. Elsevier BV, pp 39–56

Kumar K, Gambhir G, Dass A, Tripathi AK, Singh A, Jha AK, Rakshit S (2020) Genetically modified crops: current status and future prospects. Planta 251(4):91

Kumar K, Kumar B, Jha AK, Neha, Piyal P (2023) Genome editing for banded leaf and sheath blight resistance in Indian maize genotypes. Maize J 12(2):61–65

Kumar K, Kumar B, Jha AK, Neha, Pandey P, Piyal P, Jat HS (2024) Genome editing for enhancing production and productivity of maize in India. Maize J 13(2):145–150

Lacroix B, Citovsky V (2019) Pathways of DNA transfer to plants from *Agrobacterium tumefaciens* and related bacterial species. Annu Rev Phytopathol 57(1):231–251

Lax C, Tahiri G, Patiño-Medina JA, Cánovas-Márquez JT, Pérez-Ruiz JA, Osorio-Concepción M, Calo S (2020) The evolutionary significance of RNAi in the fungal kingdom. Int J Mol Sci 21(24):9348

Lei L, Pan H, Hu HY, Fan XW, Wu ZB, Li YZ (2023) Characterization of ZmPMP3g function in drought tolerance of maize. Sci Rep 13(1):7375

Li J, Zhang H, Si X, Tian Y, Chen K, Liu J, Chen H, Gao C (2017) Generation of thermosensitive male-sterile maize by targeted knockout of the ZmTMS5 gene. J Genet Genomics 44(9):465–468

Li Q, Wu G, Zhao Y, Wang B, Zhao B, Kong D, Wang H (2020) CRISPR/Cas9-mediated knockout and overexpression studies reveal a role of maize phytochrome C in regulating flowering time and plant height. Plant Biotechnol J 18(12):2520–2532

Li P, Li Z, Xie G, Zhang J (2021) Trihelix transcription factor ZmThx20 is required for kernel development in maize. Int J Mol Sci 22(22):12137

Li P, Ma H, Xiao N, Zhang Y, Xu T, Xia T (2023) Overexpression of the ZmSUS1 gene alters the content and composition of endosperm starch in maize (*Zea mays* L.). Planta 257(5):97

Liang Z, Zhang K, Chen K, Gao C (2014) Targeted mutagenesis in *Zea mays* using TALENs and the CRISPR/Cas system. J Genet Genomics 41(2):63–68

Liu C, Li X, Meng D, Zhong Y, Chen C, Dong X, Xu X, Chen B, Li W, Li L, Tian X (2017) A 4-bp insertion at ZmPLA1 encoding a putative phospholipase A generates haploid induction in maize. Mol Plant 10(3):520–522

Liu S, Jaouannet M, Dempsey DMA, Imani J, Coustau C, Kogel KH (2020) RNA-based technologies for insect control in plant production. Biotechnol Adv 39:107463

Liu L, Gallagher J, Arevalo ED, Chen R, Skopelitis T, Wu Q (2021) Enhancing grain-yield-related traits by CRISPR–Cas9 promoter editing of maize CLE genes. Nat Plants 7(3):287–294

Liu C, Kong M, Yang F, Zhu J, Qi X, Weng J, Di D, Xie C (2022) Targeted generation of null mutants in ZmGDIα confers resistance against maize rough dwarf disease without agronomic penalty. Plant Biotechnol J 20(5):803

Liu Y, Chen Z, Zhang C, Guo J, Liu Q, Chen Y, Liu X (2024) Gene editing of ZmGA20ox3 improves plant architecture and drought tolerance in maize. Plant Cell Rep 43(1):18

Lu X, Zhou Z, Yuan Z, Zhang C, Hao Z, Wang Z, Li M, Zhang D, Yong H, Han J, Li X (2020) Genetic dissection of the general combining ability of yield-related traits in maize. Front Plant Sci 11:788

Luo M, Zhang Y, Li J, Zhang P, Chen K, Song W, Zhao J (2021) Molecular dissection of maize seedling salt tolerance using a genome-wide association analysis method. Plant Biotechnol J 19(10):1937–1951

Ma M, Sun S, Zhu J, Qi X, Li G, Hu J, Xie C, Liu C (2024) Engineering high amylose and resistant starch in maize by CRISPR/Cas9-mediated editing of starch branching enzymes. Crop J 12(4):1252–1258

Mdoda L, Tamako N, Gidi LS, Naidoo D (2025) Evaluating the impact of improved maize varieties on agricultural productivity and technical efficiency among smallholder farmers in the Eastern Cape, South Africa: an empirical analysis. GM Crops Food 16(1):272–304
Metje-Sprink J, Sprink T, Hartung F (2020) Genome-edited plants in the field. Curr Opin Biotechnol 61:1–6
Mezzetti B, Smagghe G, Arpaia S, Christiaens O, Dietz-Pfeilstetter A, Jones H, Kostov K, Sabbadini S, Opsahl-Sorteberg HG, Ventura V, Taning CNT (2020) RNAi: what is its position in agriculture? J Pest Sci 93(4):1125–1130
Munkácsy G, Tulassay Z, Győrffy B (2007) RNA interference and its clinical applications. Orv Hetil 148(47):2235–2240
Ning Q, Jian Y, Du Y, Li Y, Shen X, Jia H, Zhang Z (2021) An ethylene biosynthesis enzyme controls quantitative variation in maize ear length and kernel yield. Nat Commun 12(1):5832
Noack F, Engist D, Gantois J, Gaur V, Hyjazie BF, Larsen A, M'Gonigle LK, Missirian A, Qaim M, Sargent RD, Souza-Rodrigues E, Kremen C (2024) Environmental impacts of genetically modified crops. Science 385(6712):eado9340. https://doi.org/10.1126/science.ado9340
Omolehin O, Raruang Y, Hu D, Han ZQ, Promyou S, Brown RL, Chen ZY (2024) Host-induced gene silencing of the *Aspergillus flavus* O-methyl transferase gene enhanced maize aflatoxin resistance. Toxins 17(1):8. https://doi.org/10.3390/toxins17010008
Pal A (2021) RNA interference. In: Springer protocols handbooks. Springer
Pan Z, Liu M, Zhao H, Tan Z, Liang K, Sun Q, Qiu F (2020) ZmSRL5 is involved in drought tolerance by maintaining cuticular wax structure in maize. J Integr Plant Biol 62(12):1895–1909
Pathi KM, Rink P, Budhagatapalli N, Betz R, Saado I, Hiekel S, Kumlehn J (2020) Engineering smut resistance in maize by site-directed mutagenesis of LIPOXYGENASE 3. Front Plant Sci 11:543895
Pathirana R, Carimi F (2024) Plant biotechnology—an indispensable tool for crop improvement. Plants 13(8):1133
Qi X, Gao H, Lv R, Mao W, Zhu J, Liu C, Xie C (2023) CRISPR/dCas-mediated gene activation toolkit development and its application for parthenogenesis induction in maize. Plant Commun 4(2):100449
Qi HY, Zhang DD, Liu B, Chen JY, Han D, Wang D (2024) Leveraging RNA interference technology for selective and sustainable crop protection. Front Plant Sci 15:1502015
Qing X, Li J, Lin Z, Wang W, Yi F, Chen J, Yang Z (2025) Maize transcription factor ZmEREB167 negatively regulates starch accumulation and kernel size. J Genet Genomics 52:411–421
Rajput M, Choudhary K, Kumar M, Vivekanand V, Chawade A, Ortiz R, Pareek N (2021) RNA interference and CRISPR/Cas gene editing for crop improvement: paradigm shift towards sustainable agriculture. Plants 10(9):1914
Sattar S, Sultana R, Hiddayatullah, Ullah I (2025) Modification of conventional methods and modern plant breeding techniques to enhance genetic gain for future food security. In: Crop biofortification: biotechnological approaches for achieving nutritional security under changing climate. Wiley, pp 429–438
Saurabh S, Vidyarthi AS, Prasad D (2014) RNA interference: concept to reality in crop improvement. Planta 239(3):543–564
Scott P, Pratt RC, Hoffman N, Montgomery R (2019) Specialty corns. In: Corn. AACC International Press, pp 289–303
Shi J, Gao H, Wang H, Lafitte HR, Archibald RL, Yang M, Hakimi SM, Mo H, Habben JE (2017) ARGOS 8 variants generated by CRISPR-Cas9 improve maize grain yield under field drought stress conditions. Plant Biotechnol J 15:207–216
Shukla VK, Doyon Y, Miller JC, DeKelver RC, Moehle EA, Worden SE (2009) Precise genome modification in the crop species *Zea mays* using zinc-finger nucleases. Nature 459(7245):437–441
Silver K, Cooper AM, Zhu KY (2021) Strategies for enhancing the efficiency of RNA interference in insects. Pest Manag Sci 77:2645–2658

Simmons CR, Weers BP, Reimann KS, Abbitt SE, Frank MJ, Wang W, Habben JE (2020) Maize BIG GRAIN1 homolog overexpression increases maize GRAIN yield. Plant Biotechnol J 18:2304–2315
Song Y, Li C, Zhu Y, Guo P, Wang Q, Zhang L, Di H (2022) Overexpression of ZmIPT2 gene delays leaf senescence and improves grain yield in maize. Front Plant Sci 13:963873
Subburayalu S, Panda D, Daware V, Rajeshvari I, Rathava PB, Sootrakar K, Venkatesan S, Joshi KB (2025) Transgenic plants: from laboratory to field. Int J Adv Biochem Res 8:1279–1286
Svitashev S, Young JK, Schwartz C, Gao H, Falco SC, Cigan AM (2015) Targeted mutagenesis, precise gene editing, and site-specific gene insertion in maize using Cas9 and guide RNA. Plant Physiol 169:931–945
Tian J, Wang C, Xia J, Wu L, Xu G, Wu W, Li D, Qin W, Han X, Chen Q, Jin W (2019) Teosinte ligule allele narrows plant architecture and enhances high-density maize yields. Science 365:658–664
Tiwari R, Chauhan S, Kumar K, Reddy VS, Rajam MV (2024) Genetically modified crops for sustainable agriculture and food security in Indian scenario. In: Biotechnology in India—reworking a strategy. Springer
Vélez AM, Fishilevich E, Rangasamy M, Khajuria C, McCaskill DG, Pereira AE (2020) Control of western corn rootworm via RNAi traits in maize: lethal and sublethal effects of Sec23 dsRNA. Pest Manag Sci 76(4):1500–1512
Voytas DF, Gao C (2014) Precision genome engineering and agriculture: opportunities and regulatory challenges. PLoS Biol 12:e1001877
Wang C, Burzio LA, Koch MS, Silvanovich A, Bell E (2015) Purification, characterization and safety assessment of the introduced cold shock protein B in DroughtGard™ maize. Regul Toxicol Pharmacol 71:164–173
Wang H, Wang M, Xia Z (2019) Overexpression of a maize SUMO conjugating enzyme gene (ZmSCE1e) increases sumoylation levels and enhances salt and drought tolerance in transgenic tobacco. Plant Sci 281:113–121
Wang Y, Liu X, Zheng X, Wang W, Yin X, Liu H, Ma C, Niu X, Zhu JK, Wang F (2021) Creation of aromatic maize by CRISPR/Cas. J Integr Plant Biol 63:1664–1670
Wang Y, Tang Q, Pu L, Zhang H, Li X (2022) CRISPR-Cas technology opens a new era for the creation of novel maize germplasms. Front Plant Sci 13:1049803
Wang C, Zhou Y, Wang Y, Jiao P, Liu S, Guan S, Ma Y (2025) CRISPR-Cas9-mediated editing of ZmPL1 gene improves tolerance to drought stress in maize. GM Crops Food 16:1–16
Wu J, Lawit SJ, Weers B, Sun J, Mongar N, Van Hemert J, Habben JE (2019) Overexpression of zmm28 increases maize grain yield in the field. Proc Natl Acad Sci USA 116:23850–23858
Xie K, Wu S, Li Z, Zhou Y, Zhang D, Dong Z (2018) Map-based cloning and characterization of *Zea mays* male sterility33 (ZmMs33) gene, encoding a glycerol-3-phosphate acyltransferase. Theor Appl Genet 131(6):1363–1378
Xing HL, Dong L, Wang ZP, Zhang HY, Han CY, Liu B (2014) A CRISPR/Cas9 toolkit for multiplex genome editing in plants. BMC Plant Biol 14(1):327
Xu R, Li Y, Sui Z, Lan T, Song W, Zhang M, Zhang Y, Xing J (2021) A C-terminal encoded peptide, ZmCEP1, is essential for kernel development in maize. J Exp Bot 72:5390–5406
Yang Y, Li A, Liu Y, Shu J, Wang J, Guo Y, Wu J (2024) ZmASR1 negatively regulates drought stress tolerance in maize. Plant Physiol Biochem 211:108684
Yao J, Rotenberg D, Afsharifar A, Barandoc-Alviar K, Whitfield AE (2013) Development of RNAi methods for *Peregrinus maidis*, the corn planthopper. PLoS One 8(8):e70243
Yin K, Gao C, Qiu JL (2017) Progress and prospects in plant genome editing. Nat Plants 3:1–6
Zhai X, Yan X, Zenda T, Wang N, Dong A, Yang Q, Duan H (2024) Overexpression of the peroxidase gene ZmPRX1 increases maize seedling drought tolerance by promoting root development and lignification. Crop J 12:753–765
Zhang ZY, Fu FL, Gou L, Wang HG, Li WC (2010) RNA interference-based transgenic maize resistant to maize dwarf mosaic virus. J Plant Biol 53:297–305

Zhang Z, Hu F, Sung MW, Shu C, Castillo-Gonzalez C, Koiwa H, Zhang X (2017) RISC-interacting clearing 3′–5′ exoribonucleases (RICEs) degrade uridylated cleavage fragments to maintain functional RISC in *Arabidopsis thaliana*. elife 6:e24466
Zhang S, Wu S, Niu C, Liu D, Yan T, Tian Y, Wan X (2021) ZmMs25 encoding a plastid-localized fatty acyl reductase is critical for anther and pollen development in maize. J Exp Bot 72:4298–4318
Zheng H, Hua M, Jiang M, Jiang C, Xi Y, Deng J, Zhou S (2025) Transgenic expression of mAChR-C dsRNA in maize confers efficient locust control. Plant Commun 6:101316
Zhong Y, Liu C, Qi X, Jiao Y, Wang D, Wang Y, Liu Z, Chen C, Chen B, Tian X, Li J (2019) Mutation of ZmDMP enhances haploid induction in maize. Nat Plants 5:575–580
Zhu J, Song N, Sun S, Yang W, Zhao H, Song W, Lai J (2016) Efficiency and inheritance of targeted mutagenesis in maize using CRISPR-Cas9. J Genet Genomics 43(1):25–36
Zhu Y, Liu Y, Zhou K, Tian C, Aslam M, Zhang B, Zou H (2022) Overexpression of ZmEREBP60 enhances drought tolerance in maize. J Plant Physiol 275:153763

Pathological Insights: Disease Resistance and Management in Maize

11

Pravin Kumar Bagaria, Navdeep Singh, K. Kamesh Krishnamoorthy, Pardeep Kumar, Mukesh Choudhary, Bahadur Singh Jat, Priya Garkoti, and Poonam Choudhary

Abstract

Maize (*Zea mays* L.) is a globally significant cereal crop, yet its productivity is often constrained by diverse fungal and bacterial diseases. This chapter provides a comprehensive overview of major maize diseases, including Maydis leaf blight, Turcicum leaf blight, Banded leaf and sheath blight, Common rust, Polysora rust, Charcoal rot, *Fusarium* stalk rot, Brown stripe downy mildew, Sorghum downy mildew, Rajasthan downy mildew, and Bacterial stalk rot. Detailed insights into symptoms, epidemiology, and disease progression across agro-climatic regions are discussed, with emphasis on pathogens such as *Cochliobolus heterostrophus*, *Exserohilum turcicum*, *Rhizoctonia solani*, *Puccinia* spp., *Macrophomina phaseolina*, *Fusarium verticillioides*, and *Dickeya zeae*. The chapter highlights the complexity of disease cycles and pathogens' survival strategies, including the role of the primary inoculum, the dissemination of sexual and asexual spores, and environmental stimuli that influence epidemics. Integrated disease management (IDM) strategies are emphasized, where host plant resistance remains central, supported by agronomic practices, fungicides, and biological control agents. Both qualitative (race-specific) and quantitative (polygenic, durable) resistance mechanisms are explored, with advances in QTL mapping, R-gene identification, and marker-assisted selection. Case studies of resistance loci such as *Ht1*, *Ht2*, *ZmCCoAOMT2*, and *ZmAPX1* illustrate progress in breeding. Finally, the chapter underscores challenges posed by pathogen variability and environmental stressors, while highlighting genomic and preci-

P. K. Bagaria (✉) · N. Singh · K. Kamesh Krishnamoorthy · P. Kumar · M. Choudhary · B. S. Jat · P. Garkoti
ICAR-Indian Institute of Maize Research, Ludhiana, Punjab, India
e-mail: pravin.bagaria@icar.org.in

P. Choudhary
ICAR-Central Institute of Post-Harvest Engineering and Technology, Ludhiana, Punjab, India

M. Choudhary et al. (eds.), *Next-Generation Maize Research*,
https://doi.org/10.1007/978-981-92-0722-0_11

sion breeding tools as promising approaches for developing resilient maize cultivars. This synthesis serves as a valuable resource for plant pathologists and breeders, enhancing understanding of disease dynamics and resistance breeding to secure maize-enabled food security.

Keywords

Host plant resistance · Mycotoxin · Chemical · Biological · Downy mildew · Rust · Blight · Stalk rot

11.1 Introduction

11.1.1 Global Significance

Maize (*Zea mays* L.) is one of the most important food crops globally, alongside wheat and rice, contributing at least 30% of food calories to over 4.5 billion people in more than 94 developing countries (Shiferaw et al. 2011). In parts of the world such as America and Africa, maize provides more than 20% of total food calories and serves as a staple food for approximately 900 million people. Given its widespread consumption, maize plays a crucial role in ensuring food security and supporting livelihoods in low- and lower-middle-income countries, which account for approximately 67% of the total maize production in developing regions. Maize is currently cultivated on nearly 208.3 million hectares across 125 developing countries and ranks among the top three most extensively grown crops in 75 of these nations (FAOStat 2023). Consistent with this exposure, the demand for maize is expected to double in the developing world by 2050, coinciding with an estimated increase in global population to approximately 9.3 billion (Rosegrant et al. 2009; Grote et al. 2021). The world area under maize cultivation has witnessed significant production areas in Sub-Saharan Africa (SSA), Latin America, and Asian countries (FAOStat 2023) (Fig. 11.1a, b). Since 1961, maize cultivation has nearly doubled, increasing from 106 million hectares to 208 million hectares, with significant growth observed since the early 2000s (Erenstein et al. 2022). Given current circumstances and declining wheat production, maize is projected to surpass wheat as the most widely grown crop by 2030. Over the same period, global maize yields have tripled from 2 tons per hectare in 1961 to 5.9 tons per hectare at present, resulting in a six-fold increase in total maize production.

Despite its predominance as a staple food, demand for maize has shifted significantly in recent decades, with a notable increase in its use as livestock feed. The rapid economic growth in densely populated areas of Asia, Latin America, and the Middle East has driven up demand for poultry and livestock products, in turn boosting maize production (Delgado 2003). Moreover, maize acts as a crucial industrial input in the production of biofuels, especially in the United States, and its feed market is growing in nations such as India and China, where economic progress allows for higher consumption of animal-derived products (Rosegrant 2008; Shiferaw et al. 2011; Grote et al. 2021).

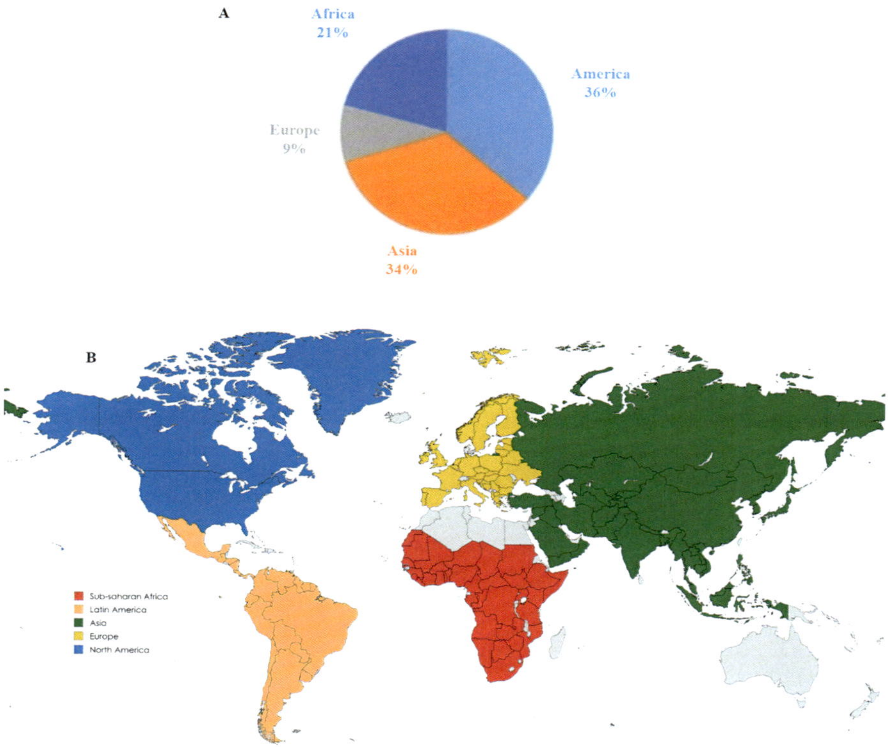

Fig. 11.1 (**a**) The percentage contribution by different continents in global maize production (FAOStat 2021); (**b**) Continental-scale description of regions under global maize cultivation, indicating the spatial distribution of maize-growing areas

11.1.2 Climate Change and Increasing Biotic Stresses in Maize Production

Climate change is significantly impacting maize cultivation, increasing its susceptibility to biotic and abiotic stresses. Changing rainfall patterns and rising temperatures are expected to shift rainfed maize mega-environments, expanding cultivation potential in higher altitudes and latitudes. However, despite this expansion, regions such as Asia and Sub-Saharan Africa are anticipated to experience increased exposure to extreme conditions, including drought, heat stress, waterlogging, and an elevated incidence of diseases and pests (Jones and Thornton 2003; Tesfaye et al. 2015; Prasanna et al. 2021).

Abiotic stressors generally intensify the effects of biotic factors, resulting in yield losses exceeding 30% in certain areas (Mittler 2006). These biotic stresses involve some of the important maize diseases, *viz.*, Turcicum leaf blight (TLB) (also known as Northern Corn Leaf Blight [NCLB]; *Exserohilum turcicum*), Maydis leaf blight (MLB) (also known as Southern Corn Leaf Blight [SCLB]; *Bipolaris maydis*), Banded leaf and sheath blight (BLSB) (*Rhizoctonia solani*), Polysora rust (*Puccinia polysora*), Common rust (*Puccinia sorghi*), and Grey Leaf Spot (GLS; *Cercospora* spp.) are among the maize diseases of global significance. Additionally, stalk, ear, and kernel rots, caused primarily by *Fusarium* and *Aspergillus* spp., pose serious threats because they are associated with mycotoxin contamination, thereby reducing both grain quality and food safety.

In Asia, post-flowering stalk rots (PFSR), downy mildews, GLS, MLB, TLB, and BLSB are among the most important diseases (Kumar et al. 2022b). The epidemiology of these diseases is more dynamic, influenced by changing climatic parameters, cropping systems, and the introductions of new germplasm. While the development of resistant varieties has managed certain diseases to some extent, others have become more prevalent due to changes in agronomic practices. For instance, the adoption of zero-tillage and crop-residue management strategies has been linked with the increasing incidence and severity of GLS (*Cercospora zeae-maydis*) (Ward et al. 1997).

11.1.3 Post-harvest Losses and Mycotoxin Contamination

Besides the pre-harvest disease management challenges, post-harvest losses due to storage fungi pose a significant concern, resulting in up to 80% losses in some areas of the world. Ear and kernel rots are particularly severe in warm, humid maize-growing environments. Approximately 55.9% of maize production in subtropical, mid-altitude, and highland zones suffers economic losses from ear rots, while such losses go up to 44% of maize produced in tropical lowlands (Compton et al. 1993). The expected increase in extreme weather conditions due to climate change is expected to exacerbate the incidence and severity of ear rots and mycotoxin contamination (Paterson and Lima 2010). Drought-prone and semi-arid areas in tropical countries are particularly vulnerable to elevated mycotoxin concentrations, posing significant risks to both animal and human health. For example, acute aflatoxicosis outbreaks were observed in Kenya in 2004 due to mycotoxin-contaminated maize, which resulted in fatal poisoning (Strosnider et al. 2006). Such a fungal attack and toxin accumulation during post-harvest phases are a major constraint in countries with poor logistic support (Stutt et al. 2023).

11.1.4 Disease Management

The development and cultivation of disease-resistant varieties remain the most effective and sustainable approach for managing diseases. Modifications in host-plant resistance through conventional breeding, genetic engineering, and molecular marker-assisted selection have facilitated the incorporation of resistance genes against major diseases of maize. In addition to integrated disease management (IDM) and genetic resistance approaches, cultural practices, biological control, and

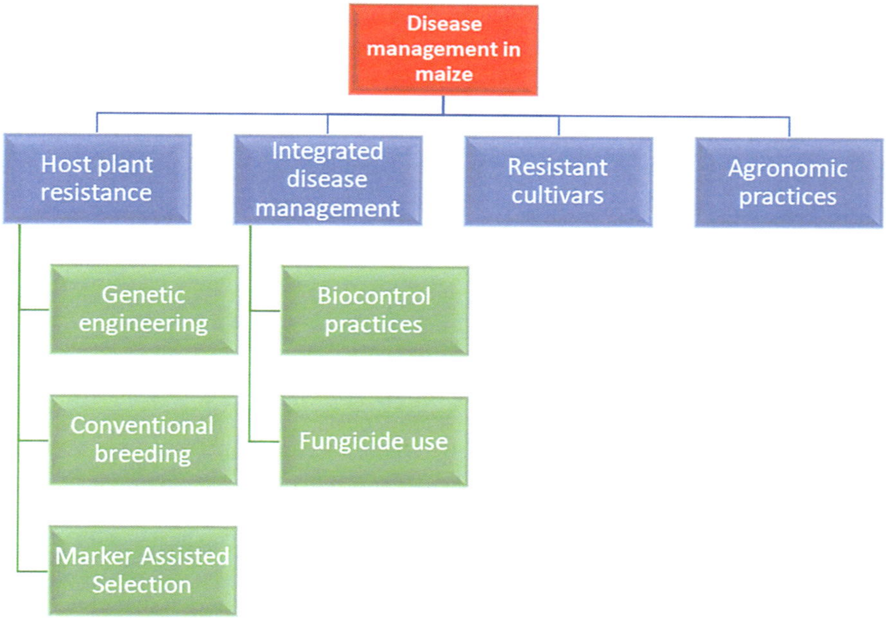

Fig. 11.2 Diagrammatic overview of the principal strategies adopted for increasing disease resistance in maize production systems

optimal fungicide use play a crucial role in reducing disease incidence. Climate-resilient maize cultivars, combined with improved agronomic practices, can further enhance disease resistance under changing environmental conditions (Fig. 11.2). Understanding the interaction between maize and its pathogens at the molecular and physiological levels is crucial for developing long-term resistance strategies and ensuring sustainable maize production in diverse agroecosystems.

11.1.4.1 Host Plant Resistance

Host plant resistance depends on the ability of plants to inherit resistance from their parents. This resistance might depend on multiple environmental factors and the inherent ability of the parent plants. It might be in multiple forms, from biochemical to structural defences. Moreover, resistance can be achieved through multiple approaches, including genetic engineering, conventional breeding, and marker-assisted selection (MAS). Genetic engineering enables the precise introduction and modification of resistance genes in plants. On the contrary, conventional breeding remains a widely used method in which susceptible and resistant genotypes are crossed to develop new varieties or genotypes with improved resistance. Furthermore, MAS supports improved breeding efficiency by using specific DNA markers associated with target resistance genes, ensuring accurate and early detection of resistant lines.

11.1.4.2 Resistant Cultivars and Agronomic Practices

The use of resistant cultivars is another important strategy for providing a long-term, cost-effective measure by incorporating naturally occurring resistance characters into commercially viable maize genotypes. Agronomic practices, including crop rotation, optimized planting in time and space, balanced fertilization, residue management, and proper irrigation, help in reducing disease prevalence under field conditions. Together, these approaches, integrated into a system, offer a holistic framework for achieving durable and sustainable disease resistance in maize.

11.2 Important Diseases of Maize in India

Maize production is significantly constrained by various diseases, which limit its yield potential across diverse agro-climatic regions worldwide. The crop is susceptible to approximately 115 diseases caused by a range of pathogens, including viruses, bacteria, fungi, and nematodes, leading to substantial yield losses (Khokhar et al. 2014). Among these, 65 diseases have been reported in India, with 11 of national significance, including one caused by a cyst nematode (AICRPM 2018). The country's maize production is systematically categorized into five major agro-climatic zones-Northern Hill Zone, North-Western Plain Zone, North-Eastern Plain Zone, Central Western Zone, and Peninsular Zone, each with designated maize-growing regions monitored under the All India Coordinated Research Project on Maize (AICRP) (Aggarwal et al. 2022).

In India, maize diseases are broadly classified into five major groups based on symptomatology: foliar diseases, seedling blights, stalk rots, ear rots, and downy mildew (Payak et al. 1973; Payak and Sharma 1985). The dynamic nature of pathogen prevalence and disease severity necessitates continuous surveillance and strategic disease management approaches to mitigate yield losses and ensure sustainable maize production. Among these are MLB/SCLB, TLB/NCLB, CLS, BLSB, FSR, ChR, BSR, Rusts, and downy mildew diseases, which are prevalent in the most maize-growing regions of India (Table 11.1). The detailed description of the primary diseases of maize has been mentioned below.

11.2.1 Maydis Leaf Blight

MLB, also known as Southern Corn Leaf Blight (SCLB), is a devastating foliar disease of maize crop caused by the ascomycetous fungus *Drechslera maydis* (Y. Nisik. & C. Miyake) Subram. and B. L. Jain [teleomorph *Cochliobolus heterostrophus* (Drechsler) Drechsler]. The pathogen is also known as *Bipolaris maydis* (Y. Nisik. & C. Miyake) Shoemaker (synonym: *Helminthosporium maydis* Y. Nisik & C. Miyake). The genus *Drechslera* (synonym *Bipolaris*) belongs to the kingdom Fungi, Phylum Ascomycota, class Dothideomycetes, order Pleosporales, family *Pleosporaceae* (Manamgoda et al. 2014; Kumar et al. 2022a). Initially considered

Table 11.1 Region-wise occurrence of important maize diseases and their associated pathogens

Disease	Casual organism	Location/state
Maydis leaf blight (MLB)/Southern Corn Leaf Blight (SCLB)	*Bipolaris maydis*	Punjab, Haryana, Bihar, Delhi, West Bengal, Uttar Pradesh, Madhya Pradesh, Odisha, Rajasthan, Gujarat
Turcicum leaf blight (TLB)/Northern Corn Leaf Blight (NCLB)	*Exserohilum turcicum*	Jammu and Kashmir, Himachal Pradesh, Uttarakhand, Bihar, Karnataka, Andhra Pradesh, Sikkim, Meghalaya, Manipur, West Bengal
Banded leaf and sheath blight (BLSB)	*Rhizoctonia solani*	Himachal Pradesh, Punjab, Uttarakhand, Haryana, Bihar, Delhi, Andhra Pradesh, Odisha, Gujarat
Charcoal rot (ChR)	*Macrophomina phaseolina*	Punjab, Delhi, Uttar Pradesh, Madhya Pradesh, Bihar, Maharashtra, Karnataka, Telangana, Tamil Nadu, Gujarat
Fusarium stalk rot (FSR)	*Fusarium verticillioides*	Rajasthan, Maharashtra, Punjab, Uttar Pradesh, Madhya Pradesh, Bihar, Karnataka, West Bengal, Gujarat, Delhi
Bacterial stalk rot (BSR)	*Dickeya zeae*	Himachal Pradesh, Uttarakhand, Punjab, Haryana, Bihar, West Bengal, Uttar Pradesh, Madhya Pradesh, Odisha, Delhi
Curvularia leaf spot (CLS)	*Curvularia lunata*	Rajasthan, Punjab, Uttar Pradesh, Madhya Pradesh, Gujarat, West Bengal, Odisha, Bihar, Haryana, Delhi
Common rust (CRust)	*Puccinia sorghi*	Haryana, Karnataka, Manipur
Polysora rust (PRust)	*Puccinia polysora*	Karnataka
Rajasthan downy mildew (RDM)	*Peronosclerospora heteropogoni*	Rajasthan
Sorghum downy mildew (SDM)	*Peronosclerospora sorghi*	Karnataka
Non-Insect Pest-Maize cyst nematode	*Heterodera zeae*	Rajasthan

of minor importance following its identification in Florida, USA, by Drechsler (1925), MLB gained considerable importance after the devastating 1970 epidemic in the United States, which was attributed to the emergence of race “T” of the pathogen. This epidemic reinforced the disease’s economic importance and susceptibility of maize hybrids carrying CMS-T cytoplasm. MLB is distributed across subtropical and tropical maize-growing regions, including Africa, the Americas, and Southeast Asia. In India, it was first reported from Maldah, West Bengal, in 1960 and has been identified in various other states, with endemic centres observed in Delhi, Haryana, Punjab, Bihar, West Bengal, Rajasthan, and Gujarat. The disease is particularly severe in warm, humid environments, resulting in yield losses of 40%–70% (Kumar et al. 2025a).

The epidemiology of MLB is closely associated with warm temperatures (20–30 °C) and high relative humidity, which enhance rapid pathogen proliferation. The mycelium, chlamydospores, or conidiophores of the pathogen overwintering in the maize debris serve as the primary inoculum for disease initiation. Conidia, which are the asexual spores of the pathogen, are disseminated by wind or rain, attach to host surfaces, and germinate leading to further perpetuation of the disease. MLB has been characterized by the formation of elongated, diamond-shaped lesions (2–6 mm wide and 3–22 mm long), featuring light brown centres with dark brown margins (Fig. 11.3a, b). Under favourable conditions, these lesions coalesce to induce comprehensive foliar blighting. Severe infection during early growth stages reduces plant vigour, lowering ear height, cob weight, grain quality, and cumulative yield potential. Furthermore, early senescence induced by the disease reduces photosynthetic efficiency, negatively affecting grain filling and fodder quality. Disease assessment is typically conducted at the soft dough stage using standardized rating scales (1–9) (Hooda et al. 2018), which are instrumental in breeding programs aimed at developing resistant maize cultivars.

11.2.2 Turcicum Leaf Blight

TLB, caused by the heterothallic ascomycetous *Setosphaeria turcica* (anamorph *Exserohilum turcicum*), is a primary foliar disease of maize. The earliest record of this disease dates back to 1878 in New Jersey, USA, with a significant epidemic occurring in Connecticut in 1889. *E. turcicum*, as a facultative parasite, exhibits peculiar pathogenic adaptability, which leads to economic yield losses ranging from 25-90% under favourable environmental conditions (Aghav et al. 2023). In India, this disease was first reported by Butler in 1907 from Bihar and the disease is prevalent in various maize-growing areas, including Jammu Kashmir, Karnataka, Himachal Pradesh, Uttarakhand, Uttar Pradesh, Odisha, Andhra Pradesh, and the North-Eastern Hill states, primarily affecting rabi maize in the plains.

The disease flourishes under moderate temperatures (18–27 °C) and high relative humidity, particularly when conditions include frequent rain showers, prolonged leaf wetness, and cool weather, which significantly intensify disease severity. Initial symptoms appear as small, light-green to greyish lesions, which later enlarge into distinctive cigar-shaped necrotic patches (2.5–15 cm in length). As the infection progresses, lesions coalesce, resulting in severe foliar blight (Fig. 11.3c). In advanced stages, *E. turcicum* sporulates heavily, producing dark grey conidia that cause scorched, brittle leaves resembling frost damage. Severe infections hinder carbohydrate movement, reducing yield and grain quality. The pathogen survives on maize debris as conidia, mycelia, or chlamydospores. Early summer moisture triggers conidial germination, with spores spread by rain (within-field) and wind (long-distance). Infection requires 6–18 h of leaf wetness, lesions develop in 7–12 days, and repeated sporulation drives rapid epidemics. IDM combining resistant varieties, cultural practices, and fungicides is vital to minimize losses and sustain productivity.

Fig. 11.3 Symptomatology of maize foliar diseases: (**a**, **b**) MLB characterized by diamond-shaped lesions with a light brown centre and dark brown margins; (**c**) TLB showing characteristic cigar-shaped necrotic lesions; (**d**, **e**) BLSB with a prominent light and dark brown banded pattern on the leaf and sheath, as well as sclerotia on the ear

11.2.3 Banded Leaf and Sheath Blight

Rhizoctonia solani f. sp. *sasakii* causes BLSB of maize. It is a soil-borne fungal disease and poses a critical economic concern, especially in areas with high relative humidity and temperature around 28 °C (Singh and Shahi 2012). This disease was first reported in Sri Lanka as a minor disease (Bertus 1927). In India, the disease was first identified in the Tarai region of Uttar Pradesh in 1960 (Payak and Renfro 1968). Since then, it has spread to various states, including Punjab, Haryana, Madhya Pradesh, West Bengal, and Assam (Rani et al. 2013). The pathogen thrives under high temperatures (optimal at 32 °C) and high relative humidity (>90%), with regular rainfall exceeding 100 mm during the early growth stages (Ahuja and Payak 1981; Singh and Shahi 2012). Disease severity is aggravated by high crop density, particularly in plants at the pre-flowering stage (approximately 40–50 days old) (Saxena 2002). The disease displays as conspicuous banded lesions on leaf sheaths, sclerotial formations, and extensive necrosis of aerial plant parts, excluding the tassel (Sharma et al. 2004) (Fig. 11.3d, e). In severe cases, BLSB can result in complete ear rot, leading to yield losses of 10-100%, with the greatest reductions occurring under continuous rainfall (Ahuja and Payak 1982; Lu et al. 2012; Hooda et al. 2017b).

The disease primarily displays as irregular, water-soaked lesions on the sheaths and bases of leaves, which swiftly expand into straw-coloured necrotic areas, accompanied by characteristic dark-coloured bands (Rani et al. 2013). Infection advances from the basal sheath to the developing ear, leading to husk leaf caking, premature drying, and complete ear deterioration. The pathogen advances through sclerotia and mycelial fragments remaining in the soil or infected crop debris, with primary inoculum sources including sclerotia (overwintered) that surface during field preparation (Ahuja and Payak 1985; Simon et al. 2014). The secondary dissemination of disease occurs through physical contact between infected and healthy plants, resulting in intra-field disease spread primarily during the cropping season (Gilligan 2002). High relative humidity (>70%) and prolonged wet conditions facilitate rapid pathogen colonization, whereas dry conditions slow disease progression.

11.2.4 Common Rust

Common rust (CRust) of maize caused by *Puccinia sorghi* Schw is a significant foliar disease. The disease is identified by oval to elongated cinnamon-brown pustules spread across the upper and lower leaf surfaces. The rupturing of these pustules releases red powdery urediniospores, which are dispersed by wind, facilitating the rapid spread of disease. As the infection progresses, the pustules turn a darker, brownish-black colour, producing teliospores that enable the survival of the pathogen between cropping seasons (Fig. 11.4a, b). Under severe outbreaks, these pustules may also develop on the ears and tassels.

Puccinia sorghi exhibits a heteroecious and macrocyclic life cycle, relying on an alternate host (*Oxalis* spp.). It undergoes five spore stages, namely: teliospores, basidiospores, spermatia, aeciospores, and urediniospores. In tropical and subtropical regions, urediospores remain on crop residue, serving as the primary inoculum for the subsequent cropping seasons. The dissemination of urediospores by wind allows the pathogen to migrate seasonally, spreading from tropical and subtropical areas into temperate maize-growing regions. The sexual stage of the fungus is predominantly observed in warmer climates. Epidemiological studies indicate that

Fig. 11.4 Symptomatology of maize diseases: (**a**, **b**) CRust with elongated cinnamon-brown pustules spread across both sides of the leaf; (**c**) PRust with circular to oval pustules that initially appear orange-red to light cinnamon brown; and (**d**, **e**) ChR with abnormal drying of the plant and a distinctive charcoal-grey to black appearance inside the stalk

common rust thrives in environments with moderate temperatures (16–25 °C) and high relative humidity, making subtropical, temperate, and highland maize-growing regions particularly susceptible to outbreaks (Bekeko 2019).

11.2.5 Polysora Rust

Polysora rust (PRust) of maize also known as Southern corn rust is caused by *Puccinia polysora* Underw. The disease was first identified in Alabama, USA

(Underwood 1897). It is a foliar disease that significantly reduces maize production under humid, warm conditions. The disease is identified by small (0.2–2 mm), circular to oval pustules that initially appear orange-red to light cinnamon brown. These pustules mainly develop on the upper leaf surface, with fewer and slower-developing lesions on the lower surface (Fig. 11.4c). Unlike common rust, the pustules of *P. polysora* are less powdery during the starting phase. The leaf epidermis remains intact over the pustules for a longer duration before eventually rupturing through a longitudinal slit. As the disease progresses, chocolate-brown to black teliospores form in a ring surrounding the uredial pustule, thereby contributing to the disease cycle and the pathogen's survival.

The primary and secondary sources of inoculum for *P. polysora* are urediniospores, which overwinter on infected crop debris and are disseminated by wind and rain splash. Upon settling on maize foliage, the spores germinate and penetrate the leaves through stomatal entry, leading to the formation of uredial pustules. These pustules eventually rupture, releasing additional urediniospores as a medium for secondary infections. The disease proliferates under warm, humid conditions, with an optimum temperature ranging between 24 and 28 °C for urediniospore germination, infection, and symptom development. A prolonged period of high humidity, dew, or light rain accelerates disease progression.

11.2.6 Charcoal Rot

Charcoal rot (ChR), caused by *Macrophomina phaseolina* (Tassi) Goid, was first reported in India in 1936 by Uppal et al. (1936). Charcoal rot is a devastating disease that poses a significant constraint to maize production, resulting in yield losses of up to 70% (Harlapur et al. 2023). The severity of this disease is more devastating under drought and dry soil conditions. Charcoal rot typically appears after flowering, especially under soil–moisture stress. The initial symptoms include abnormal drying of the upper leaves, which can result in stem lodging and premature plant death (Fig. 11.4d, e). At maturity, the disease is identified by the discolouration of the lower stem internodes, typically limited to the first five nodes, which display a distinctive charcoal-grey to black appearance.

Affected stems become fragile, losing structural integrity, and can be easily compressed between fingers. The disease's progression is closely linked to environmental factors, making its management particularly challenging in regions experiencing prolonged drought conditions. The pathogen invades the root 1–2 weeks after planting, and symptoms are visible only after plant maturity, which makes its management quite challenging (Khaledi and Taheri 2016).

11.2.7 Fusarium Stalk Rot

Fusarium stalk rot (FSR), a devastating root and stalk disease, was first documented in the United States by Pammel in 1914. The primary causal agent of *Fusarium* stalk rot is *Fusarium verticillioides*. In India, the disease was first reported in 1964 from Mount Abu, Rajasthan, by Arya and Jain. FSR causes yield losses up to 42% and may accelerate further under prolonged dry conditions (Khokhar et al. 2013). The disease is particularly prevalent under water-stressed soil conditions, where it significantly affects plant health and productivity.

Symptoms of FSR typically appear after the flowering stage, with affected plants exhibiting wilting and dull, greyish-green leaf discolouration (Fig. 11.5a, b). The lower stalk turned from dark green to straw-coloured. Rotting is observed at the root crown part and lower internodes. Upon splitting open the infected stalks, internal tissues display a significant light pink to tan discolouration. In some cases, FSR occurs concurrently with charcoal rot, further exacerbating crop damage and yield losses.

11.2.8 Brown Stripe Downy Mildew

Brown stripe downy mildew (BSDM) of maize is caused by *Sclerophthora rayssiae* var. *zeae* Payak & Renfro. This pathogen exclusively affects leaf tissues. The initial symptoms are characterized by narrow, chlorotic, or yellowish stripes, typically 3–7 mm wide, which may later develop into reddish-purple discolouration in specific maize genotypes. These lesions are clearly separated by leaf veins, which run parallel to them. As the infection advances, adjacent lesions coalesce, resulting in extensive striping and blotching, primarily on the lower leaves, where disease severity is likely to be highest (Fig. 11.5c). Unlike other downy mildews, *S. rayssiae* var. *zeae* does not induce systemic infection, thereby limiting its impact to the foliage.

The primary inoculum of the pathogen consists of soil-borne oospores and sporangia (conidia) present in infected crop residues. Oospores germinate in response to root exudates when soil temperatures exceed 20 °C, producing germ tubes that penetrate underground plant tissues and initiate infection. Subsequently, sporangia (conidia) formed on the sporangiophores (conidiophores) emerge through the stomata, predominantly at night. These are disseminated via wind and rain splash, facilitating secondary infections. The disease is favoured by high relative humidity of 80% for more than 10 h or 100% for at least 4 h at night. The optimal temperature range for disease development is between 10 and 33 °C, particularly under prolonged humid conditions.

11.2.9 Sorghum Downy Mildew

The sorghum downy mildew (SDM) disease in maize is caused by *Peronosclerospora sorghi* Weston and Uppal Shaw. The pathogen induces systemic infection in seedlings shortly after emergence. Infected seedlings exhibit characteristic symptoms, including chlorosis, stunted growth, and sterility, resulting from oospore formation. Infected leaves become narrow, pale yellow, and covered in fine downy fungal growth, particularly more significant on lower leaf surfaces near the basal region

Fig. 11.5 Symptomatology of maize diseases: (**a**, **b**) FSR showing shrunken, discoloured stalk with dark green to straw-coloured lower internodes; Upon split open, the infected stalk's internal tissues display a light pink to tan discolouration (**c**) BSDM with narrow, chlorotic or yellowish stripes; (**d**) SDM-infected seedlings exhibiting chlorosis, stunted growth, and fine downy fungal growth, especially on the lower leaf surfaces near the base; (**e**, **f**) RDM showing characteristic symptom known as the "half-diseased leaf"

(Fig. 11.5d). The systemic infection in plants leads to the development of distinctive leaf stripes, resembling extensive colonization by the pathogen.

The disease cycle closely resembles that of BSDM (*Sclerophthora rayssiae* var. *zeae*), where primary infection originates from overwintering oospores in infected plant debris. Under favourable conditions, oospores germinate and release motile zoospores, which infect young seedlings. Secondary infection occurs via wind-dispersed conidia, which penetrate maize leaves through stomata. Conidial production is optimal at 24–26 °C, with germination requiring free moisture or high humidity, favouring an environment of 21–25 °C. Systemic infection is most severe when temperatures range between 11 and 32 °C and moisture persists for at least 4 h, creating conducive conditions for rapid disease spread.

11.2.10 Rajasthan Downy Mildew

Rajasthan downy mildew (RDM) of maize is caused by *Peronosclerospora heteropogoni*. The pathogen induces systemic infection and severely hampers plant growth and yield. The disease appears early in the seedling stage and is diagnosed by the characteristic "half-diseased leaf" symptom, where pale discolouration appears at the base of the second and third leaves. Infected leaves have yellow stripes that extend from the base to the green portions (Fig. 11.5e, f). Furthermore, systemically affected plants exhibit a distinct yellowish appearance visible from a distance. Severely infected seedlings often do not survive beyond the knee-high stage, while in some cases, surviving plants exhibit upright, yellow leaves that are closely spaced along the stem.

Systemic infections inhibit cob formation, leading to small, poorly filled, or non-functional ears. In some cases, tassel development may occur but mainly resulting to sterile or infertile pollen production, further impairing reproductive success.

Peronosclerospora heteropogoni primarily survives and propagates through the production of conidia and oospores on collateral hosts such as *Heteropogon contortus* and *Heteropogon melanocarpus*, with maize serving as a secondary host where only conidia are produced. Oospores formed on these grass hosts serve as the primary inoculum, initially infecting the grasses before spreading to maize at the 2–3 leaf stage. Systemically infected leaves of both host plants produce large numbers of conidia, particularly during the early morning hours (2:30 AM–3:00 AM) on both leaf surfaces, facilitating rapid disease spread. The optimal disease development occurs at temperatures between 23.5 and 30.1 °C with an average rainfall of 154.4 mm and a relative humidity of 87.3%. Conidial germination is most efficient at 25 °C, while no germination occurs at temperatures exceeding 40 °C.

11.2.11 Bacterial Stalk Rot

In recent years, BSR has emerged as a destructive disease of *Kharif* maize in India (Kumar et al. 2016), which was first documented by Prasad (1930). The initial report of this disease was made by Sabet from Egypt followed by its first report in India by Hingorani (Sabet 1954; Hingorani et al. 1960). Initially, the incitant was identified as *Erwinia dissolves* (Sinha and Prasad 1977), but later disease symptomatology showed closer relatedness to *Erwinia chrysanthemi* pv. *zeae* (Hingorani et al. 1960; Martínez-Cisneros et al. 2014). Further, Samson et al. (2005) taxonomically renamed the pathogen as *Dickeya zeae*. The high temperatures, elevated relative humidity, and continuous rainfall are conducive to pathogen proliferation and infection.

Fig. 11.6 Symptomatology of BSR showing water-soaked lesions, rotting at the basal stem, particularly in the leaf sheaths, and rapid decay of the basal internodes

The pathogen generally infects maize at 30–35 DAS and may persist until harvest, with peak severity occurring at the tasseling stage. The initial symptoms include water-soaked lesions and basal stem rot, particularly in the leaf sheaths, followed by rapid decay of the basal internodes (Fig. 11.6). As the disease progresses, leaves begin to roll, and infected tissues lose their natural green colour, exhibiting a characteristic boiled-water appearance. Advanced infection causes severe internal tissue maceration accompanied by a strong fermenting odour. Later, these symptoms can lead to substantial yield losses in maize under conducive environmental conditions.

11.3 Disease Resistance and Management

Efficient disease management strategies in maize require a joint venture of sustainable, economical, and eco-friendly practices. The utilization of resistant cultivars is one of the most effective approaches, which not only reduces disease incidence and severity but also can be coupled with other IDM approaches to maximize crop yield and safeguard the environment (Katan 2017). Notably, no single cultivar/variety shows complete resistance to every pathogen. Cultivars/varieties labelled as resistant exhibit a higher level of defence against certain diseases than those labelled as tolerant (Baysal-Gurel et al. 2012).

Plant diseases are classified into two groups, viz., qualitative and quantitative resistance. Qualitative resistance is primarily race-specific, meaning it provides resistance against a primary set of pathogens. It is conferred by a single gene (dominant or recessive) and provides high levels of resistance. In contrast, quantitative resistance is governed by an oligogenic or polygenic mode of inheritance. It is regarded as race-nonspecific resistance. It is often more durable than the qualitative resistance due to its supplementary or moderately dominant gene action. While qualitative resistance is more suitable for use in breeding programs, quantitative resistance is generally preferred in field conditions due to its wide effectiveness and long-term stability (Lindhout 2002; Parlevliet 2002). Maize retains substantial potential to accommodate genes conferring resistance to major diseases (Kumar et al. 2025a). In maize, most deployed resistance is quantitative, with qualitative resistance genes (R-genes) mapped through Mendelian genetics. R-genes have been extensively studied, with their signalling pathways and mechanisms of action elucidated (Glazebrook 2001; Hammond-Kosack and Parker 2003).

A synthesis of 50 publications on maize disease resistance mapping has documented 437 dQTL, 17 R-genes, and 25 R-gene analogues (Wisser et al. 2006). Over the past decade, extensive research has explored the genomic organization of disease resistance genes in maize and other crops. McMullen and Simcox (1995) reviewed the distribution of disease and insect resistance genes in maize, organizing the genome into 100 bins of approximately 20 cM each (Gardiner et al. 1993). Their analysis, based on 16 studies, revealed a non-random distribution of resistance loci, synthesizing data from 34 dQTL and 19 significant resistance genes. These findings underscore the importance of understanding the genetic basis of disease resistance in maize, thereby facilitating the development of more resilient cultivars for sustainable crop production. The integrated breeding framework (Fig. 11.7) emphasizes how molecular techniques, high-throughput phenotyping, and gene-editing programs work together to speed up the development of disease-resistant maize varieties. Ongoing pathogen surveillance and close stakeholder collaboration further refine resistance strategies and help ensure the successful adoption of improved cultivars. Numerous disease-resistant hybrids across different agroclimatic zones have been released from 2016 to the present by various private and public sector institutes (Table 11.2).

11.3.1 Disease Resistance Against Primary Maize Disease

11.3.1.1 Maydis Leaf Blight

Maydis leaf blight (MLB) remains an important constraint in maize production. MLB causes yield losses up to 40% under favourable environmental conditions (Kumar et al. 2022a, 2025a; Aggarwal et al. 2024). Several studies have demonstrated that no maize cultivars exhibit complete immunity to MLB disease. However, specific inbred lines exhibited substantial resistance (Kumar et al. 2016). The absence of fully resistant varieties has necessitated a shift toward polygenic and quantitative resistance breeding, which offers the most viable approach for managing MLB disease. Numerous genetic studies have demonstrated that a dominant genetic effect is associated with MLB resistance, characterized by distinct additive × additive interactions. However, maturity-related traits show significant dominance × dominance interactions. These findings showed that hybrid breeding practices, leveraging resistant inbred lines such as CML269-1 and P72c1Xbrasil1177-2 in combination with susceptible lines HKIPC4B and ESM 113, hold promise for improving disease resistance. Moreover, heritability assessment for MLB disease resistance at both broader-sense (91.9%) and narrower-sense (84.3%) levels highlights the potential for genetic improvement through selection.

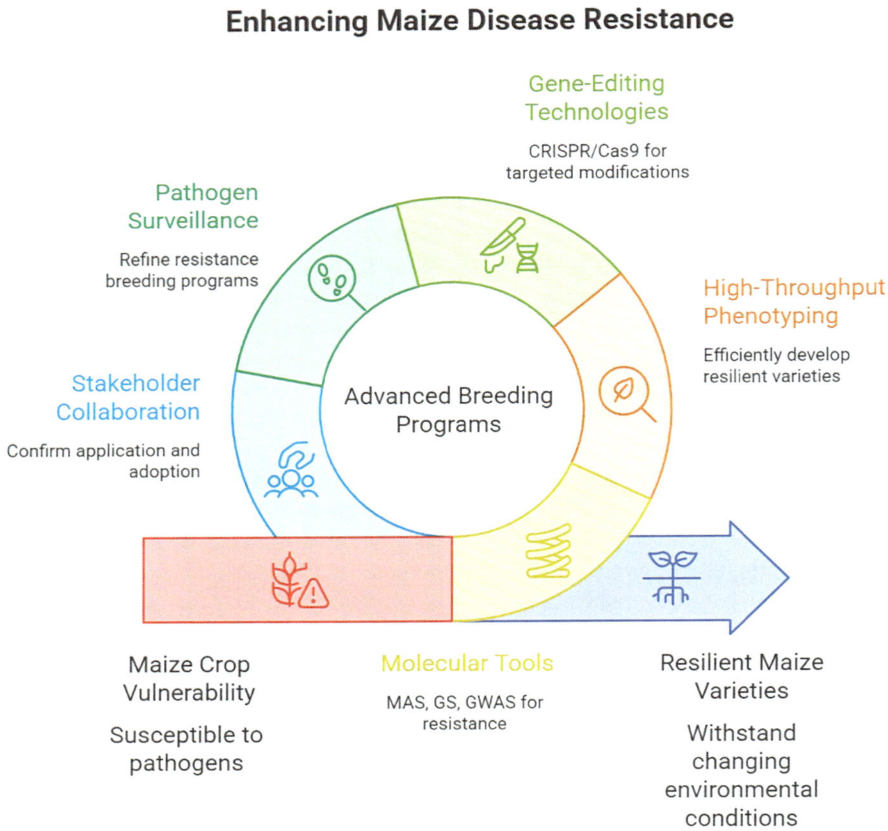

Fig. 11.7 Synergistic approaches to strengthen maize disease resistance against important maize diseases

Various historical studies from the mid-twentieth century identified MLB resistance as especially recessive (Smith and Hooker 1973). The advancements in molecular genetics include restriction fragment length polymorphism (RFLP) and amplified fragment length polymorphism (AFLP) markers. These techniques facilitated the mapping of MLB-resistant quantitative trait loci (QTLs), such as *rhm1*, on the short arm of chromosome 6 (Zhao et al. 2012; Choudhary et al. 2019). Further improvement of this genetic region has pinpointed a key contender gene, *lysine histidine transporter 1* (*LHT1*), within an interval of 8.56 kb. Several QTL studies from different scientific communities worldwide have confirmed the presence of resistance loci with additive, partially dominant, or epistatic effects (Choudhary et al. 2019). Among these, *ZmCCoAOMT2*, which encodes a caffeoyl-CoA O-methyltransferase, plays a crucial role in MLB disease resistance by participating in lignin biosynthesis and regulating the hypersensitive response. Variations in gene expression and amino acid sequences within *ZmCCoAOMT2* influence metabolite

Table 11.2 Catalogue of different maize hybrids released with documented disease resistance against major maize diseases across diverse agro-climatic regions of India

Disease	Zones				
	NHZ	NWPZ	NEPZ	PZ	CWZ
Maydis leaf blight (MLB)	HPC-4, DKC-9126, IBCH 402, CP Sweet 2, Megha Maize-1 (RCM 1-61), Shalimar Maize Hybrid-5 (IMH-221) (DMRH-1417)	SMH 4555, Pusa Super Sweet Corn-3, Pusa Waxy Maize Hybrid-1, Pusa Biofortified Maize Hybrid-6, Pusa Biofortified Maize Hybrid-7, P 34407 (PM 21111L), Pusa Biofortified Maize hybrid-4, Punjab Baby Corn 1, IMH 225, GK 3302, GK 3303, IMH 222, JC 4, JC 12, PMH 13, ADV 7132 (ADV 765), PAC 745 Gold	ADV 785 (ADV-7733), Pusa Super Sweet Corn-3, Pusa Biofortified Maize Hybrid-8, Pusa Biofortified Maize Hybrid-6, IMH 230, IMH 231, Top Class (KMH 1411), QPMH 6 (EDV of PMH 6), IMH 228, KMH 8121 (HM 20105), HT 5110 (HT519074), Jawahar Maize-1014 (HMM 1014), PAC 751	Pusa Super Sweet Corn-3, DKC-9126, Rajendranagar Makka-182, P 34407 (PM 21111L), Pusa Biofortified Maize hybrid-4, JKMH 4243, ADV-756 (ADV 0990296), HTMH 5108, JKMH-4848 (IMR 498), JKMH 4243, IMH 225	TMMH-2882, SMH 4555, Pusa Super Sweet Corn-3, Pusa Biofortified Maize Hybrid-6, Pusa Biofortified Maize Hybrid-7, DKC-9133, Pusa Biofortified Maize hybrid-4 (APH 4), ADV-756 (ADV 0990296), Pratap Hybrid-6, Pratap QPM Hybrid-5

(continued)

Table 11.2 (continued)

Disease	Zones				
	NHZ	NWPZ	NEPZ	PZ	CWZ
Turcicum leaf blight (TLB)	HPC-4, DKC-9126, IBCH 402, Pant Composite Maize 4 (PCM-4) (DOP-339), VL VitA (FPVH 1), Pant Composite Maize 4 (PCM-4) (DOP-339), VL VitA (FPVH 1), Shalimar Kishenganga-3 (KDM-30), CP Sweet 2, DKC 8211 (IU8229), Him Palam Maize Composite 2 (L 316), Megha Maize-1 (RCM 1-61), Shalimar Maize Hybrid-5 (IMH-221) (DMRH-1417)	NK-7660 (SYN-207660), P 34407 (PM 21111L), CP 999, Pusa Biofortified Maize hybrid-4, Pusa Biofortified Maize hybrid-5, IMH 225, IMH 222, ADV 7132 (ADV 765), PAC 745 Gold	ADV 785 (ADV-7733), IMH 230, IMH 231, Pusa Biofortified Maize hybrid-5, VL Madhurima (FSCH 144), DMRH 1301, VL Madhurima (FSCH 144), IMH 228, Jawahar Maize-1014 (HMM 1014), Siata Mimpui (MZM 11), Mizo Popcorn 1 (MZM 17), NMH-51, PAC 751	NK-7660 (SYN-207660), ADV 752 (ADV-7211), DKC-9126, KMH-8333, Rajendranagar Makka-182, VNR 4597, PM 20205L, P 34407 (PM 21111L), Bio 207, Pusa Pop Corn Hybrid-2, Pusa Biofortified Maize hybrid-4, Pusa Biofortified Maize hybrid-5, JKMH 4243, ADV 768 (ADV 7251), MM 9344 (DMH 192), ADV-756 (ADV 0990296), HTMH 5108, HTMH 5109 (HT 51412616), JKMH-4848 (IMR 498), Pusa HM-8 Improved, JKMH 4243, ADV 768 (ADV 7251), R 3414 (RMH 3414), IBCH 401, IPCH 501 (IPCH 1901)	DKC-9133, DKC-9165, Pusa Biofortified Maize hybrid-4 (APH 4), Pusa Biofortified Maize hybrid-5 (APTQH-5), ADV-756 (ADV 0990296), DMRH 1301, DMRH 1308, IBCH 401 (IMHSB 19KB-2), Pratap Hybrid-6 (EH-2936), Pratap Raj Hybrid Maize-1010 (WH-1010)

(continued)

Table 11.2 (continued)

Disease	Zones				
	NHZ	NWPZ	NEPZ	PZ	CWZ
Banded leaf and sheath blight (BLSB)	HQPM-29, IBCH 402, Pant Composite Maize 4 (PCM-4) (DOP-339), Pant Composite Maize 4 (PCM-4) (DOP-339), CP Sweet 2, DKC 8211 (IU8229), Megha Maize-1 (RCM 1-61)	SMH 4555, P 34407 (PM 21111L), CP 999, Pusa Biofortified Maize hybrid-4, GK 3302, PAC 745 Gold	Top Class (KMH 1411), LG 34.05 (BL 900), Siata Mimpui (MZM 11), Mizo Popcorn 1 (MZM 17)	P 34407 (PM 21111L), Pusa Biofortified Maize hybrid-4 (APH 4), HTMH 5108, LG 34.05 (BL 900), R 3414 (RMH 3414)	TMMH-2882, SMH 4555, DKC-9133, DKC-9165, Pusa Biofortified Maize hybrid-4 (APH 4), LG 34.05 (BL 900)
Charcoal rot (ChR)	IBCH 402, CP Sweet 2	SMH 4555, Pusa HQPM4 Improved, CP 999, Pusa Pop Corn Hybrid-1, Pusa Biofortified Maize hybrid-5, IMH 225, JC 4, JC 12, PMH 13	IMH 230, Pusa Biofortified Maize hybrid-5, Pusa HM 4 (Male Sterile Baby Corn-2), VL Madhurima (FSCH 144), DMRH 1301, VL Madhurima (FSCH 144), IMH 228, Pusa Jawahar Hybrid Maize-2, NMH-51	ADV 752 (ADV-7211), KMH-8333, Rajendranagar Makka-182, VNR 4597, Bio 207, Pusa Pop Corn Hybrid-1, Pusa Biofortified Maize hybrid-5, Pusa HM 4 (Male Sterile Baby Corn-2), JKMH 4243, ADV 768 (ADV 7251), MM 9344 (DMH 192), ADV-756 (ADV 0990296), HTMH 5109 (HT 51412616), JKMH-4848 (IMR 498), Pusa HM-8 Improved, JKMH 4243, ADV 768 (ADV 7251), R 3414 (RMH 3414), IBCH 401, IMH 225	TMMH-2882, SMH 4555, Pusa HQPM4 Improved, Pusa Biofortified Maize hybrid-5, Pusa HM 4 (Male Sterile Baby Corn-2), ADV-756 (ADV 0990296), DMRH 1301, DMRH 1308, IBCH 401, Pratap Raj Hybrid Maize-1010 (WH-1010)

(continued)

Table 11.2 (continued)

Disease	Zones NHZ	NWPZ	NEPZ	PZ	CWZ
Sorghum downy mildew (SDM)		KMH 8322, ADV 7132 (ADV 765), PAC 745 Gold	LG 34.05 (BL 900), PAC 751	ADV-756 (ADV 0990296), LG 34.05 (BL 900), JKMH-4848 (IMR 498), NMH-51	ADV-756 (ADV 0990296), LG 34.05 (BL 900), Pratap Hybrid-6 (EH-2936)
Common rust (CRust)			LG 34.05 (BL 900)	MM 9344 (DMH 192), HTMH 5108, HTMH 5109 (HT 51412616), LG 34.05 (BL 900), Pusa HM-8 Improved	LG 34.05 (BL 900)
Fusarium stalk rot (FSR)	DKC-9126, Megha Maize-1 (RCM 1-61)	Pusa HQPM4 Improved, P 34407 (PM 21111L), DMRHP 1402, IMH 225, IMH 222, ADV 7132 (ADV 765)	IMH 231, VL Madhurima (FSCH 144), VL Madhurima (FSCH 144)	ADV 752 (ADV-7211), DKC-9126, DKC-9144, PM 20205L, P 34407 (PM 21111L), ADV 768 (ADV 7251), ADV-756 (ADV 0990296), HTMH 5108, ADV 768 (ADV 7251), IBCH 401	R 8050 (R-3437), Pusa Male Sterile Baby Corn Hybrid-3, Pusa HQPM4 Improved, DKC-9133, ADV-756 (ADV 0990296), DMRHP 1402, IBCH 401, DKC 9215 (IU8636), Pratap Hybrid-6 (EH-2936), Pratap Raj Hybrid Maize-1010 (WH-1010), Pratap Raj Hybrid Maize-1095 (WH-1095), Pratap QPM Hybrid-5 (EHQ-64)

(continued)

Table 11.2 (continued)

	Zones				
Disease	NHZ	NWPZ	NEPZ	PZ	CWZ
Bacterial stalk rot (BSR)	HQPM-29, DKC 8211 (IU8229), Him Palam Maize Composite 2 (L 316)		Pusa Jawahar Hybrid Maize-2	DKC-9144, PM 20205L, ADV-756 (ADV 0990296), HTMH 5108	DKC-9133, ADV-756 (ADV 0990296)
Curvularia leaf spot (CLS)		Pusa HQPM4 Improved	VL Madhurima (FSCH 144), VL Madhurima (FSCH 144)	ADV-756 (ADV 0990296), HTMH 5108, JKMH-4848 (IMR 498), IBCH 401	R 8050 (R-3437), Pusa Male Sterile Baby Corn Hybrid-3, Pusa HQPM4 Improved, ADV-756 (ADV 0990296), IBCH 401, IMH 225, DKC 9215 (IU8636), Pratap Raj Hybrid Maize-1010 (WH-1010)

synthesis in the phenylpropanoid pathway, thereby contributing to disease resistance. More recently, the *Zea mays ascorbate peroxidase 1* (*ZmAPX1*) gene has been identified as a positive regulator of MLB resistance by regulating hydrogen peroxide (H_2O_2) levels and activating the jasmonic acid-mediated defence pathway. The identification of these resistance-associated genes provides valuable genetic resources for breeding more resilient maize varieties.

11.3.1.2 Turcicum Leaf Blight

The recognition of resistance sources against *Exserohilum turcicum*, the cause of turcicum leaf blight, has been a global research focus. The comprehensive germplasm screening has led to the recognition of highly resistant inbred lines. These lines offer an encouraging thoroughfare for maize advancement. The development of new varieties with enhanced disease resistance and tolerance to numerous abiotic and biotic stresses represents a viable strategy to enhance maize productivity and profitability, particularly for small-scale farmers.

An important aspect of TLB management is identifying physiological races, enabling targeted breeding efforts. These races have been differentiated based on lesion development and post-inoculation of the pathogen. Over the past two decades, different physiological races have been identified globally, which include races 0 and 1 in China, 23 N, 23, and 2 N in Mexico, races 23, 23 N, and 0 in Zambia, and 0, 2 N, and 23 N in Uganda (Jenkins and Robert 1952; Lević et al. 2008). Kumar et al. (2022b) conducted a genetic analysis for TLB resistance using 288 assorted maize inbred lines. A total of 89 SSR markers were used to screen the lines for resistance. Disease screening was carried out across two hot-spot locations, showing a diverse resistance pattern. Using highly polymorphic markers leads to confirming a strong genetic diversity and purity among lines. The structure analysis of population grouped the inbred lines into 3–5 clusters consistent with their origin, with substantial genetic variation perceived within and among populations. GWAS analysis revealed 15 SSR markers associated with TLB resistance across chromosomes, representing 11 candidate genes, which offer valuable resources for molecular breeding. Such variability underscores the importance of developing durable resistance sources that remain effective across diverse agroclimatic conditions and pathogen populations. Another study conducted by Kumar et al. (2025b) screened 384 maize inbred lines for finding candidate genes against TLB via GWAS analysis at four hot-spot locations. Sixteen marker trait associations (MTAs) were identified for resistance on all the chromosomes except 1, 9, and 10, showing 20-30% phenotypic variation. Furthermore, five Ortho-CGs and 18 CGs were identified in the genomic regions of MTAs. Some of these identified CGs, including *Zm00001eb226290*, *Zm00001eb293590*, and *Zm00001eb293670*, are responsible for TLB disease resistance in maize and sorghum.

Both qualitative and quantitative genetic factors characterize TLB resistance in maize. Various studies indicate that resistance is particularly polygenic, involving multiple genes. These genes collectively diminish lesion numbers while driving minimal effects on pathogen sporulation. Research conducted by Mallikarjuna et al. (2007) in India enlightened the genetic makeup of resistance across six maize generations. Their findings marked the involvement of additive, dominance, and epistatic interactions, along with dominance and dominance × dominance effects being frequent in most crosses. These genetic interactions play a crucial role in shaping resistance, particularly at anthesis, and are instrumental in breeding superior hybrids through double-cross and top-cross breeding strategies.

Introducing significant resistance genes has also been a key path in maize breeding programs. Classical studies conducted by Hooker and colleagues (Hooker 1963; Hooker et al. 1970) identified six dominant genes, namely *Ht1*, *Ht2*, *Ht3*, *Htm1*, *Htn1*, and *NN*, in addition to a single recessive gene, *ht4*, which confers resistance to various *S. turcica* races. However, homozygosity for *Ht* genes provides lasting resistance. Despite this, polygenic resistance often offers more durable protection, especially during severe epidemics. The evolving nature of the pathogen, incorporating both major-gene and polygenic resistance, remains a strategic breeding approach to achieve long-term management of TLB disease.

The extended evolution of *E. turcicum* necessitates ongoing research to develop resilient maize genotypes. Future breeding programs should focus on MAS and genomic tools to accelerate the deployment of resistance genes against emerging TLB races globally.

11.3.1.3 Banded Leaf and Sheath Blight

BLSB remains a significant problem for maize production, particularly devastating in humid and subtropical regions. Traditional disease management strategies, which include fungicide applications and other cultural practices, have been insufficient or economically unattainable for large-scale management practices. Furthermore, the development of resistant hybrids through traditional breeding programs has been hindered by the scarcity of resistant germplasm and challenges associated with artificial inoculation. Despite the economic importance of BLSB, studies on the genetic basis of resistance in maize remain limited (Sharma 1999).

Quantitative Trait Loci (QTL) mapping has emerged as a powerful tool for diagnosing the genetic basis of complex disease resistance (Choudhary et al. 2019). This method enables researchers to locate specific resistance loci, analyse gene-environment interactions, and facilitate MAS to improve resistance traits. An important landmark study by Zhao et al. (2006) mapped 77 QTLs associated with BLSB resistance across chromosomes 1, 2, 3, 4, 5, 6, and 10 in an F_2 population derived from a cross between resistant (R15) and susceptible (478) inbred lines. The findings of this study provide a crucial basis for future breeding efforts for the introgression of resistance genes into elite maize varieties.

Along with QTL mapping, combining ability analysis has been an important tool in evaluating resistance inheritance (Choudhary et al. 2019). Vimla et al. (1988) assessed 15 maize crosses involving six inbred lines across two locations and reported both general combining ability (GCA) and specific combining ability (SCA) to have a substantial effect on disease response. Nevertheless, GCA effects were predominant, indicating that the additive genetic effects play a key role in resistance. Notably, the inbred line CM 104 showed strong potential as a resistance donor, making it a valuable resource for breeding programs. Ongoing efforts aim to identify and tag additional resistance QTLs to improve the genetic resilience of modern maize hybrids (Zhao et al. 2006). The incorporation of MAS in advanced breeding programs offers an encouraging avenue for intensifying resistance development.

11.3.1.4 Post-flowering Stalk Rot

Macrophomina phaseolina, the incitant of charcoal rot in maize, is a devastating pathogen in maize production. The resistance to charcoal rot (ChR) is a polygenic trait in nature governed by both additive and non-additive gene action, with significant environmental interaction (Singh and Kaiser 1991; Krishna et al. 2013; Mir et al. 2018; Hooda et al. 2017a). Accustomed to its quantitative inheritance and the significant effects of various environmental parameters, breeding for charcoal rot resistance requires incorporating advanced molecular practices to enhance genetic gains over time.

Advanced breeding programs utilize molecular technologies to enhance resistance breeding practices in maize. The use of genotypic data enables early-stage selection, allowing the development of breeding populations without proportionally increasing phenotyping efforts (Cooper et al. 2014; Hooda et al. 2017a). QTL analysis is an important tool for identifying genomic regions linked to disease resistance against BLSB. Moreover, various QTL mapping studies for other stalk rot diseases, such as *Gibberella* stalk rot (GSR) and Anthracnose stalk rot (ASR), have been reported in maize, including the fine-mapping of a major QTL for *Fusarium graminearum* resistance (Yang et al. 2010; Zhang et al. 2012; Ma et al. 2017) and the cloning of an NBS-LRR resistance gene for ASR on chromosome 4 (Jung et al. 1994; Abad et al. 2006; Broglie et al. 2011), molecular mapping for charcoal rot resistance in maize remains limited. This gap may be attributed to the complex nature of the disease, limited availability of resistant germplasm, and challenges in achieving consistent disease expression under natural conditions due to potential co-infections with other stalk rot pathogens. However, QTLs for *M. phaseolina* resistance have been identified in other crops such as sorghum (Mahmoud et al. 2018), soybean (da Silva et al. 2019) and sesame (Chowdhury et al. 2017), providing valuable insights for maize research.

Resistance to *Fusarium* stalk rot (FSR) in maize is a complex trait and was investigated using genomic prediction in three doubled-haploid populations derived from two crosses (Lohithaswa et al. 2025). Six genomic models were analysed and found that prediction accuracy has been increased with increasing the size of the training population along with the density of markers. The prediction accuracy also increased when the TS:VS ratios totalled 80:20 or 75:25. The study exhibited that F2 populations revealed higher genetic variation. The positive correlation and cross-validation using test cross values confirmed the efficiency of FSR resistance using GE-BV.

A genome-wide association study (GWAS) was conducted by Rashid et al. (2021) to identify genetic variants associated with charcoal rot using the CIMMYT Asia association mapping panel, comprising 396 tropical-adapted maize lines. The researchers identified approximately 19 significantly associated SNPs that are substantially associated with charcoal rot resistance, and verified through QTL mapping in two $F_{2:3}$ populations. Notably, the two overlapping QTLs, *qMSR6* and *qFMSR6*, located on chromosome 6 at 103.51 Mb, have been identified in different mapping populations (resistance donors), underscoring their potential role in resistance breeding.

Further phenotypic screening of 37 maize genotypes by Sravya et al. (2023) using the toothpick inoculation method at the Agricultural College, Polasa, Jagtial, has provided important insights into resistance sources against BLSB disease in maize. The study revealed the disease responses ranging from highly resistant (score 1) to moderately susceptible (score 7). Among all the lines, one inbred line (PFSR 135), three testers (CML 286, CML 451, and BML 7), and eight hybrids (PFSR 51 × BML 6, PFSR 132 × CML 286, PFSR 29 × CML 451, PFSR 70 × BML 6, PFSR 76 × CML 286, PFSR 135 × CML 286, PFSR 70 × BML 7, and PFSR 76 × BML 7) showed resistance against charcoal rot, and may serve as promising candidates for charcoal resistance breeding programs.

11.3.2 Integrated Disease Management (IDM) Strategies

IDM embodies different disease control strategies on the basis of multidisciplinary techniques. Some of the important IDM practices contributing towards disease reduction and control are listed in Table 11.3. One of the important agronomic strategies is crop rotation, which reduces the carryover of pathogens, thus providing a barrier to perpetuation. The cultivation of resistant/tolerant cultivars and hybrids sustains inherent defence mechanisms against selective phytopathogens. Sowing certified pathogen-free seeds reduces the initial spread into the crop system. Under temperate conditions, removing plant debris/residue from the soil reduces pathogen survival. Appropriate weed management plays an essential role in disease management, as most weeds are alternate hosts for various phytopathogens. Furthermore, soil fertility management increases plant vigour by stimulating growth, thereby enhancing resistance to disease pressure.

Table 11.3 Integrated disease management practices in maize

Name of disease	IDM practices
MLB, TLB, Brown leaf spot	*Cultural Practices* 1. Implement crop rotation to break the disease cycle and reduce pathogen inoculum in the field 2. Adopt proper field sanitation, including removal and destruction of infected crop residues, to minimize primary inoculum sources *Resistant Varieties* Use the resistant varieties mentioned in Table 11.2 *Biological Control* Seed treatment with *Azospirillum* (ACD 15 or ACD 20 Stran) @ 25 g and *Trichoderma* @ 6 g per kg seeds *Chemical Control* 1. For *MLB*, Seed treatment with Thiram @ 3 g/kg seed + foliar application of Mancozeb (Dithane M-45, Indofil M-45) or zineb (Dithane Z-78) at the first appearance of symptoms @ 2–2.5 g/L of water after 45 days of sowing + foliar spray of azoxystrobin 18.2% w/w + difenoconazole 11.4% w/w SC @ 1 mL/L at days after sowing is effective. 2. For TLB, seed treatment with Thiram 75 WP or Captan 50 WP @ 2 g/kg seed. Under severe infection, foliar application of azoxystrobin 18.2% w/w + difenoconazole 11.4% w/w SC @ 1 mL/L of water or mancozeb (Indofil M-45) @ 2.5 g/L of water. 3. To effectively control Brown Spot disease in maize, treating seeds with 20 g *Trichoderma* chalk formulation + mancozeb 63% or Thiram 40 F.S. @ 6 g/kg seed or use fungicides such as Benlate, Bavistin, and Plantavax at a dosage of 2.5 g/kg seed is recommended
BLSB	*Cultural Practices* Removal of the two basal leaves, along with the leaf sheath, to reduce disease incidence *Chemical Control* Foliar application of azoxystrobin 18.2% w/w + difenoconazole 11.4% w/w SC @ 1 mL/L of water immediately after appearance of first symptom, repeat the application at 10–15 day's interval *IDM Module* Use IDM module, soil amended with *Trichoderma harzianum* formulation @ 6 t/acre farm yard manure + seed treatment with Salicylic acid (100 ppm) + foliar spray of azoxystrobin 18.2% w/w + difenoconazole 11.4% w/w SC @ 1 mL/L using 500 L water/ha

(continued)

Table 11.3 (continued)

Name of disease	IDM practices
Charcoal rot	*Cultural Practices* 1. Deep summer ploughing 2. Field sanitation 3. Avoid water stress at the flowering stage 4. Adopt crop rotation with a non-cereal crop 5. If charcoal rot appears in the field, harvest the crop early to reduce ear loss *Resistant Varieties* Use of resistant varieties mentioned in Table 11.2 *Biological Control* Seed treatment with 25 g PSB (*Pseudomonas striata* H-21) and 6 g *Trichoderma harzianum* *Chemical Control* 5 mL Thiram 40 FS per kg seeds before sowing
Fusarium stalk rot	*Cultural Practices* 1. Field sanitation 2. Crop rotation with a non-cereal crop 3. Apply a balanced dose of fertilizers (low dose of nitrogen and high dose of potassium) 4. If stalk rot appears in the field, harvest the crop early to reduce ear loss *Resistant Varieties* Use of resistant varieties mentioned in Table 11.2 *Biological Control* FYM-100 kg + 2.5 kg *Trichoderma* chalk powder and slightly wet + Incubate it for 15–20 days, and after green growth of *Trichoderma*, mix it with 500 kg FYM + Vermicompost 100 kg + 10 kg ash of dung. Apply at the time of seed sowing as a prophylactic measure *Chemical Control* 1. Foliar spray after 40 and 55 DAS or immediately after the appearance of symptoms by using 2 g/L of carbendazim 12% + mancozeb 63% thoroughly covering the entire foliage. 2. If the disease severity is more than 50 DAS, give a second spray of tebuconazole @ 2 g/L solution *IDM Module* Seed Treatment: It should be done by using 6 g carbendazim/kg seed or 6 g tebuconazole + 20 g *Trichoderma chalk* formulation + 100 g dung ash. Mix all with the seed thoroughly in a plastic container or steel container by shaking it well (prophylactic measure)
Bacterial stalk rot	*Cultural Practices* 1. Destroy diseased plant debris 2. Adequate drainage avoids stagnation of water for a longer duration *Resistant Varieties* Use of resistant varieties mentioned in Table 11.2 *Chemical Control* Drenching of bleaching powder, $Ca(ClO)_2$, with 1.5 g/15 L of water immediately after the appearance of the first visible symptom

(continued)

Table 11.3 (continued)

Name of disease	IDM practices
Common rust, Polysora rust	*Cultural Practices* Cultivation of early-maturing varieties to escape severe infections *Resistant Varieties* Use of resistant varieties mentioned in Table 11.2 *Chemical Control* 1. Foliar spray of propiconazole 25% EC @ 1 mL/L of water 2. Foliar spray of mancozeb (Indofil M-45) @ 2.5–4 g/L of water upon the first appearance of pustules. If required, three spray applications at 10-day intervals are recommended
Brown stripe downy mildew	*Cultural Practices* 1. Roughing and burning of infected crop debris 2. Remove collateral host Takri grass (*Digitaria sanguinalis*) *Chemical Control* 1. Foliar application of mancozeb (Indofil M-45) @ 200 g/acre in 100 L of water after a fortnight of sowing
Rajasthan downy mildew (RDH)	*Cultural Practices* Roughing and burning of infected crop debris *Chemical Control* 1. Seed treatment with metalaxyl (Apron 35 SD) @ 8 g/kg seed 2. Foliar application of Ridomil MZ 68 (0.1%) after the appearance of the first symptom
Sorghum downy mildew (SDM)	*Cultural Practices* Late *Kharif* plantation must be avoided *Chemical Control* 1. Seed treatment with metalaxyl 35% WS or metalaxyl 8% + mancozeb 64% (WP) @ 3 g/kg seed 2. Foliar application of Amistar top (azoxystrobin 18.2% w/w + difenoconazole 11.4% w/w SC) @ 0.1% after appearance of first symptom

11.4 Conclusion

Maize, as a cornerstone of global food security, is continually threatened by diverse fungal and bacterial pathogens, including *Cochliobolus heterostrophus*, *Exserohilum turcicum*, *Rhizoctonia solani*, *Puccinia* spp., *Macrophomina phaseolina*, *Fusarium verticillioides*, and *Dickeya zeae*. These pathogens exploit favourable environmental conditions to initiate epidemics, while their variability complicates control. Moisture, temperature fluctuations, and crop residues drive infection cycles, enabling rapid spread. Complementary strategies—crop rotation, residue management, optimized planting densities, fungicides, and biological control—reduce inoculum pressure and strengthen resilience. The sustainable management of major maize diseases, MLB, TLB, BLSB, and charcoal rot, remains a formidable challenge for global maize cultivation. Yet, notable progress has been achieved through the identification and deployment of resistant germplasm in large-scale breeding programs. Advances in molecular breeding, quantitative trait locus (QTL) mapping, and genomic tools have broadened opportunities for durable resistance. Genes such as *Ht1*, *Ht2*, and *Ht3* have provided effective protection against TLB, while polygenic resistance mechanisms offer more enduring safeguards against evolving pathogen races. Resistance to BLSB and charcoal rot involves complex genetic interactions, requiring integrated breeding strategies that combine conventional and biotechnological innovations. The discovery of key QTLs and single-nucleotide polymorphism (SNP) markers has further strengthened the foundation for future breeding programs. Innovative approaches, such as biochar application, have shown promise in mitigating soil-borne diseases like charcoal rot. Biochar improves soil health and, when combined with biocontrol agents such as *Trichoderma viride*, enhances disease suppression.

The convergence of modern biotechnological tools with conventional breeding practices will be pivotal in developing climate-resilient maize hybrids. A multi-pronged approach that integrates genetic resistance, agronomic practices, and pathogen surveillance will remain essential for sustainable disease management. Future priorities include advancing research on pathogenic variability, host-pathogen interactions, and polygenic resistance, particularly against *M. phaseolina*. Emerging tools such as genomic selection, GWAS, and CRISPR/Cas9 gene editing promise to accelerate breeding progress. By uniting innovation, surveillance, and farmer-centred dissemination, resilient maize cultivars can be developed and adopted widely, ensuring sustainable productivity and global food security.

References

Abad L, Wolters P, Stucker D, Davis P (2006) Advances in anthracnose stalk rot resistance. In: The Fifth National IPM Symposium "Delivering on a Promise". https://www.ipmcenters.org/ipmsymposiumv/posters/037.pdf

Aggarwal SK, Hooda KS, Bagaria PK, Debnath S, Mallikarjuna N, Gogoi R, Kaur HJ (2022) Surveillance of major maize diseases and their distribution scenario in different agroclimatic zones of India. Maize J 11(2):88–94

Aggarwal SK, Hooda KS, Kaur H, Gogoi R, Chauhan P, Bagaria PK, Lal MK (2024) Comparative evaluation of management modules against Maydis leaf blight disease in maize (*Zea mays*). Eur J Plant Pathol 168(3):485–495

Aghav MA, Shinde VS, Khaire P, Latake SB (2023) Morpho-cultural traits of the pathogen *Exserohilum turcicum* responsible for turcicum leaf blight in maize. J Plant Dis Sci 18(2):93–98

Ahuja SC, Payak MM (1981) A laboratory method for evaluating maize germplasm to banded leaf and sheath blight. CABI Digital Library, pp 34–37

Ahuja SC, Payak MM (1982) Symptoms and signs of banded leaf and sheath blight of maize. Phytoparasitica 10:41–49

Ahuja SC, Payak MM (1985) Comparative biology, pathology and karyology of rice and maize isolates of *Rhizoctonia solani* f.sp. *sasakii*. Int Rice Res Newsl 10:5–6

AICRPM (2018) Annual progress report on *Kharif* maize, pp 1075

Baysal-Gurel F, Gardener BM, Miller SA (2012) Soil-borne disease management in organic vegetable production. Organic Agriculture. https://eorganic.org/node/7581

Bekeko Z (2019) Distribution and importance of common rust of maize (*Puccinia sorghi* Schw.) in Hararghe Highlands, Eastern Ethiopia. East Afr J Sci 13(2):135–144

Bertus LS (1927) A sclerotial disease of maize (*Zea mays* L.) due to *Rhizoctonia solani* Kühn. CABI Digital Library, pp 44–46

Broglie KE, Butler KH, Butruille MG, da Silva Conceição A, Frey TJ, Hawk JA, Wolters PJCC (2011) U.S. Patent No. 8,062,847. U.S. Patent and Trademark Office, Washington, DC

Choudhary M, Wani SH, Kumar P, Bagaria PK, Rakshit S, Roorkiwal M, Varshney RK (2019) QTLian breeding for climate resilience in cereals: progress and prospects. Funct Integr Genomics 19(5):685–701

Compton JAF, Tyler PS, Hindmarsh PS, Golob P, Boxall RA, Haines CP (1993) Reducing losses in small farm grain storage in the tropics. Trop Sci 33:283–318

Cooper M, Messina CD, Podlich D, Totir LR, Baumgarten A, Hausmann NJ, Graham G (2014) Predicting the future of plant breeding: complementing empirical evaluation with genetic prediction. Crop Pasture Sci 65(4):311–336

Chowdhury S, Basu A, Kundu S (2017) Biotrophy-necrotrophy switch in pathogen evoke differential response in resistant and susceptible sesame involving multiple signaling pathways at different phases. Sci Rep 7(1):17251

da Silva MP, Klepadlo M, Gbur EE, Pereira A, Mason RE, Rupe JC, Chen P (2019) QTL mapping of charcoal rot resistance in PI 567562A soybean accession. Crop Sci 59(2):474–479.

Delgado CJ (2003) The rising consumption of meat and milk in developing countries has sparked a new food revolution. J Nutr 133:3907S–3910S

Drechsler C (1925) Leafspot of Maize caused by *Ophiobolus heterostrophus*, the ascigerons stage of a *Helminthosporium* exhibiting bipolar germination. J Agric Res 31:701–726

Erenstein O, Jaleta M, Sonder K, Mottaleb K, Prasanna BM (2022) Global maize production, consumption and trade: trends and R&D implications. Food Secur 14(5):1295–1319

FAOStat (2021) FAO Stat. FAO, Rome. http://www.fao.org/faostat

FAOStat (2023) Statistical databases and data sets of the Food and Agriculture Organisation of the United Nations. http://faostat.fao.org/default.aspx

Gardiner JM, Coe EH, Melia-Hancock S, Hoisington DA, Chao S (1993) Development of a core RFLP map in maize using an immortalized F2 population. Genetics 134(3):917–930

Gilligan CA (2002) An epidemiological framework for disease management. Adv Bot Res 38:1–64

Glazebrook J (2001) Genes controlling expression of defence responses in *Arabidopsis*-2001 status. Curr Opin Plant Biol 4(4):301–308

Grote U, Fasse A, Nguyen TT, Erenstein O (2021) Food security and the dynamics of wheat and maize value chains in Africa and Asia. Front Sustain Food Syst 4:617009

Hammond-Kosack KE, Parker JE (2003) Deciphering plant-pathogen communication: fresh perspectives for molecular resistance breeding. Curr Opin Biotechnol 14(2):177–193

Harlapur SI, Khokhar MK, Hooda KS (2023) Epidemiology and management of charcoal rot of maize caused by Macrophomina phaseolina (Tassi.) Goid. In: Macrophomina phaseolina. Academic Press, pp 83–89

Hingorani MK, Grant UJ, Singh NH (1960) *Erwinia carotovora* f.sp. *zeae*, a destructive pathogen of maize in India. CABI Digital Library, pp 151–157

Hooda KS, Khokhar MK, Sekhar JC, Bagaria PK, Sharma SS, Kaur H, Reddy RR (2017a) Identification of resistance sources against post flowering stalk rot in maize. CABI Digital Library, pp 432–439

Hooda KS, Khokhar MK, Shekhar M, Karjagi CG, Kumar B, Mallikarjuna N, Yadav OP (2017b) Turcicum leaf blight-sustainable management of a re-emerging maize disease. J Plant Dis Protect 124:101–113

Hooda KS, Bagaria PK, Khokhar M, Kaur H, Rakshit S (2018) Mass screening techniques for resistance to maize diseases. ICAR-Indian Institute of Maize Research, Ludhiana, p 93

Hooker AL (1963) Inheritance of chlorotic-lesion resistance to *Helminthosporium turcicum* in seedling corn. Phytopathology 53:660–662

Hooker AL, Smith DR, Lim SM, Musson MD (1970) Physiological races of *Helminthosporium maydis* and disease resistance. Plant Dis Rep 54:1109–1110

Jenkins MT, Robert AL (1952) Inheritance of resistance to the leaf blight of corn caused by *Helminthosporium turcicum*. Agron J 44:136–140

Jones PG, Thornton PK (2003) The potential impacts of climate change on maize production in Africa and Latin America in 2055. Glob Environ Chang 13:51–59. https://doi.org/10.1016/S0959-3780(02)00090-0

Jung M, Weldekidan T, Schaff D, Paterson A, Tingey S, Hawk J (1994) Generation-means analysis and quantitative trait locus mapping of anthracnose stalk rot genes in maize. Theor Appl Genet 89(4):413–418

Katan J (2017) Diseases caused by soil-borne pathogens: Biology, management and challenges. J Plant Pathol 99:305–315

Khaledi N, Taheri P (2016) Biocontrol mechanisms of *Trichoderma harzianum* against soybean charcoal rot caused by *Macrophomina phaseolina*. J Plant Protect Res 56(1):21–31

Khokhar MK, Hooda KS, Sharma SS, Singh VIMLA, Saini ANITA (2013) *Fusarium* stalk rot: a major threat to maize production in India. Maize J 1:1–6

Khokhar MK, Hooda KS, Sharma SS, Singh V (2014) Post-flowering stalk rot complex of maize: present status and prospects. Maydica 59:226–242

Krishna KM, Kulakarni N, Kumar RS (2013) Genetic analysis of post-flowering stalk rot resistance in maize (*Zea mays* L.). Indian J Genet Plant Breed 73(3):328–331

Kumar B, Hooda KS, Gogoi R, Kumar V, Kumar S, Abhishek A, Yadav OP (2016) Inheritance study and stable sources of maydis leaf blight (*Cochliobolus heterostrophus*) resistance in tropical maize germplasm. Cereal Res Commun 44(3):424–434

Kumar B, Choudhary M, Kumar K, Kumar P, Kumar S, Bagaria PK, Rakshit S (2022a) Maydis leaf blight of maize: update on status, sustainable management and genetic architecture of its resistance. Physiol Mol Plant Pathol 121:101889

Kumar B, Choudhary M, Kumar P, Kumar K, Kumar S, Singh BK, Rakshit S (2022b) Population structure analysis and association mapping for turcicum leaf blight resistance in tropical maize using SSR markers. Genes 13(4):618

Kumar B, Pushpendra, Kumar S, Kumar P, Singh R, Yankanchi S, Jat HS (2025a) Understanding the genetic basis of resistance to maydis leaf blight and maturity-related traits in corn. Front Plant Sci 16:1551940

Kumar B, Singh R, Lahkar C, Dar ZA, Mallikarjuna N, Devlash R, Jat HS (2025b) Genomic insights for turcicum leaf blight resistance in maize (*Zea mays* L.). J Agric Food Res 24:102340

Lević J, Stanković S, Petrović T (2008) The determination of *Exserohilum turcicum* virulence factors in Serbia. Genetika 40(3):271–281

Lindhout P (2002) The perspectives of polygenic resistance in breeding for durable disease resistance. Euphytica 124:217–226

Lohithaswa HC, Showkath Babu BM, Sowmya MS, Banakar SK, Mallikarjuna N, Jadesha G, Mallikarjuna MG, Balasundara DC, Anand P (2025) Assessment of the potential for genomic selection to improve resistance to *fusarium* stalk rot in maize. Front Plant Sci 16:1631408

Lu Y, Xu J, Yuan Z, Hao Z, Xie C, Li X, Xu Y (2012) Comparative LD mapping using single SNPs and haplotypes identifies QTL for plant height and biomass as secondary traits of drought tolerance in maize. Mol Breed 30:407–418

Mahmoud AF, Abou-Elwafa SF, Shehzad T (2018) Identification of charcoal rot resistance QTLs in sorghum using association and in silico analyses. J Appl Gen 59(3):243–251.

Ma C, Ma X, Yao L, Liu Y, Du F, Yang X, Xu M (2017) *qRfg3*, a novel quantitative resistance locus against *Gibberella* stalk rot in maize. Theor Appl Genet 130(8):1723–1734

Mallikarjuna N, Gowda KP, Manjunath B, Kumar KK, Kumar NS (2007) Turcicum leaf blight disease severity of maize in relation to meteorological factors. Environ Ecol 25:778–784

Manamgoda DS, Rossman AY, Castlebury LA, Crous PW, Madrid H, Chukeatirote E, Hyde KD (2014) The genus bipolaris. Stud Mycol 79(1):221–288

Martínez-Cisneros BA, Juárez-López G, Valencia-Torres N, Durán-Peralta E, Mezzalama M (2014) First report of bacterial stalk rot of maize caused by *Dickeya zeae* in Mexico. Plant Dis 98(9):1267–1267

McMullen MD, Simcox KD (1995) Genomic organization of disease and insect resistance genes in maize. Mol Plant-Microbe Interact 8:811–815

Mir ZR, Singh PK, Zaidi PH, Vinayan MT, Sharma SS, Krishna MK, Vemula AK, Rathore A, Nair SK (2018) Genetic analysis of resistance to post flowering stalk rot in tropical germplasm of maize (*Zea mays* L.). Crop Prot 106:42–49

Mittler R (2006) Abiotic stress, the field environment and stress combination. Trends Plant Sci 11(1):15–19

Parlevliet JE (2002) Durability of resistance against fungal, bacterial and viral pathogens: the present situation. Euphytica 124:147–156

Paterson RRM, Lima N (2010) How will climate change affect mycotoxins in food? Food Res Int 43(7):1902–1914

Payak MM, Renfro BL (1968) Combating maize disease. Indian Farmer Dis 1:53–58

Payak MM, Sharma RC (1985) Maize diseases and approaches to their management in India. Int J Pest Manag 31(4):302–310

Prasad HH (1930) A bacterial stalk rot of Maize. CABI Digital librar, 25:pp–72.

Payak MM, Sharma RC, Lilaramani J (1973) How to control maize diseases. Indian Farm 23:20–23

Prasanna BM, Cairns JE, Zaidi PH, Beyene Y, Makumbi D, Gowda M, Magorokosho C, Zaman-Allah M, Olsen M, Das A, Worku M, Gethi J, Vivek BS, Nair SK, Rashid Z, Vinayan MT, Issa AB, San Vicente F, Dhliwayo T, Zhang X (2021) Beat the stress: breeding for climate resilience in maize for the tropical rainfed environments. Theor Appl Genet 134(6):1729–1752. https://doi.org/10.1007/s00122-021-03773-7

Rani VD, Reddy PN, Devi GU (2013) Banded leaf and sheath blight of maize incited by *Rhizoctonia solani* f. sp. *sasakii* and its management. A review. Int J Pharm Biol Arch 4:52–60

Rashid Z, Kaur H, Babu V, Singh PK, Harlapur SI, Nair SK (2021) Identification and validation of genomic regions associated with charcoal rot resistance in tropical maize by genome-wide association and linkage mapping. Front Plant Sci 12:726767

Rosegrant MW (2008) Biofuels and grain prices: impacts and policy responses. Testimony for the U.S. Senate Committee on Homeland Security and Governmental Affairs. International Food Policy Research Institute, Washington, DC

Rosegrant MR, Ringler C, Sulser TB, Ewing M, Palazzo A, Zhu T (2009) Agriculture and food security under global change: prospects for 2025/2050. International Food Policy Research Institute, Washington, DC

Sabet KA (1954) A new bacterial disease of Maize in Egypt. Empire J Exp Agric 22:64–65

Samson R, Legendre JB, Christen R, Saux MFL, Achouak W, Gardan L (2005) Transfer of *Pectobacterium chrysanthemi* (Burkholder et al., 1953) Brenner et al. (1973) and *Brenneria paradisiaca* to the genus *Dickeya* gen. nov. as *Dickeya chrysanthemi* comb. nov. and *Dickeya paradisiaca* comb. nov. and delineation of four novel species, *Dickeya dadantii* sp. nov., *Dickeya dianthicola* sp. nov., *Dickeya dieffenbachiae* sp. nov. and *Dickeya zeae* sp. nov. Int J Syst Evol Microbiol 55(4):1415–1427

Saxena SC (2002) Bio-intensive integrated disease management of banded leaf and sheath blight of maize. In: Proceedings at the Asian regional maize workshop: new technologies for the new millennium, Bangkok, Thailand

Sharma RR (1999) Pathological and biological attributes of *Rhizoctonia solani* f.sp. *sasakii* Exner, the causal organism of banded leaf and sheath blight of maize. MSc thesis, RAU, RCA, Udaipur

Sharma RR, Gour HN, Rathore RS (2004) Etiology of banded leaf and sheath blight symptoms on maize. Crop Sci 46:1039–1045

Shiferaw B, Prasanna BM, Hellin J, Bänziger M (2011) Crops that feed the world 6. Past successes and future challenges to the role played by maize in global food security. Food Secur 3:307–327

Simon TE, Le Cointe R, Delarue P, Morlière S, Montfort F, Hervé MR, Poggi S (2014) Interplay between parasitism and host ontogenic resistance in the epidemiology of the soil-borne plant pathogen *Rhizoctonia solani*. PLoS One 9(8):105159

Singh RDN, Kaiser SAKM (1991) Genetic analysis of resistance to charcoal rot of maize. J Mycopathol Res 29:103–109

Singh A, Shahi JP (2012) Banded leaf and sheath blight: an emerging disease of maize (*Zea mays* L). Maydica 57(3):215–219

Sinha SK, Prasad M (1977) Bacterial stalk rot of maize, its symptoms and host-range. Zentralbl Bakteriol Parasitenkd Infektionskr Hyg 132(1):81–88
Smith DR, Hooker AL (1973) Monogenic chlorotic-lesion resistance in corn to *Helminthosporium maydis*. Crop Sci 13(3):330–331
Sravya TL, Seshu G, Goud TY, Kumar MN (2023) Screening of some maize cultivars for charcoal rot in the Northern Region of Telangana, India. Int J Plant Soil Sci 35:1542–1548
Strosnider H, Azziz-Baumgartner E, Banziger M, Bhat RV, Breiman R, Brune MN (2006) Work group report: public health strategies for reducing aflatoxin exposure in developing countries. Environ Health Perspect 114:1898–1903
Stutt RO, Castle MD, Markwell P, Baker R, Gilligan CA (2023) An integrated model for pre-and post-harvest aflatoxin contamination in maize. NPJ Sci Food 7(1):60
Tesfaye K, Gbegbelegbe S, Cairns JE, Shiferaw B, Prasanna BM, Sonder K, Boote K, Makumbi D, Robertson R (2015) Maize systems under climate change in sub-Saharan Africa: potential impacts on production and food security. Int J Clim Change Strateg Manag 7:247–271. https://doi.org/10.1108/IJCCSM-01-2014-0005
Underwood LM (1897) Some new fungi chiefly from Alabama. Bull Torrey Bot Club 24(2):81–86
Uppal BN, Kolhatkar KG, Patel MK (1936) Blight and hollow stem of sorghum. Indian J Agric Sci 6:1323–1334
Vimla B, Mukherjee BK, Ahuja VP (1988) Combining ability analysis for resistance to banded leaf and sheath blight of maize. Indian J Genet Plant Breed 48:75–79
Ward JMJ, Laing MD, Cairns ALP (1997) Management practices to reduce grey leaf spot in maize. Crop Sci 37:1257–1262
Wisser RJ, Balint-Kurti PJ, Nelson RJ (2006) The genetic architecture of disease resistance in maize: a synthesis of published studies. Phytopathology 96(2):120–129
Yang Q, Yin G, Guo Y, Zhang D, Chen S, Xu M (2010) A major QTL for resistance to *Gibberella* stalk rot in maize. Theor Appl Genet 121(4):673–687
Zhang Y, Crous PW, Schoch CL, Hyde KD (2012) Pleosporales. Fungal Divers 53:1–221
Zhao M, Zhang Z, Zhang S, Li W, Jeffers DP, Rong T, Pan G (2006) Quantitative trait loci for resistance to banded leaf and sheath blight in maize. Crop Sci 46(3):1039–1045
Zhao Y, Lu X, Liu C, Guan H, Zhang M, Li Z, Lai J (2012) Identification and fine mapping of *rhm1* locus for resistance to southern corn leaf blight in maize. J Integr Plant Biol 54(5):321–329

Entomological Insights and Genetic Interventions for Sustainable Pest Management in Maize

12

S. B. Suby, P. Lakshmi Soujanya, K. R. Yathish, Jawala Jindal, Mukesh Choudhary, and Chikkappa Gangadhar Karjagi

Abstract

The invasion of fall armyworm (*Spodoptera frugiperda*) in India has dramatically reshaped the maize pest landscape, previously dominated by native stemborers, presenting a formidable challenge to the crop's stress-mitigation potential. Historically, the trade-off between native genetic resistance and yield drove hybrid development to prioritize productivity, with insect resistance largely conferred by transgenic traits in major maize-producing regions. In India, where transgenic food crops remain under moratorium and hybrid breeding programs initiated in the late 1980s relied on indigenous germplasm supported by international collaborations, varietal releases have emphasized agronomic superiority while retaining moderate insect and disease tolerance. When combined with improved agronomic practices, including integrated pest management (IPM), these hybrids have demonstrated resilience to native insect pests that impose comparatively lower stress. However, *S. frugiperda* causes direct ear damage and increases mycotoxin risk, highlighting the urgent need to incorporate multiple resistance traits through pre-breeding programs. Double haploid (DH)–based line development provides a powerful platform to rapidly fix and evaluate complex, quantitative resistance traits. The success of these strategies increasingly depends on precise and scalable phenotyping, as conventional field screening is labour-intensive and often confounded by genotype × environment interactions.

S. B. Suby (✉) · K. R. Yathish · M. Choudhary · C. G. Karjagi
ICAR-Indian Institute of Maize Research, Ludhiana, Punjab, India

P. L. Soujanya
Winter Nursery Centre, ICAR-Indian Institute of Maize Research, Hyderabad, Telangana, India

J. Jindal
Punjab Agricultural University, Ludhiana, Punjab, India
e-mail: jawalajindal@pau.edu

M. Choudhary et al. (eds.), *Next-Generation Maize Research*,
https://doi.org/10.1007/978-981-92-0722-0_12

Advances in high-throughput and precision phenotyping now enable rapid, objective, and comprehensive assessment of insect damage, stress responses, and post-damage recovery, thereby enhancing selection efficiency and accelerating genetic gain for resistance. Pioneering efforts by organizations such as the USDA and CIMMYT illustrate the feasibility of combining robust insect and disease resistance with tolerance to abiotic stresses in tropical maize. As maize assumes a growing role in biofuel production under India's National Biofuel Policy 2018, breeding strategies must balance high grain yield, industrial utility, and durable stress tolerance-a balance increasingly achievable through DH-enabled precision breeding integrated with genomic tools, IPM, and sustainable agricultural practices.

Keywords

Pest · IPM · Breeding · Resistance · Management

12.1 Introduction

Maize production in India faces significant challenges from both abiotic and biotic stressors, with insect pests consistently causing yield losses of 10%–20%, which can escalate up to 80% in severe infestations. Historically, the spotted stem borer [*Chilo partellus* (Swinhoe)], pink stem borer [*Sesamia inferens* (Walker)], and shoot fly (*Atherigona* spp.), were the key insect pests economically impacting maize in India, challenging rainy (*kharif*), autumn (*rabi*), and spring season maize crops, respectively. In addition to this, sporadic infestation of black cutworm [*Agrotis ipsilon* (Hüfnagel)] causes economic damage to *rabi* maize in eastern plains and early *kharif* maize in northern hills. The cob borer *Helicoverpa armigera* (Hübner) was the only insect causing direct damage to corn, impacting the marketability of sweet corn, particularly during the *rabi* harvest. However, the invasion of the fall armyworm (*Spodoptera frugiperda*) in May 2018 drastically altered the pest dynamics, inflicting damage on all growth stages of maize and emerging as the most destructive pest currently affecting maize across India, except in the late *rabi* and spring-sown maize regions of northwest India. An integrated pest management (IPM) approach, imposing a combination of resistant cultivars, biological control, agroecological approaches, and targeted pesticide use, is pursued in India to keep the pest populations below the economic threshold. This chapter discusses host plant resistance (HPR) and its deployments as the first layer of defence in maize IPM.

12.2 Maize Stemborers: Global Diversity and the Indian Context

Globally, maize production is affected by many economically damaging stalk borers like European corn borer [*Ostrinia nubilalis* (Hübner)] prevalent in North America, Europe and North Africa; southwestern corn borer [*Diatraea grandiosella* (Dyar)], American sugarcane borer [*Diatraea saccharalis* (Fabricius)], and, neotropical corn borer [*Diatraea lineolata* (Walker)] in Americas; mediterranean corn borer (MCB) or pink stem borer [*Sesamia nonagrioides* (Lefebvre)] in Africa, west Asia and southern Europe; greater sugarcane borer [*Sesamia cretica* (Led)] in northern Africa, Asia and Europe, African pink stem borer [*Sesamia calamistis* (Hmps)], African sugarcane borer [*Eldana saccharina* (Walker)], and African maize stalk borer [*Busseola fusca* (Fuller)] in Africa; Asian corn borer or oriental corn borer [*Ostrinia furnacalis* (Guenee)] in south-east Asia; *C. partellus* in Asia and Africa, and, *S. inferens* in south and southeast Asia. The stemborers that economically damage maize crops in India are *C. partellus* and *S. inferens*, typically infesting *kharif* and *rabi* maize, respectively, during the V3–6 growth stages. Unlike the notorious *O. nubilalis* and *B. fusca*, *C. partellus* and *S. inferens* were rarely reported to cause ear damage in their native range. *C. partellus* lays over 500 eggs on the lamina of upper leaves, while *S. inferens* deposits over 300 eggs within the lower leaf sheaths, in more than 10 egg masses in 4–5 plants (Taneja and Nwanze 1990; Kaur et al. 2015). Upon hatching within 3–4 days, *C. partellus* neonates congregate into the leaf whorl and feed in groups, whereas *S. inferens* larvae feed at their hatching site within the leaf sheath. After about a week, *C. partellus* larvae exit the whorl, bore the stalk base, and tunnel upward toward the meristem. In contrast, *S. inferens* larvae feed on leaf sheath tissue and tunnel inward towards the meristem. Damage to the meristem by both pests leads to withering of the central leaf whorl, leading to the 'dead heart' symptom. Such plants might give rise to non-productive tillers. Since the older larval stages of both *C. partellus* and *S. inferens* feed by tunnelling the stalk, it directly restricts the translocation of nutrients, thus significantly limits the potential yield of maize.

12.2.1 Stalk Resistance: The Key Host Plant Resistance Trait to Stemborers and Stalk Rot

Stalk tunnelling by maize borers globally can cause 30% yield losses, equivalent to worldwide losses of 311.3 million tons every year (López-Malvar et al. 2021). In addition to directly damaging the stalk, stem borers can predispose maize to stalk rot pathogens by creating entry wounds and serving as vectors, particularly for *Fusarium* spp. (Gatch and Munkvold 2002). Thus, stalk resistance plays a major role in HPR to stem borers, primarily due to tissue toughness, which forms a difficult-to-penetrate rind and hard pith, effectively delaying larval penetration and subsequent tunnelling within the stem (Martin et al. 2004; Suby et al. 2020; Guo et al. 2022). Thus, the most vulnerable stage of stalk borers to maize is from early to mid-vegetative stages. However, *O. nubilalis*, the most destructive of them all, also damages maize during the reproductive stage, causing stalk breakage, tassel, ear, and kernel damage, as well as ear drop (Chiang and Hodson 1950). Thus, the majority of work in HPR to stemborers is demonstrated in *O. nubilalis*. Studies indicate that resistance to *O. nubilalis* stalk tunnelling is controlled by multiple Quantitative trait locus (QTLs), each contributing a portion of the observed phenotypic variance (Bohn et al. 2000; Krakowsky et al. 2002; Cardinal and Lee 2005). These studies have progressively advanced toward the efficient use of marker-assisted selection (MAS), with a transition from QTL-based approaches to a more comprehensive genome-wide selection strategy for resistance against *O. nubilalis* and *S. nonagrioides* (Foiada et al. 2015; Jiménez-Galindo et al. 2019; López-Malvar et al. 2022).

Genetic diversity, explored through germplasm screening, is vital for developing resilient maize cultivars with improved stalk resistance. Although phenotypic selection has been effective for increasing resistance to maize borers like *O. nubilalis* and *S. nonagrioides*, genetic gains have often been limited due to the unfavourable genetic correlation between reduced tunnel length (an indicator of stalk resistance) and grain yield (Papst et al. 2004; Butrón et al. 2009). However, research using diallel crossbreeding among maize populations has proven effective for evaluating and selecting for resistance traits across various environments, whether under natural infestations or controlled conditions. This approach helps to identify genetic combinations that can optimize both resistance and yield despite the complexities of their genetic interplay. Stalk resistance to *S. calamistis* and *E. sacharina* resistance and agronomic traits work by additive gene action. In contrast, non-additive gene action tends to affect traits like dead heart, a key indicator of stem borer damage, while a combination of both gene actions is crucial for traits such as stalk breakage and grain yield (Ajala et al. 2008; Sandoya et al. 2010). Generation mean analysis of a cross between the *S. inferens* resistant inbred line DMR E62 and CML 451 revealed negative additive and dominance effects, along with positive additive × dominance, dominance × dominance, and epistatic interactions (Santosh et al. 2012), suggesting the complex genetic architecture of *S. inferens* resistance. Thus, the identification of promising population pairs for reciprocal recurrent selection has emerged as a strategic approach to enhancing genetic resistance while improving yield-related characteristics. This strategy leverages the combined effects of general combining ability and specific combining ability to optimize resistance against multiple stem borer species while simultaneously improving yield potential (Ajala et al. 2008; Sandoya et al. 2010). Furthermore, Huang et al. (2025) developed a machine learning model showing that maize stalk bending strength can be accurately predicted, with the seventh internode's crushing strength identified as the key trait for lodging resistance. Studies show that genomic selection (GS) achieved higher genetic gains for yield, while phenotypic selection (PS) was more successful in reducing tunnel length (a measure of borer resistance) but did not significantly boost yield. The findings suggest that while PS is effective for enhancing specific resistance traits, GS may offer more balanced improvements for both resistance and yield, although the gains per breeding cycle for both traits might be relatively modest (Gesteiro et al. 2023). Gene mapping and expression studies provide clues for stalk rot resistance in maize. For example, Wang et al. (2022) employed proteomic studies in an inbred line Lx9801 (*Fusarium graminearum* inoculated vs mock-inoculated) to identify antioxidant protein *ZmPrx5* as a positive regulator of maize resistance to *Fusarium* stalk rot. Later, Hou et al. (2023) used 295 inbred lines in a GWA study to identify 69 candidate genes for maize resistance to *Pythium aristosporum* stalk rot, with functional validation of *Zm00001d051313* as a positive regulator of antifungal defense. Similarly, in GWAS study, Wang et al. (2024) identified

29 SNPs linked to maize stalk lodging resistance, highlighting *Zm00001d014449* for cell wall formation and *Zm00001d005300 (MYB44)* for stress response as key genes enhancing stalk strength. Recently, Wang et al. (2025) used BSA-Seq approach to identify a candidate region on chromosome 6, with KASP marker linked two key genes (*Zm00001eb260120*, *Zm00001eb260260*) for *Fusarium* stalk rot resistance.

While stalk strength against lodging and insect resistance is directly assessed through penetration resistance, it can also be indirectly evaluated by quantifying cell wall–fortifying hydroxycinnamates, particularly *p*-coumaric acid (PCA) and ferulic acid (FA), which serve as key biochemical markers (Liu et al. 2022; López-Malvar et al. 2019). PCA also plays the main antibiosis factor to *C. partellus* (Gundappa et al. 2013), and its higher concentrations in both rind and pith tissues have been correlated with increased resistance to *S. nonagrioides* (Gesteiro et al. 2021). When the role of PCA was investigated using Recombinant Inbred Lines (RILs) derived from a Multiparent Advanced Generation Intercross (MAGIC) population, it was observed that resistant lines exhibited higher PCA concentrations. However, these lines also displayed agronomically undesirable traits, such as reduced plant height and delayed maturity, highlighting a trade-off between resistance and yield (López-Malvar et al. 2021). Similarly, selective breeding of maize inbred lines to achieve up to a threefold variation in diferulate concentrations revealed that while these changes did not impact pest resistance, forage digestibility, or ethanol production, higher diferulate concentrations were associated with reduced cellulose and lower sugar release during enzymatic hydrolysis (López-Malvar et al. 2025). These findings suggest that breeding strategies should prioritize grain yield and industrial utility, while considering resistance traits like PCA as secondary selection criteria, particularly in light of maize's growing role as a biofuel source in India.

12.3 Shoot Fly, *Atherigona* spp. and Its Management

The shoot fly complex (*Atherigona* spp.), including *A. orientalis* (Schiner), *A. naqvii* (Steyskal), *A. soccata* (Rondani), etc., is a significant pest of spring maize, particularly in the northwestern plains of India. Infestation by *Atherigona* spp. begins as early as 2 days after seedling emergence and persists for up to 3 weeks, making it the most challenging stress in maize. Since its infestation can cause plant death as early as 7 days after germination, insecticidal seed treatment remains the only effective management tool to date. Severe infestations are commonly observed in crops sown between the third week of February and the first week of March in regions such as Punjab, Haryana, Uttarakhand, and Uttar Pradesh. This pest can lead to up to 60% plant loss and approximately 21% grain yield reduction. In Punjab, *A. naqvii* is the predominant species, incurring a loss of 27.93%–45.04% in grain yield (Jindal 2013), while a species complex primarily consisting of *A. soccata* and *A. orientalis*, along with a smaller proportion of *A. naqvii*, is found in neighbouring states like Haryana, Delhi, Uttarakhand, and Uttar Pradesh, though their relative abundance remains unclear. The eggs of *A. soccata* and *A. orientalis* are boat-shaped with reticulated wing-like projections on either side, whereas those *of A. naqvii* is rice grain-shaped with fine surface ridges. The female lays 2–30 eggs on the lower surface of basal leaves, the shoot, or in the soil near the plant base. Eggs hatch within 3 days, and the creamy-yellow maggots burrow into the shoot, destroying the growing point, ultimately causing 'dead heart' within 2 weeks of germination.

Resistance to shootfly is influenced by a combination of morphological, biochemical, and genetic factors. Morphologically, genotypes that exhibited higher seedling vigour, smaller leaf area, fewer leaves, and thinner stems showed reduced incidence of dead hearts. These traits likely make the plants less favourable for shoot fly larvae to feed on and develop, thereby providing a degree of tolerance. In contrast, none of the genotypes demonstrated antixenosis, i.e., they did not deter shoot fly oviposition. Tolerance, assessed by a lower proportion of dead hearts relative to overall damage (including leaf injury), and plant vigour emerged as a key resistance mechanism (Jindal et al. 2007). Additionally, antibiosis was identified, as indicated by its negative impact on larval and pupal weight, as well as on the duration and survival of the insect's life stages (Goyal et al. 2020). Biochemically, the study found that susceptible genotypes had higher levels of total soluble sugars, proteins, and amino acids, which could support shoot fly growth. In contrast, resistant genotypes had increased levels of phenolic compounds, tannins, and defensive enzymes like phenylalanine ammonia lyase (PAL) and polyphenol oxidase (PPO), which inhibit insect development (Goyal et al. 2020). On the genetic front, resistance to shoot fly was found to be polygenic, with 19 putative QTLs identified, indicating a complex genetic control mechanism. Major QTLs, such as *qDH9.1* (associated with dead hearts) and *qEC9.1* (associated with oviposition), were mapped to chromosome 9, explaining significant phenotypic variance (15.03% and 18.89%, respectively). Candidate genes in these QTL regions included those involved in allelochemical synthesis (e.g., cytochrome P450), protease inhibitors, receptor kinases, and components of the ubiquitin-proteasome pathway, all of which likely contribute to the plant's defence mechanisms. Moreover, these QTLs were found to be syntenic with regions on chromosome 10 in sorghum, where the shoot fly is the most destructive pest, suggesting an evolutionarily conserved genetic basis for resistance (Vikal et al. 2020). Further identification of additional QTLs and genomic regions influencing resistance can facilitate the development of effective resistance breeding strategies to combat shootfly, a challenge that remains largely confined to basic research.

12.4 Fall Armyworm (*Spodoptera frugiperda*) and Its Management

The fall armyworm (*Spodoptera frugiperda*, FAW) is an invasive pest that inflicts severe damage to maize throughout its growth stages, from early vegetative (V2–3) to the reproductive (R3) phase, particularly during the *kharif* season across most regions of India. A prominent sign of the FAW's success as a species is its ability to lay over 1000 eggs in several masses, covered with protective scales, distributed across multiple plants, and positioned on both sides of the leaves. In the *kharif* season, these eggs typically hatch within 3 days, after which the neonate larvae aggregate in the leaf whorl and begin feeding within. Unlike stem borers and defoliators, which typically infest less than 50% of the maize plant population, FAW infestations often cover 100% of plants. This widespread infestation is primarily attributed to the high fecundity and cannibalistic behaviour of older FAW larvae, which promote local dispersal, while the moths' long-distance migratory capacity further accelerates rapid spread across fields and regions. FAW's voracious feeding behaviour often exceeds the maize plants' ability to recover from foliage loss, causing substantial yield reductions and directly impacting yield through damage to the corn ears. In the absence of effective control measures, the fall armyworm can cause yield losses ranging from 17% to 40% in *kharif* maize across different locations in India. However, maize sown between late *rabi* and spring in northwestern India has experienced minimal impact from fall armyworm, as prolonged temperatures below 10 °C combined with aridity are likely to hinder its survival (Dahi et al. 2020; Maino et al. 2021). While FAW's long-distance migration causes re-infestation every year, its local dispersal and overlapping generations contribute to rapid re-infestation within a week of pesticide application, often compelling farmers to spray repeatedly. To counter this, a coordinated, area-wide management strategy is essential, where synchronized sowing of tolerant genotypes serves as the first line of management, helping to reduce pest pressure. Since FAW larvae feed on both vegetative and reproductive stages of the crop, incorporating both foliar and ear resistance is critical for strengthening host plant resistance (HPR).

12.4.1 Vegetative and Ear Resistance to Fall Armyworm

Defoliation by fall armyworm at both early and later growth stages reduces maize yield, with early-stage damage affecting grain number and later-stage damage further impacting dry matter accumulation and forage quality (Tashiro et al. 2024). Foliar resistance to *S. frugiperda* and other leaf-feeding insects during the vegetative stage in maize is primarily governed by antibiosis, through defensive metabolites, and tolerance, which collectively reduce foliar damage (Williams et al. 1995). An early transition from the vegetative to reproductive stage, along with increased leaf toughness, also contributes to resistance (Williams et al. 1998). Ear damage by *S. frugiperda*, and other ear-feeding insect pests, such as *Helicoverpa zea*, *O. nubilalis*, *O. furnacalis*, and *B. fusca*, not only reduces yield but also significantly increases the risk of mycotoxin contamination by facilitating infection from ear rot pathogens like *Aspergillus flavus* and *Fusarium* spp. (Guthrie et al. 1981; McMillian et al. 1985; Flett and Van Rensburg 1992; Windham et al. 1999; Farhan et al. 2020; Barton et al. 2024). However, maize employs a range of resistance mechanisms against ear-feeding pests, governed by both physical and biochemical factors. Physical barriers, such as tight and extended husk coverage along with an increased number of husk layers, help prevent or delay ear-feeding pests and the subsequent invasion of ear rot pathogens (Widstrom et al. 1970; Moore and Tracy 2021; Magarini et al. 2024). The USDA-ARS released germplasm population Zapalote Chico 2451F contains maysin, a flavone glycoside that provides resistance to silk-feeding insects, viz., *H. zea*, *S. frugiperda*, and *Euxesta stigmatias*, reducing ear damage (Wiseman and Carpenter 1995; Widstrom et al. 2003).

The presence of multiple resistance factors suggests the involvement of QTLs, as demonstrated in several studies using F_2-derived maize populations subjected to artificial infestation by defoliating insect pests and subsequent genetic mapping. QTLs on chromosomes 6, 9, and 10 were identified as key contributors to resistance against *D. grandiosella* and *S. frugiperda* leaf feeding. The candidate insect resistance genes, including the *mir* gene family on chromosome 6 and the *glossy15* locus on chromosome 9, were mapped within these QTL regions (Brooks et al. 2005). Similarly, analysis of Mp705 (resistant) × Mp719 (susceptible) $F_{2:3}$ families identified two major QTLs in bins 4.06 and 9.03 associated with resistance to *S. frugiperda* leaf feeding (Womack et al. 2020). GWAS using 423 tropical maize lines identified 56 significant marker–trait associations for *S. frugiperda* resistance, governing foliar damage scores, ear rot, and ear damage traits, with chromosome 4 accounting for the highest number of SNP markers associated with foliar damage (Kamweru et al. 2022). Similarly, GWAS identified 15 SNP markers and 7 genes associated with *S. frugiperda* resistance using 289 maize lines, followed by metabolic pathway analysis, which revealed key pathways such as hormone signalling, carotenoid biosynthesis, known antibiosis agents, and cuticular wax formation (Warburton et al. 2023).

12.5 Genetic Trade-Offs Between Yield and Defence: Constitutive Versus Induced Resistance

The underlying mechanisms of genetic resistance comprise both constitutive and induced defence systems that operate against insect damage. Well-known constitutive defence metabolites in maize, such as PCA, DIMBOA (2,4-dihydroxy-7-methoxy-1,4-benzoxazin-3-one), and maysin, function as antinutritional factors for insects. However, their continuous production is energy-intensive for the plant, as evidenced by the yield penalty observed when breeding for high PCA levels (López-Malvar et al. 2021). While research has primarily focused on these constitutive defences, emerging evidence suggests that induced defences may be more efficient, as they require fewer resources and are activated only in response to insect attack. Generally, defence response in plants to defoliation is triggered by jasmonic acid (JA), a key phytohormone involved in insect defence. The immediate response to JA signalling is the emission of herbivore-induced plant volatiles (HIPVs), which alert the neighbouring plants, regulate the behaviour of herbivores, and serve as an SOS signal to natural enemies (Hu et al. 2021). Maize landraces were found to emit HIPVs in response to *C. partellus* attack, attracting both egg and larval parasitoids, whereas commercial hybrids lack this defence trait (Tamiru et al. 2011). Notably, the landrace Braz1006 exhibited significantly higher production of the HIPV

(E)-caryophyllene and its associated transcript *TPS23* following *C. partellus* egg deposition compared to elite breeding lines, thereby attracting the parasitoid *Cotesia sesamiae* (Tamiru et al. 2017). The *S. frugiperda-resistant* inbred Mp708 exhibits approximately threefold higher JA levels before herbivore feeding, constitutive expression of JA-inducible genes, and the release of (E)-β-caryophyllene, suggesting strong constitutive defences and a heightened ability to respond to *S. frugiperda* attack (Shivaji et al. 2010; Smith et al. 2012). Interestingly, *S. frugiperda* relies on host plant-derived volatiles for host location, with the constitutive methyl salicylate and the induced (E)-α-bergamotene acting as oviposition attractants for adult moths, while the induced (E)-4,8-dimethyl-1,3,7-nonatriene (DMNT) serves as an oviposition deterrent (Yactayo-Chang et al. 2021).

In addition to volatiles, other defence metabolites are induced in response to insect feeding, both in the short term and long term. In the short term, Mp708 induces the defensive protein Mir1-CP (Maize insect resistance 1-cysteine protease) upon herbivory (Lopez et al. 2007). Similarly, the Chinese maize inbred Xi502 accumulates high levels of HDMBOA-Glc upon *S. frugiperda* feeding (Malook et al. 2021). Significant differences have been observed in maize's responses to short-term versus long-term feeding by stemborers (Rodríguez et al. 2021; Soujanya et al. 2021). In response to *S. nonagrioides* feeding, the susceptible inbred line (EP42) exhibited higher constitutive antibiosis, which weakened over time, whereas the resistant inbred line (PB130) progressively induced its defences (Rodríguez et al. 2021). In response to *S. inferens* feeding or JA treatment, the resistant genotypes (DMRE 63, CM 500) and moderately resistant genotype (WNZ ExoticPool) accumulated significantly higher levels of PCA and ferulic acid FA, compared to the susceptible genotypes (CM 202, BML 6) (Soujanya et al. 2021).

These findings suggest that maize has lost some of its inducible defensive traits due to a trade-off for higher yield during selective breeding. However, these traits could potentially be reintroduced to improve pest resistance. In this process, its wild ancestor, teosinte, serves as one such genetic reservoir of defence mechanisms against insect pests and pathogens that could be utilized for introgression. (Rosenthal and Dirzo 1997; de Lange et al. 2014). *S. frugiperda* prefer maize 1.8 times greater than teosinte, and force-fed larvae on it had 1.2 times less weight compared to maize (Takahashi et al. 2012). Experiments in India with inbred lines LM 24, CML 72, along with teosinte and its backcross derivatives, demonstrated moderate resistance to FAW infestations. Teosinte exhibited superior stress tolerance, largely due to elevated peroxidase (POD), phenylalanine ammonia lyase (PAL), tyrosine ammonia lyase (TAL), and β-1,3-glucanase enzyme activities, which led to increased phenolic content and enhanced lignification of leaves under FAW feeding. The backcross line (LM 24 × teosinte) inherited some of these HPR traits, displaying comparable activities of POD, β-glucanase, chitinase, and protease inhibitors, as well as similar levels of total phenols and lignin, surpassing those observed in LM 24. On the other hand, the well-known resistant line CML 72 exhibited a distinct defence strategy by achieving the highest activities of β-glucanase, chitinase, and protease inhibitors (Anwer et al. 2024).

12.6 Transgenic Resistance: The Promise, Pitfalls and the Way Forward

Genetically modified maize (GM-maize) expressing *Bacillus thuringiensis* (Bt) toxins, such as Cry and Vip proteins, has significantly improved pest management in maize cultivation by reducing reliance on chemical insecticides and enhancing crop yields. The first genetically modified maize was specifically engineered to express the Cry1Ab protein, which targets the *O. nubilalis*, and registered for field release in the 1990s in USA (US Environmental Protection Agency 2010). Observing the growing success of the technology, many single Cry protein maize cultivars were registered, like Herculex I, expressing Cry1F protein targeting *S. frugiperda* (Siebert et al. 2008). Cry proteins work by binding to receptors in the gut of the insect, disrupting digestion and causing death. Use of a single Cry gene put selection pressure on target pests to evolve against it, over time, as exemplified by Cry1Ab resistance of *B. fusca* in South Africa (Van Rensburg 2007), Cry1F resistance of *S. frugiperda* in the Americas (Storer et al. 2010; Huang et al. 2014; Farias et al. 2016; Chandrasena et al. 2018) and Cry3Bb1 resistance of *Diabrotica virgifera* LeConte (western corn rootworm) in the USA (Gassmann et al. 2011).

Engineering diverse Cry proteins not only helps prevent or delay resistance development in primary target pests but also provides broader protection against multiple insect species. Thus, new Bt corn genotypes have been developed to express multiple Cry proteins, such as Cry1F, Cry2Ab2, engineered Cry1 and Cry2 variants, along with Vip proteins like Vip3Aa20, offering enhanced defence against a wide range of major and minor pests like FAW, *D. grandiosella*, *Diabrotica v. zeae*, *Diatraea saccharalis*, *Spodoptera exigua*, *Helicoverpa zea*, *Agrotis ipsilon*, *Striacosta albicosta*, etc. (Siebert et al. 2012; Rule et al. 2014). The latest advancement is the DBN3601T event, expressing Cry1Ab and Vip3Aa19, released in China to combat FAW (Zhao et al. 2023), which invaded the country in 2019. Similarly, RNA interference (RNAi) by knocking down the expression of crucial genes for the growth and development of the target pest in turn, results in lethal phenotypic deformities or mortality. As a species-specific technology, RNAi can be deployed through sprayable formulations or transgenesis, offering a promising next-generation alternative for insect pest management. The first effective RNAi-based transgenic corn expressed dsSNF7, targeted the *Diabrotica v. virgifera* and caused significant mortality (Baum et al. 2007). To prevent or slow the development of resistance to genetically modified (GM) insect-resistant traits, strategies such as refuge planting are implemented alongside rigorous resistance monitoring. This approach ensures that a portion of the pest population remains susceptible to GM traits, preserving the long-term efficacy of the technology. Additionally, IPM strategies help manage pest populations and extend the durability of GM traits.

The promise of GM traits like Bt lies not only in their ability to confer resistance against insect pests throughout both the vegetative and reproductive stages but also in their unique advantage of reducing stalk rot and grain mycotoxin contamination, a benefit rarely achieved through native resistance (Gatch and Munkvold 2002; Barton et al. 2024). By 2023, GM crops were cultivated in 76 countries, covering

13.38% of the world's total farmland, representing a 121-fold increase since 1996, with maize expressing single or pyramided Bt traits ranking as the third most widely grown GM crop (Cheng et al. 2024). However, in India, policy decisions on transgenic corn cultivation remain suspended, making the direct use of Bt and RNAi technologies unfeasible in current agricultural practices. Nevertheless, the core principles of these technologies can still be leveraged through alternative approaches, such as developing efficient formulations like Dipel®, a widely used Bt formulation in India. Similarly, RNAi-based formulations offer a promising alternative, enabling their benefits while avoiding reliance on GM crops. Recently in India, RNAi targeting *COPIα*, *COPIβ*, and *GSTU1* genes of *S. frugiperda* demonstrated strong insecticidal potential when administered through diet and maize leaves, suggesting that sprayable RNAi formulations could be developed as a 'green strategy' for managing this pest (Bera et al. 2025).

12.7 Breeding Strategies for Sustainable Insect Pest Management in Maize

Internationally, collaborative research efforts during the pre-transgenic era, led by CIMMYT (Mexico) and USDA-ARS (Mississippi) from the 1970s to the 1990s, played a crucial role in identifying genetic variation in maize that could support breeding for native resistance to FAW. CIMMYT's breeding programs extensively utilized tropical, subtropical, and temperate maize germplasm, including USDA-ARS lines, leading to the development of two major populations: Multiple Insect Resistant Tropical (MIRT) and Multiple Borer Resistant (MBR). These populations served as a foundation for breeding improved maize inbred lines with partial resistance to FAW and maize borers. Several FAW-resistant tropical germplasm developed through these programs have been widely disseminated, including CML71, CML124, CML125, CML338, CML333, CML334, CML370, CML372, and CML574 (Prasanna et al. 2019). Meanwhile, USDA's breeding efforts resulted in the development of temperate maize inbred lines such as Mp496, Mp701, Mp708, Mp713, Mp714, and Mp716, which were primarily derived from CIMMYT's tropical maize germplasm through multiple generations of selection for resistance to FAW and *D. grandiosella* (Williams et al. 1990). Additionally, landrace-derived lines such as ZC-2451F, originating from the Zapalote Chico landrace of Mexico, were released by USDA-ARS as a source of improved resistance to silk and ear feeding by FAW and other ear feeding pests (Widstrom et al. 2003).

However, with transgenic maize now dominating major maize-producing regions worldwide, breeding efforts have increasingly prioritized yield enhancement, often at the expense of developing native genetic resistance. In India, where active hybrid breeding programs began in the late 1980s with the support of CIMMYT germplasm, and in the absence of transgenic food crops, varietal releases have traditionally focused on agronomic superiority combined with moderate tolerance to insects and diseases. These hybrids, when cultivated under improved agricultural practices and IPM, have contributed to a significant reduction in insecticide use. Nevertheless, the recent invasion of FAW has introduced new challenges, necessitating the incorporation of ear resistance into India's maize breeding program. Moreover, the strong association between increasing insect-derived ear injury, rising yield loss, and heightened aflatoxin accumulation is of particular concern in geographies where high temperatures, insect pressure, and aflatoxin stress are prevalent (Pruter et al. 2020). Thus, the continuous selection and development of multiple stress tolerant maize lines remain vital for mitigating pest impacts on crop productivity. Importantly, native resistance was explored using MIRT and Antigua group-1 populations, which led to the development of inbred lines resistant to *S. inferens* through recurrent selection, with the DMR E63 (WNZPBTL 9) (IC0594373; INGR14014) genotype emerging as particularly promising (Sekhar et al. 2010, 2014; Santosh et al. 2012). This genotype has also exhibited vegetative-stage resistance to FAW (Soujanya et al. 2022) and is currently being utilized in breeding programs. However, incorporating multiple resistance traits such as foliage, stalk, and ear resistance against insect pests and pathogens into maize breeding programs requires a systematic and multidisciplinary approach. Pioneering efforts by organizations like the USDA and CIMMYT have set the benchmark for achieving robust resistance against pests such as *O. nubilalis*, FAW, and for mitigating stalk lodging. USDA's research on enhancing stalk strength and increasing lignin content to reduce lodging while enhancing borer resistance (Martin et al. 2004), along with CIMMYT's successful integration of multi-stress resistance through combined phenotypic and genotypic selection in high stress-prone environments (Prasanna et al. 2022), represent strategies worth emulating. Given the complexity of insect resistance, genomic selection presents a promising approach by leveraging genome-wide dense markers for predictive modelling, as demonstrated in a genome-wide association study (GWAS) conducted on a panel of over 400 tropical CIMMYT maize lines against *S. frugiperda* (Kamweru et al. 2022). Similarly, a GWAS identified nine SSR markers and 19 QTL loci associated with multiple resistance traits including the defensive metabolite DIMBOA against *O. furnacalis* (Niu et al. 2023). This strategy not only enhances selection accuracy against multiple biotic stresses but also supports association analyses to uncover the genetic basis of key traits (Badji et al. 2020). More recently, building on CIMMYT's long-standing focus on multi-stress resilience, breeding strategies have increasingly utilized doubled haploid (DH) technology to rapidly fix favourable allelic combinations from multi-parent, stress-tolerant elite germplasm, thereby accelerating the development of genetically uniform inbred lines for quantitative resistance traits. For example, DH and pedigree lines derived from three-parent populations of elite lines, tolerant to maize lethal necrosis (MLN), maize streak virus (MSV), and drought, and exhibiting high general combining ability for grain yield, have demonstrated that complex, polygenic stress-resistance traits can be efficiently stabilized without compromising hybrid performance (Aleri et al. 2025).The integration of genomic selection with DH-derived training populations further enhances selection accuracy and genetic gain for quantitative resistance traits. Recent evidence demonstrates the growing use of DH lines for evaluating fall armyworm resistance in maize, particularly in populations derived from tropical germplasm introgressed to broaden the narrow genetic base of temperate maize. Using the BS39 population and its DH derivatives, BS39 and especially BS39_DH_012 exhibited moderate resistance to fall armyworm leaf injury, with growth stage–dependent variation suggesting the involvement of distinct resistance mechanisms (Abel et al. 2025). By adopting and refining these approaches, Indian corn breeding programs can develop high-yielding, pest-resistant hybrids tailored to the region's specific challenges and opportunities.

12.7.1 Breeding Strategies to Integrate Insect and Associated Pathogen Resistance in Corn and Its Deployment in IPM

12.7.1.1 Pre-breeding as a Foundation for Insect and Pathogen Resistance

Landraces, exotic germplasm, and wild relatives constitute important reservoirs of resistance to major maize insect pests and fungal pathogens. Nevertheless, these genetic resources frequently exhibit undesirable agronomic attributes, necessitating systematic pre-breeding efforts prior to their incorporation into elite breeding programs (Cobb et al. 2019). Pre-breeding represents a crucial interface between genetic resource utilization and applied crop improvement, facilitating the identification, validation, and introgression of novel resistance alleles into adapted genetic backgrounds while minimizing linkage drag. By expanding the genetic base of elite germplasm, pre-breeding provides durable resistance options that support medium- to long-term breeding objectives (Ragot et al. 2018). Recent GWAS and QTL mapping efforts have identified multiple loci associated with resistance to insect pests and ear rot pathogens, underscoring the polygenic nature of these traits and highlighting the necessity of integrating marker-assisted introgression with genomic prediction approaches (Kamweru et al. 2022). Collectively, these findings emphasize the central role of pre-breeding in the development of resilient maize cultivars capable of withstanding diverse biotic stresses.

12.7.1.2 Precision Phenotyping and High-Throughput Trait Assessment

Accurate and reproducible phenotyping underpins genetic improvement for resistance to insect pests and associated diseases in maize. Conventional multi-environment field screening is labour-intensive, costly, and frequently constrained by strong genotype × environment interactions, particularly for complex resistance traits targeting insect pests and insect-mediated pathogen infections (Chawade et al. 2019; Song et al. 2021). High-throughput plant phenotyping (HTPP) technologies address these limitations by enabling rapid, objective, and scalable assessment of complex traits. Advances in satellite-based imaging, unmanned aerial vehicles (UAVs), and ground-based sensor platforms have substantially enhanced field-based maize phenotyping. UAV-mounted multispectral, hyperspectral, and thermal sensors facilitate canopy-level quantification of defoliation, plant vigour, and stress responses associated with insect damage, while proximal sensing platforms allow high-resolution measurement of plant architecture, lesion development, and disease progression (Chawade et al. 2019). Precision phenotyping further supports indirect assessment of resistance-related traits, including stay-green, canopy temperature, biomass accumulation, and post-injury recovery, which are commonly linked to tolerance against insect pests and stalk- and ear-rot pathogens. Integration of HTPP-derived data with advanced statistical and genomic models improves phenotypic resolution, enhances selection efficiency, and accelerates genetic gain in resistance-oriented maize breeding programs (Song et al. 2021).

12.7.1.3 Trait Introgression Via Conventional Breeding

(i) Backcross breeding: Resistant donor lines are crossed with elite Indian maize inbred lines, including HKI 1105, HKI 161, HKI 163, HKI 193–1, BML6, BML7, LM 13, LM-14, IML418–1, IML23, and IML24. Successive backcrossing is then employed to recover the superior agronomic performance and adaptation of the recurrent parent while concurrently selecting for resistance traits. Phenotypic and, where available, marker-assisted selection can be applied at each backcross generation to enhance selection efficiency and minimize linkage drag. DH technology may be integrated at the F_1 or early backcross stages (BC_1F_1 or BC_2F_1) to rapidly fix resistance alleles and generate completely homozygous resistant inbred lines, thereby substantially shortening the breeding cycle.

(ii) Recurrent selection: Recurrent selection remains a cornerstone of conventional resistance breeding in maize, particularly for the improvement of quantitatively inherited traits. This approach enables the simultaneous accumulation of favourable alleles conferring resistance to multiple insect pests and associated diseases. Selection cycles should be designed to improve resistance across key plant components, including stalk strength, foliar resistance, and ear protection, while maintaining overall agronomic performance. Emphasis should be placed on both antibiosis and post-infestation recovery capacity, thereby enhancing durable, broad-spectrum resistance across successive breeding generations.

12.7.1.4 Modern Breeding Approaches for Insect and Disease Resistance in Maize

(i) Developing maize cultivars with durable resistance to insect pests and associated pathogens requires the integration of conventional, molecular, genomic, and biotechnological approaches. Modern breeding platforms leverage complementary strategies, viz., molecular markers, genomic selection, doubled haploids (DH), and transgenic or genome-editing technologies to accelerate genetic gain while maintaining broad-spectrum and long-lasting resistance (Fig. 12.1).

(ii) Molecular breeding approaches: Molecular breeding has substantially enhanced the efficiency of maize improvement by enabling precise selection for resistance traits. Functional markers derived from QTL mapping, GWAS, and allele mining are directly associated with genes controlling resistance mechanisms such as antibiosis, antixenosis, and tolerance (Mandal 2018; Salgotra and Stewart 2020). Marker-assisted selection (MAS), including forward breeding and marker-assisted backcrossing, allows early identification of resistant genotypes, accelerates genetic gain, and minimizes linkage drag (Moose and Mumm 2008). While numerous markers for FAW, stemborers, and disease tolerance have been reported, extensive validation across diverse germplasm and environments remains essential prior to routine deployment.

(iii) Genomic Selection (GS) for multi-trait resistance: Genomic selection (GS) enables the prediction of genomic estimated breeding values (GEBVs) for complex, polygenic traits, such as resistance to FAW, stem borers, and ear rot pathogens (Lenz et al. 2020). By using genome-wide markers and well-phenotyped training populations, GS captures both major- and minor-effect loci, allowing selection of superior individuals even when phenotyping resources are limited or environments are variable. Multi-trait GS models can predict resistance across foliar, stalk, and ear tissues, guiding parental selection and optimized crossing strategies. GS can also predict the performance of untested lines, accelerating the development of multi-stress-resistant maize.

(iv) Doubled haploids (DH) and accelerated breeding: Conventional maize breeding requires multiple generations of inbreeding and multi-environment testing, extending cultivar development timelines (Moose and Mumm 2008; Cobb et al. 2019). Accelerated breeding approaches, including DH, shuttle breeding, and speed breeding, substantially reduce cycle time. DH technology is particularly impactful, rapidly generating completely homozygous lines within one or two generations. This allows rapid fixation and pyramiding of resistance alleles for multiple pests and associated diseases, facilitating faster hybrid develop-

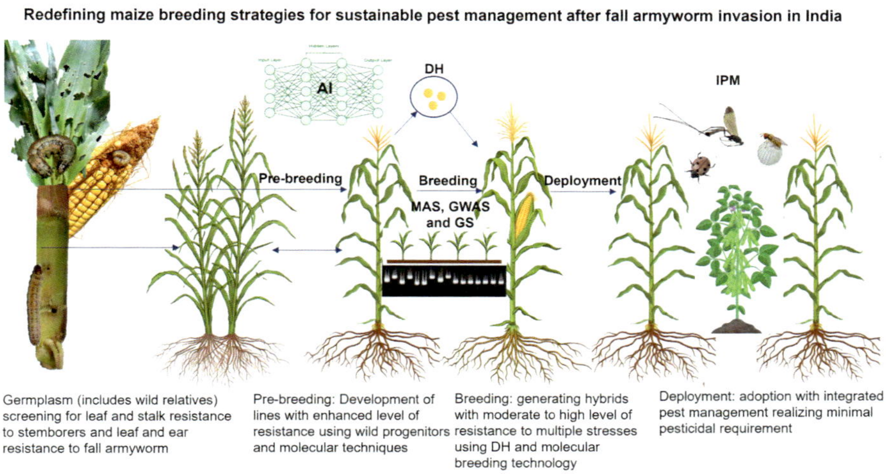

Fig. 12.1 Conceptual framework illustrating entomology-led, genetics-enabled maize breeding for sustainable pest management. The figure depicts the continuum from wild relatives and pre-breeding to advanced breeding and field deployment, integrating host plant resistance, plant–insect interactions, and genetic interventions (MAS, GWAS, genomic selection, and AI-assisted decision support) to develop pest-resilient maize hybrids, culminating in deployment under integrated pest management for reduced pesticide dependence

ment. DH lines also provide stable, uniform genetic backgrounds for the integration of transgenic or genome-edited traits, supporting consistent performance across environments. DH-based breeding is now widely applied in global maize programs and is increasingly adopted in India through ICAR- and CIMMYT-supported facilities.

(v) Trait- and tissue-specific resistance: Targeting tissue-specific traits enhances durability and efficacy of resistance. Foliar resistance is improved by selecting for lower leaf damage and elevated antifeedant compounds, such as PCA and DIMBOA. Stalk resistance is strengthened through markers for penetration resistance, rind thickness, lignin content, and PCA. Ear resistance is enhanced via maysin, husk tightness and husk length, which restrict pest entry, reduce damage by FAW, *H. armigera*, secondary pests, and associated ear rot pathogens. DH technology enables rapid fixation and pyramiding of these tissue-specific resistance alleles into uniform genetic backgrounds, creating maize lines with broad-spectrum, durable resistance.

(vi) Where regulatory frameworks permit, transgenic strategies deploying Cry, Vip, and RNAi constructs can specifically target FAW and stemborers. Genome-editing technologies, particularly CRISPR/Cas systems, allow precise modifications of regulatory elements controlling inducible defence responses and defence-related pathways, such as jasmonic acid signalling and benzoxazinoid biosynthesis. Compared with conventional GM crops, genome-edited maize can be developed more rapidly, provides multi-stress resistance, and in some regions faces fewer regulatory barriers (Spencer-Lopes et al. 2018; Prasanna et al. 2022). Combining transgenic or genome-edited traits with native resistance and DH lines allows pyramiding of multiple defence mechanisms, enhancing durability, reducing selection pressure on pest populations, and enabling environmentally sustainable control of insect pests and associated pathogens.

12.7.1.5 Integration of IPM Practices, Stakeholder Engagement, and Capacity Building

(i) Integration with Integrated Pest Management (IPM): Host-plant resistance is most effective when deployed as a core component of IPM. Resistant hybrids should be combined with complementary IPM tools, including biological control agents such as parasitoids, predators, microbial biopesticides, and botanicals, as well as cultural practices like optimized planting dates, intercropping, crop rotation, and residue management. The synergy between genetic resistance and IPM not only enhances pest suppression but also delays the evolution of pest resistance to both chemical and transgenic interventions, ensuring sustainable, environmentally safe crop protection.

(ii) Stakeholder engagement and capacity building: Effective deployment of resistant hybrids and IPM practices requires collaboration among research institutions, seed companies, extension agencies, and farming communities. Training programs should focus on the adoption of improved hybrids, proper agronomic management, pest monitoring, and IPM implementation. Strengthening farmer capacity ensures that the benefits of modern breeding technologies, viz., molecular, genomic, DH, transgenic, and genome-edited approaches, are fully realized at the field level, facilitating rapid dissemination, adoption, and sustained impact of resistant maize cultivars.

12.8 Conclusion

The invasion of FAW has fundamentally altered maize pest dynamics in India, intensifying yield losses and challenging established management strategies. While native stem borers and shoot fly were previously managed through moderate host-plant resistance integrated with IPM, FAW's high fecundity, migratory capacity, and ability to damage both vegetative and reproductive stages necessitate a paradigm shift in breeding and management approaches. Effective strategies must also address ear rot and associated mycotoxin risks, which can compromise grain quality and food safety. The future resilience of India's maize production depends on the strategic integration of host-plant resistance with next-generation breeding technologies, including DH pipelines, MAS, GS, genome editing, and precision phenotyping. These approaches accelerate allele fixation, improve trait evaluation, and shorten breeding cycles, while maintaining high yield potential. Key target traits include stalk strength, inducible defensive metabolites mediated by JA, such as PCA, DIMBOA, and HIPVs. Valuable resistance sources, including teosinte and resistant inbred lines, can be rapidly introgressed and stabilized using these technologies. Complementary agronomic innovations such as mechanization, minimal tillage, precision agriculture, UAVs, AI-based pest monitoring, and farmer-centric digital advisory platforms are essential to scale sustainable pest management practices. Smallholder farmers must be empowered with access to pest-resistant hybrids developed through DH-enabled pipelines, supported by robust extension services to ensure adoption and effective management. Although transgenic solutions have proven effective globally, India's regulatory framework currently emphasizes conventional and molecular breeding strategies, including genome editing. Ensuring long-term productivity and resilience will require the convergence of host-plant resistance, advanced breeding technologies, IPM, mechanization, and digital agronomy tools. Such integrated approaches are critical to safeguard maize productivity, resilience, and food, feed, and energy security for generations to come.

References

Abel CA, Woolfolk SW, Frei UK (2025) Response of doubled haploid lines derived from maize population BS39 to leaf injury by the fall armyworm, *Spodoptera frugiperda*. Southwest Entomol 50(4):1612–1621

Ajala SO, Odiyi AC, Thé C, Olaoye G (2008) Population cross diallel of maize genotypes with varying levels of resistance to the pink stalk borer (Sesamia calamistis Hampson) and the sugarcane borer (*Eldana sacharina*). Maydica 53(2):79

Aleri I, Beyene Y, Chavangi A, Gowda M, Burgueño J, Chaikam V (2025) Comparison of hybrid maize (*Zea mays* L.) performance developed through conventional pedigree and doubled haploid breeding methods. Plant Breed 25:1–18. https://doi.org/10.1111/pbr.70024

Anwer M, Kaur K, Jindal J, Suneja Y, Garg T (2024) Peroxidase, β-1,3-glucanase, chitinase and protease inhibitor activities in conjunction with phenols confer tolerance to wild and backcross-derived maize genotypes against *Spodoptera frugiperda*. Cereal Res Commun 52(3):993–1000

Badji A, Kwemoi DB, Machida L, Okii D, Mwila N, Agbahoungba S, Kumi F, Ibanda A, Bararyenya A, Solemanegy M, Odong T (2020) Genetic basis of maize resistance to multiple insect pests: integrated genome-wide comparative mapping and candidate gene prioritization. Genes 11(6):689

Barton WY, Buntin GD, Toews MD (2024) Bt trait efficacy against corn earworm (Helicoverpa zea) for preserving grain yield and reducing mycotoxin contamination of field corn. Insects 15(12):914

Baum JA, Bogaert T, Clinton W, Heck GR, Feldmann P, Ilagan O, Johnson S, Plaetinck G, Munyikwa T, Pleau M, Vaughn T, Roberts J (2007) Control of coleopteran insect pests through RNA interference. Nat Biotechnol 25:1322–1326

Bera P, Suby SB, Dixit S, Vijayan V, Kumar N, Sekhar JC, Vadassery J (2025) Identification of novel target genes for RNAi-mediated management of fall armyworm (*Spodoptera frugiperda*). Crop Prot 187:106972

Bohn M, Schulz B, Kreps R, Klein D, Melchinger AE (2000) QTL mapping for resistance against the European corn borer (*Ostrinia nubilalis* H.) in early-maturing European dent germplasm. Theor Appl Genet 101:907–917

Brooks TD, Willcox MC, Williams WP, Buckley PM (2005) Quantitative trait loci conferring resistance to fall armyworm and southwestern corn borer leaf feeding damage. Crop Sci 45(6):2430–2434

Butrón A, Sandoya G, Revilla P, Malvar RA (2009) Genetics of resistance to the pink stem borer (*Sesamia nonagrioides*) in maize (*Zea mays*). Ann Appl Biol 154(2):205–217

Cardinal AJ, Lee M (2005) Genetic relationships between resistance to stalk tunneling by the European corn borer and cell-wall components in maize population B73 × B52. Theor Appl Genet 111:1–7

Chandrasena DI, Signorini AM, Abratti G, Storer NP, Olaciregui ML, Alves AP, Pilcher CD (2018) Characterization of field-evolved resistance to *Bacillus thuringiensis*-derived Cry1F δ-endotoxin in *Spodoptera frugiperda* populations from Argentina. Pest Manag Sci 74(3):746–754

Chawade A, Van Ham J, Blomquist H, Bagge O, Alexandersson E, Ortiz R (2019) High-throughput field phenotyping tools for plant breeding and precision agriculture. Agronomy 9(5):258. https://doi.org/10.3390/agronomy9050258

Cheng X, Li H, Tang Q, Zhang H, Liu T, Wang Y (2024) Trends in the global commercialization of genetically modified crops in 2023. J Integr Agric 23(12):3943–3952

Chiang HC, Hodson AC (1950) Stalk breakage caused by European corn borer and its effect on harvesting of field corn. J Econ Entomol 43(4):415–422

Cobb JN, Juma RU, Biswas PS, Arbelaez JD, Rutkoski J, Atlin G, Hagen T, Quinn M, Ng EH (2019) Enhancing the rate of genetic gain in public-sector plant breeding programs: lessons from the breeder's equation. Theor Appl Genet 132(3):627–645

Dahi HF, Salem SA, Gamil WE, Mohamed HO (2020) Heat requirements for the fall armyworm *Spodoptera frugiperda* as a new invasive pest in Egypt. Egypt Acad J Biol Sci A Entomol 13(4):73–85

de Lange ES, Balmer D, Mauch-Mani B, Turlings TCJ (2014) Insect and pathogen attack and resistance in maize and its wild ancestors, the teosintes. New Phytol 204:329–341

Farhan Y, Smith JL, Limay-Rios V, Schaafsma AW (2020) Effect of simulated lepidopteran ear feeding injury on mycotoxin accumulation in grain corn. J Econ Entomol 113(5):2187–2196

Farias JR, Andow DA, Horikoshi RJ, Bernardi D, Ribeiro RDS, Nascimento ARD, Santos ACD, Omoto C (2016) Frequency of Cry1F resistance alleles in *Spodoptera frugiperda* (Lepidoptera: Noctuidae) in Brazil. Pest Manag Sci 72(12):2295–2302

Flett BC, Van Rensburg JB (1992) Effect of *Busseola fusca* on maize ear rot incidence caused by *Fusarium moniliforme* and *Stenocarpella maydis*. S Afr J Plant Soil 9(4):177–179

Foiada F, Westermeier P, Kessel B, Ouzunova M, Wimmer V, Mayerhofer W, Presterl T, Dilger M, Kreps R, Eder J, Schön CC (2015) Improving resistance to the European corn borer: a comprehensive study in elite maize using QTL mapping and genome-wide prediction. Theor Appl Genet 128(5):875–891

Gassmann AJ, Petzold-Maxwell JL, Keweshan RS, Dunbar MW (2011) Field-evolved resistance to Bt maize by western corn rootworm. PLoS One 6(7):e22629

Gatch EW, Munkvold GP (2002) Fungal species composition in maize stalks relative to European corn borer injury and transgenic insect protection. Plant Dis 86(10):1156–1162

Gesteiro N, Butrón A, Estévez S, Santiago R (2021) Role of maize cell-wall phenylpropanoids in stem-borer resistance. Phytochemistry 185:112683

Gesteiro N, Ordás B, Butrón A, De la Fuente M, Jiménez-Galindo JC, Samayoa LF et al (2023) Genomic versus phenotypic selection to improve corn borer resistance and grain yield. Front Plant Sci 14:1162440

Goyal S, Jindal J, Kaur K (2020) Physico-chemical bases of resistance in maize against shoot fly *Atherigona naqvii*. Euphytica 216:1–20

Gundappa, Kumar P, Suby SB (2013) Antibiosis effects of phenolic acids on maize spotted stem borer *Chilo partellus*. Indian J Entomol 75(3):247–250

Guo J, He K, Meng Y, Hellmich RL, Chen S, Lopez MD, Lauter N, Wang Z (2022) Asian corn borer damage is affected by rind penetration strength of corn stalks in a spatiotemporally dependent manner. Plant Direct 6(2):e381

Guthrie WD, Lillehoj EB, McMillian WW, Barry D, Kwolek WF, Franz AO, Catalano EA, Russell WA, Widstrom NW (1981) Effect of hybrids with different levels of susceptibility to second-generation European corn borers on aflatoxin contamination in corn. J Agric Food Chem 29(6):1170–1172

Hou M, Cao Y, Zhang X, Zhang S, Jia T, Yang J, Han S, Wang L, Li J, Wang H, Zhang L (2023) Genome-wide association study of maize resistance to *Pythium aristosporum* stalk rot. Front Plant Sci 14:1239635

Hu L, Zhang K, Wu Z, Xu J, Erb M (2021) Plant volatiles as regulators of plant defense and herbivore immunity. Curr Opin Insect Sci 44:82–88

Huang F, Qureshi JA, Meagher RL Jr, Reisig DD, Head GP, Andow DA, Ni X, Kerns D, Buntin GD, Niu Y, Yang F (2014) Cry1F resistance in fall armyworm (Spodoptera frugiperda): single gene versus pyramided Bt maize. PLoS One 9(11):e112958

Huang G, Guo Y, Tan W, Zhang M, Li Z, Zhou Y, Duan L (2025) Development of a model for maize stalk lodging resistance. Field Crop Res 325:109828

Jiménez-Galindo JC, Malvar RA, Butrón A, Santiago R, Samayoa LF, Caicedo M, Ordás B (2019) Mapping resistance to corn borers in a maize MAGIC population. BMC Plant Biol 19:1–7

Jindal J (2013) Incidence of insect pests and management of shoot fly *Atherigona* spp. in maize. PhD thesis, Punjab Agricultural University, Ludhiana, India

Jindal J, Hari NS, Grewal MS, Chawla JS (2007) Evaluation of maize genotypes against shoot fly *Atherigona naqvii*. Crop Improv 34(2):160–165

Kamweru I, Anani BY, Beyene Y, Makumbi D, Adetimirin VO, Prasanna BM, Gowda M (2022) Genomic analysis of resistance to fall armyworm in CIMMYT maize lines. Genes 13(2):251

Kaur JA, Kumar P, Singh J, Suby SB, Bajya DR (2015) Egg-laying pattern of *Sesamia inferens* on maize. Indian J Agric Sci 85:109–113

Krakowsky MD, Brinkman MJ, Woodman-Clikeman WL, Lee M (2002) Genetic components of resistance to European corn borer stalk tunneling. Crop Sci 42(4):1309–1315

Lenz PR, Nadeau S, Azaiez A, Gérardi S, Deslauriers M, Perron M, Isabel N, Beaulieu J, Bousquet J (2020) Genomic prediction for hastening and improving efficiency of forward selection in conifer polycross mating designs: an example from white spruce. Heredity 124(4):562–578

Liu L, Liu S, Lu H, Tian Z, Zhao H, Wei D et al (2022) Integration of transcriptome and metabolome analyses reveals lodging-resistance genes in maize. Front Genet 13:1001195

Lopez L, Camas A, Shivaji R, Ankala A, Williams P, Luthe D (2007) Mir1-CP accumulation in maize vascular tissues in response to herbivory. Planta 226:517–527
López-Malvar A, Butrón A, Samayoa LF, Figueroa-Garrido DJ, Malvar RA, Santiago R (2019) Genome-wide association analysis for maize stem cell wall hydroxycinnamates. BMC Plant Biol 19:1–12
López-Malvar A, Reséndiz Z, Santiago R, Jiménez-Galindo JC, Malvar RA (2021) Relationships between stalk resistance and corn borers, agronomic traits, and cell wall hydroxycinnamates in a set of recombinant inbred lines from a maize MAGIC population. Agronomy 11(6):1132
López-Malvar A, Malvar RA, Butrón A, Revilla P, Jiménez-Galindo JC, Souto XC, Santiago R (2022) SNP identification for maize cell wall hydroxycinnamates. Phytochemistry 193:113002
López-Malvar A, Malvar RA, Butrón A, Souto XC, Santiago R (2025) Assessing breeding success for contrasting diferulate concentrations. BMC Plant Biol 25:18
Magarini A, Colombo F, Cassani E, Ghidoli M, Landoni M, Sangiorgio S, Pilu R (2024) The role of husk traits in maize susceptibility to Fusarium verticillioides. Food Energy Secur 13(2):e537
Maino JL, Schouten R, Overton K, Day R, Ekesi S, Bett B, Barton M, Gregg PC, Umina PA, Reynolds OL (2021) Regional and seasonal activity predictions for fall armyworm in Australia. Curr Res Insect Sci 1:100010
Malook SU, Liu XF, Ma C, Qi J, Liu W, Zhou S (2021) Transcriptomic responses of fall armyworm feeding on resistant maize with high benzoxazinoid content. Agronomy 11(12):2503
Mandal L (2018) Potential applications of molecular markers in plants. Curr Trends Biomed Eng Biosci 12(4):85–87
Martin SA, Darrah LL, Hibbard BE (2004) Divergent selection for rind penetrometer resistance and effects on European corn borer damage. Crop Sci 44(3):711–717
McMillian WW, Wilson DM, Widstrom NW (1985) Aflatoxin contamination of preharvest corn in Georgia. J Environ Qual 14(2):200–202
Moore VM, Tracy WF (2021) Combining ability of husk extension, maysin content, and corn earworm resistance. J Am Soc Hortic Sci 146(1):14–23
Moose SP, Mumm RH (2008) Molecular plant breeding as the foundation for 21st century crop improvement. Plant Physiol 147(3):969–977
Niu Y, Zhao X, Chao W, Lu P, Bai X, Mao T (2023) Genetic variation, DIMBOA accumulation, and candidate gene identification in maize multiple insect resistance. Int J Mol Sci 24(3):2138
Papst C, Bohn M, Utz HF, Melchinger AE, Klein D, Eder J (2004) QTL mapping for European corn borer resistance and agronomic traits in early-maturing maize. Theor Appl Genet 108:1545–1554
Prasanna BM, Bruce AY, Beyene Y, Makumbi D, Gowda M, Asim M, Martinelli S, Parimi S (2019) Host plant resistance in maize to fall armyworm. In: Fall armyworm in Asia: a guide for integrated Pest management. CIMMYT, pp 99–120
Prasanna BM, Burgueño J, Beyene Y, Makumbi D, Asea G, Woyengo V, Tarekegne A, Magorokosho C, Wegary D, Ndhlela T, Zaman-Allah M (2022) Genetic trends in CIMMYT's tropical maize breeding pipelines. Sci Rep 12(1):20110
Pruter LS, Weaver M, Brewer MJ (2020) Risk factors and strategies for insect-derived ear injury and aflatoxin accumulation in maize. J Integr Pest Manage 11(1):13
Ragot M, Bonierbale M, Weltzien E (2018) From market demand to breeding decisions: a framework. CGIAR Gender and Breeding Initiative Working Paper, No. 2, pp 1–32
Rodríguez VM, Velasco P, Cao A, Santiago R, Malvar RA, Butrón A (2021) Systemic maize responses modulate resistance to stem borers. Front Plant Sci 11:627468
Rosenthal JP, Dirzo R (1997) Effects of life history and domestication on plant defence against insects. Evol Ecol 11:337–355
Rule DM, Nolting SP, Prasifka PL, Storer NP, Hopkins BW, Scherder EF, Siebert MW, Hendrix WH III (2014) Efficacy of pyramided Bt proteins Cry1F, Cry1A.105, and Cry2Ab2 expressed in SmartStax corn hybrids against lepidopteran insect pests in the northern United States. J Econ Entomol 107(1):403–409
Salgotra RK, Stewart CN (2020) Functional markers for precision plant breeding. Int J Mol Sci 21(13):4792

Sandoya G, Malvar RA, Santiago R, Alvarez A, Revilla P, Butrón A (2010) Effects of selection for resistance to *Sesamia nonagrioides* on maize yield and stability. Ann Appl Biol 156(3):377–386
Santosh HB, Sekhar JC, Rakshit S, Gadag RN, Dass S (2012) Epistatic interactions for susceptibility to pink borer in maize. Indian J Genet Plant Breed 72(3):284–289
Sekhar JC, Rakshit S, Kumar P, Venkatesh S, Sharma RK, Anuradha M et al (2010) Improvement of resistance in maize through selection against pink borer. Indian J Genet Plant Breed 70(2):204–206
Sekhar JC, Kumar P, Kumar RS, Kaul J, Rakshit S, Dass S (2014) DMRE 63 maize germplasm with resistance to pink borer. Indian J Plant Genet Resour 27(2):191–193
Shivaji R, Camas A, Ankala A, Engelberth J, Tumlinson JH, Williams WP, Wilkinson JR, Luthe DS (2010) Plants on constant alert: elevated levels of jasmonic acid and jasmonate-induced transcripts in caterpillar-resistant maize. J Chem Ecol 36(2):179–191
Siebert MW, Tindall KV, Leonard BR, Van Duyn JW, Babcock JM (2008) Evaluation of Cry1F maize against fall armyworm. J Entomol Sci 43(1):41–51
Siebert MW, Nolting SP, Hendrix W, Dhavala S, Craig C, Leonard BR, Stewart SD, All J, Musser FR, Buntin GD, Samuel L (2012) Evaluation of corn hybrids expressing Cry1F, Cry1A.105, Cry2Ab2, Cry34Ab1/Cry35Ab1, and Cry3Bb1 against southern United States insect pests. J Econ Entomol 105(5):1825–1834
Smith WEC, Shivaji R, Williams WP, Luthe DS, Sandoya GV, Smith CL, Sparks DL, Brown AE (2012) A maize line resistant to herbivory constitutively releases (E)-β-caryophyllene. J Econ Entomol 105(1):120–128
Song P, Wang J, Guo X, Yang W, Zhao C (2021) High-throughput phenotyping in future crop breeding. Crop J 9(3):633–645
Soujanya PL, Sekhar JC, Ratnavathi CV, Karjagi CG, Shobha E, Suby SB, Yathish KR, Sunil N, Rakshit S (2021) Induction of cell wall phenolic monomers as part of direct defense response in maize to pink stem borer (*Sesamia inferens* Walker) and non-insect interactions. Sci Rep 11(1):14770
Soujanya PL, Sekhar JC, Yathish KR, Karjagi CG, Rao KS, Suby SB, Jat SL, Kumar B, Kumar K, Vadessery J, Subaharan K (2022) Leaf damage-based phenotyping technique and its validation against fall armyworm, *Spodoptera frugiperda* (J.E. Smith), in maize. Front Plant Sci 13:906207
Spencer-Lopes MM, Forster BP, Jankuloski L (2018) Manual on mutation breeding, 3rd edn. Food and Agriculture Organization of the United Nations
Storer NP, Babcock JM, Schlenz M, Meade T, Thompson GD, Bing JW, Huckaba RM (2010) Discovery and characterization of field resistance to Bt maize: *Spodoptera frugiperda* (Lepidoptera: Noctuidae) in Puerto Rico. J Econ Entomol 103(4):1031–1038
Suby SB, Jha SK, Karjagi CG, Kumar P, Sekhar JC, Kaur J, Cholla AK, Soujanya PL, Sharma RK, Rakshit S (2020) Penetration resistance of second above-ground internode in V6–10 stage maize plants confers resistance to stalk boring larvae of Chilo partellus (Swinhoe) in maize. Phytoparasitica 48(3):455–469
Takahashi CG, Kalns LL, Bernal JS (2012) Plant defense against fall armyworm in maize and teosinte. Entomol Exp Appl 145(3):191–200
Tamiru A, Bruce TJ, Woodcock CM, Caulfield JC, Midega CA, Ogol CK, Mayon P, Birkett MA, Pickett JA, Khan ZR (2011) Maize landraces recruit egg and larval parasitoids in response to egg deposition by a herbivore. Ecol Lett 14(11):1075–1083
Tamiru A, Bruce TJ, Richter A, Woodcock CM, Midega CA, Degenhardt J, Kelemu S, Pickett JA, Khan ZR (2017) A maize landrace that emits defense volatiles in response to herbivore eggs possesses a strongly inducible terpene synthase gene. Ecol Evol 7(8):2835–2845
Taneja SL, Nwanze KF (1990) Mass rearing of Chilo spp. for resistance screening. Int J Trop Insect Sci 11(4–5):603–616
Tashiro K, Ishitani M, Murai S, Niimi M, Tobisa M, Idota S et al (2024) Effects of defoliation simulating fall armyworm damage in maize. AgriEngineering 6(2):1847–1858
U.S. Environmental Protection Agency (2010) Cry1Ab and Cry1F Bt plant-incorporated protectants. Biopesticides Registration Action Document

Van Rensburg JB (2007) First report of field resistance by Busseola fusca to Bt maize. S Afr J Plant Soil 24(3):147–151
Vikal Y, Kaur A, Jindal J, Kaur K, Pathak D, Garg T, Singh A, Singh P, Yadav I (2020) Identification of genomic regions associated with shoot fly resistance in maize and their syntenic relationships in the sorghum genome. PLoS One 15(6):e0234335
Wang S, Liu W, Chen Z, Zhang J, Jia X, Gou M et al (2022) The antioxidant protein ZmPrx5 contributes resistance to maize stalk rot. Crop J 10(4):1049–1058
Wang B, Yang M, Guo H, Wang J, Wang Z, Lu H et al (2024) Genome-wide association study for stalk lodging resistance in maize. BMC Genomics 25(1):19
Wang B, Wang S, Xu L, Li Z, Ding X, Ma L et al (2025) Gene mapping and genetic analysis of maize resistance to stalk rot. Int J Mol Sci 26(24):11866
Warburton ML, Woolfolk SW, Smith JS, Hawkins LK, Castano-Duque L, Lebar MD, Williams WP (2023) Genes and genetic mechanisms contributing to fall armyworm resistance in maize. Plant Genome 16(2):e20311
Widstrom NW, McMillian WW, Wiseman BR (1970) Resistance in corn to corn earworm and fall armyworm. J Econ Entomol 63(3):803–808
Widstrom NW, Wiseman BR, Snook ME, Nuessly GS, Scully BT (2003) Registration of maize population Zapalote Chico 2451F. Crop Sci 42:444–445
Williams WP, Davis FM, Windham GL (1990) Registration of Mp708 germplasm line of maize. Crop Sci 30:757
Williams WP, Buckley PM, Davis FM (1995) Combining ability for fall armyworm and southwestern corn borer resistance. Theor Appl Genet 90:275–278
Williams WP, Davis FM, Buckley PM, Hedin PA, Baker GT, Luthe DS (1998) Resistance to fall armyworm and southwestern corn borer at different maize growth stages. J Econ Entomol 91(6):1471–1480
Windham GL, Williams WP, Davis FM (1999) Effects of southwestern corn borer on Aspergillus flavus infection and aflatoxin accumulation. Plant Dis 83(6):535–540
Wiseman BR, Carpenter JE (1995) Growth inhibition of corn earworm larvae on resistant corn silk diets. J Econ Entomol 88(4):1037–1043
Womack ED, Williams WP, Smith JS, Warburton ML, Bhattramakki D (2020) Mapping QTL for fall armyworm resistance in maize inbred Mp705. J Econ Entomol 113(2):956–963
Yactayo-Chang JP, Mendoza J, Willms SD, Rering CC, Beck JJ, Block AK (2021) Maize volatiles influencing oviposition and feeding of *Spodoptera frugiperda*. J Chem Ecol 47(8):799–809
Zhao S, Yang X, Liu D, Sun X, Li G, Wu K (2023) Performance of the domestic Bt corn event expressing pyramided Cry1Ab and Vip3Aa19 against *Spodoptera frugiperda* in China. Pest Manag Sci 79(3):1018–1029

Optimizing Agronomic Practices for Sustainable Maize Production

13

S. D. Bamboriya, D. M. Mahala,
and Sumitra Devi Bamboriya

Abstract

Maize is the backbone of India's feed and fuel industries and contributes immensely to the nation's economic growth and development. It has emerged as a major player in India's agriculture sector in recent years due to its raising demand in the feed, ethanol, and starch sectors. A significant shift from conventional practices to technology-driven sustainable approaches is vital for boosting maize productivity to cater to the increasing industrial demand of this crop. This chapter addresses the vital aspect of sustainable maize production, focusing on improved production technologies like precision land levelling, zero tillage, bed planting, mechanized crop cultivation, drip irrigation, and fertigation. These cutting-edge technologies, often found to be more productive, profitable, and sustainable, and leave lesser environmental impacts. The strategic integration of these innovative technologies saves precious natural resources, reduces dependency on chemical inputs, provides ecological balance and resilience against weather abnormalities, preserves biodiversity, improves soil health, and overall well-being of the agri-ecosystem. Hence, the transition from conventional methods to technology-driven approaches is imperative for boosting maize productivity and profitability and ensuring the long-term viability and sustainability of agricultural systems.

Keywords

Bed planting · Conservation agriculture · Drip irrigation · Mechanized farming · Soil health · Zero tillage

S. D. Bamboriya (✉) · D. M. Mahala
ICAR-IIMR, Ludhiana, Punjab, India

S. D. Bamboriya
SKNAU, Jobner, Rajasthan, India

M. Choudhary et al. (eds.), *Next-Generation Maize Research*,
https://doi.org/10.1007/978-981-92-0722-0_13

13.1 Introduction

Maize (*Zea mays* L.) is the third leading cereal crop in India with respect to area coverage (12.4 mha) and production (42.4 mt) (MoA&FW 2025). More recently, maize has taken the centre stage of India's feed and energy security, as around 51% of maize output is being consumed in the poultry sector and 18% in the ethanol industry. Maize has got the title of India's fastest growing cereal as its cultivation is gaining momentum in recent years due to its raising demand in ethanol, starch, and feed sectors. Additionally, the Indian government is also promoting its cultivation in northwest India to address the residue burning, water, and electricity crisis. As a result, maize has witnessed a 36% surge in area and up to 87% bloom in production over the previous decade (2015–16 to 2024–25), as shown in Fig. 13.1. In spite of a record production surge, the demand-supply gap is widening as the domestic demand has overtaken the supplies. India emerged as a net importer of maize, and its imports touched 9.7 lakh tonnes during the last fiscal year (2024–25). Projections show that maize production will have to be doubled (86 mt) by 2047 to meet India's food and feed demand and E-30 ethanol target. With a little to no possibility of maize area expansion, these production targets have to be achieved solely by increasing productivity, that too with limited resource use, considering the environment in view. Currently, India's average maize productivity has stagnated at 3–3.5 t/ha and remains far below the global average (6 t/ha). The major drivers of poor yield include pest and disease outbreaks, moisture stress, waterlogging, and poor agronomic management. These critical yield gaps point to a huge scope for increasing maize yield and urge revising the maize sector to ensure food security, meet industrial demand, and raise farmers' income.

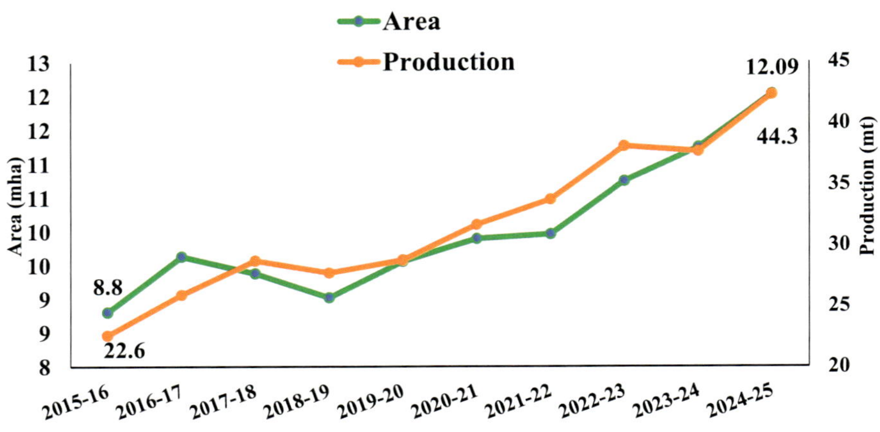

Fig. 13.1 Surge in all India maize area and production between 2015–16 and 2024–25 (DAC &FW 2021). The graph shows the dramatic growth in both maize cultivation area (~1.5 fold) and maize production (~two fold) during the previous decade

13.1.1 Pathways to Sustainable Maize Production

Agrarian growth and development in India are facing numerous challenges, namely, soil degradation, climate vagaries, looming water and energy crisis, resource dwindling, rising cultivation cost, etc., threatening the overall agricultural sustainability of the nation in the long run. The ever-increasing population and increased food demand further amplify the pressure on already scarce natural resources and necessitate immediate action for efficient and sustainable use of farm resources. In this section, we will explore the possible agronomic strategies for enhancing maize productivity and profitability in sustainable ways.

13.1.2 Precision Land Leveling (PLL)

Most cultivated fields have uneven topography and field slopes of varying degrees that led to poor crop stand establishment, uneven irrigation depth, runoff of irrigation water, inefficient use of agri-inputs, and poor crop yield (Jat et al. 2009). PLL facilitates the production of uniform field surfaces, minimizes the spatial heterogeneity and micro-topography by removing unneeded depressions and higher contours. It ensures uniformity in irrigation water and fertilizer application, subsequently enhancing crop productivity and inputs (water, nutrients, and energy) use efficiency. Studies showed that PLL improves crop stand and farm productivity, saves a considerable amount of irrigation water, and reduces labour deployment in weeding (Jat et al. 2006). Jat et al. (2009) compared PPL with traditional land leveling (TLL) and reported that PPL helped in saving irrigation water by 10%–14% in rice and wheat. In addition, the PLL enhanced rice-wheat system productivity (7.4%) and profitability (54%) as compared to TLL.

13.1.3 Precision Planting

Maize planting is generally accomplished by hand, where heterogeneous plant spacing and depth, non-synchronous crop emergence, and sub-optimal planting density are quite common phenomena. The non-equidistant plant distribution modifies canopy structure and light interception, thereby affecting photosynthesis, resource-use efficiency, crop yield, and net profitability. Extreme heterogeneous plant distribution may decrease crop productivity up to 30% (Wade et al. 1988). Precision planting of a single seed/hill at fixed spacing and depth is crucial in achieving better crop yield and quality (Fig. 13.2). A pneumatic planter is third-generation seeding machinery that ensures uniformity in seed placement and plant-to-plant spacing, favouring simultaneous emergence and uniform establishment of crop (Singh et al. 2018). Studies have shown that pneumatic planter minimizes mechanical damage to seed and avoids placement of double seed per hill, thereby reducing seed wastage as well as seed requirement per unit area (Chandel et al. 2024). Singh et al. (2018) compared the performance of different planting

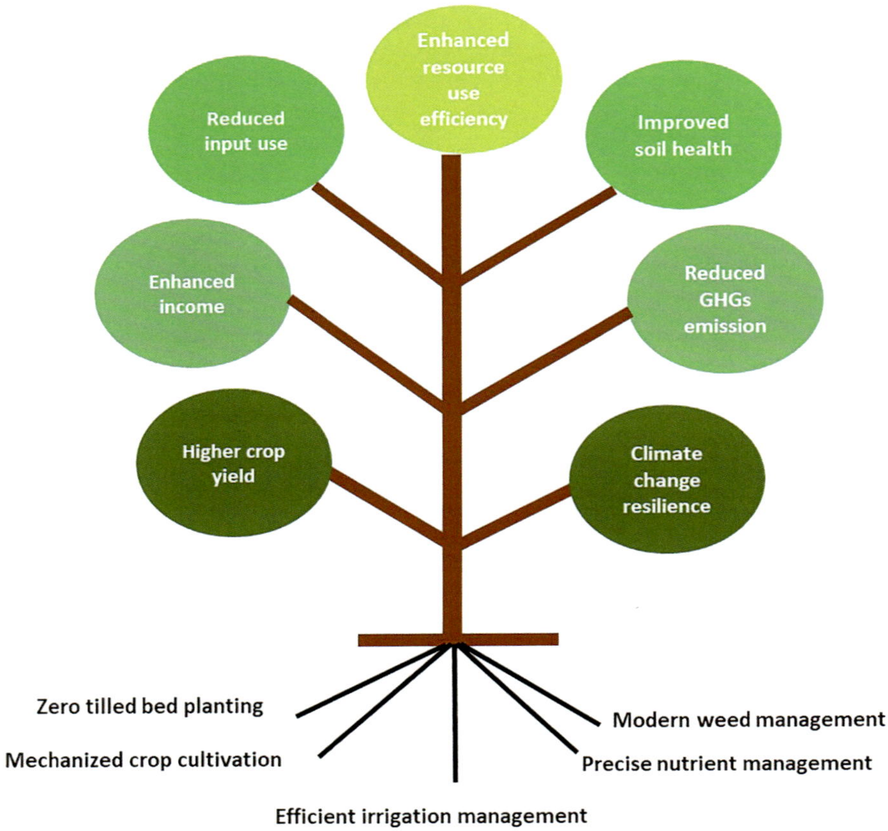

Fig. 13.2 Improved maize production technologies enhance productivity, resilience, and sustainability while conserving natural resources

machineries and reported that pneumatic planters improved maize yield by 3.8% and saved the operational cost by 70% over manual planting on ridges. Chandel et al. (2024) compared pneumatic planters with various raised-bed/flat-bed planters. They noted the highest quality feed index, lowest missing and multiple indexes, and the least operational cost with pneumatic planter (Table 13.1). Owing to appropriate seeding depth and perfect equidistance spacing, the pneumatic planter produced maximum yield and also recorded the highest energy productivity. Mechanized maize sowing with a vacuum precision planter saved 32% and 20% cost of planting compared to the conventional method and seed-cum-ferti drill, respectively (Manjulatha et al. 2022). The reduced planting cost was due to lower seed requirements and the elimination of thinning operations, as only one seed was placed per hill.

Table 13.1 The machinery operational parameters, maize yield, and energy productivity for various row crop planters/sowing techniques

Parameters	Raised bed planter		Pneumatic planter		Manual sowing
	Inclined plate	Vertical plate	Raised bed	Flat bed	
Quality of feed index (%)	67.44[a]	48.04[a]	87.99[a]	85.25[a]	55.22[a]
Missing index (%)	19.00[ab]	17.52[a]	7.64[ab]	10.797[a]	44.78[ab]
Multiple index (%)	13.63[ab]	34.33[b]	4.37[a]	3.79[a]	0.00[c]
Operation cost (Rs./ha)	1170.0.	1646.3	1008.1	746.4	6209.7
Yield (t/ha)	7.71	6.75	8.61	8.34	4.56
Energy productivity (kg MJ^{-1})	0.13	0.11	0.14	0.14	0.08

In the Table, means with the same letter are not significantly different.

13.2 Mechanized Crop Cultivation

After globalization in 1991, India has witnessed a dramatic migration of labour from agriculture to industry and service sectors in search of better employment options. Government-sponsored schemes like Mahatma Gandhi National Rural Employment Guarantee Act (MNREGA) added fuel to the fire and further raise the labour scarcity and forced wages to touch a peak. In this scenario, farm mechanization holds the key to sustain agricultural productivity. Machinery use in agriculture enables timely field preparation, seeding, weeding, pesticide spray and harvesting, resulting in increased farm productivity (Bamboriya et al. 2020). It speeds up agricultural tasks, reduces the time window between previous crop harvesting and succeeding crop seeding, and facilitates crop intensification and diversification. Mechanization reduces labour deployment in crop production and save agri-input use, thus reducing production cost and making agriculture economically viable (Manjulatha et al. 2022). The reduced drudgery and workload and improved quality of life associated with farm mechanization may attract youth in agriculture. Under the changing climatic scenarios, farm machinery is immensely important in the adaptation process by reducing weather risk.

Mechanization in maize offers a pathway towards more efficient, profitable, and sustainable maize production, especially in the era of climate change. The majority of the maize production in India comes from rainfed areas, where agricultural activities are rain-dependent, thereby time-bound and need to be carried out in a short time span. Intensive cropping in the irrigated zones necessitates quick planting and harvesting to integrate more crops per year. Hence, mechanization is vital for sustaining maize production in the irrigated as well as non-irrigated agro-ecologies. Wider inter-row spacing and vast scale cultivation of single-cross hybrids with uniform maturity and ear placement height offer an opportunity for seed-to-seed mechanization in maize (Bamboriya et al. 2020). As per the impact evaluation studies, agriculture mechanization reduced labour requirement (30%), time of operations (33%), diesel consumption (22.4%), fertilizer requirement (12.7%), seed rate (11%), and weed infestation (26.6%) and improved crop productivity (17.9%) (DAC&FW, 2021).

13.3 Zero till Farming

Soil is the most important natural resource, ensuring food security, supporting the livelihood of large masses, and sustaining ecosystem services. Recent estimates suggest that almost one-third of the total geographical area of India is hardly hit by various forms of soil degradation. Intensive tillage, negligence of organic matter addition, and adoption of unsustainable cropping systems contributed to the depletion of soil organic carbon and soil fertility, and caused widespread micronutrient deficiency. Soil degradation implications include: reduced agricultural output and resilience against environmental shocks, increased dependency on off-farm resources, higher production costs, damage to biodiversity, threat to the nation's food security and socio-economic stability, and climate change amplification.

Zero tillage (ZT), residue retention (+R), and diversified crop rotation are widely advocated as options for reversing soil degradation, restoring soil fertility, increasing resource use efficiency, system resilience, and farm profitability (Lal 2019). Conventional method of field preparation includes multiple (5–6) passes of tillage machinery like disc plough, cultivator/harrow, and planker to make a crumbly seedbed prior to crop planting. The repetitive and intensive tillage operations consume a large amount of fuel and emit significant amounts of greenhouse gases, favour soil erosion, reduce soil organic matter, and also disturb soil structure and biological life (Lal 2019). These tillage operations alone account for 20%–30% cost of the maize production, leading to lower profit margins. ZT provides several benefits, like mitigating CO_2 emissions, enhancing carbon sequestration and improving soil structure, reducing fuel consumption, timely crop sowing, reducing soil erosion, etc. (Lal 2019). Likewise, surface residue retention offers multi-fold benefits, including reduced runoff, soil erosion, and water evaporation, increased moisture absorption and retention, suppression of weeds, improvement in soil organic matter and C-sequestration and soil structure, improved microbial diversity and activities, and enhanced nutrient storage and cycling (Jat et al. 2013; Parihar et al. 2016; Lal 2019).

Zero till planting in maize-wheat systems reduced production cost (33.7%), and irrigation water use (10.8%), and also improved water productivity (26.2%), crop productivity (11.3%), and net returns (US$ 281/ha) compared to the conventional tillage (CT) system (Jat et al. 2013). A decade-long adoption of ZT recorded significantly higher mean grain yield (8 t/ha) of winter maize over CT (5.9) (Jat et al. 2019a). Parihar et al. (2016) tested ZT and CT under various maize-based rotations for six consecutive years. ZT systems provided 15% higher system productivity, US$ 403/ha higher income, and reduced variable cost by 8% than CT. The ZT systems improved soil physical quality by improving the infiltration rate and soil aggregation (>0.25 mm) and lowering bulk density and penetration resistance over the CT system (Jat et al. 2013). The ZT + R minimized the risks of drought stress through conservation of soil moisture and reduction in soil temperature, therefore, improved the system resilience during low rainfall years and provided better yield over CT (Parihar et al. 2016; Lal 2019).

13.4 Bed Planting

Conventional flatbed planting of maize faces many challenges, namely, waterlogging, excessive water use, reduced water productivity, soil compaction, and increased weed infestation, affecting maize productivity, profitability, and long-term sustainability (Gupta et al. 2023). Maize is a highly water-sensitive cereal, and temporary water ponding, especially at V_3 stage, causes 58.8%–69.8% yield reduction (Yu et al. 2015). A shift from flat to raised bed planting creates ideal growing conditions for water-sensitive crops by preventing soil compaction, improving drainage and aeration (Gupta et al. 2023). Bakker et al. (2005) studied the effect of raised beds on soil physical properties and noted that the bed top surface had higher infiltration rates and hydraulic conductivity and a lower incidence of waterlogging over flat beds.

Naresh et al. (2014) reported that raised bed + PLL saved irrigation water by 16.5%–20.4% over conventional practices. Maize produced 11.8% more yield under raised bed + PLL as compared to flatbed + TLL. Devkota et al. (2021) compared laser levelled raised beds with laser levelled flat beds and reported that raised bed planting improved maize productivity (16%), net profit (24%), and nitrogen (N)-use efficiency (33%) (Table 13.2). In addition, the irrigation amount was reduced by 66 mm (18%), and irrigation water productivity was increased by 42%. With a sufficient decrease in water use (9%–20%), broad bed improved maize water productivity by 11%–26% compared to ridge planting (Singh et al. 2025).

Permanent raised bed (PRB) planting is a novel technique spreading its roots in conservation agriculture (CA)-based maize farming where the bed is left undisturbed except for bed reshaping every year, and previous crop residues are kept on

Table 13.2 Crop yield, input use efficiency, and economics under various land levelling and crop establishment methods

Treatment	Grain yield (t/ha)	Net profit (Rs. /ha)	N recovery efficiency (%)	Water use (mm/ha)	WPi (kg/m^3)
Wide raised bed + PLL	4.41	24,545	46	311	1.42
Wide raised bed + TLL	4.08	22,708	41	351	1.16
Flatbed + PLL	4.05	21,391	42	352	1.15
Flatbed + TLL	3.8	19,853	33	377	1.0

Source: Devkota et al. (2021

the soil surface. CA-based PRBs have been shown to increase maize yield, save water (11%), and improve water productivity (16%) over flat beds (Jat et al. 2013). Additional benefits include reduced crop lodging, rainwater harvesting, better drainage, less water stagnation, and so far. It was reported that adopting zero-tillage with PRBs improved soil physical properties by increasing aggregate formation, water infiltration, and reducing soil hardness compared with traditional practices (Jat et al. 2009). Permanent raised beds support zero till seeding, accompanied by reduced tillage cost and energy saving, and reduced gaseous emission (Ram et al. 2005). Stubble retention on permanent beds delivers a variety of benefits, including improved soil organic matter and reduced soil evaporation.

13.5 Precision Irrigation Management

India is witnessing an all-time high-water scarcity as half of India's population is surviving under high to extreme high-water stress (NITI Aayog 2018). Agriculture is the major consumer of fresh water resources; therefore, the looming water crisis is the biggest threat to agricultural sustainability. Widespread cultivation of water guzzling crops like rice, faulty irrigation practices, negligence of water conservation technologies, lack of suitable water pricing and undesirable subsidy on the electricity are major contributing factors for agricultural water shortage. As per estimates, by 2030, Indian agriculture could face a 570 BCM shortage in irrigation water supply (NITI Aayog 2018). The continuous population explosion, climate change, and rapid industrialization will make this scenario even worse. In this context, precise irrigation management is utmost important for agriculture intensification, food security, and agricultural sustainability.

Among the various improved versions of irrigation technologies developed so far, drip irrigation tops the list for better irrigation water management (Gupta et al. 2023). Under a traditional irrigation system, a substantial amount of the precious resource is wasted through runoff, evaporation, and deep percolation, which leads to poor yield and higher irrigation costs (Singh et al. 2025). Drip irrigation eliminates such water losses by providing a small amount of water directly into the root zone. Integrating CA with surface drip irrigation (SDI) saved a considerable amount of irrigation water (168 mm) and improved crop water productivity (259%) and grain yield (14%) of maize over the furrow irrigation system with CT (Sandhu et al. 2019). The sub-surface drip irrigation (SSDI) + CA saved 48%–53% irrigation water in rice and 42–53% in wheat compared to flood irrigation + CT, mainly by minimizing losses through percolation and evaporation (Sidhu et al. 2019). Likewise, Jat et al. (2019a, b) reported 22.7% water saving and 33.3% improved water productivity in maize crop with SSDI + CA over flood irrigation + CA. Gupta et al. (2023) reported 11.2% yield improvement, 55% water saving, and 140% improvement in water productivity in maize-wheat rotation with SSDI over furrow/flood irrigation. Drip irrigation has been shown to reduce water and nutrient losses through runoff (37%–49%), percolation (61%–74%), soil evaporation (8%–36%), nitrate leaching (58%–62%), and ammonia volatilization (33%–55%) as compared to flood irrigation (Leghari et al. 2024). SDI with a broad bed helped in saving of 26%–27% irrigation water and increased crop water productivity up to 44%, over broad bed with furrow irrigation (Singh et al. 2025).

13.6 Efficient Nutrient Management

Plant nutrients had a profound impact on boosting agricultural productivity and safeguarding food supply for the world population. Non-judicious use of fertilizers, especially over-reliance on N-fertilizer and negligence of potassium (K)-fertilizer, has led to soil fertility depletion, nutrient imbalances, soil biodiversity loss, environmental pollution, and poor crop yields. Negligence of organic manure addition, along with intensive tillage, reduced soil organic carbon and increased the need for fertilizer application. Therefore, integrated nutrient management is vital to improve crop productivity, reduce fertilizer waste and costs, minimize environmental footprints, and support agricultural sustainability. Application of 10–15 t/ha farm yard manure along with 100% recommended fertilizer dose has been shown to improve maize productivity and profitability (Mahala et al. 2023). Five splits application of N (20% Basal, 25% at V4, 30% at V8, 20% at VT, and 5% at GF) has been found to improve maize yields up to 22%, compared to the three-split method (33% at basal, 33% at V8, and 33% at VT). Likewise, three-splits application of K (at sowing, knee-high, and tasselling stages) has been shown to improve maize grain yield and K use efficiency over basal application alone (Channakeshava et al. 2022). Single-time band application of phosphorus (P) fertilizer directly into the root zone is desirable to enhance fertilizer use efficiency (Mahala et al. 2023). Soil application of 20–25 kg ZnSO4/ha has been shown to increase maize grain yield in field trials (Shivay and Prasad 2014). In deficit soils, application of 30 kg sulphur and 1.5 kg boron/ha had a significant positive effect on the maize productivity (Mahala et al. 2023).

Fertigation, through precise and simultaneous application of two agri-inputs, i.e. water and nutrients, enhances crop yield, net profits, and fertilizer use efficiency and minimizes fertilizer doses and their negative environmental impacts. Precise N application close to the crop root zone in multiple splits via drip irrigation reduced N leaching and volatilization losses, improved N use efficiency, and saved ~20% of the N over the conventional method (Sidhu et al. 2019). (Jat et al. 2019a) and Gupta et al. (2023) reported 20–25% N saving with SSDI over flood irrigation in maize. Fertigation through drip irrigation in ZT maize improved N-use efficiency by 31.3% over flood irrigation + CA (Jat et al. 2019a). Fertigation in maize at 10-day intervals enhanced the N recovery by 17% in comparison to furrow irrigation (Sandhu et al. 2019). Likewise, through right placement (crop root zone), right dose (small amounts), and right time (more splits) of fertilizer application, 30.4% higher agronomy efficiency of N was recorded in maize under CA + SSDI over CT + Flood irrigation (Gupta et al. 2023). Multi-split applications of P via fertigation enhanced P uptake by reducing its soil fixation, leading to improved maize yield, and P use efficiency over its single-time basal application (Kang et al. 2021).

13.7 Modern Weed Management

Weeds are the major biotic stress threatening maize yield, particularly in the rainy season. Wider inter-row space, sluggish initial growth, and delayed ground cover make the crop more prone to weed infestation (Kumar et al. 2021). Weeds, if left uncontrolled, cause 50% grain yield losses in maize (Radheshyam et al. 2021). To minimize yield losses from weed interference, the maize crop must be kept weed-free, typically between emergence and the eight-leaf stage.

Atrazine is commonly used as a pre-emergence (PE) herbicide in maize, which provides effective control of broadleaf and some narrowleaf weeds. Recently, a new generation pre-emergence herbicide-pyroxasulfone has been launched for broad-spectrum and selective weed control in maize. It is a soil-active, systemic herbicide that works by disrupting the fatty acid production in weed plants. Pyroxasulfone@127 g/ha was found more effective PE herbicide than atrazine in maize and showed greater efficacy even at lower application rates (Kumar et al. 2021). This systemic, low-dose herbicide with a more specific and targeted formulation is ideal for minimizing herbicidal buildup in the ecosystem and ensuring sustainable crop productivity. For late-season weed control in maize, new generation herbicides, namely, tembotrione (34.4% SC) and topramezone, were introduced during the last decade. These HPPD (4-hydroxyphenylpyruvate dioxygenase) inhibiting herbicides effectively control weeds through disrupting the carotene biosynthesis process. Sequential use of atrazine as PE fb tembotrione (120 g/ha)/ topramezone (25 g/ha) as post-emergence (PoE) gave effective control of weeds and improved net profitability of maize farming over maintaining a weed-free environment through manual weeding (Sreelatha et al. 2020). The post-emergence application of these systemic herbicides at reduced doses limits herbicide load into the agro-ecosystem and reduces environmental hazards.

The simultaneous application of two herbicides with different modes of action increases the weed control spectrum, therefore vital to check the growth of difficult-to-control weeds. The tank mix application of 75% dose of atrazine (750 g/ha) with either tembotrione (120 g/ha) or topramezone (25.2 g/ha) was found highly effective for broad-spectrum weed control in maize when applied as early PoE (15 DAS) (Radheshyam et al. 2021). Likewise, the pre-mix formulation of atrazine (PS II-inhibitor) + mesotrione (HPPD-inhibitor) as a PoE herbicide was found promising in limiting weed growth and improving maize productivity (Singh et al. 2024). The synergistic relationship between these two herbicides improves herbicide efficiency, reduces the effective dose, and increases sustainability in herbicide use.

Herbicides are vital for weed management in modern agriculture; however, their use has posed a serious threat to the environment, biodiversity, and food safety, and therefore highlighted the need to adopt some alternative weed management strategies. Recent trends in farming like organic farming, and looming labour shortage have increased the scope of mechanized weeding. Mechanical weed control using tractor-mounted implements is feasible for maize crops owing to its wide row planting that provide adequate space for mobility of the tractor. This operation not only kills the weeds but also loosen the soil surface, ensuring better soil aeration and water infiltration (Bamboriya et al. 2020). Cultivators are commonly used agricultural implements for inter-row weeding, as their tine spacing can be adjusted as per crop requirements. Singh and Bhosale (2014) compared the performance of cultivator and manual weeding in sugarcane and noted that the weeding cost was 97.16% lower with higher field efficiency (0.94 h/h) with cultivator over manual weeding (0.0018 h/h). Nowadays, the self-propelled power weeder is increasingly being used for inter-row weeding operations in wide-spaced crops. Power weeder in maize reduced the operational cost (Rs. 1350/ha) and improved the field capacity (0.067 ha/h) over manual weeding (Rs. 4051/ha and 0.002 ha/h) (Shekhar et al. 2010). The mechanical weeding should be performed up to the knee-high stage of the crop; otherwise, the limited ground clearance of machinery may damage the crop foliage at later stages. Always run the machinery at shallow depth to minimize the mechanical damage to plant roots.

13.8 Conclusion

This chapter has explored a range of innovative maize production practices, including zero tillage, crop establishment methods, precision nutrient and water management, mechanization, and the use of low-dose herbicides. These scientifically proven technologies minimize natural resource degradation, reduce agri-input use, and enhance production potentials and farm earnings. Widespread adoption of these sustainable practices is immensely important for achieving desirable production targets, making agriculture economically viable, addressing the burning agriculture issues, and safeguarding the future of agriculture.

References

Bakker D, Hamilton G, Houlbrooke DJ, Spann C (2005) The effect of raised beds on soil structure, waterlogging, and productivity on duplex soils in Western Australia. Aust J Soil Res 43(5):575–585. https://doi.org/10.1071/SR03118

Bamboriya SD, Jat SL, Sreelatha D, Mahala DM, Rakshit S (2020) Mechanized maize production for enhanced productivity and profitability. IIMR technical bulletin 2020/1. Indian Institute of Maize Research, Ludhiana

Chandel R, Kumar N, Singh M, Thakur S (2024) Scaling mechanization and profitability in maize cultivation through innovative maize planters along with agroforestry approach: sustainable and climate smart approach to diversify rice based cereal systems in various regions. Intech Open. https://doi.org/10.5772/intechopen.111766

Channakeshava S, Siddapur VS, Naveena DV, Mamatha B (2022) Effect of different levels and split application of potassium on growth and yield of maize (*Zea mays* L.). Mysore J Agric Sci 56(4):130–141

DAC &FW (2021) Press release. Available at: https://www.pib.gov.in/PressReleasePage.aspx?-PRID=1705515. Accessed: 11 Jan 2026

Devkota KP, Yadav S, Humphreys E, Kumar A, Kumar P, Kumar V, Malik RK, Srivastava AK (2021) Land gradient and configuration effects on yield, irrigation amount and irrigation water productivity in rice–wheat and maize–wheat cropping systems in eastern India. Agric Water Manag 255:107036. https://doi.org/10.1016/j.agwat.2021.107036

Gupta N, Singh Y, Jat HS, Singh LK, Sidhu HS, Gathala MK, Jat ML (2023) Precise irrigation water and nitrogen management improve water and nitrogen use efficiencies under conservation agriculture in the maize–wheat systems. Sci Rep 13:12060

Jat ML, Chandna P, Gupta R, Sharma S, Gill M (2006) Laser land levelling: a precursor technology for resource conservation. Rice–wheat consortium technical bulletin series 7. Rice–Wheat Consortium for the Indo-Gangetic Plains, New Delhi

Jat ML, Gathala MK, Ladha JK, Saharawat YS, Jat AS, Kumar V, Sharma SK, Kumar V, Gupta R (2009) Evaluation of precision land leveling and double zero-till systems in the rice–wheat rotation: water use, productivity, profitability and soil physical properties. Soil Tillage Res 105:112–121

Jat ML, Gathala MK, Saharawat YS, Tetarwal JP, Gupta R, Singh Y (2013) Double no-till and permanent raised beds in maize–wheat rotation of north-western Indo-Gangetic plains of India: effects on crop yields, water productivity, profitability and soil physical properties. Field Crop Res 149:291–299

Jat HS, Sharma PC, Datta A, Choudhary M, Kakraliya SK, Singh Y, Sidhu HS, Gerard B, Jat ML (2019a) Re-designing irrigated intensive cereal systems through bundling precision agronomic innovations for transitioning towards agricultural sustainability in North-West India. Sci Rep 9:17929. https://doi.org/10.1038/s41598-019-54086-1

Jat RK, Singh RG, Kumar M, Jat ML, Parihar CM, Bijarniya D, Sutaliya JM, Jat MK, Parihar MD, Kakraliya SK, Gupta RK (2019b) Ten years of conservation agriculture in a rice–maize rotation of eastern Gangetic Plains of India: yield trends, water productivity and economic profitability. Field Crop Res 232:1–10. https://doi.org/10.1016/j.fcr.2018.12.004

Kang L, Zhang G, Chu G (2021) Split delivering phosphorus via fertigation to a calcareous soil increased P availability and maize yield (*Zea mays* L.) by reducing P fixation. J Soils Sediments 21(6):1–14

Kumar RM, Yamanura, Boraiah B (2021) Assessment of bio-efficacy of novel pyroxasulfone for controlling weeds in summer maize. Indian J Weed Sci 53(1):99–103

Lal R (2019) Promoting "4 per thousand" and "adapting African agriculture" by south–south cooperation: conservation agriculture and sustainable intensification. Soil Tillage Res 188:27–34. https://doi.org/10.1016/j.still.2017.12.015

Leghari SJ, Hu K, Wei Y, Wang T, Laghari Y (2024) Modelling the effects of cropping systems and irrigation methods on water consumption, N fates and crop yields in the North China Plain. Comput Electron Agric 218:108677

Mahala DM, Jat SL, Bamboriya SD, Singh AK, Sepat S, Kumar B, Sharma RP (2023) Best crop management practices in maize cultivation. IIMR technical bulletin 02/2023. ICAR–Indian Institute of Maize Research, Ludhiana

Manjulatha G, Rajanikanth E, Sowjanya B, Rani U (2022) Impact of mechanization under maize cultivation in Karimnagar district of Telangana state, India for enhanced profitability. Theor Biol Forum 27–33. https://doi.org/10.5281/zenodo.7466071

Ministry of Agriculture & Farmers Welfare (2025). All India: crop-wise area, production & yield. Available at: https://upag.gov.in/dash-reports/allindiaapy. Accessed: 7 Oct 2025

Naresh RK, Rathore RS, Kumar P, Singh SP, Singh A, Shahi UP (2014) Effect of precision land leveling and permanent raised bed planting on soil properties, input use efficiency, productivity and profitability under maize (*Zea mays*)–wheat (*Triticum aestivum*) cropping system. Indian J Agric Sci 84(6):725–732

Parihar CM, Jat SL, Singh AK, Kumar B, Singh Y, Pradhan S, Pooniya V, Dhauja A, Chaudhary V, Jat ML, Jat RK, Yadav OP (2016) Conservation agriculture in irrigated intensive maize-based systems of North-Western India: effects on crop yields, water productivity and economic profitability. Field Crop Res 193:104–116

Radheshyam R, Jat SL, Parihar CM, Singh AK, Pooniya V, Singh R (2021). Post-emergence herbicides enhance yield and profitability in Kharif maize (*Zea mays L.*). Indian J Agric Sci 91(11):1566–1570

Ram H, Singh Y, Timsina J, Humphreys E, Dhillon SS, Kumar K, Kler DS (2005) Performance of upland crops on raised beds in northwestern India. In: Roth CH, Fischer RA, Meisner CA (eds)

Evaluation and performance of permanent raised bed cropping systems in Asia, Australia and Mexico. ACIAR Proceedings No. 121. ACAIR

Sandhu OS, Gupta RK, Thind HS, Jat ML, Sidhu HS, Singh Y (2019) Drip irrigation and nitrogen management for improving crop yields, nitrogen use efficiency and water productivity of maize–wheat system on permanent beds in north-West India. Agric Water Manag 219:19–26. https://doi.org/10.1016/j.agwat.2019.03.040

Shekhar S, Chandra S, Roy DK (2010) Performance evaluation of different weeding tools in maize. Indian J Weed Sci 42(1–2):95–97

Shivay YS, Prasad R (2014) Effect of source and methods of zinc application on corn productivity, nitrogen and zinc concentrations and uptake by high quality protein corn (*Zea mays*). Egypt J Biol 16(1):72–78. https://doi.org/10.4314/ejb.v16i1.10

Sidhu HS, Jat ML, Singh Y, Sidhu RK, Gupta N, Singh P, Singh P, Jat HS, Gerard B (2019) Subsurface drip fertigation with conservation agriculture in a rice–wheat system: a breakthrough for addressing water and nitrogen use efficiency. Agric Water Manag 216:273–283

Singh TP, Bhosale AT (2014) Comparative performance evaluation of different mechanical equipment for weed control in sugarcane crop in northern-Western Tarai region of Uttarakhand. Afr J Agric Res 9(43):3226–3232

Singh M, Kumar M, Prakash A, Sharma K, Mishra PK (2018) Comparative field performance of pneumatic planters for planting of maize crop. Agric Eng Today 42(3):12–18

Singh S, Yadav RA, Prajapati SK, Kumar P, Yadav PK, Khan N, Sachan P (2024) Impact of different doses of herbicides on weed density and weed control efficiency in maize. Int J Plant Soil Sci 36(5):425–431

Singh K, Sindhu VK, Brar A, Lamba JS, Deol JS, Jakhar RS, Singh R (2025) Optimizing water use in spring maize through integrated agro-techniques in groundwater-stressed regions of north-western India. Agric Water Manag 317:109635

Sreelatha D, Swarnalatha V, Ramprakash T, Reddy MLK (2020) Evaluation of new post-emergence herbicides in rainfed maize. Maize J 9(1):50–55

Wade LJ, Norris CP, Walsh PA (1988) Effects of suboptimal plant density and non-uniformity in plant spacing on grain yield of rain-grown sunflower. Aust J Exp Agric 28:617–622

Yu WD, Feng LP, Hu CD, Peng JY (2015) Effects of waterlogging during seedling stage on the growth and yield of summer maize in Huang–Huai Region. Chin J Ecol 34(8):2161–2166

Soil Management for Sustainable Maize Production

14

Deep Mohan Mahala, Shanti Devi Bamboriya,
Shankar Lal Jat, and Hanuman Sahay Jat

Abstract

The sustainable maize production is facing multiple challenges such as deteriorating soil fertility, inefficient use of water and nutrients, and increased climatic variability. Maize has high demands for both nutrients and moisture, and imbalanced fertilizer application, depletion of soil organic matter, and suboptimal irrigation practices have contributed to yield instability and soil degradation. The chapter combines essential soil management techniques that help maintain sustainable maize production through combined approaches for soil fertility and moisture management. Balanced fertilization guided by the "4R" nutrient stewardship framework is highlighted as a means to improve nutrient-use efficiency and reduce environmental losses. The role of organic amendments such as farmyard manure, compost, green manures, poultry manure, and vermicompost is discussed for their impact on soil health and their ability to cycle nutrients and sustain agricultural productivity. In addition, the water requirements for maize growth at various developmental stages are discussed, along with efficient irrigation practices. Methods to enhance irrigation system performance and maintain soil moisture and water drainage. Integrated soil-centric management is essential to improve maize resilience, stabilizing yields, and promoting environmental sustainability.

Keywords

Soil fertility · Nutrient use efficiency · Organic amendments · Irrigation management · Sustainable maize production

D. M. Mahala (✉) · S. D. Bamboriya · S. L. Jat · H. S. Jat
ICAR-Indian Institute of Maize Research, Ludhiana, Punjab, India

M. Choudhary et al. (eds.), *Next-Generation Maize Research*,
https://doi.org/10.1007/978-981-92-0722-0_14

14.1 Introduction

Maize (*Zea mays*), commonly known as corn, is one of the world's most vital cereal crops and is ranked alongside rice and wheat in terms of global production and socio-economic importance (Erenstein et al. 2022). Maize cultivation occurs in over 160 countries around the world and accounts for more than one-third of all grain output at the global level. Additionally, it serves as staple food for hundreds of millions of people, feed for livestock and raw material for various industrial products. Further, the versatility and ability of maize to thrive in diverse agro-climatic conditions have earned it the title "Queen of Cereals." In India, maize is the third most produced cereal after rice and wheat, with cultivation expanding rapidly to meet rising demand. In the year 2024–25, the maize cultivated in 12 million hectares with a production of 44 million tonnes, showing a two-fold increase from 2014–15 levels. While 20%–22% of this production is consumed directly as food, most of it goes to poultry and livestock feed as well as industrial sectors such as starch and ethanol production. Therefore, maize has a vital role in enhancing food security, rural livelihoods, and industrial growth in India.

However, irrespective of its growing importance, there are several pressing challenges, some of them, like climate change, declining soil fertility, and unsustainable management practices (agronomic and soil management practices), which are a major threat to sustainable maize cultivation in India. Out of the soil management practices, one major issue is nutrient imbalance in application, resulting in soil fertility decline and environmental pollution. The prevalence of nutrient imbalance is evident from fertilizer consumption patterns; for instance, India's overall N:P:K consumption ratio in recent years has deviated significantly (exceeding 10:4:1 in 2023–24) from the recommended 4:2:1. Such disproportional fertilizer use can lead to issues like soil acidification, trace nutrient deficiencies, and lower nutrient-use efficiency, ultimately threatening sustainable production. Research shows that uneven fertilizer application of nitrogen, phosphorus, and potassium results in decreased agricultural production. Further, the practice of planting maize consecutively with wheat rotations that lack proper organic matter return has caused nutrient depletion (Timsina et al. 2010). Therefore, a lack of balanced nutrient availability in soil can be a leading cause of stagnant or decreasing maize yields.

Another challenging issue is that the maize sector is grappling with poor resource-use efficiency, particularly in terms of water and nutrient use, which means that a significant portion of inputs like irrigation water and fertilizers does not get effectively converted into crop output. On one hand, in irrigated pockets (such as winter maize in Bihar or Andhra Pradesh), the inefficient water management approach, like flood irrigation, can lead to water wastage and nutrient leaching (Kumar et al. 2013). On the other hand, large portions of rainfed areas suffer from water scarcity due to a lack of conservation measures. Further, low nutrient use efficiency, which is a consequence of the nutrient imbalances and suboptimal fertilizer application practices (such as the kind of fertilizer, time, and place of application), is also widespread. This not only raises production costs but also causes environmental pollution (such as nitrate runoff and greenhouse gas emissions) and threatens the sustainability of maize cultivation. Hence, improving the efficiency of input use is crucial to reducing the negative externalities associated with maize cultivation.

The above-discussed soil management issues are critical factors responsible for stagnating yields in some regions and pose risks to long-term productivity, urging the need for solutions centred on improved soil management practices, which can address multiple challenges simultaneously. The benefits of soil-centric management are two-fold: productivity gains and environmental sustainability. One thing is that healthier soils directly translate to better maize crop growth and yield stability. Soils rich in organic matter, for instance, have higher nutrient-supplying power and water holding capacity, which buffer maize against drought and nutrient stress (Lal 2006; Oldfield et al. 2019). This resilience is increasingly crucial under climate variability. And the second thing is that sustainable soil management reduces agriculture's environmental footprint. Practices that improve input efficiency mitigate negative impacts such as nutrient leaching and greenhouse gas emissions (Robertson and Vitousek 2009). In essence, good soil management in maize systems produces dual benefits because it enables high crop yields today while protecting soil resources for future generations, which fulfils both production targets and environmental protection needs. This chapter focuses on soil management approaches to support sustainable maize production. It discusses how soil conditions like nutrients, organic matter, and moisture affect plant growth and explores how to practically manage them. Drawing on recent research and examples from India and South Asia, the chapter discusses how these practices can increase yields, strengthen maize crop resilience, and support long-term soil health.

14.2 Soil Fertility and Nutrient Management

14.2.1 Essential Nutrients for Maize

Maize, in fact, every crop requires 17 essential nutrients for proper growth and development. These include carbon, hydrogen, and oxygen (obtained from air and water) and 14 mineral nutrients supplied by soil. The mineral nutrients are classified as macronutrients (needed in large amounts) and micronutrients (needed in trace amounts). The six macronutrients are nitrogen (N), phosphorus (P), potassium (K)—often called primary macronutrients—and the secondary macronutrients calcium (Ca), magnesium (Mg), and sulfur (S). The eight essential micronutrients needed for development include boron (B), chlorine (Cl), copper (Cu), iron (Fe), manganese (Mn), molybdenum (Mo), nickel (Ni) and zinc (Zn). The different nutrients in plant growth serve distinct functions, which result in particular signs of deficiency that cause damage to plant production. We have briefly described the physiological roles of macro and micronutrients in the maize crop and some common symptoms with which the deficiencies can be identified in Table 14.1.

Table 14.1 Essential nutrients for maize, their primary functions, and deficiency symptoms

Nutrient	Primary function	Deficiency symptoms in maize
Nitrogen (N)	Component of chlorophyll, proteins, and enzymes; promotes vegetative growth and green foliage	Older leaves turn pale yellow or light green, starting at the tips (V-shaped chlorosis); overall stunted growth
Phosphorus (P)	Energy transfer (ATP/ADP), nucleic acids; essential for root development and early growth	Stunted plants with dark green foliage; older leaves and stems may develop a purple-red tint (anthocyanin accumulation)
Potassium (K)	Regulates osmotic balance, enzyme activation, and stomatal function; improves drought tolerance and stalk strength	Yellowing and brown scorching along margins of older leaves; leaf curling and eventual necrosis at tips/edges (leaf "burn")
Calcium (ca)	Structural role in cell walls (calcium pectate); aids cell division in meristems; membrane stability	Newest leaves deformed or "hooked," with curled edges; fine yellowing along margins and growing tips die back (brittle leaves)
Magnesium (Mg)	Central part of the chlorophyll molecule; cofactor for many enzymes in carbohydrate metabolism	Interveinal chlorosis (yellow-white stripes between veins) on older leaves; in maize, lower leaves show striping and may develop purple underside
Sulfur (S)	Constituent of certain amino acids (cysteine, methionine) and vitamins; needed for protein synthesis	Uniform pale yellowing of young leaves and growing points; overall delayed growth and late flowering (resembles N deficiency but on newest leaves)
Zinc (Zn)	Enzyme cofactor (e.g. auxin synthesis, protein synthesis); important for growth hormones	"White bud" syndrome—Broad white or yellow bands on young leaves, starting at base; short internodes and little leaf size; leaf tips may die in severe cases
Iron (Fe)	Required for chlorophyll formation (though not in chlorophyll itself); involved in redox enzymes and electron transport	Interveinal chlorosis on newest leaves, veins stay green while tissue in between turns almost white; affects full length of leaf, similar to Mn deficiency
Manganese (Mn)	Activates enzymes in photosynthesis and N assimilation; part of the water-splitting system in photosynthesis	Interveinal chlorosis on young leaves (whitish stripes), often with small dark or necrotic specks; can be confused with Fe deficiency
Copper (Cu)	Enzyme activator; involved in photosynthesis, respiration, and lignin synthesis in cell walls	Young leaves fail to fully expand (twisted or rolled), yellowing from tip; leaf tips and edges die (necrosis); older leaves may show bluish-green tinge and edge necrosis
Boron (B)	Essential for cell wall formation, cell division, sugar transport, and pollen tube growth	New leaves distorted with thick, brittle texture and wrinkling; poor tassel and ear development (due to pollination failure and kernel abortion)

(continued)

Table 14.1 (continued)

Nutrient	Primary function	Deficiency symptoms in maize
Molybdenum (Mo)	Component of enzymes for nitrate reduction and nitrogen fixation; aids synthesis of amino acids	General nitrogen deficiency-like symptoms (pale, lower leaves yellowing) when nitrate is abundant but cannot be utilized; in some cases, marginal leaf burn and cupping of leaves
Chlorine (Cl)	Involved in osmotic regulation, stomatal activity, and the oxygen-evolution step of photosynthesis	Rarely deficient in field; can cause wilting, chlorotic blotches, and bronzing of leaves; roots stunted and thickened when cl is extremely low

Ni (Nickel) is also essential in trace amounts for enzyme function (especially urease), but maize Ni deficiencies are practically unknown in soils with normal organic matter. In general, micronutrient deficiencies in maize are less common than N, P, or K deficiencies, but Zn deficiency is notably widespread in maize-growing regions

14.2.2 Fertilizer Application Practices

An important part of managing soil fertility is effective fertilization. In order to meet crop demands, it entails not only selecting the appropriate kind of fertilizer but also applying the proper quantity at certain times and depositing it in specific locations (commonly referred to as the “4Rs”—right source, right rate, right time, right place). Due to maize’s high nutrient requirements, farmers must apply NPK fertilizers many times, including a basal treatment at planting and subsequent top-dress applications as the crop develops. Fertilizer sources range from straight fertilizers (which include urea for N application and single superphosphate for P and muriate of potash for K) to compound fertilizers (e.g. DAP—diammonium phosphate 18–46–0, for N and P). The effectiveness of crop nutrient absorption depends on when and how farmers choose to apply their nutrients (Sun et al. 2022).

14.2.2.1 Types of Fertilizer and Timing

For maize, a common practice is to apply most or all of the phosphorus and potassium at planting (basal) since these nutrients are needed early for root and seedling development and they do not move readily in soil (Goulding 1989; Hinsinger 2001). Nitrogen, on the other hand, should be split-applied: a portion at planting and the remainder in one or two top-dressings during vegetative growth. To stop soil N loss from leaching and volatilization, N supply should be matched with the plant’s peak growth period (Debele and Taressa 2023). Matching the plant N need through split N application can significantly increase nitrogen use efficiency (NUE) and maize yields (Singh et al. 2024). However, farmers need to determine their best fertilizer application time as the site-specific soil characteristics and weather patterns decide when to use N fertilizer (Zamukulu et al. 2024). Beyond N, secondary nutrients like sulfur (if needed) are often applied at planting or in early growth, and micronutrients (like zinc, iron) can be given as soil application (e.g. zinc sulphate or iron sulphate at planting) or foliar spray on the appearance of deficiency symptoms as discussed in Table 14.1.

14.2.2.2 Fertilizer Placement Methods

How fertilizer is placed in relation to the seed or plant affects nutrient availability and losses. Table 14.2 presents a comparison between different fertilizer placement techniques, which shows their standard application depth/position, required equipment/tool, nitrogen use efficiency, labour requirements, and appropriate situations.

The selection of placement methods needs to happen based on what resources farms have and what their environmental situation looks like. For instance, an experienced farmer will use drip fertigation to get greater results, whereas smallholder farmers without machinery will broadcast fertilizer. The objective is to minimize losses to the air or water while ensuring that the maize plant has access to nutrients at the right time. In order to increase the nutrient use efficiency, banding, splitting the N dose, and incorporating surface-applied urea are being used. The worldwide average NUE for cereal crops reaches only 33%, which results in the loss of two-thirds of all N applications. The farmers can enhance NUE through improved placement and timing methods, which include side-dressing and fertigation, while reducing environmental harm from fertilizer loss and gas emissions.

14.2.3 Organic Amendments for Maize

Organic amendments are essential components of sustainable maize cultivation as they improve both soil fertility and soil health. In contrast to synthetic fertilizers, organic sources promote soil structure, water retention, and microbial development by releasing nutrients gradually and increasing soil organic matter. Vermicompost, compost, green manure crops, farmyard manure (FYM), and poultry manure are the common organic amendments. The different materials contain different amounts of nutrients while having different release patterns.

14.2.3.1 Farmyard Manure (FYM)

The decomposed cattle dung, urine, and bedding materials make up this heavy manure substance. The product contains typically 0.5% N, and 0.2% P_2O_5, and 0.5% K_2O on a fresh weight basis (Kumar et al. 2021) (values can vary with bedding and handling). A common field application rate is 5–10 tons/ha (higher rates up to 15–20 t/ha may be used in organic systems, but handling becomes difficult). At ~10 t/ha, FYM supplies roughly 50 kg N, 20 kg P_2O_5, 50 kg K_2O total, though not all is immediately available. The primary benefit of FYM becomes evident because it builds up soil organic carbon which acts as microbial food sources and improves cation exchange properties for enhanced nutrient storage (Rasool et al. 2008). The release of nutrients from FYM happens slowly, so the first crop gets only 30% of its accessible nitrogen; the rest becomes available to subsequent crops. Thus, the long-term fertility of soil improves through FYM application because it establishes a slow process to increase soil nitrogen availability (as evidenced by crop yields increasing when farmers use manure at regular intervals) (Moharana et al. 2012).

Table 14.2 Comparison of common fertilizer placement methods in maize

Placement method	Position/depth	Equipment	Approx. N use efficiency	Labor and cost	Recommended situations	References
Broadcast and incorporate (basal)	Evenly spread on surface; mixed into top 10 cm soil by ploughing or harrow	Tractor-mounted spreader or manual broadcasting, followed by tillage tool	Low-moderate (40%–60% N recovery if incorporated; less if left on surface)	Low labour if mechanized (one pass); low cost	Pre-plant application of P and K	Ladha et al. (2005)
Band placement (starter or at planting)	Concentrated band 5 cm beside and below seeds (at planting) or below the seed zone	Planter or drill with fertilizer attachment (for seed-fertiliser placement)	Moderate-high (can improve N/P uptake; higher efficiency than broadcast)	Done with planting—moderate labour; requires appropriate equipment	Cooler soils or no-till (helps early growth); applying P or a small N dose as starter	Alam et al. (2018)
Side-dressing (post-emergence)	Band applied alongside rows, 5–10 cm deep, ~15–30 cm from maize plants, at V6-V8 stage.	Side-dress toolbar with knives or injection tines, or hand placement in small farms	High (~60%–70% + if well-timed and incorporated). Supplies N when crop can utilize it	Additional field pass required—higher labour/fuel; moderate cost for equipment	Soils prone to N loss (sandy or wet climates); high N requirement systems	Wu et al. (2017)
Fertigation (through irrigation)	Dissolved fertilizer delivered in irrigation water directly to root zone (depth depends on drip depth or furrow infiltration)	Drip lines or sprinkler with fertilizer injector/pump system	Very high (up to ~80%–90% with careful scheduling). Minimal volatilization or leaching if properly managed	Low labour after setup; high initial cost for the system; requires soluble fertilizers	Irrigated maize fields, especially under drip. Ideal for frequent small N doses matching crop uptake	Çetin and Akalp (2019)

14.2.3.2 Compost

Compost refers to organic matter (crop residues, farm wastes, sometimes city compost) that has been piled and decomposed by microbes under managed conditions. The nutrient content of properly prepared compost includes 1.0%–1.5% N, 0.5%–1% P_2O_5, and 1%–2% K_2O by dry weight, which exceeds the nutrient levels found in raw FYM. Recommended rates for compost might be 5–8 t/ha for general soil improvement (often used in conjunction with fertilizers). The main advantage of compost lies in its ability to provide stable humus, which enhances soil structure and improves clay soil aeration and its ability to retain nutrients (Garbowski et al. 2023). Typically, only ~10%–20% of compost's N might be plant-available in the season of application (depending on its maturity and C:N ratio). The long-term fertility and biological health of soil benefits from composting, but farmers need to add additional nutrients, which maize plants require for their high growth needs.

14.2.3.3 Green Manure

Green manuring involves growing a quick-growing plant (often a legume) and ploughing it into the soil fresh, to add organic matter and biologically fix nitrogen. Common green manure crops for maize systems include sunnhemp (*Crotalaria juncea*), dhaincha (*Sesbania aculeata*), and cowpea. They fix large amounts of atmospheric N and can generate 4–6 tons of dry biomass per hectare when they reach maturity. Sunnhemp plants can add between 60–133 kg N/ha to the soil if grown for 8 weeks (Garzon et al. 2023), though the average values range between 50–60 kg N/ha during 2–3-month period. The nutrient content of a green manure legume might be ~2%–3% N in dry matter, 0.2%–0.3% P, and 1%–2% K. The recommended practice is to incorporate the green manure crop into the soil at least 2–3 weeks before planting maize, giving time for initial decomposition. Similar to other manures, the green manure improves the soil structure and organic matter content. As the plant material is tender with a low C:N ratio, N release is relatively fast depending on soil moisture and temperature, about 50% or more of the N could mineralize in the first month (Giller 2001). So, the application of this manure can match the initial to middle stages of N requirements for the maize crop when done correctly. Additionally, green manure crops help suppress weeds and prevent soil erosion (Teasdale 1996).

14.2.3.4 Poultry Manure

The combination of chicken waste with bedding materials such as rice hulls or sawdust makes up poultry litter, which functions as a dense organic fertilizer. The nutrient content of this product exceeds FYM because it contains 3% N, 2.5%–3% P_2O_5, and 1.3%–1.5% K_2O (Ijoyah 2007). In other terms, one ton of poultry manure might contain ~30 kg N, 25–26 kg P_2O_5, 14 kg K_2O. Because of this, application rates are lower; typically, 2–5 t/ha of poultry manure is applied to maize fields. Even 2 t/ha of manure can supply ~60 kg N. The nutrient release from poultry manure is relatively quick, since a significant fraction of N in fresh poultry litter is in readily available forms (uric acid, ammonium). About 50% or more of its N may be available in the first season (Ruiz Diaz and Sawyer 2008), thus it is often referred to as a *quick-acting* organic fertilizer. Poultry manure contains N and P nutrients together with micronutrients, which include zinc and copper. It contains high amounts of ammoniacal N, which needs proper composting or immediate soil application after use to stop ammonia from releasing into the atmosphere. It's also "hotter," so the excessive rates will damage seedlings while creating nutrient imbalances, so users need to use it with caution. Poultry manure functions like other manures to build up soil organic matter during extended periods yet it provides immediate nutrient enrichment to the soil.

14.2.3.5 Vermicompost

Vermicompost is the product of organic waste processed by earthworms (often *Eisenia fetida*). The product consists of small particles that contain humus and have a nutrient profile that shows moderate levels of all essential elements. The nutrient content of Vermicompost includes 1.2%–1.8% N, 0.8%–1.1% P, and 1.5%–2.0% K. The nutrient content of vermicompost exceeds that of ordinary compost because the vermicomposting process both concentrates nutrients and enhances their availability in the final product. Field application rates are usually 1–3 t/ha for cash crops (because vermicompost is often produced in limited quantities and is more expensive). In a maize context, 2 t/ha vermicompost might supply ~30 kg N, among other nutrients. The worm digestion process in vermicompost production creates two essential benefits, which improve soil structure and microbial activity through its delivery of helpful microbes and enzymes (Chatterjee et al. 2020).

The plant takes in more than 60% of its total N during its first 8 weeks of growth (just before tasselling) (Nasielski et al. 2019). The application of green manures and poultry manure can provide enough N to meet crop requirements during this period. The slow N release from FYM and compost makes these materials suitable for long-term fertility maintenance because they do not meet the high N needs of crops at their peak growth stage. The rate at which organic N becomes available for plant uptake depends on how decomposition conditions, i.e. soil moisture, temperature, and C:N ratio, affect the process (Chiriac et al. 2025). Organic sources need to be included in sustainable maize production because they help maintain soil fertility, which will benefit future agricultural needs. The practice also leads to higher soil organic carbon levels, which produce better soil structure and more diverse microbial communities. The process leads to better nutrient cycling (Rai et al. 2025) and enhances the efficiency of fertilizers added (Ali et al. 2019). Organic sources need to be included in sustainable maize production because they help maintain soil fertility, which will benefit future agricultural needs.

14.3 Irrigation and Soil Moisture Management for Maize

14.3.1 Importance of Soil Moisture for Maize Growth and Productivity

Water is a vital component of plant growth, influencing nearly every aspect of its physiology. The water content in fresh maize plants reaches 70%–90%, and plants need proper soil moisture to perform cell expansion, nutrient transport, and photosynthesis. Water in the soil solution dissolves plant nutrients before it transports these nutrients through the roots. The roots of maize plants function best at optimal soil moisture levels because they can then extract and move nutrients, which leads to strong plant development and maximum crop production. Water stress creates problems for these biological processes because drought conditions force stomata to shut down, which blocks CO_2 from entering plants for photosynthesis and decreases the flow of nutrients into plant roots. The extended period of dry conditions results in nutrient deficiencies, which can cause wilted leaves and poor growth, thus showing that maize plants need consistent soil moisture to thrive.

The different growth stages of maize have varied water needs and moisture sensitivities over the season. The total water requirements for the maize crop throughout an entire growing season under optimal conditions range between 500 and 800 mm (Akinmutimi 2015). The crop requires minimal water during its initial growth period. It needs only 2–3 mm of water each day due to its small canopy and root system (Kranz et al. 2008). The crop requires more water during its fast-growing vegetative stage and reproductive development period. Around tasselling (VT) and silking (R1) (the onset of flowering), the maize crop needs roughly 0.3 inches (~8 mm) of water per day in hot, dry conditions (Kranz et al. 2008). This mid-season period is when the crop's canopy is fully developed and evaporative demand is high. The plant needs 4–5 mm of water daily to produce grain until it reaches the dough and dent stages, before its water needs decrease to less than 2 mm until it reaches physiological maturity (Kranz et al. 2008). Thus, mid-season (flowering to early grain fill) is the critical window of highest water demand for maize.

Drought stress around flowering can cause delayed silk emergence and silk desiccation, thus poor pollination and ultimately kernel abortion (Otegui et al. 1995). The yield losses of 5%–8% per day have been observed during severe drought at tasselling/silking (Shaw and Newman 1988). The process of early grain development, which includes the blister to dough stage, remains vulnerable to stress. The application of moderate to severe stress during grain filling results in smaller kernel sizes and weights, which leads to a 3%–4% yield reduction for each day of stress (Shaw and Newman 1988). Maize shows better resistance to short-term water deficits during the early vegetative growth period. During early growth (e.g. 4–6 leaf stage), mild drought stress may cause temporary growth restriction but their overall production remains unaffected when sufficient water becomes available during flowering time. Agronomists have observed that plants that face moderate water shortages at the beginning of their growth will create deeper root structures, which will protect them later from droughts.

Just as water scarcity, excessive soil moisture can also affect plant development. The roots of maize plants need oxygen for respiration, but roots cannot access oxygen saturated/waterlogged soils, which results in damaged roots, restricted nutrient uptake, and increased disease vulnerability. Research shows that waterlogged soil conditions lead to a 20%–30% decrease in maize yields, but at the reproductive stage, waterlogging can cause greater losses (~40% or more) (Tian et al. 2021). In addition, the heavy rainfall combined with insufficient water drainage not only decreases root oxygen levels but also leaches away vital nutrients such as N from the soil (Acharya et al. 2025). Thus, the sustainable production of maize requires farmers to maintain proper soil moisture levels by avoiding both severe drought and waterlogged conditions.

14.3.2 Efficient Irrigation Techniques for Maize

When rainfall is inadequate or unreliable, irrigation is used to supply water to crops. Several irrigation techniques are used for maize cultivation, each presents distinct operational characteristics that depend on the size of the farm and available resources. The major irrigation methods consist of surface irrigation through furrow and flood methods, sprinkler irrigation, and micro-irrigation systems like drip irrigation with surface or subsurface drip lines. The process of selecting appropriate irrigation methods needs to balance between achieving water efficiency and practical feasibility.

14.3.2.1 Surface (Furrow) Irrigation

Surface irrigation distributes water through furrows, which function as small channels to supply water between crop rows across the entire field. Maize is often planted on ridges or beds, and water flows by gravity down the furrows to infiltrate into the root zone. It is popular among smallholder farmers and in large canal-irrigated schemes alike because it does not require advanced equipment. However, furrow irrigation tends to be the least water-efficient method. A large fraction of the water can be lost to deep percolation beyond the root zone or runoff at the end of the field. The operational efficiency of surface irrigation systems operates at a rate between 50 and 60% (Bayisa et al. 2021), meaning nearly half the water may be wasted if management is poor. The practice of furrow irrigation requires farmers to perform extensive manual work because they need to operate the water distribution and determine the optimal time for irrigation. The system has various drawbacks, yet its budget-friendly cost makes it suitable for small farmers. The method becomes practical when farmers have access to abundant water at affordable costs.

14.3.2.2 Sprinkler Irrigation

In sprinkler systems, the water is pumped through pipes to nozzles that spray droplets over the crop. Common setups for maize include centre pivot systems, which function as automated sprinklers that operate on large farms to irrigate fields through circular movements. Sprinklers have a moderate to high efficiency, typically around 65%–75% in standard setups (Howell 2003) and up to ~85%–90% with modern low-pressure, drop-nozzle designs (Evans and Sadler 2008). Sprinklers distribute water equally across different land elevations without requiring any land levelling process. These systems enable farmers to maintain precise control of water application rates because they can deliver small amounts of irrigation water at regular intervals, which matches crop requirements better than flood irrigation methods. Sprinklers are well-suited to coarse-textured (sandy) soils where furrow water would just percolate too quickly. The system allows farmers to perform fertigation by adding fertilizers to their irrigation water and chemigation for pest management. From a cost point of view, the system requires initial expenses for the pump, piping systems, and operating costs. Sprinkler irrigation systems are an effective method for large landholding farmers. The US Great Plains region, along with other semi-arid areas, uses centre pivot sprinklers to deliver water efficiently to their commercial maize fields.

14.3.2.3 Drip Irrigation

The drip irrigation system delivers water to plant roots through plastic tubes, which have small emitters. Drip lines in maize fields run parallel to crop rows, which exist either on top of the soil or at a minimal depth below the surface. The application of water occurs through low-pressure systems, which produce a slow drip that percolates into the soil. This method is the most water-efficient form of irrigation—field application efficiency is typically 90% or higher because water is placed only where needed and surface evaporation is minimized. Research shows that drip irrigation not only helps obtain water savings along with better crop production. For example, a 7-year study demonstrated that drip irrigation systems consumed 25% less water than border flood irrigation systems while producing 14% more maize grain output and achieving 54% better water usage efficiency (Liu et al. 2022, 2023). By minimizing losses, drip irrigation is ideal for water-scarce areas and can greatly reduce the irrigation requirement of maize.

Despite its efficiency advantages, drip irrigation has a higher cost and management demands. The installation of drip lines (tubing, filters, emitters) represents a significant investment, often feasible for larger farms or high-value crops. For smallholders, drip kits are sometimes used on smaller plots, but widespread adoption may require financial support or cost-sharing. When operated properly, however, drip allows precise control of water and even nutrients (via fertigation) to match the crop's needs at different growth stages. Labour requirements during the season are low (mostly monitoring), though installing and removing surface drip lines each season can be labour-intensive. Subsurface drip irrigation (SDI) is a variant where drip laterals are permanently buried below the soil surface (commonly at 15–30 cm depth). This method eliminates surface evaporation completely and delivers water directly in the root zone. SDI combined with precise scheduling has shown remarkable results in Indian maize systems. A 3-year ICAR study in the Trans-Gangetic Plains (NW India) under conservation agriculture reported that subsurface drip (at 100% of recommended N fertilizer) increased maize grain yields by ~15.8% over conventional flood irrigation, while reducing irrigation water use by about 55% (Gupta et al. 2023). In summary, drip irrigation (surface or subsurface) is a promising technology for sustainable intensification in water-limited regions.

14.3.2.4 Rainfed Systems with Supplemental Irrigation

A large portion of the world's maize is grown under rainfed conditions, meaning the crop relies on rainfall for its water supply. In rainfed farming, maize can often yield well in regions with reliable rainfall distribution; however, if dry spells occur during critical periods (e.g. around flowering), significant yield losses may result. To mitigate this risk, farmers may use supplemental irrigation—providing additional water during drought stress periods to ensure key growth stages have adequate moisture. Supplemental irrigation can be as simple as a smallholder using a watering can or pump to irrigate a maize field during a mid-season dry spell, or as complex as a large farm using irrigation equipment only when rainfall fails. The goal is not to fully irrigate the crop throughout the season, but to "bridge" rainfall deficits at crucial times. Even a relatively small amount of irrigation can have outsized benefits under rainfed conditions. For instance, studies in semi-arid regions show that adding one or two irrigations during drought can significantly increase maize yields and stability compared to purely rainfed crops (Jian et al. 2024). In the US Midwest (which is traditionally rainfed), model projections indicate that introducing limited irrigation could raise maize yields by 10%–30% in dry years and reduce yield variability (Ferin and Kucharik 2024). Similarly, in parts of Africa and South Asia, capturing rainwater (through ponds or soil water conservation structures) and using it for supplemental irrigation during dry spells has proven to boost maize yields and help cope with rainfall variability (Zeleke 2020).

The advantages of supplemental irrigation are that it requires less water than full-season irrigation and can often use simpler, low-cost methods (since irrigation is needed only occasionally). Farmers can prioritize watering during the crop's moisture-sensitive stages (like tasselling and grain fill) to get the best return on limited water. The water-use efficiency of supplemental irrigation is very high in terms of yield per unit of irrigation water, because rainfall is doing most of the work and irrigation is only topping up the deficit. From a cost perspective, rainfed systems with supplemental irrigation are appealing for smallholders since they might only need a small pump or manual labour for a few irrigations, rather than an extensive irrigation infrastructure. The challenge is that if drought is extreme or prolonged, rainfed fields may require more water than is available, so supplemental irrigation is a risk management tool rather than a guarantee of full irrigation. Nonetheless, integrating even a little irrigation into rainfed maize systems can dramatically improve food security by preventing total crop failures in drought years.

14.3.3 Managing Soil Moisture Stress in Maize

Ensuring that maize crops have neither too little nor too much water requires active soil moisture management. Farmers and agronomists employ various strategies to cope with drought (water deficits) and excess moisture conditions. Key approaches include proper irrigation scheduling, soil moisture monitoring, agronomic practices to conserve water and drainage, or other measures for waterlogged fields. Below, we discuss how to manage moisture stress for sustainable maize production.

14.3.4 Irrigation Scheduling and Crop Water Requirements

One of the most important aspects of irrigation management is *when* and *how much* to irrigate. Proper scheduling aims to supply water in sync with the crop's needs, avoiding stress at critical stages while minimizing waste. In practice, this means monitoring the crop and soil and timing irrigations to maintain soil moisture in the root zone within an optimal range. Maize's water requirement follows a predictable pattern—low in early growth, peaking at mid-season, then declining. Farmers can use this knowledge along with weather data to anticipate the crop's water needs. A common tool is calculating evapotranspiration (ET): using local weather (temperature, humidity, etc.) to estimate daily water use by maize, often adjusted with a crop coefficient (Kc) for the growth stage. By replacing only the water the crop has used (and factoring in any rainfall), farmers keep soil moisture within the target range and avoid both drought stress and over-irrigation.

In scheduling irrigation, attention to maize growth stages is crucial. Avoiding moisture stress during tasselling and silking is the top priority; thus even farmers practicing deficit irrigation will ensure the crop is well-watered during the flowering period (Maazou et al. 2016). If water supplies are limited, it is often recommended to concentrate irrigation around the reproductive stage and early grain fill since these have the largest impact on yield. For example, under limited water, a farmer might skip some early vegetative irrigations (letting the crop use stored soil moisture) to save water for critical later stages—research in Ethiopia showed that using alternate-furrow irrigation and cutting back to 75% of full water still produced a good yield when water was focused on critical periods. Modern tools like soil moisture sensors are a great aid in scheduling decisions. Devices such as tensiometers, gypsum blocks, or electronic probes can be placed in the field to measure soil water content or tension. These sensors provide real-time feedback on whether the soil in the root zone is drying out. For instance, a tensiometer might indicate when soil suction exceeds a threshold (say for e.g. 40 centibars), signalling that irrigation is needed to prevent stress. Even simpler methods like the "feel method" (checking soil by hand at various depths) or observing plant indicators (leaf rolling or slight wilting in the afternoon) can guide smallholders on when to irrigate. The overall goal is to *schedule irrigation to meet the evapotranspiration demand of maize at each stage without significant gaps*. This typically results in more frequent irrigations during peak use (often every 3–5 days around silking in hot climates) and less frequent during early and late season. By keeping the soil moisture within the crop's requirement zone, irrigation scheduling thus prevents the yield losses associated with water stress.

14.3.5 Soil Moisture Conservation Practices

Apart from adding water via irrigation, soil moisture can be managed by conserving water that is already in the soil. For this purpose, agronomic practices that improve the soil's water holding capacity or reduce evaporation can be very effective, especially for rainfed or limited irrigation areas. Mulching is one widely recommended practice that involves covering the soil surface with organic residues (crop stover, straw) or plastic film. Mulches act as an insulating layer, reduce direct evaporation from the soil, and suppress weeds that would otherwise compete for water. In maize fields, leaving the previous crop's residues (e.g. maize stalks or wheat straw) on the surface can significantly decrease soil moisture loss. Research has shown that maize under straw mulch experienced cooler soil temperatures and retained higher moisture in the root zone during dry spells to support better crop growth and yield under drought conditions (Liu et al. 2022). Organic mulches also improve infiltration of rainwater and irrigation by breaking the impact of raindrops and preventing crusting of soil (Lal 1998). Similarly, conservation tillage techniques (like minimum till or no-till farming) help build soil structure and organic matter over time, which can increase the soil's capacity to store water (Farahani et al. 2022). No-till maize with residue cover often shows greater available moisture in the soil profile compared to conventionally tilled soil because there is less evaporation and improved soil porosity for water to percolate (Farahani et al. 2022; Liebhard et al. 2022). These practices not only conserve moisture but also reduce erosion and improve soil health, offering sustainability co-benefits.

Another water-conserving practice is optimizing the root zone conditions. Deep and healthy root systems enable maize to tap moisture from a larger soil volume. Farmers can encourage deeper rooting by avoiding soil compaction (which can create hardpans that restrict roots) and by ensuring any early irrigation does not keep moisture too shallow (which could discourage roots from exploring deeper). It is observed that maize does not extract water uniformly with depth—roughly 40% of water uptake is from the upper quarter of the root zone, 30% from the next quarter, and less from deeper layers (Sharp and Davies 1985). If irrigation is always shallow, roots may proliferate only near the surface. By contrast, slightly drying the surface early on or using practices like alternate wetting and drying (allowing the topsoil to dry intermittently) can stimulate roots to grow downward. However, one must be careful that too much early-season stress could reduce ear development. That's why key is a balanced approach where roots are incentivized to grow deep by avoiding overwatering the topsoil, while still supplying enough moisture to not impair vegetative growth significantly (Vera Muñoz et al. 2019; Pardossi et al. 2009). When the maize root system fully explores down to 1 m or more, it can utilize subsoil moisture reserves during dry periods, effectively buffering the crop against drought. In irrigation management, this translates to practices like less frequent, deeper irrigation (for example, applying a larger irrigation depth but less often, so water penetrates deeply). That ensures moisture throughout the root zone and keeps deeper soil layers available as a bank for the plant to draw from in between irrigations.

14.3.6 Drainage and Management of Excess Moisture

On the other end of the spectrum, managing excess water is essential in humid regions or irrigated areas with poor drainage. Maize is reasonably tolerant of short-term soil saturation (a day or two), but beyond that, waterlogging stress sets in quickly. Without adequate oxygen, maize roots can die within days, and, as noted earlier, yield losses averaging ~30% have been recorded in waterlogged conditions (Behera et al. 2025). To prevent yield losses from waterlogging, farmers implement drainage solutions and careful water management. In high-rainfall areas or low-lying fields, installing field drains or ditches helps remove standing water. Surface drainage can be achieved by shaping the field with gentle slopes to outlet channels, or simply making small drain furrows across the field to guide water off after heavy rains. In tropical smallholder contexts, farmers sometimes plant maize on raised beds or ridges specifically to elevate the root zone above waterlogged soil in the furrow bottoms. This ridge/furrow system allows excess rain to collect in furrows away from the plant base. Certain agronomic choices can also mitigate the effects of temporary waterlogging. Ensuring good soil structure through organic matter and avoiding compaction helps water infiltrate more uniformly (Hamza and Anderson 2005). Rotating with deep-rooted crops or cover crops can improve soil porosity and drainage over time (Chen et al. 2014). Nonetheless, the primary defence against excess water is providing drainage. Striking the right moisture balance will support healthy root development, efficient nutrient uptake, and robust maize growth.

14.3.7 Rainwater Harvesting and Moisture Conservation in Drylands

In many dryland farming areas, large-scale irrigation is unavailable, so farmers focus on capturing and conserving rainwater in situ. Rainwater harvesting techniques can significantly improve soil moisture for maize without external irrigation. These include bunding, terracing, and constructing small earth dams to impound runoff, which can later infiltrate into fields. For example, contour bunds or ridges across a slope slow down rain runoff, allowing more water to soak into the soil profile where crops can use it. In parts of sub-Saharan Africa, simple pits or "zai" holes are dug and filled with organic matter so when rain comes, water collects in the pits and sinks, which creates moist pockets for the maize's root zone. Such methods have enabled farmers to cultivate maize in regions with erratic rainfall by maximizing the fraction of rain that becomes available soil moisture.

Alongside water harvesting, fallow management is a traditional strategy in dry areas, leaving land fallow for a season to build up soil moisture for the following crop. While continuous cultivation is often necessary for food production, rotating in a bare or weedy fallow (or preferably a cover crop that requires minimal water) can replenish deeper soil layers with moisture that the subsequent maize crop can use. Researchers and extension agents recommend replacing bare fallow with cover crops that protect the soil but are drought-tolerant or terminated early, so they do not deplete stored moisture. The combined effect of soil cover, enhanced infiltration, and reduced evaporation under cover cropping can increase the plant-available water for maize. Thus, managing soil moisture for maize involves a range of techniques: timely irrigation (or rainwater use) to meet crop demands, agronomic practices to retain and conserve soil moisture, and infrastructure to remove excess water. By understanding maize's water needs and stress responses, farmers can implement an integrated moisture management plan i.e. irrigating efficiently with suitable methods, conserving rainwater and soil moisture, and protecting the crop from both drought and waterlogging.

14.4 Conclusion

Sustainable maize production is governed by effective soil management, particularly soil fertility and moisture. Fertilization under the guidance of "4Rs" principle (right source, rate, time, and place ensures nutrient availability that could match maize growth stages. Organic amendments can help to improve soil organic matter while developing better soil structure, nutrient cycling, and long-term fertility. The management of soil moisture needs to be precise because maize plants show extreme vulnerability to drought and waterlogging conditions, especially during the reproductive stage. The efficient irrigation techniques, along with proper scheduling, moisture conservation practices, and sufficient drainage, collectively lead to better water productivity and yield stability. Water-limited and rainfed systems can use supplemental irrigation and rainwater harvesting as effective methods to reduce their risks. Overall, an integrated, soil-centric management approach enhances maize resilience, sustains productivity, and supports environmental sustainability under increasing climatic and resource pressures.

References

Acharya B, Dodla S, Tubana B, Gentimis T, Rontani F, Adhikari R, Setiyono T (2025). Characterizing optimum N rate in waterlogged maize (*Zea mays* L.) with unmanned aerial vehicle (UAV) remote sensing) Agronomy 15:434

Akinmutimi AL (2015) Estimation of water requirements of early and late season maize using penman's equation. J Soil Sci Environ Manag 6:24–28

Alam MK, Bell RW, Salahin N, Pathan S, Mondol ATMAI, Alam MJ, Shil NC (2018) Banding of fertilizer improves phosphorus acquisition and yield of zero tillage maize. Sustainability 10:3234

Ali Q, Ashraf S, Kamran M, Ijaz M, Rehman A, Tahir M, Ahmad S (2019) Organic manuring for agronomic crops. In: Agronomic crops, vol 2. Elsevier, pp 163–193

Bayisa GD, Hordofa T, Tezera K, Tesfaye A, Ashame G, Wondimu T (2021) Maize yield and water use efficiency under different irrigation levels and furrow irrigation methods. Air Soil Water Res 14:11786221211058177

Behera S, Kar I, Yadav A, Sahu A (2025) Exploring waterlogging challenges, causes and mitigating strategies in maize (*Zea mays* L.). J Agron Crop Sci 211:e70109

Çetin Ö, Akalp E (2019) Efficient use of water and fertilizers in irrigated agriculture: drip irrigation and fertigation. Acta Horticult Regiotect 2:97–102

Chatterjee R, Debnath A, Mishra S (2020) Vermicompost and soil health. In: Soil health. Springer International Publishing, pp 69–88

Chen G, Weil RR, Hill RL (2014) Effects of compaction and cover crops on soil least limiting water range and air permeability. Soil Tillage Res 136:61–69

Chiriac OP, Pittarello M, Moretti B, Zavattaro L (2025) Factors influencing nitrogen derived from soil organic matter mineralisation. Agric Ecosyst Environ 381:109444
Debele M, Taressa B (2023) Urea split application effects on maize (*Zea mays* L.) growth and grain yield. Int J Agron 2023:6673773
Erenstein O, Jaleta M, Sonder K, Mottaleb K, Prasanna BM (2022) Global maize production, consumption and trade. Food Secur 14:1295–1319
Evans RG, Sadler EJ (2008) Methods and technologies to improve efficiency of water use. Water Resour Res 44:1–15
Farahani E, Emami H, Forouhar M (2022) Effects of tillage systems on soil organic carbon and physical properties. Land Degrad Dev 33:1307–1320
Ferin KM, Kucharik CJ (2024) Irrigation expansion effects on maize yield and nitrogen leaching. Agric Syst 219:104055
Garbowski T, Bar-Michalczyk D, Charazińska S, Grabowska-Polanowska B, Kowalczyk A, Lochyński P (2023) Natural soil amendments in agriculture. Soil Tillage Res 225:105462
Garzon J, Vendramini JM, Silveira ML, Dubeux JC, Liao HL, Sollenberger LE, Alencar NM (2023) Residue management and genotype effects on sunn hemp decomposition. Agron J 115:261–272
Giller KE (2001) Nitrogen fixation in tropical cropping systems. CABI Publishing
Goulding KWT (1989) Potassium fixation and release. In: Methodology in soil-K research. International Potash Institute, pp 137–154
Gupta N, Singh Y, Jat HS, Singh LK, Choudhary KM, Sidhu HS, Gathala MK, Jat ML (2023) Precision irrigation and nitrogen management in maize–wheat systems. Sci Rep 13:12060
Hamza MA, Anderson WK (2005) Soil compaction in cropping systems: a review. Soil Tillage Res 82:121–145
Hinsinger P (2001) Bioavailability of soil inorganic phosphorus: a review. Plant Soil 237:173–195
Howell TA (2003) Irrigation efficiency. In: Encyclopedia of water science. CRC Press, pp 467–500
Ijoyah MO (2007) Poultry manure effects on muskmelon yield. Afr J Biotechnol 6:1884
Jian H, Gao Z, Guo Y, Xu X, Li X, Yu M, Du X (2024) Supplemental irrigation reduces maize yield loss under heat stress. Agric Water Manag 299:108888
Kranz WL, Irmak S, Van Donk SJ, Yonts CD, Martin DL (2008) Irrigation management for corn. University of Nebraska–Lincoln Extension
Kumar A, Jat SL, Kumar R, Yadav OP (2013) Maize production systems for improving resource-use efficiency. Directorate of Maize Research, New Delhi
Kumar S, Dhar S, Barthakur S, Rajawat MVS, Kochewad SA, Meena LR (2021) Farmyard manure as K-fertilizer in wheat. Front Environ Sci 9:764489
Ladha JK, Pathak H, Krupnik TJ, Six J, van Kessel C (2005) Fertilizer nitrogen efficiency in cereal production. Adv Agron 87:85–156
Lal R (1998) Soil erosion impacts on productivity. Crit Rev Plant Sci 17:319–464
Lal R (2006) Soil organic carbon restoration for crop yields. Land Degrad Dev 17:197–209
Liebhard G, Klik A, Neugschwandtner RW, Nolz R (2022) Tillage effects on soil water and soybean evapotranspiration. Agric Water Manag 269:107719
Liu M, Wang G, Liang F, Li Q, Tian Y, Jia H (2022) Optimal irrigation levels for maize under drip irrigation. Water 14:3822
Liu M, Liang F, Li Q, Wang G, Tian Y, Jia H (2023) Drip irrigation improves maize water use efficiency. Sci Rep 13:8392
Maazou ARS, Tu J, Qiu J, Liu Z (2016) Breeding for drought tolerance in maize (*Zea mays* L.). Am J Plant Sci 7:1858–1872
Moharana PC, Sharma BM, Biswas DR, Dwivedi BS, Singh RV (2012) Nutrient management effects on soil carbon pools. Field Crop Res 136:32–41
Nasielski J, Earl H, Deen B (2019) Luxury nitrogen uptake in maize under stress. Front Plant Sci 10:318
Oldfield EE, Bradford MA, Wood SA (2019) Soil organic matter and crop yields: a meta-analysis. Soil 5:15–32

Otegui ME, Andrade FH, Suero EE (1995) Maize growth and kernel abortion under drought. Field Crop Res 40:87–94
Pardossi A, Incrocci L, Incrocci G, Malorgio F, Battista P, Bacci L, Balendonck J (2009) Root zone sensors for irrigation management. Sensors 9:2809–2835
Rai A, Mishra R, Patra A, Singh AK, Sharma S, Bahuguna A, Rai S, Rajput J, Gangwar A, Jha S, Tripathi SK (2025) Microbes and compost in climate resilient agriculture. In: Sustainable management of agro-food waste. Academic Press, pp 105–125
Rasool R, Kukal SS, Hira GS (2008) Soil organic carbon under long-term fertilization. Soil Tillage Res 101:31–36
Robertson GP, Vitousek PM (2009) Nitrogen in agriculture. Annu Rev Environ Resour 34:97–125
Ruiz Diaz DA, Sawyer JE (2008) Plant-available nitrogen from poultry manure. Agron J 100:1318–1326
Sharp RE, Davies WJ (1985) Root growth and water uptake in maize. J Exp Bot 36:1441–1456
Shaw RH, Newman JE (1988) Weather stress in the corn crop. Oklahoma Cooperative Extension Service
Singh A, Rudnick D, Snow DD, Proctor C, Puntel L, Iqbal J (2024) Split nitrogen application effects on nitrate leaching. Agrosyst Geosci Environ 7:e20554
Sun X, Li Y, Heinen M, Ritzema H, Hellegers P, van Dam J (2022) Fertigation strategies for water and nitrogen efficiency. Agriculture 13:17
Teasdale JR (1996) Cover crops and weed management. J Prod Agric 9:475–479
Tian LX, Zhang YC, Chen PL, Zhang FF, Li J, Yan F, Feng BL (2021) Waterlogging effects on crop yield: a meta-analysis. Front Plant Sci 12:634898
Timsina J, Jat ML, Majumdar K (2010) Rice–maize systems of South Asia. Plant Soil 335:65–82
Vera Muñoz J, Conejero Puente W, Conesa MÁ, Ruiz Sánchez MC (2019) Irrigation factor approach based on soil water content. Irrig Sci 37:1–12
Wu M, Li G, Li W, Liu J, Liu M, Jiang C, Li Z (2017) Nitrogen fertilizer deep placement in rice. Front Plant Sci 8:1227
Zamukulu PM, Bagula EM, Mondo JM, Chuma GB, Bulonza GM, Mwaluke S, Mushagalusa GN (2024) Nitrogen split application effects on maize yield. CABI Agric Biosci 5:22
Zeleke Z (2020) Supplementary irrigation effects on maize (*Zea mays* L.) yield. Int J Res Stud Agric Sci 6:4–9

Advancements in Machinery Solutions for Sustainable Maize Cultivation Across Different Agro-ecologies

15

Manpreet Singh, H. S. Sidhu, Manish Kakraliya, Manish Kumar, Shankar Lal Jat, and Hanuman Sahay Jat

Abstract

Maize-based cropping systems such as maize–wheat (MW), rice-maize (RM), and maize-maize (MM) are gaining importance across the different Indian agro-ecologies due to their high productivity, water-use efficiency, and adaptability to Conservation Agriculture (CA) practices. However, the sustainability of these systems depends heavily on the integration of appropriate mechanization solutions that support minimal soil disturbance, residue retention, and precise input management. Innovative machinery such as laser land levelling enhances irrigation uniformity and water-use efficiency by 20%–25%, whereas residue management technologies such as Turbo Happy Seeder and Super Straw Management System facilitate direct seeding in high-residue conditions, curbing residue burning and improving soil carbon dynamics in the rice–maize system. Zero-till and precision planters, equipped with disc openers and pneumatic metering systems, ensure optimal seed and fertilizer placement, enhancing resource-use efficiency in maize–maize and maize–wheat system. Additionally, raised-bed and permanent-bed planters have demonstrated higher maize productivity (up to 13 t ha^{-1}), improved water-use efficiency THS by 30%–40%, and better nutrient-use efficiency (15%–20%) under CA conditions. Adoption of CA-based mechanization in maize systems not only enhances productivity and profitability but

M. Singh · H. S. Sidhu · H. S. Jat (✉)
Punjab Agricultural University (PAU), Ludhiana, Punjab, India
e-mail: sidhuhs@pau.edu

M. Kakraliya · S. L. Jat
ICAR-Indian Institute of Maize Research, Ludhiana, Punjab, India

M. Kumar
International Maize and Improvement Centre (CIMMYT), BISA Farm, Ludhiana, Punjab, India
e-mail: ma.kumar@cgiar.org

M. Choudhary et al. (eds.), *Next-Generation Maize Research*,
https://doi.org/10.1007/978-981-92-0722-0_15

also reduces greenhouse gas emissions and improves resource-use efficiency. Wider adoption can be achieved through custom-hiring centres, policy support, and capacity building of farmers and service providers.

Keywords

Conservation agriculture · Maize-based cropping systems · Mechanization · Resource-use efficiency · Residue management · Indo-Gangetic Plains

15.1 Introduction

Maize is the third most important cereal crop in India after rice and wheat, cultivated across approximately 12 million hectares with diverse agro-ecological conditions. The major maize-growing states include Karnataka, Madhya Pradesh, Maharashtra, Telangana, Bihar, and Rajasthan. Maize is grown in all three seasons, viz., kharif, rabi, and spring, and supports a wide range of maize-based cropping systems. Prominent systems include maize–wheat, maize–mustard, maize-chickpea, rice-maize-potato, and maize-vegetable rotations in northern India; maize–ragi, maize-red gram, maize-maize, and maize-sunflower systems in southern states; and rice–maize systems in eastern India. The popular intercropping maize-based systems are maize + sugarcane, maize + vegetables, maize + redgram, maize + soybean, maize + moong, maize + urd, maize + cowpea, etc., Increasing demand from feed, food, ethanol and industrial sectors has encouraged farmers to adopt high-yielding hybrids and more intensive, resource-efficient maize-based cropping systems across the country. Maize is increasingly preferred in conservation agriculture (CA)–based systems because it responds well to practices such as minimum soil disturbance, residue retention, and diversified crop rotations, which together enhance soil health and resource efficiency. CA improves soil structure, moisture availability, and microbial activity more than conventional tillage systems resulting in strong yield stability, especially in rainfed and moisture-stress environments. Residue mulching reduces soil erosion and suppresses weeds, while maize's compatibility with crops like wheat, legumes, and vegetables makes it ideal for sustainable, low-input rotations. As a result, farmers adopting CA often achieve better soil organic carbon, reduced production costs, and higher long-term productivity compared to conventional management systems, making maize a key driver in scaling conservation agriculture across India.

Successful implementation of CA demands an integrated approach encompassing precise seeding, irrigation, fertilization, residue management, and weed control, all tailored to site-specific soil and climatic conditions. The development and adoption of suitable machinery for these operations are crucial for scaling up CA in India's major maize-based cropping systems (sequential and intercropping). In-field residue retention and incorporation under CA can significantly improve soil organic matter and nutrient cycling, thereby enhancing the sustainability of any cereal-based systems. Innovative practices such as intercropping, relay cropping with short-duration legumes (e.g., mungbean, urdbean, cowpea, lathyrus) and vegetables (spinach, cauliflower, cabbage, potato, etc.) integration within maize-based systems have shown potential to improve cropping intensity, soil fertility, and farm profitability. In maize-based systems such as rice–maize, maize–wheat, and maize–maize rotations, resource conservation technologies like zero tillage (ZT), minimum tillage (MT), and precise seeding have demonstrated high potential for improving resource-use efficiency and sustainability when supported by advanced CA-compatible machinery like bed planters (narrow; 67 cm and wide beds; 120 cm), pneumatic planters (wide beds; 120 cm), Happy Seeder, and Zero-till Multi-crop planters.

Despite these proven advantages, the adoption of CA machinery in India remains constrained by several factors, including small and fragmented landholdings, limited capital investment capacity, irregular field shapes, lack of skilled operators, and competition between machinery and human labour in maize domain areas. Addressing these challenges through technological innovation, custom-hiring models, and policy interventions is vital for large-scale adoption of CA practices. Accelerating mechanization in maize-based cropping systems can thus play a transformative role in advancing sustainable intensification, ensuring resource conservation, and improving the climate resilience of Indian agriculture.

15.2 Machinery Solutions for Maize Systems

Maize is increasingly being recognized as a promising alternative for agricultural diversification in upland areas due to its higher water productivity and adaptability to varying climatic conditions (Jat et al. 2013; Adak et al. 2023). Traditionally, seedbed preparation for maize involves several conventional tillage operations, which increase production costs and delay sowing. However, maize can be efficiently established without preparatory tillage using zero-tillage technology equipped with an inclined plate planter attachment. Adoption of zero tillage (ZT) offers multiple benefits- savings in diesel and time, reduction in greenhouse gas emissions, better conservation of irrigation water during the first irrigation, ultimately leading to lower production costs and timely planting over large areas (Gathala et al. 2015; Jat et al. 2013). Machineries for varied maize systems under different agro-ecosystems are discussed below.

15.2.1 Laser Land Leveller

Laser land levelling (LLL) is a crucial prerequisite before adopting conventional/conservation agriculture (CA) practices in any cropping system (Fig. 15.1). In India, irrigation remains a vital input, with nearly 40% of cropland being irrigated, contributing about 60%–80% of total food production (Singh et al. 2014). Efficient water management through proper irrigation scheduling, along with advanced crop management techniques such as laser-assisted precision land

Fig. 15.1 Laser land leveller

levelling and improved management practices, can significantly improve water productivity and resource use efficiency in maize-based systems. Studies have demonstrated that LLL can save irrigation water by 20%–25% and reduce irrigation time by about 30%, while enhancing the productivity of major crops like maize, rice, wheat, and sugarcane by 10%–15% (Jat et al. 2006; Sidhu et al. 2007; Jat et al. 2015a). Latif et al. (2013) reported reduced weed intensity and an approximately 11% lower herbicide cost in wheat fields levelled using laser technology compared to traditional land levelling (TLL). Additionally, LLL increases the cultivable area by 1.5%–10.0% due to the elimination of unnecessary bunds and channels compared to TLL (Sidhu et al. 2007; Jat et al. 2009). Uneven fields and unlined irrigation channels cause substantial water losses, making laser-assisted precision levelling vital for improving irrigation efficiency and water productivity. As a foundational technology for CA, precision land levelling enhances crop yields and optimizes input use efficiency, particularly for water and nutrients (Jat et al. 2006; Lohan et al. 2014).

This technique employs laser guidance (light amplification by stimulated emission of radiation) to achieve a highly accurate levelling within ±2 cm of the field's mean elevation. The use of an automatic survey system (ASS) for levelling has proven to be more time-efficient, cost-effective, and economical compared to manual survey methods (MSM) (Manpreet-Singh et al. 2020). The comparative analysis (Table 15.1) revealed that survey time, labour, and survey costs were reduced by 57.9%, 79.2%, and 38.2%, respectively, for ASS compared to MSM. The field capacity of LLL (including survey time) was significantly higher for ASS due to the reduced labour requirements and faster operation. Although the operating cost of ASS was about 7.4% higher than MSM because of the added expenses for the control box, power mast, and bucket attachment, the total laser levelling and system costs were significantly lower ($P < 0.05$) for ASS compared to MSM. Overall, the performance indicators strongly favoured the automatic survey system over the manual approach. The laser leveller, being a custom-operated implement, not only

Table 15.1 Comparative economics of an automatic survey system (ASS) and manual survey method (MSM)

Parameter	ASS	MSM	% change over MSM	*P* value
Survey diesel (l/ha)	0.3	–	–	–
Survey time (min/ha)	12.1 ± 0.9	28.7 ± 1.1	−57.9	<0.0001
Survey labour cost (Rs/ha)	7.5 ± 1.1	36.2 ± 1.3	−79.2	<0.0001
Total survey cost (Rs/ha)	22.4 ± 1.6	36.2 ± 1.3	−38.2	<0.0001
System capacity (ha/hr)	0.3 ± 0.0	0.2 ± 0.0	53.0	0.04
Laser unit cost (Rs/hr)	657.8	612.6	7.4	–
Laser cost (Rs/ha)	2121.1 ± 196.7	3028.4 ± 318.73	−30.0	0.03
Total system cost (Rs/ha)	2143.5 ± 197.5	3064.6 ± 319.12	−30.1	0.03

Source: Manpreet-Singh et al. (2020)

saves time and labour for operators but also provides precision levelling for farmers. Reduced fuel consumption contributes to lower greenhouse gas emissions, supporting environmental sustainability. LLL thus holds immense potential to enhance agricultural productivity, promote water conservation, ensure food security, and reduce the environmental footprint of irrigated production systems in India. For instance, in Punjab alone, laser land levelling has been adopted over 1.75 million hectares with more than 7850 operational units. Although the initial investment in a laser leveller is high, its adoption through custom hiring services has made it economically feasible for small and marginal farmers. The efficiency and cost-effectiveness of LLL improve with larger farm sizes, allowing better utilization of machinery and maximizing its operational benefits.

15.2.2 Happy Seeder

Minimum and zero-tillage (ZT) technologies have emerged as essential components of Conservation Agriculture (CA) for sustainable intensification of maize-based systems such as maize–wheat, maize–maize, and rice–maize. These systems offer multiple benefits, including lower production costs, reduced water and fuel use, improved soil health, and enhanced timeliness of sowing compared to conventional tillage practices (Malik et al. 2004; Singh et al. 2008; Erenstein and Laxmi 2008; Lohan et al. 2018). However, the direct seeding of maize or wheat into residues after mechanized harvesting often faces several operational challenges, such as residue accumulation in furrow openers, poor traction of the seed metering drive wheel due to loose straw or stover, and the need for frequent lifting of implements in heavy residue conditions. These limitations can lead to uneven seed depth, poor crop establishment, and reduced early vigour in maize and wheat crops (Fig. 15.2).

Fig. 15.2 Happy Seeder

In north-western (NW) India, where maize-based systems are expanding rapidly as an alternative to rice, a large volume of maize and rice residues poses both an opportunity and a challenge. Burning of these residues leads to significant air pollution, nutrient losses, and greenhouse gas (GHG)emissions. To address this, residue management technologies such as the Happy Seeder (HS), Smart Seeder, and Super Seeder have been developed to facilitate direct seeding in heavy residue conditions. Originally designed for sowing wheat into rice residues (Sidhu et al. 2007; Singh et al. 2009), it has now been successfully adapted for any cereal systems, including maize-based systems. Happy Seeder integrates a straw management rotor with flail-type blades and a zero-till seed drill, enabling simultaneous residue management and precise seed placement. The flail blades chop crop stover and push the residues as mulch between seeded rows, improving moisture conservation, reducing weed emergence, and ensuring uniform plant establishment.

Field evaluations in rice–wheat systems across Punjab and Haryana have demonstrated yield gains of 3%–8% and improved profitability by 15%–20% compared to conventional tillage, owing to better soil moisture retention, reduced evaporation, and enhanced soil thermal conditions under surface residues (Gathala et al. 2011; Singh et al. 2011). In maize–maize systems, studies by Jat et al. (2019, 2020, 2025) and Kumar et al. (2022) reported 10%–12% higher system productivity, 25% improvement in water-use efficiency, and up to 20% reduction in fuel consumption under ZT-residue retention. Similarly, for rice–maize systems, on-station and on-farm trials in Bihar and Odisha have shown 0.8–1.2 t ha^{-1} higher maize yields under residue-retained ZT conditions, coupled with a 30%–35% reduction in irrigation water use and a 40% decline in operational energy inputs. Retention of maize and rice residues enhances soil organic carbon, moderates soil temperature, and reduces soil moisture loss by 30–50 mm, often saving one irrigation event for the succeeding crop. It also minimizes weed pressure, thereby reducing herbicide use and costs. Studies have shown that the HS reduces energy consumption by up to 80% and CO_2 emissions by 60% compared to conventional systems, while providing 1.5–2 times higher net returns due to savings in labour, fuel, and input costs (Sidhu et al. 2015).

To cater to small and marginal farmers cultivating maize-maize and rice-maize systems in eastern India, a smaller version of the Turbo Happy Seeder (THS), an improved version of Happy Seeder, was developed that can be operated using a two-wheel tractor. This compact model enables direct seeding of wheat or maize into heavy residues on small plots, while also being suitable for sowing short-duration summer crops like mungbean or maize fodder after wheat harvest, enhancing overall system profitability. Additionally, the use of a knife roller or residue shredder mounted on the tractor front effectively cuts and chops maize stalks into 10–15 cm pieces before seeding, improving machine performance in maize-based CA systems. Despite its clear advantages and government subsidies (up to 60% of the purchase price), THS adoption in maize-based systems remains moderate due to factors such as a short operational window, lower seeding capacity (0.3 ha h^{-1} compared to 0.5 ha h^{-1} under conventional systems), and the requirement for uniform residue spreading. However, recent policy measures such as residue management legislation, custom hiring centres (CHCs), and integration of residue-handling attachments have accelerated THS adoption in maize-dominated areas. Current estimates suggest that THS and related CA machinery cover >1.5 Mha across eastern India, contributing to a 30%–40% reduction in residue burning and improving air quality across the Indo-Gangetic Plains.

15.2.3 Zero-Till Multi-Crop Planter/Seed Drill

Planters equipped with multi-crop inclined-plate metering mechanisms are highly suitable for diversified maize-based cropping systems (Fig. 15.3). These planters

Fig. 15.3 Zero-till multi-crop planter/seed drill

utilize an inclined plate seed-metering mechanism with notched cells that ensure precise seed placement and uniform distribution. The four-wheel tractor-driven design enables consistent seed-to-seed and row-to-row spacing with minimal seed injury, reducing skips caused by machine vibration compared to other metering systems. The major advantage of these planters lies in their versatility, i.e. they can be easily adjusted for different crops such as maize, rice, cotton, soybean, and groundnut by simply replacing the crop-specific inclined plates and modifying the row spacing. Additionally, an inclined plate box can be attached to an existing zero-till drill as a cost-effective alternative to purchasing a separate machine. These planters are compatible with tractors of around 35 hp., making them well-suited for small and medium-scale farmers.

Recent research has validated the performance and economic benefits of inclined-plate planters under Indian and South Asian field conditions. Sahu et al. (2017) reported that a tractor-drawn multi-crop inclined-plate planter developed at IGKV, Raipur, achieved a field capacity of 0.70 ha h^{-1} with a field efficiency of about 80%, reducing sowing costs by nearly 50% compared to traditional practices. Similarly, Nandini et al. (2018) found that a manually operated inclined-plate planter for maize and pulses achieved 91%–93% field efficiency and reduced sowing costs by approximately 94% relative to manual planting.

15.2.4 Straw Management System (SMS)

The combine harvester collects cereal crops (rice, wheat, maize) across a swath equal to its cutter bar width and deposits straw residues from the straw walkers into the centre of the harvested area, forming windrows of loose straw parallel to the direction of operation. Uniform spreading of these residues is essential for the effective functioning of second-generation drills such as the Happy Seeder, Smart Seeder, and Super Seeder; however, manual spreading requires substantial labour input, ranging from 8 to 13 man-hours per hectare. To overcome this challenge, the Super Straw Management System (Super SMS) was developed as an attachment for self-propelled combine harvesters equipped with a 4.27 m cutter bar and a 110 hp. engine (Manpreet-Singh et al. 2020). The Super SMS uniformly chops and spreads straw by collecting it from one side of the combine, cutting it with a rotor fitted with serrated blades, and blowing it tangentially through a deflector across the entire cutter bar width (Fig. 15.4).

Performance evaluation of the combine harvester with and without the Super SMS demonstrated significant improvements in residue management without compromising machine performance (Manpreet-Singh et al. 2020). The use of Super SMS markedly increased the proportion of chopped straw in the 0–20 cm range, reducing the weighted mean chop size to 13.7 cm compared to 27.0 cm under conventional harvesting. Detailed analysis of chop size distribution revealed that the proportion of very fine (<2.5 cm), fine (2.5–7.5 cm), and medium (7.5–12.5 cm) fractions increased from 8.0% to 16.2%, 6.3% to 13.8%, and 5.3% to 15.9%, respectively, while the coarse fraction (12.5–20.0 cm) increased from 11.3% to 22.7%. In

Fig. 15.4 Combine harvester with Super SMS

contrast, the proportion of very coarse residues (>20 cm) was substantially reduced from 31.4% to 69.1%, leading to a significant reduction in average chop size from 26.9 to 13.7 cm (Singh et al. 2023).

Despite a slight reduction in operating speed from 3.6 to 3.4 km h^{-1}, the field capacity remained nearly unchanged (~1.2 ha h^{-1}), while fuel consumption increased by 23.4% (from 8.9 to 11.9 L ha^{-1}) with the use of Super SMS. Machine performance parameters indicated that straw walker losses increased marginally from 0.3% to 0.7%, whereas sieve losses remained constant at 0.2%. Importantly, cleaning efficiency was maintained at 96%, and total grain losses remained well below the permissible limit of 2.5% as per IS 8122 (1994) standards. The Super SMS significantly enhances residue management by uniformly spreading finely chopped straw, thereby facilitating the efficient operation of conservation agriculture implements such as the Happy Seeder and zero-till drills. Its adoption not only reduces the need for residue burning but also contributes to improved soil organic matter and operational efficiency. Owing to these advantages, the Super SMS has been widely recommended across Punjab and other Indian states as a key intervention for sustainable residue management and mitigation of air pollution (Singh et al. 2023; Sidhu et al. 2015).

15.2.5 Raised-Bed (Ridge) Planter

Raised-bed planting is widely recognised as an optimal method for maize cultivation during both the monsoon and winter seasons, particularly under conditions of excess moisture or limited water availability in rainfed areas. Sowing on the southern side of east–west–oriented ridges or beds enhances germination. Using raised-bed planters equipped with inclined-plate, cupping or roller-type seed-metering systems allows for precise placement of seeds and fertilisers in a single operation, resulting in a good crop stand, higher productivity and improved resource-use efficiency. Earlier work (e.g., Australian Centre for International Agricultural Research; ACIAR 2008) showed that raised-bed planting could reduce irrigation water usage by 20%–30% (and in some cases 30%–50%) in north-western India, while achieving higher yields. Additional benefits and synergies emerge when raised bed planting is combined with conservation agriculture (CA) practices such as permanent beds (PBs) and zero-tillage (ZT) within maize-based cropping systems, which improve overall system productivity and economic returns compared to conventional tillage (CT) methods (Parihar et al. 2016). Recent multi-year research across the Indo-Gangetic Plains (IGP) during the rabi seasons of 2020–21 to 2023–24 has demonstrated the remarkable advantages of raised bed planting (RBP) in maize cultivation. Maize yields under RBP ranged from 9.28 to 10.53 t ha^{-1}, markedly surpassing the 5.70–8.29 t ha^{-1} achieved under conventional flat-bed planting. Beyond yield enhancement, RBP boosted net returns by nearly 20%, improved water use efficiency (WUE) by about 35%, and enhanced nutrient use efficiency (NUE) expressed as grain yield per unit of applied nitrogen by approximately 25% (Jat et al. 2018, 2025). However, a participatory on-farm study in Uttar Pradesh further revealed that precision land levelling combined with permanent raised beds reduced irrigation water use by ~20.4%, while increasing yields by ~13.5% in wheat and ~11.8% in maize than flat-bed systems (Jat et al. 2015a; Naresh et al. 2014).

15.2.6 Permanent Raised Bed Planter

To maximize the benefits of bed planting technology under conservation agriculture, maize can be sown on permanent raised beds in a single pass without the need for preparatory tillage (Figs. 15.5 and 15.6). The permanent bed planter is designed with double-disc furrow openers and a bed shaper that enable precise seed placement and maintenance of bed geometry over multiple cropping cycles. This machine places one maize row at the centre of each bed with a seed-to-seed spacing of 67.5 × 20.0 cm, ensuring optimal plant population and uniform crop stand. The double-disc furrow openers create a narrow furrow, which causes minimal disturbance to the established beds compared to conventional zero-tillage (ZT) tine openers, thereby preserving soil structure and residue cover. Field studies conducted in the Indo-Gangetic Plains have demonstrated that permanent raised beds (PRB) in maize–wheat systems significantly improve crop productivity, soil physical properties, and irrigation water-use efficiency compared to conventional tillage (CT). According to Jat et al. (2013, 2018), maize yields increased by up to 73% under PRB over CT, while irrigation water savings were about 25%, enhancing water productivity and profitability (Jat et al. 2015b, 2018). Similarly, Gathala et al. (2015) reported that CA-based tillage and crop establishment methods such as permanent

Fig. 15.5 Permanent raised bed planter

Fig. 15.6 Permanent raised bed former in maize-based systems

beds improved maize–wheat system productivity and soil quality, resulting in higher net returns and better resource-use efficiency. Recent long-term studies from the North China Plain further confirm these advantages, showing that permanently raised beds enhance soil hydrologic properties, reduce surface compaction, and promote stable infiltration rates over 6 years of continuous maize-wheat rotation (Han et al. 2022). Moreover, maize yield under raised-bed planting (RBP) ranged from 9.28 to 10.53 t ha^{-1}, significantly higher than flat-bed systems (5.70–8.29 t ha^{-1}), with 35% higher water-use efficiency and 25% higher nitrogen-use efficiency (Jat et al. 2025).

Fig. 15.7 Pneumatic planter

15.2.7 Pneumatic Planter

Pneumatic planters are advanced precision seeding machines designed to enhance seed placement accuracy, uniform spacing, and crop stand establishment across various crops, including maize, rice, cotton, soybean, and groundnut (Fig. 15.7). These planters use a pneumatic seed metering mechanism that employs air pressure to pick and release individual seeds through suction holes, ensuring minimal seed damage and uniform seed-to-seed and row-to-row spacing (Singh et al. 2005). Compared to mechanical planters, pneumatic planters offer higher operational efficiency, improved seed singulation, and reduced seed skipping, particularly under field vibrations or irregular terrain conditions (Mahapatra et al. 2024). The pneumatic seed metering system consists of a rotating disc with seed cells, a vacuum chamber, and a delivery system that allows precise metering and placement of seeds even at higher forward speeds, resulting in consistent plant populations and better crop uniformity. Recent technological advancements, such as adjustable vacuum pressure, variable seed discs, and electronic seed monitoring systems, have further improved the performance of pneumatic planters under conservation agriculture conditions (Mahapatra et al. 2024). These planters can be operated efficiently with tractors of 35 hp. and above, offering versatility for small and large farms alike.

15.2.8 Double-Disc Planter

The double-disc planter is equipped with dual disc openers that efficiently cut through crop residues in advance of the seed row, ensuring uniform seed placement (Fig. 15.8). This passive, anti-blocking double-disc mechanism, typically powered by four-wheel tractors, is widely utilized on medium-sized farms to plant four rows of maize in wheat residues as part of a maize–wheat rotation. The inclusion of press wheels behind the double-disc openers ensures proper closure of the seed furrow, improving soil–seed contact and enhancing germination rates. Moreover, the planter incorporates an inclined plate seed metering system, ensuring accurate seed distribution and minimal seed injury during planting (Jat et al. 2013).

Fig. 15.8 Double-disc planter

15.2.9 Tractor Front-Mounted Knife Roller for Managing Crop Residues

The front-mounted knife roller comprises two rollers fitted with straight knives along their circumference, rotating passively with ground contact (Fig. 15.9). These knives efficiently shear maize or wheat residues between the soil surface and the cutting edge, facilitating effective residue management. Mounted on the front of the tractor, the roller operates in conjunction with a rear-mounted seeding implement, allowing for simultaneous residue management and seed drilling in a single tractor pass. This configuration enables direct drilling of subsequent crops, such as wheat, immediately after maize harvest, enhancing operational efficiency and residue utilization (Adak et al. 2023). The knife roller is suitable for 45-hp tractors and achieves a field capacity of approximately 0.8 ha h^{-1} under field conditions.

15.2.10 Bed Former

Raised-bed planting through a tractor-drawn ridger provides an effective and economical solution for irrigation management, particularly under dry and high-temperature environments (Figs. 15.6 and 15.10). The ridger can be modified with a seed drill attachment, enabling simultaneous furrow creation and sowing by

Fig. 15.9 Tractor front-mounted knife roller

Fig. 15.10 Bed former

adjusting tine positions. Its design includes a frame with staggered tines fitted with miniature furrow openers that form furrows spaced 33.5 cm apart, converting the remaining soil into a seedbed. This tractor-operated implement, typically requiring a 45-hp tractor, demonstrates a field capacity of 0.4 ha h^{-1}, and is commonly utilized in maize–wheat rotations for shaping and maintaining permanent raised beds (Han et al. 2022). Additionally, the equipment can perform mechanical weeding, ridge reshaping, and earthing-up of maize plants, contributing to improved soil aeration and moisture conservation.

15.3 CA Machinery for Small Farmers

Manual and animal traction seeders are typically small, lightweight, and simple in design, making them easy to manufacture, use, and maintain. They are commonly employed on small farms and in hilly areas where mechanized options are limited. Manual equipment represents the basic level of mechanization, and advancements in equipment design aim to improve productivity, energy efficiency, and ergonomics. Studies have shown that low-cost, human- and animal-powered planters can significantly enhance seeding precision, reduce drudgery, and improve time efficiency for smallholders under CA systems (Sims and Kienzle 2015). The adoption of improved human- and animal-pulled equipment requires a demand-driven innovation system that integrates farmer feedback and local manufacturing capacity (Jin 2014). Manually operated planters generally have a field capacity of around 0.05–0.08 ha h^{-1}, making them suitable for fragmented and sloped terrains typical of Indian hill regions.

15.3.1 Single-Row Punch Planter

The punch planter is suitable for planting maize, soybean, sunflower, and cowpea under normal/ZT conditions (Fig. 15.11). It features two compartments for seed and

Fig. 15.11 Single-row punch planter

fertilizer, a punching unit for dibbling seeds into residues, a furrow opener for fertilizer placement, and a press wheel to regulate planting depth. Field trials have shown that single-row punch planters improve seed placement uniformity and reduce seed losses compared with traditional broadcasting methods, thereby enhancing crop stand establishment and yield stability in smallholder maize systems (Sims and Kienzle 2015). The operator walks behind the planter and operates it using handlebars. This single-row planter typically weighs about 30 kg.

15.3.2 Manual Direct Planter

This multi-crop planter generally includes a zero-till tine for opening a small seed row and features two separate hoppers for seed and fertilizer, with individual drop tubes delivering them into the slot. The rear wheel drives the seed metering device and acts as both a seed-covering and press wheel. According to Jin (2014), such small-scale direct planters have been successfully adapted in South and Southeast Asia, leading to up to 35% time savings and 20%–25% labour cost reduction in maize-based systems. The operator walks behind the seed drill, controlling it with handlebars.

15.3.3 Manual Double-Disc Planter

The manual double-disc planter, designed for smallholder CA, uses double-disc openers to cut through crop residues and open narrow seed rows. It includes separate hoppers for seed and fertilizer with individual drop tubes for placement accuracy. Behind the opener, a wheel-driven metering device ensures uniform seed distribution and adequate soil cover. Field evaluations have demonstrated that manual double-disc planters improve seed-to-soil contact and early crop establishment, especially in residue-retaining maize–wheat rotations (Sims and Kienzle 2015).

15.4 Intercropping with Vegetables, Pulses, and Oilseeds

Maize is highly suitable as an intercrop with vegetables, pulses, and oilseeds because of its adaptable growth habit, wider row spacing, and efficient use of sunlight and soil nutrients. Its upright architecture allows sufficient light penetration for companion crops such as beans, peas, cowpea, soybean, groundnut, and various short-duration vegetables (Fig. 15.12). Legumes grown alongside maize enrich the soil through nitrogen fixation, improving overall system productivity and reducing fertilizer needs. Similarly, oilseeds and vegetables benefit from the microclimate created by maize, which helps in moisture conservation and weed suppression. Intercropping maize with these crops not only enhances total land-use efficiency and system profitability but also reduces production risks by diversifying income sources for farmers.

Fig. 15.12 Intercropping with vegetables, pulses, and oilseeds

15.5 Suggestions for Large-Scale Adoption of CA Machinery in India

To advance the adoption of conservation agriculture (CA) in India, it's essential to enhance the development and uptake of high-performance no/minimum-till seeders. Current seeders are limited in size and uptake, necessitating several key recommendations.

(i) Policy support is critical, requiring adequate research funding to develop new machinery for CA implementation.
(ii) Existing seeders need further refinement to accommodate various crops, field sizes, and soils across different regions and power sources.
(iii) To address the constraints faced by marginal farmers with limited resources, seeders should be light-weight, simple, affordable, and suitable for low-horsepower tractors.
(iv) For heavy clayey soils, tines should be used instead of discs in no-till seeders to facilitate planting and simplify design and operation.

(v) All no-till seeders should incorporate depth-control mechanisms and contour-following capabilities, particularly in rough or hilly terrains, to overcome natural obstacles and ensure optimal seed placement.
(vi) Anti-blocking mechanisms are crucial, particularly in high-residue conditions requiring further development to ensure efficient performance and adoption.
(vii) Closer cooperation among research institutions, enterprises, and farmers is necessary to design suitable no/minimum-till seeders tailored to different cropping areas, ensuring their effective implementation and widespread adoption.

15.6 Conclusion and Future Perspectives

Conservation agriculture (CA) is increasingly seen as a sustainable alternative to conventional maize-based farming systems, offering improved agricultural production, enhanced nitrogen-use efficiency, and reduced GHG emissions. This approach also strengthens system resilience to climate variability and resource constraints. The successful adoption of CA in these systems relies heavily on the availability of suitable machinery capable of residue management, precise seed placement, and efficient fertilizer application across diverse soil and climatic conditions. Machines that can simultaneously seed and manage residues are vital for the effective implementation of CA in maize-based cropping systems. Machines play a pivotal role in improving resource-use efficiency, soil health, and residue management. Similarly, CA machinery minimizes residue burning, conserves soil moisture, and enhances system profitability and sustainability while improving the soil carbon sequestration and nutrient-use efficiency, contributing to sustainable intensification.

Furthermore, there is an increasing need for machinery that enables precise fertilizer placement and residue incorporation, thereby enhancing nutrient-use efficiency and maintaining soil structure under continuous maize-based systems. To accelerate the adoption of CA and promote diversification, there is an urgent requirement to develop and evaluate multi-crop, multi-utility machines suited to varying field conditions and power sources. Investment in capacity building, training, and demonstration programs is equally crucial to ensure the effective operation and maintenance of machinery among small and marginal farmers. Ultimately, sustainable mechanization, combined with sound agronomic management and strong policy support, will determine the long-term success of CA in maize-based cropping systems, ensuring enhanced productivity, profitability, and environmental sustainability.

References

ACIAR (2008) Permanent beds and rice residue management for rice–wheat systems in the Gangetic Plains. In: Humphreys E, Roth CH (eds) ACIAR proceedings no. 127. Australian Centre for International Agricultural Research, Canberra

Adak S, Ghosh PK, Tiwari R, Jat ML, Kumar S, Singh M, Singh S, Kumar R (2023) Impact of contrasting tillage, residue mulch and nitrogen on crop rotation and soil quality in maize–

wheat systems in north-western Indo-Gangetic Plains. Front Sustain Food Syst 7:1230207. https://doi.org/10.3389/fsufs.2023.1230207

Erenstein O, Laxmi V (2008) Zero tillage impacts in India's rice–wheat systems. Soil Tillage Res 100:1–14

Gathala MK, Kumar V, Saharawat YS, Jat HS, Singh M, Kumar A, Sharma PC, Kumar V, Ladha JK (2011) Happy Seeder technology: a solution for residue management for the sustainability and improved production of the rice–wheat system of the Indo-Gangetic Plains. In: Resilient food Systems for a changing world: proceedings of the 5th world congress on conservation agriculture (Brisbane, Australia, 25–29 Sept 2011)

Gathala MK, Timsina J, Islam MS, Rahman MM, Hossain MI, Harun-Ar-Rashid M, Ghosh AK, Krupnik TJ, Tiwari TP, McDonald A (2015) Conservation agriculture based tillage and crop establishment options can maintain farmers' yields and increase profits in South Asia's rice–maize systems: evidence from Bangladesh. Field Crop Res 172:85–98. https://doi.org/10.1016/j.fcr.2014.12.003

Han L, He J, Li H, Wang Q, Lu C, Yang H, Yang H (2022) Soil hydrologic properties in permanent raised beds—a field study experiment on wheat–maize cropping systems. Land Degrad Dev 34(3):698–709. https://doi.org/10.1002/ldr.4487

Jat ML, Chandna P, Gupta RK (2006) Laser land leveling: a precursor technology for resource conservation, Technical bulletin series 7. Rice–Wheat Consortium for the Indo-Gangetic Plains, New Delhi

He J, Zhang Z, Li H, Wang Q (2014) Development of small/medium size no-till and minimum-till seeders in Asia: A review. Int. J. Agric. Biol. Eng. 7(4): 1–12

Jat ML, Gupta R, Ramasundaram P, Saharawat YS, Singh RG (2009) Laser-assisted precision land leveling: A potential technology for resource conservation in irrigated intensive production systems of the Indo-Gangetic Plains. In: Ladha JK et al (eds) Integrated crop and resource management in the rice–wheat system in South Asia. International Rice Research Institute, Los Baños, pp 223–238

Jat ML, Gathala MK, Saharawat YS, Tetarwal JP, Gupta R, Yadvinder-Singh. (2013) Double no-till and permanent raised beds in maize–wheat rotation of north-western Indo-Gangetic plains of India: effects on crop yields, water productivity, profitability and soil physical properties. Field Crop Res 149:291–299. https://doi.org/10.1016/j.fcr.2013.05.010

Jat ML, Singh Y, Gill G, Sidhu HS, Gupta R (2015a) Laser-assisted precision land leveling impacts in irrigated intensive production systems of South Asia. Adv Soil Sci 22:323–352

Jat HS, Singh G, Singh R, Choudhary M, Gathala MK, Jat ML, Sharma DK (2015b) Management influence on maize–wheat system performance, water productivity and soil biology. Soil Use Manag 31(4):534–543

Jat RD, Jat HS, Nanwal RK, Yadav AK, Bana A, Choudhary KM, Kakraliya SK, Sutaliya JM, Sapkota TB, Jat ML (2018) Conservation agriculture and precision nutrient management practices in maize-wheat system: effects on crop and water productivity and economic profitability. Field Crop Res 222:111–120

Jat HS, Sharma PC, Datta A, Choudhary M, Kakraliya SK, Sidhu HS, Gerard B, Jat ML (2019) Re-designing irrigated intensive cereal systems through bundling precision agronomic innovations for transitioning towards agricultural sustainability in north-West India. Sci Rep 9(1):1–14

Jat HS, Kumar V, Datta A, Choudhary M, Kakraliya SK, Poonia T, McDonald AJ, Jat ML, Sharma PC (2020) Designing profitable, resource use efficient and environmentally sound cereal based systems for the Western Indo-Gangetic plains. Sci Rep 10(1):1–16

Jat RK, Meena VS, Reddy IR, Sohane RK, Singh RN, Durgude S, Kumar S, Pazhanismy S, Meena SK, Sharda K, Singh S, Singh RK, Kumari S, Singh KM, Kumar G, Lenka AK, Chaurasiya A, Sahu R, Choudhary GL, Nath P, Yadav PK, Kumar A, Jha RK, Kumar U, Das A, Jha AK, Tripathi DP, Sagar S (2025) Optimizing maize systems with raised beds: boosting productivity, profitability, and sustainability. Front Sustain Food Syst 9:1484079. https://doi.org/10.3389/fsufs.2025.1484079

Kumar R, Mishra JS, Naik SK, Mondal S, Meena RS, Kumar S, Dubey AK, Makarana G, Jha BK, Mali SS, Biswas AK, Choudhary AK, Hans H, Dubey R, Kumar S, Sundaram PK, Raman RK, Monobrullah M, Kumar S, Kumar U, Bhatt BP (2022) Impact of crop establishment and residue management on soil properties and productivity in rice-fallow ecosystems in India. Land Degradation & Development: 1–15

Latif A, Shakir AS, Rashid MU (2013) Appraisal of economic impact of zero tillage, laser land leveling and bed-furrow interventions in Punjab, Pakistan. Pak J Eng Appl Sci 1:65–81

Lohan SK, Singh M, Sidhu HS (2014) Laser guided land leveling and grading for precision farming. In: Ram T, Lohan SK, Singh R, Singh P (eds) Precision farming—a new approach. Astral International Pvt. Ltd., New Delhi, pp 148–158

Lohan SK, Jat HS, Yadav AK, Sidhu HS, Jat ML, Chauhan BS (2018) Burning issues of paddy residue management in north-west states of India. Renew Sust Energ Rev 81:693–706

Mahapatra J, Tiwari PS, Singh KP, Nandede BM, Sahni RK, Pagare V, Singh J, Shrinivasa DJ, Mandal S (2024) An adjustable pneumatic planter with reduced source vibration for better precision in field seeding. Sensors 24(11):3399. https://doi.org/10.3390/s24113399

Malik RK, Yadav A, Gill GS, Singh S (2004) Evolution and acceleration of no-till farming in rice–wheat cropping system of the Indo-Gangetic Plains. In: Proceedings of the 4th international crop science congress (Brisbane, 29 Sept–3 Oct 2004)

Manpreet-Singh, Sidhu HS, Yadvinder-Singh (2020) Performance evaluation of automatic vis-à-vis manual topographic survey for precision land leveling. Precis Agric 21:300–310

Nandini Y, Moses SC, Aalam RN (2018) Design of manually operated multi-crop inclined plate planter for farm women. Int J Curr Microbiol App Sci 7(6):931–940. https://ijcmas.com/7-6-2018/Y.%20Nandini%2C%20et%2-0al.pdf

Naresh RK, Rathore RS, Kumar P, Singh SP, Singh A, Shahi UP (2014) Effect of precision land leveling and permanent raised bed planting on soil properties, input use efficiency, productivity and profitability under maize–wheat cropping system. Indian J Agric Sci 84(6):725–732. https://doi.org/10.56093/ijas.v84i6.41464

Parihar CM, Jat SL, Singh AK, Kumar B, Yadvinder-Singh, Pradhan S, Pooniya V, Dhauja A, Kakraliya SK, Jat ML, Jat RK (2016) Conservation agriculture in irrigated intensive maize-based systems of North-Western India: effects on crop yields, water productivity and economic profitability. Field Crop Res 193:104–116

Sahu M, Verma A, Dave AK (2017) Performance evaluation of tractor–drawn multi-crop inclined plate planter. J Plant Dev Sci. https://jpds.co.in/wp-content/uploads/2019/07/10.-Re-PDF-Manisha-Sahu-1147.pdf

Sidhu HS, Mahal JS, Dhaliwal IS (2007) Laser land leveling—a boon for sustaining Punjab agriculture, Farm machinery bulletin-2007/01:13. Punjab Agricultural University, Ludhiana

Sidhu HS, Singh S, Singh Y, Blackwell J, Humphreys E, Singh V, Singh P (2015) Development and evaluation of the turbo Happy Seeder for sowing wheat into heavy rice residues in NW India. Field Crop Res 184:201–212

Sims B, Kienzle J (2015) Mechanization of conservation agriculture for smallholders. Environments 2(2):139–156. https://doi.org/10.3390/environments2020139

Singh RC, Singh G, Saraswat DC (2005) Optimisation of design and operational parameters of a pneumatic seed metering device for planting cottonseeds. Biosyst Eng 92(4):429–438. https://doi.org/10.1016/j.biosystemseng.2005.08.002

Singh RP, Dhaliwal HS, Humphreys E (2008) Economic evaluation of the Happy Seeder for rice–wheat systems in Punjab, India. In: Proceedings of the 52nd Australian agricultural and resource economics society conference (Canberra, Australia, 5–8 Feb 2008)

Singh B, Eberbach PL, Humphreys E (2011) The effect of rice straw mulch on evapotranspiration, transpiration and soil evaporation of irrigated wheat in Punjab, India. Field Crop Res 98:1847–1855

Singh Y, Sidhu HS, Singh M, Humphreys E, Kukal SS, Singh B, Blackwell J (2009) Happy seeder: A conservation agriculture technology for managing rice residues. Technical Bulletin No. 01. Department of Soils, Punjab Agricultural University, Ludhiana

Singh Y, Kukal SS, Jat ML, Sidhu HS, Gupta R (2014) Improving water productivity of wheat-based cropping systems in South Asia for sustained productivity. Adv Agron 127:157–258

Singh M, Sidhu HS, Goyal R, Verma A, Jat ML, Singh SK, Singh R, Singh Y (2023) Super straw management system-an innovative and complementing mechanised solution for in-situ recycling of Rice residue in Rice-wheat system of South Asia. AMA-Agric Mech Asia Afr Latin Am 54(3):25–34

Smart Maize Farming: Enhancing Precision Agriculture

16

N. Sunil, U. Srinivasulu Reddy, Deep Mohan Mahala,
B. Jyothi, Mukesh Choudhary, and Hanuman Sahay Jat

Abstract

Smart maize farming integrates advanced technologies such as the Internet of Things (IoT), artificial intelligence (AI), drones, big data analytics, and geospatial tools to enhance productivity, sustainability, and decision-making in maize cultivation. This chapter presents an in-depth overview of technological innovations transforming maize farming, including sensor-based monitoring, UAV-enabled field analysis, machine learning applications for disease and pest detection, and data-driven yield prediction models. Precision agriculture tools such as GPS, GIS, variable rate technology, and precision irrigation enable site-specific management that optimizes water, fertilizer, and pesticide use while reducing environmental impact. Case studies from Africa, Europe, and Asia demonstrate real-world benefits of these technologies in improving yield estimation, enhancing chlorophyll content assessment, and strengthening resource management. Despite the promise of smart farming, smallholder farmers face challenges including high costs, limited technical skills, and infrastructural constraints. The chapter discusses practical solutions, including cooperative technology sharing models and mobile-based advisory systems. Emerging innovations such as blockchain, hyperspectral imaging, and AI-powered robotics indicate a future where maize production becomes increasingly efficient, resilient, and sustainable. Overall, smart maize farming represents a transformative pathway

N. Sunil (✉) · B. Jyothi · M. Choudhary
Winter Nursery Centre, ICAR-Indian Institute of Maize Research, Hyderabad, India
e-mail: mukesh.choudhary1@icar.org.in

U. Srinivasulu Reddy
Machine Learning and Data Analytics Lab, Department of Computer Applications, National Institute of Technology, Tiruchirappalli, India

D. M. Mahala · H. S. Jat
ICAR-Indian Institute of Maize Research, PAU Campus, Ludhiana, India

M. Choudhary et al. (eds.), *Next-Generation Maize Research*,
https://doi.org/10.1007/978-981-92-0722-0_16

toward meeting global food security demands under changing climatic and resource conditions.

Keywords

Smart maize farming · Smart maize · AI · IoT · Machine learning

16.1 Introduction

Smart farming integrates technologies such as the Internet of Things (IoT), drones, artificial intelligence (AI), and big data analytics to improve productivity, sustainability, and precision in crop management. In maize systems, these tools enable real-time monitoring of crop health and environmental conditions, supporting data-driven management decisions (Sharma and Shivandu 2024). In maize-dependent countries such as Rwanda, where post-harvest losses threaten food security, IoT-based storage systems using temperature–humidity sensors, gas detectors, and GSM/GPRS modules provide real-time alerts via mobile interfaces, allowing remote storage management. In addition, machine-learning–based maize grain image classification enhances quality assessment and reduces spoilage and aflatoxin contamination (Ndayikengurukiye et al. 2023; Viviane et al. 2023). Globally, maize is a major cereal crop for food, feed, and industrial use, consumed by over 900 million people, with annual production exceeding 1.2 billion tons (FAOSTAT 2023). Its adaptability across diverse agroecological zones makes it a critical crop for addressing challenges related to population growth, climate variability, and resource scarcity (Shiferaw et al. 2011; Prasanna et al. 2021). Conventional maize cultivation relies on uniform input application, often ignoring spatial variability in soil, moisture, and pest pressure. Precision agriculture overcomes these limitations through site-specific management, improving input-use efficiency, yields, and risk reduction (Li et al. 2020; Dhau et al. 2022).

IoT forms the backbone of smart maize farming, with sensor networks continuously monitoring soil moisture, temperature, pH, electrical conductivity, rainfall, canopy growth, and nutrient status. These real-time datasets support precise irrigation and fertilizer scheduling, minimizing waste and enhancing crop performance (Cicioğlu and Çalhan 2021; Jawad et al. 2020). AI further strengthens decision-making by analyzing large datasets to predict pest outbreaks, classify maize diseases, and optimize irrigation and nutrient management. High diagnostic accuracy has been reported for diseases such as leaf blight, rust, and fall armyworm, enabling timely interventions (Ghosh et al. 2019; Wang et al. 2022). Big data analytics converts sensor-derived information into actionable insights by integrating spatial and temporal datasets, facilitating yield forecasting, nutrient requirement assessment, and climate risk evaluation (Zhang and Kovacs 2019; Chlingaryan et al. 2018). Drones complement these technologies by providing high-resolution multispectral and thermal imagery to assess crop vigor, water stress, nutrient status, and early disease symptoms. Vegetation indices such as NDVI and GNDVI are widely used to monitor canopy health, biomass accumulation, and yield potential in maize (Wahab et al. 2018, Torres-Sánchez et al. 2018). Collectively, IoT, AI, drones, and big data analytics are transforming maize production by improving irrigation efficiency, optimizing nutrient use, and enabling rapid pest and disease detection. These technologies significantly reduce water, fertilizer, and pesticide inputs, lowering environmental impacts. Precision irrigation alone can reduce maize water use by up to 40%, while targeted pest management decreases chemical application and greenhouse gas emissions (Balafoutis et al. 2020). Overall, the adoption of smart farming technologies promotes climate-resilient, resource-efficient maize production systems, enhancing productivity, profitability, and sustainability to support global food security under changing climatic conditions.

16.2 Technological Innovations in Maize Farming

16.2.1 Role of IoT and Sensors in Crop Monitoring

The Internet of Things (IoT) has become a core component of modern agriculture by enabling real-time field monitoring through interconnected sensors and wireless sensor networks. IoT refers to the integration of devices and sensors with the internet to enable seamless data transmission and analysis. While scalability and communication challenges exist, such as device identification and data routing—the potential of IoT to enhance efficiency and productivity is substantial. In precision farming, sensor-generated data support timely, site-specific interventions, such as GPS-guided fertilizer and pesticide application, thereby reducing resource use and waste. However, successful adoption requires farmers to manage sensors, cloud platforms, and user interfaces, necessitating technical capacity building. The emerging concept of the Internet of Everything (IoE), combined with big data and cloud computing, further strengthens spatiotemporal decision-making in agriculture (Alahmad et al. 2023). In maize cultivation, IoT-based monitoring relies on multiple sensor types to assess crop health and environmental conditions. Soil moisture sensors support precise irrigation scheduling, while nutrient (NPK) sensors enable targeted fertilization. Temperature and humidity sensors help characterize microclimatic conditions relevant to planting and disease management. Volatile organic compound (VOC) sensors provide early stress signals for timely crop protection (Gbiliy et al. 2025). Ultrasound sensors track plant movement and growth dynamics (Maginga et al. 2024) and passive infrared (PIR) sensors detect pest activity, supporting integrated pest management strategies (Syafiq Muhammad et al. 2023; Prasetyo et al. 2024). Together, these sensor networks enhance precision agriculture, leading to improved resource efficiency, yield stability, and sustainability.

The integration of IoT with machine learning (ML) represents a major advancement in smart maize farming. Sharma et al. (2025) demonstrated an IoT-enabled system combining sensor-based data acquisition with ML analysis of satellite and drone imagery to assess crop health, estimate yield, and optimize resource use. Soil and environmental sensors captured field-level data, while remote sensing imagery generated vegetation indices for stress detection. Comparative analyses of ML models used in maize prediction indicate that random forest models combined with satellite imagery achieved up to 95% accuracy in detecting Maize Streak Virus, enabling early disease intervention. Nevertheless, the dependence on large training datasets remains a key limitation affecting scalability across regions.

16.2.2 Use of Drones for Field Analysis and Data Collection

Unmanned Aerial Vehicles (UAVs), or drones, have emerged as a central tool in digital precision agriculture due to advances in robotics, control systems, electronics, and artificial intelligence. Agricultural drones offer high-resolution imaging, rapid data acquisition, and operational flexibility at declining costs. Sensor miniaturization and nanotechnology advancements allow drones to perform multiple tasks simultaneously without compromising flight endurance, enhancing efficiency in large-scale maize monitoring (Toscano et al. 2024).

Recent progress in maize disease detection increasingly relies on spectral analysis, Raman spectroscopy, and remote sensing to identify pathogen-induced stress at early stages. Distinct spectral signatures among maize genotypes, potentially linked to unknown metabolites, improve sensitivity to biotic stress detection (Murray et al. 2025). Integrating Plant Reactome knowledge graphs with spectral data further aids in distinguishing disease-specific signals from general stress responses. Remote sensing platforms—including RGB, near-infrared (NIR), multispectral, and hyperspectral imaging—are deployed via drones, satellites, and ground-based robotic systems (Thomas et al. 2018). While satellite imagery offers broad spatial coverage, resolution limitations restrict its effectiveness. Drones equipped with hyperspectral and thermal sensors bridge this gap by delivering high-resolution, real-time field-scale monitoring. Ground-based tools such as Raman spectrometers and PhotosynQ devices provide localized biochemical insights, although challenges remain in achieving pathogen-specific detection at large scales (Thomasson et al. 2022). Equipped with multispectral sensors, drones enable rapid identification of spatial variability in crop health, early detection of drought and disease stress, and precise intervention planning, including targeted input application or replanting. Compared with conventional scouting methods, drone-based monitoring significantly improves speed, accuracy, and labor efficiency. Applications include early diagnosis of nutrient deficiencies and pest infestations, canopy-based yield estimation, and large-scale field assessment. In Kenya, drone-assisted surveys over maize fields achieved approximately 95% accuracy in yield prediction, supporting improved market planning, storage management, and overall farm profitability.

16.3 Data-Driven Decision Making

16.3.1 Big Data Analytics for Yield Prediction

Big data analytics integrates information from diverse sources, including weather forecasts, historical yield records, soil health indicators, and real-time sensor data, to generate actionable insights for maize production. Predictive models developed from these datasets enable accurate yield forecasting, supporting informed decisions on harvest timing, logistics, and resource allocation. Big data in agriculture encompasses structured, semi-structured, and unstructured data, ranging from terabytes to zettabytes, and requires advanced analytical tools for effective processing. Structured data includes organized databases, while unstructured data consists of text, imagery, and sensor outputs without a predefined format. Semi-structured data, such as XML files and RFID-generated information, combine features of both. Accurate yield prediction is increasingly important due to the declining availability of experienced farmers, particularly in resource-limited regions, highlighting the need for data-driven decision-support systems. Advanced machine-learning approaches, including N-stage Convolutional Neural Networks (CNNs), have been developed to support precision agriculture by optimizing irrigation scheduling, disease detection, and yield prediction. Experimental results indicate high predictive accuracy and practical applicability. Future advancements may incorporate Generative Adversarial Networks (GANs) to generate synthetic training data, improving model robustness and enhancing disease detection and yield forecasting capabilities.

16.3.2 Machine Learning Applications in Pest and Disease Management

Machine learning has become a transformative tool in maize pest and disease management by enabling early diagnosis and targeted interventions. Integrating UAV-based hyperspectral imaging with satellite remote sensing has significantly improved maize Leaf Chlorophyll Content (LCC) estimation, overcoming limitations of satellite data caused by mixed pixels from soil variability and intercropping. A hyperspectral maize database developed using a random linear mixed model (RLMM) employed spectral similarity measures—including Euclidean, Manhattan, Minkowski, and Cosine distances—to align Sentinel-2 data with UAV spectra. Among multiple vegetation indices tested, the Normalized Area Over Reflectance Curve (NAOC) showed the strongest correlation with field measurements, demonstrating its value for precision nutrient management and yield prediction (Yang et al. 2024).

Image-based weed identification is another major ML application in precision agriculture. The DRONEWEED dataset provides high-resolution UAV RGB imagery for early-stage weed classification in maize, enabling targeted weed control during critical growth stages (BBCH14–BBCH17). Deep learning models such as

CNNs and Vision Transformers trained on structured datasets containing over 67,000 labeled images have significantly improved weed mapping accuracy, reducing herbicide use and promoting sustainable crop management (Mesias-Ruiz et al. 2025). Real-time disease detection has been further advanced through hybrid deep learning models. Thakur et al. (2024) introduced a CondConViT framework combining Vision Transformers and CNNs with a conditional convolutional attention mechanism for maize disease classification using drone imagery. The incorporation of a GAN-based data augmentation strategy (C3GAN) enhanced model generalization under limited training data conditions. Continuous model refinement through farmer feedback and expert validation highlights the role of adaptive, AI-driven decision-making in smart maize farming.

16.3.3 Georeferencing Technologies in Precision Agriculture

Georeferencing technologies, particularly GPS, underpin precision agriculture by providing accurate spatial information for variable rate technology (VRT), allowing site-specific application of fertilizers and pesticides (Pandey et al. 2021). Geographic Information Systems (GIS) complement GPS by integrating spatial datasets related to soil properties, nutrient status, moisture, pH, topography, and pest pressure to generate detailed decision-support maps (Mani et al. 2021). These tools enable precise input optimization and improved field-level management. GPS- and GNSS-based guidance systems provide real-time positioning for farm machinery, enhancing efficiency in planting, harvesting, and input application. Automated steering systems reduce labor costs, minimize overlaps and gaps, and support the development of autonomous agricultural machinery. Despite limitations related to signal availability, GNSS has transformed manual farming practices into data-driven, mechanized systems central to Agriculture 4.0 (Sharma and Shivandu 2024). Differential GPS (DGPS) further improves positional accuracy by correcting satellite errors, making it essential for high-precision operations. Remote sensing technologies, using aircraft and satellites to capture reflected electromagnetic radiation across visible, infrared, and ultraviolet wavelengths, provide critical insights into crop health, soil conditions, weed infestation, nutrient deficiencies, and water stress. These geospatial tools collectively strengthen data-driven decision-making, enabling timely interventions and enhancing productivity and sustainability in maize farming (Padro et al. 2019).

16.4 Resource Management and Sustainability

Resource management and sustainability are central to smart maize farming, ensuring long-term productivity while addressing environmental protection and food security. Effective resource management focuses on the optimal use of water, soil, nutrients, and energy through technologies such as IoT sensors, satellite imagery, and data analytics (Fig. 16.1). These tools enable precise monitoring and timely

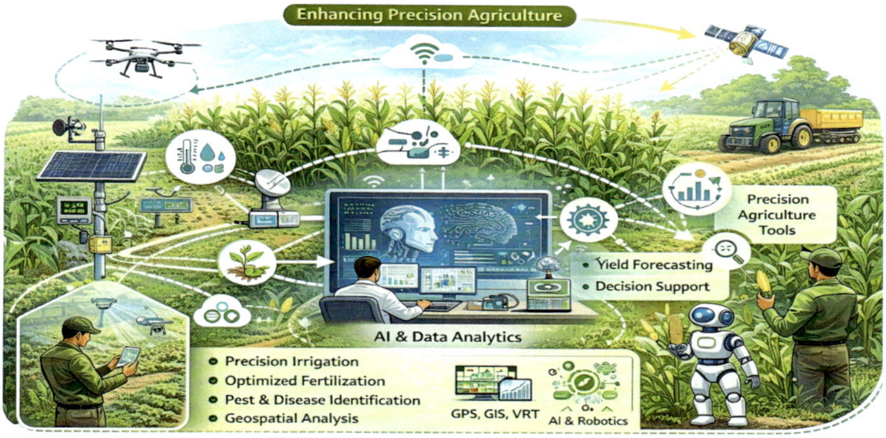

Fig. 16.1 Smart maize farming framework illustrating the integration of IoT sensors, remote sensing, drones, AI-driven analytics, and precision machinery to enable real-time monitoring, decision support, and site-specific management of water, nutrients, pests, and crops for enhanced productivity, sustainability, and resource-use efficiency

interventions, for example, precision irrigation systems that maintain optimal soil moisture while minimizing water waste, and soil nutrient sensors that guide fertilizer application. Sustainability-oriented practices further aim to reduce the ecological footprint of maize cultivation without compromising yield. Importantly, smart farming solutions also have social and economic dimensions, as scalable digital tools can empower smallholder farmers, promote equitable growth, and support resilient agricultural systems globally.

16.4.1 Precision Irrigation Techniques

Maize is a water-intensive crop, with growth stages such as germination, tasseling, and grain filling requiring consistent moisture. Precision irrigation delivers water exactly when and where it is needed, improving water-use efficiency and preventing yield losses caused by over- or under-irrigation. Reinforcement learning–based irrigation systems have demonstrated water savings of up to 40% without yield penalties by integrating historical and real-time environmental data (Abioye et al. 2022). The integration of artificial intelligence with IoT sensors, remote sensing, and machine learning allows real-time analysis of soil moisture, weather conditions, and crop health (Fig. 16.1). This data-driven approach ensures optimal irrigation scheduling at critical growth stages, significantly enhancing water efficiency (Adewusi et al. 2024).

16.4.2 Technological Innovations in Precision Irrigation

IoT-enabled sensors continuously monitor soil moisture, temperature, and humidity, transmitting data through WSNs for automated irrigation control. UAV imagery and GIS-based analysis further identify water-stressed zones, enabling targeted irrigation across large maize fields (Abioye et al. 2022).

16.4.3 Machine Learning for Irrigation Management

Machine learning models predict maize irrigation requirements using historical and real-time environmental data. Supervised models such as decision trees, random forests, and support vector machines estimate water demand based on soil and weather variables (Yamac et al. 2024). Regression models guide daily irrigation needs (Pulido-Calvo et al. 2007), while deep learning approaches, including RNNs and LSTM networks, forecast weather patterns and evapotranspiration, enabling dynamic irrigation adjustments and reduced water use (Chlingaryan et al. 2018; Karunathilake et al. 2023).

16.4.4 Remote Sensing and GIS Integration

GIS tools delineate management zones and monitor spatial variability in soil moisture, while satellite and UAV-based remote sensing capture vegetation indices and crop water stress. UAV multispectral imagery enables precise irrigation interventions and reduces wastage across large maize fields (Abioye et al. 2022). The integration of AI with drone and satellite data further enhances stress detection and timely management decisions (Lachgar et al. 2023).

16.4.5 Digital Tools and Applications

Mobile and web-based platforms support real-time irrigation and fertigation decisions in maize farming. Mobile applications such as the Smart Farm App enable remote monitoring and control of irrigation systems, while CFertigUAL integrates irrigation scheduling with nutrient delivery in controlled environments (Abioye et al. 2022). Adapted tools like the Cotton App provide soil water balance alerts applicable to maize systems (Vellidis et al. 2016). Web-based platforms such as ThingSpeak offer real-time data visualization, and ARSPivot DSS integrates GIS data to automate center-pivot irrigation systems for large-scale maize production (Abioye et al. 2022). When combined with center-pivot or drip irrigation systems, these digital tools enhance water-use efficiency and yield stability (Chlingaryan et al. 2018).

16.4.6 Efficient Use of Fertilizers and Pesticides

Excessive fertilizer and pesticide use in conventional maize farming leads to nutrient losses, environmental pollution, and increased production costs. Smart farming technologies—such as IoT sensors, nanotechnology, and data analytics—enable site-specific input management by continuously monitoring soil and crop conditions (Anand et al. 2023). AI-driven geospatial analysis supports precise input application tailored to field variability, reducing overuse and environmental contamination (Choudhary et al. 2024; Obeidat et al. 2024). Nanotechnology has significantly improved fertilizer efficiency through nanofertilizers that provide controlled nutrient release. Nanocarriers such as hydroxyapatite, chitosan nanoparticles, and mesoporous silica enhance nutrient use efficiency, increase root growth, chlorophyll content, and yield, while reducing leaching losses in maize and other crops (Fincheira et al. 2023). IoT-based smart systems further optimize fertilizer delivery by integrating nutrient application with irrigation (fertigation), using real-time sensor feedback and machine learning recommendations to align inputs with crop demand and environmental conditions. The combined use of nanotechnology and smart farming tools offers substantial potential for improving productivity, reducing costs, and maintaining soil health, although widespread adoption depends on affordable technologies, farmer training, and supportive policies (Manjunatha et al. 2016).

16.4.7 Variable Rate Technology (VRT) and UAV-Based Input Application

Variable Rate Technology (VRT) enables site-specific fertilizer and pesticide application and is increasingly integrated with UAV platforms. UAV-mounted VRT systems dynamically adjust application rates using real-time field data, optimizing input distribution while minimizing wastage (Fig. 16.1). Drone sprayers and automated machinery improve uniform coverage, reduce labor requirements, and enable rapid response to spatial variability in crop stress and nutrient deficiency (Sagheer et al. 2020). IoT sensors provide continuous feedback on soil moisture and nutrient status, allowing adaptive fertilization and pest management strategies (Chen et al. 2022). The flexibility of UAVs allows effective operation in difficult terrains, making VRT-equipped drones suitable for diverse agricultural landscapes. By reducing both over- and under-application of inputs, these systems enhance economic returns while minimizing environmental impacts and supporting sustainable maize production (Song et al. 2021).

16.5 Challenges and Solutions

16.5.1 Addressing Technological Adoption Barriers

The adoption of smart farming technologies, including precision tools for efficient fertilizer use, faces several barriers despite their potential benefits. Key challenges include high implementation costs, lack of technical knowledge among farmers, limited infrastructure in rural areas, and concerns about data security and privacy. These barriers often prevent small and medium-scale farmers, who constitute the majority of agricultural producers globally, from leveraging smart farming solutions effectively.

16.5.1.1 High Costs and Limited Access

Advanced technologies like IoT sensors, drones, and nanotechnology-based fertilizers require substantial initial investment. Small-scale farmers often lack access to capital, making these innovations inaccessible to them. Moreover, the maintenance and operational costs of such systems can add to the financial burden. To address this, government subsidies, microfinance schemes, and public-private partnerships are essential. For instance, India's PM-Kisan scheme provides financial support to small farmers, enabling them to invest in smart agricultural tools.

16.5.1.2 Knowledge and Skill Gaps

Many farmers lack the technical expertise needed to operate smart farming equipment or interpret data from digital platforms. This gap hampers the effective utilization of technologies like precision irrigation or IoT-enabled nutrient delivery systems. Extension services, capacity-building workshops, and user-friendly interfaces can bridge this gap. For example, the Food and Agriculture Organization conducts farmer field schools to enhance technical know-how.

16.5.1.3 Infrastructure Deficiencies

Inadequate rural infrastructure, such as unreliable internet connectivity and electricity supply, limits the deployment of smart farming technologies. Rural broadband initiatives, such as the USDA's ReConnect Program in the United States, aim to extend high-speed internet access to underserved areas, fostering the adoption of digital farming tools.

16.5.1.4 Data Security and Privacy Concerns

The integration of IoT devices and cloud-based platforms raises concerns about data ownership, misuse, and privacy. Farmers may be reluctant to adopt these technologies without clear guidelines on data protection. Establishing transparent regulations and promoting trusted third-party platforms can alleviate these concerns. The European Union's General Data Protection Regulation (GDPR) provides a robust framework for data privacy, which can serve as a model for agricultural technologies and farmer incentives. With a collaborative approach involving governments, private companies, and non-governmental organizations, the barriers to adopting smart farming technologies can be overcome, ensuring their benefits reach even the most resource-constrained farmers.

16.5.2 Solutions for Smallholder Farmers

Community-based approaches, such as shared drone services and cooperative data platforms, make advanced technologies accessible. Mobile-based decision-support systems tailored for small-scale maize farmers can also drive adoption.

16.6 Case Studies and Real-World Applications

Case Study 1: UAV-Derived GNDVI for Maize Vigor and Yield Estimation (Wahab et al. 2018)
This study evaluated the effectiveness of UAV-derived Green Normalized Difference Vegetation Index (GNDVI) in assessing maize vigor and predicting yield in smallholder systems of Sub-Saharan Africa. Conducted in two villages in Ghana, it covered 87 maize plots from 60 households. UAV flights with RGB and NIR sensors were performed at 100 m altitude, and GNDVI was calculated and correlated with field measurements, including SPAD chlorophyll content, visual vigor scores, and crop-cut yields at 5, 9, and 13 weeks after planting. Results showed GNDVI peaked at 9 WAP (~0.72) and declined by 13 WAP (~0.65), with early-season GNDVI significantly correlating with yield ($r \approx 0.39$ for maximum and $r \approx 0.37$ for mean). SPAD had moderate correlations, while visual vigor scores were weaker. The study concluded that UAV-derived GNDVI provided more reliable early yield predictions than traditional field assessments, supporting timely nutrient and irrigation interventions in resource-limited smallholder systems.

Case Study 2: Precision Agriculture Tools for Monitoring Maize Yield in Hungary (Széles et al. 2024)
This research assessed the relationship among SPAD, NDVI, and maize grain yield under varying nitrogen and irrigation regimes to identify optimal nutrient strategies. Conducted at the University of Debrecen, the factorial experiment involved three maize hybrids under irrigated and rainfed conditions, with nitrogen applied at different rates and stages. SPAD and NDVI were measured across six growth stages, supported by UAV multispectral imagery. Results showed SPAD values increased during vegetative stages and declined during reproductive stages, with higher values under irrigation. NDVI peaked around V12–Vn and effectively distinguished irrigated from rainfed plots. The highest yield (11.86 Mg ha^{-1}) was achieved with 120 kg N ha^{-1} under irrigation, and NDVI at R3 strongly predicted yield ($R^2 \approx 0.75$). The study concluded that integrating SPAD and NDVI enables adaptive nitrogen and irrigation management, enhancing efficiency and reducing production risk.

Case Study 3: Enhancing Satellite-Based Chlorophyll Estimation Using UAV Hyperspectral Data (Yang et al. 2024)
This study aimed to improve satellite-derived leaf chlorophyll content (LCC) estimates by integrating high-resolution UAV hyperspectral data using a random linear mixed model. Conducted in a 100-ha maize–soybean intercropping system in Shandong Province, China, UAV hyperspectral images were collected at jointing, tasseling, and milking stages, while Sentinel-2A/B data were preprocessed. LCC was measured at 56 field locations using a SPAD meter, and various spectral matching methods and vegetation indices were evaluated. Cosine similarity achieved the highest accuracy, with 75% of Sentinel-2 pixels showing >0.95 similarity with UAV spectra. NAOC had the strongest relationship with LCC (R^2 = 0.83), while NDVI suffered from saturation effects. Integrating UAV hyperspectral data improved satellite-based LCC predictions (R^2 = 0.90), outperforming existing products like MuSyQ in intercropping conditions. The study concluded that high-resolution hyperspectral integration enhances nutrient diagnostics and improves the reliability of satellite-based monitoring platforms for smart maize farming.

16.7 Future Trends and Developments

As the evolution of agriculture takes place, there is a vast transformation in the way food is produced by the new technologies that emerge in smart farming and precision agriculture. These changes in advancements are significant for maize production, where precision-driven solutions tend to enhance efficiency, optimize resource utilization, and improve the overall sustainability. The following emphasizes the latest trends in precision agriculture and the future of smart farming in maize cultivation (Table 16.1).

16.7.1 Emerging Technologies in Precision Agriculture

16.7.1.1 Blockchain Technology

Blockchain enhances traceability in maize supply chains by creating an immutable ledger that records every transaction from farm to table. This transparency ensures that all stakeholders, including farmers, suppliers, and consumers, can verify the source and journey of maize products. As a result, it promotes fair trade practices by assuring consumers of the quality and authenticity of the product, while also helping farmers obtain better prices for their crops.

16.7.1.2 AI-Powered Robots

AI-powered robots are revolutionizing maize farming by automating critical tasks such as seed planting, weed control, and harvesting. These intelligent machines use computer vision and machine learning to navigate fields, identify crop rows, and perform operations with high precision. In maize breeding programs, robots can assist in detasseling, cob sorting, and phenotyping, reducing labor costs and human

Table 16.1 Impact of precision agriculture technologies on crop productivity, resource use, and economic benefits

Technology	Yield increase (%)	Water use reduction (%)	Fertilizer use reduction (%)	Pesticide reduction (%)	Economic benefit (%)	Reference
Variable rate technology (VRT)	12–25	15–30	Up to 46%	Up to 51%	Up to 38% cost savings	Sanyaolu and Sadowski (2024)
Precision irrigation	10–20	30–50	5%–15%	10%–20%	10%–25% economic return	Monteiro et al. (2021)
Remote sensing and GIS	12–30	20–40	15%–20%	10%–15%	15%–30% increase in production efficiency	Getahun et al. (2024)
Drones for crop monitoring	15–25	10–25	10%–15%	20%–35%	12%–30% higher profitability	Mgendi (2024)
Automated machinery (AI and robotics)	15–30	20–40	20%–35%	25%–40%	20%–40% reduction in labor costs	Domingues et al. (2022)

error. Their ability to optimize input use—like fertilizers and pesticides—enhances resource efficiency, boosts yields, and supports environmentally sustainable maize production (Bala et al. 2021).

16.7.1.3 Hyperspectral Imaging

Hyperspectral imaging provides a powerful tool for monitoring maize crop health by capturing reflectance data across hundreds of light wavelengths. This technology enables early detection of nutrient deficiencies, pest infestations, and diseases—often before symptoms are visible. In maize breeding, hyperspectral data support high-throughput phenotyping by identifying stress responses and growth patterns at the molecular level. When integrated with AI models, it allows for precise diagnostics and targeted interventions, improving yield potential and resilience under variable environmental conditions (Shu et al. 2021).

16.7.2 The Future of Smart Farming in Maize Production

The adoption of smart farming technologies in maize production is expected to expand rapidly, driven by advances in AI, the IoT, and big data analytics. These technologies enable farmers to collect and analyze large volumes of data on soil health, weather conditions, and crop performance, supporting data-driven decision-making and optimized farm operations. AI-based models can predict optimal planting windows and monitor crop health using satellite imagery and sensor data, while IoT devices, such as soil moisture and climate sensors, provide real-time field insights. Big data analytics further enhances resource management by identifying patterns that reduce input costs and improve efficiency. Realizing the full potential of smart farming requires coordinated efforts among key stakeholders. Government policies that promote innovation—such as subsidies for technology adoption and funding for agricultural research—can accelerate farmer uptake. Collaboration with private companies supports the development of customized digital tools, while partnerships with research institutions foster innovation and the dissemination of best practices. Together, these collaborative approaches can scale smart farming technologies globally, improving productivity, sustainability, and resilience in maize-based food systems to meet the needs of a growing population.

16.8 Conclusion

The integration of GPS- and GNSS-based guidance systems has fundamentally transformed maize cultivation by improving operational efficiency, accuracy, and precision. Real-time positioning enables automated planting, harvesting, and field management, reducing human error during critical operations. Given maize's global importance as a staple crop, these technological advancements play a crucial role in enhancing productivity and supporting farmer livelihoods. Variable Rate Application further strengthens precision maize farming by enabling site-specific management of seeds, fertilizers, and water based on spatial variability within fields. GNSS-derived data allow farmers to align input application with soil and crop requirements, maximizing yield while minimizing resource wastage and environmental impact. This targeted approach contributes directly to sustainable maize production systems. Despite these benefits, challenges such as limited satellite coverage and signal instability can affect the accuracy of GPS/GNSS-based tools. Addressing these constraints through continuous technological improvements is essential to ensure reliable precision farming across diverse production environments. As maize farming becomes increasingly commercialized, efficient machinery management has also gained importance. GNSS-enabled automated steering systems reduce overlap, improve row spacing, lower labor requirements, and enhance overall operational efficiency. Looking ahead, GNSS technology will continue to underpin future innovations in maize farming, including the development of autonomous agricultural machinery and robots that depend on precise navigation. As these technologies mature, maize production systems are expected to become more productive, resource-efficient, and environmentally sustainable, ensuring resilience and food security in the face of growing global demand.

References

Abioye EA, Hensel O, Esau TJ, Elijah O, Abidin MSZ, Ayobami AS, Yerima O, Nasirahmadi A (2022) Precision irrigation management using machine learning and digital farming solutions. AgriEngineering 4(1):70–103. https://doi.org/10.3390/agriengineering4010005

Adewusi AO, Asuzu OF, Olorunsogo T, Iwuanyanwu C, Adaga E, Daraojimba DO (2024) AI in precision agriculture: a review of technologies for sustainable farming practices. World J Adv Res Rev 21(1):2276–2285

Alahmad T, Neményi M, Nyéki A (2023) Applying IoT sensors and big data to improve precision crop production: a review. Agronomy 13(10):2603. https://doi.org/10.3390/agronomy13102603

Anand J, Yusoff N, Ab Ghani H, Thoti KK (2023) Technological applications in smart farming. A bibliometric analysis. Adv Sustain Technol 2(2):30–40

Bala JA, Olaniyi OM, Folorunso TA, Daniya E (2021) An IoT-based autonomous robot system for maize precision agriculture operations in sub-Saharan Africa. In: Singh KK, Nayyar A, Tanwar S, Abouhawwash M (eds) Emergence of cyber physical system and IoT in smart automation and robotics. Springer, Cham, pp 63–75. https://doi.org/10.1007/978-3-030-66222-6_5

Balafoutis AT, Beck B, Fountas S, Vangeyte J, van der Wal T, Soto I, Gómez-Barbero M, Barnes A, Eory V (2020) Precision agriculture technologies positively contributing to GHG emissions mitigation. J Clean Prod 244:118852. https://doi.org/10.1016/j.jclepro.2019.118852

Chen Q, Li L, Chong C, Wang X (2022) AI-enhanced soil management and smart farming. Soil Use Manag 38(1):7–13. https://doi.org/10.1111/sum.12745

Chlingaryan A, Sukkarieh S, Whelan B (2018) Machine learning approaches for crop yield prediction and climate-smart farming practices. Comput Electron Agric 151:61–69. https://doi.org/10.1016/j.compag.2018.05.023

Choudhary R, Mantri SP, Barse MVA, Chitnis S (2024) Leveraging AI in smart agro-informatics: a review of data science applications. Int Res J Adv Eng Manag 2(6):1964–1975

Cicioğlu M, Çalhan A (2021) Smart agriculture with internet of things in cornfields. Comput Electr Eng 90:106982. https://doi.org/10.1016/j.compeleceng.2020.106982

Dhau I, Namara RE, Nyang'au, J. (2022) Precision agriculture for maize production: a review. Agric Syst 196:103343. https://doi.org/10.1016/j.agsy.2021.103343

Domingues T, Brandão T, Ferreira JC (2022) Machine learning for detection and prediction of crop diseases and pests: a comprehensive survey. Agriculture 12(9):1350. https://doi.org/10.3390/agriculture12091350

FAOSTAT (2023) FAOSTAT database. Food and Agriculture Organization of the United Nations (FAO), Rome. Available at: https://openknowledge.fao.org/server/api/core/bitstreams/df90e6cf-4178-4361-97d4-5154a9213877/content

Fincheira P, Hoffmann N, Tortella G, Ruiz A, Cornejo P, Diez MC, Seabra AB, Benavides-Mendoza A, Rubilar O (2023) Eco-efficient systems based on nanocarriers for the controlled release of fertilizers and pesticides. Nano 13(13):1978. https://doi.org/10.3390/nano13131978

Gbiliy A, Senosy I, Saadan MA, Di X, Sayed M, Gao YY, Hao GF (2025) Nanoarchitectures-powered volatile organic compound sensors for sustainable agriculture. Coord Chem Rev 548:217169. https://doi.org/10.1016/j.ccr.2024.217169

Getahun S, Kefale H, Gelaye Y (2024) Application of precision agriculture technologies for sustainable crop production and environmental sustainability. Sci World J 2024:2126734. https://doi.org/10.1155/2024/2126734

Ghosh S, Das P, Chakraborty S (2019) Maize leaf disease detection using deep learning techniques. Inf Process Agric 6(3):396–410. https://doi.org/10.1016/j.inpa.2019.07.003

Jawad HM, Nordin R, Gharghan SK, Jawad AM, Ismail M (2020) Energy-efficient wireless sensor networks for precision agriculture. Sensors 20(6):1523. https://doi.org/10.3390/s20061523

Karunathilake EMBM, Le AT, Heo S, Chung YS, Mansoor S (2023) The path to smart farming: innovations and opportunities. Agriculture 13(8):1593. https://doi.org/10.3390/agriculture13081593

Lachgar M, Hrimech H, Kartit A (2023) Unmanned aerial vehicle-based applications in smart farming: a systematic review. Int J Adv Comput Sci Appl 14(6):1–16. https://doi.org/10.14569/IJACSA.2023.0140608

Li B, Xu X, Zhang L, Feng Z (2020) Applications of precision agriculture technologies for maize production. Field Crop Res 245:107659. https://doi.org/10.1016/j.fcr.2019.107659

Maginga TJ, Masabo E, Bakunzibake P, Kim KS, Nsenga J (2024) Using wavelet transform and hybrid CNN–LSTM models on VOC and ultrasound IoT sensor data for non-visual maize disease detection. Heliyon 10(4). https://doi.org/10.1016/j.heliyon.2024.e28162

Mani PK, Mandal A, Biswas S, Sarkar B, Mitran T, Meena RS (2021) Remote sensing and geographic information system: a tool for precision farming. In: Geospatial technologies for crops and soils. Springer, pp 49–111. https://doi.org/10.1007/978-981-15-6024-8_3

Manjunatha SB, Biradar DP, Aladakatti YR (2016) Nanotechnology and its applications in agriculture: a review. J Farm Sci 29(1):1–13

Mesias-Ruiz GA, Peña JM, de Castro AI, Dorado J (2025) Drone imagery dataset for early-season weed classification in maize and tomato crops. Data Brief 58:111203. https://doi.org/10.1016/j.dib.2024.111203

Mgendi G (2024) Unlocking the potential of precision agriculture for sustainable farming. Discov Agric 2:87. https://doi.org/10.1007/s44321-024-00087-5

Monteiro A, Santos S, Gonçalves P (2021) Precision agriculture for crop and livestock farming: a brief review. Animals 11(8):2345. https://doi.org/10.3390/ani11082345

Murray SC, Verhoef A, Adak A, Sen D, Salzman R, Jaiswal P, Naithani S (2025) Detecting novel plant pathogen threats to food system security by integrating the plant Reactome and remote sensing. Curr Opin Plant Biol 83:102684. https://doi.org/10.1016/j.pbi.2024.102684

Ndayikengurukiye S, Twahirwa E, Nshimiyimana L (2023) IoT-based postharvest monitoring system for maize storage. J Stored Prod Res 105:102372. https://doi.org/10.1016/j.jspr.2023.102372

Obeidat MA, Abdallah J, Hamadneh T, Qawaqneh H, Mansour AM (2024) Enhancing agricultural operations through AI-driven agent communication in smart farming systems. Ing Syst d'Inf 29(3):917–926. https://doi.org/10.18280/isi.290322

Padro JC, Muñoz FJ, Planas J, Pons X (2019) Comparison of four UAV georeferencing methods for environmental monitoring purposes. Int J Appl Earth Obs Geoinf 75:130–140. https://doi.org/10.1016/j.jag.2018.10.011

Pandey PC, Tripathi AK, Sharma JK (2021) An evaluation of GPS opportunity in market for precision agriculture. In: GPS and GNSS Technology in Geosciences. Elsevier, Amsterdam, pp 337–349. https://doi.org/10.1016/B978-0-12-818617-6.00018-2

Prasanna BM, Cairns JE, Zaidi PH, Beyene Y, Gowda M, Magorokosho C, Makumbi D (2021) Beat the stress: breeding for climate-resilient maize. Crop J 9(3):430–445. https://doi.org/10.1016/j.cj.2021.01.006

Prasetyo N, Syamsu M, Arman S (2024) Design of pest detection and repellent device based on Arduino Uno using PIR sensor. J Comput Sci Adv 2(5):285–296. https://doi.org/10.70177/jsca.v2i5.1326

Pulido-Calvo I, Montesinos P, Roldán J, Ruiz-Navarro F (2007) Linear regressions and neural approaches to water demand forecasting. Biosyst Eng 97(2):283–293. https://doi.org/10.1016/j.biosystemseng.2007.03.013

Sagheer A, Mohammed M, Riad K, Alhajhoj M (2020) A cloud-based IoT platform for precision control of soilless greenhouse cultivation. Sensors 21(1):223. https://doi.org/10.3390/s21010223

Sanyaolu M, Sadowski A (2024) The role of precision agriculture technologies in enhancing sustainable agriculture. Sustainability 16(15):6668. https://doi.org/10.3390/su16156668

Sharma K, Shivandu SK (2024) Integrating artificial intelligence and internet of things for enhanced crop monitoring. Sensors International 5:100292. https://doi.org/10.1016/j.sintl.2024.100292

Sharma A, Gowda D, Nagarajan SKS, Shah K, Prasanth P, Saxena R (2025) IoT-enhanced machine learning for remote sensing and image processing in agriculture. In: Advances in smart agriculture. Springer. https://doi.org/10.1007/978-981-96-7289-9_45

Shiferaw B, Prasanna BM, Hellin J, Bänziger M (2011) Past successes and future challenges of maize in global food security. Food Secur 3(3):307–327. https://doi.org/10.1007/s12571-011-0140-5
Shu M, Shen M, Zuo J, Yin P, Wang M, Xie Z, Tang J, Wang R, Li B, Yang X (2021) UAV-based hyperspectral imaging to estimate crop traits in maize inbred lines. Plant Phenomics 2021:9845201. https://doi.org/10.34133/2021/9845201
Song C, Zhou Z, Zang Y, Zhao L, Yang W, Luo X, Zhu Q (2021) Variable-rate control system for UAV-based granular fertilizer spreader. Comput Electron Agric 180:105832. https://doi.org/10.1016/j.compag.2020.105832
Syafiq Muhammad A, Mandala S, Gunawan PH (2023) IoT-based pest detection in maize plants using machine learning. In: Proceedings of the international conference on data science and its applications. Bandung, IEEE, pp 254–258. https://doi.org/10.1109/ICoDSA58501.2023.10277633
Széles A, Huzsvai L, Mohammed S, Nyéki A, Zagyi P, Horváth É, Tamás A (2024) Precision agricultural technology for advanced monitoring of maize yield. J Agric Food Res 15:100967. https://doi.org/10.1016/j.jafr.2023.100967
Thakur PS, Chaturvedi S, Khanna P, Sheorey T, Ojha A (2024) Real-time plant disease identification using vision transformers. IEEE Trans AgriFood Electron 2(4):269–283. https://doi.org/10.1109/TAFE.2024.3389127
Thomas S, Kuska MT, Bohnenkamp D, Brugger A, Alisaac E, Wahabzada M, Mahlein AK (2018) Benefits of hyperspectral imaging for plant disease detection. J Plant Dis Prot 125:5–20. https://doi.org/10.1007/s41348-017-0124-6
Thomasson A, Kolomiets M, Kurouski D, Mullet J, Murray S, Simpson S, Vann T (2022) Automated precision phenotyping greenhouses. Authorea Preprints. https://doi.org/10.22541/au.165642112.47383477/v1
Torres-Sánchez J, Peña JM, López-Granados F (2018) Assessing maize nitrogen status using drone-based multispectral imagery. Remote Sens 10(3):520. https://doi.org/10.3390/rs10030520
Toscano F, Fiorentino C, Capece N, Erra U, Travascia D, Scopa A, D'Antonio P (2024) Unmanned aerial vehicle for precision agriculture: a review. IEEE Access. https://doi.org/10.1109/ACCESS.2024.3387462
Viviane I, Emmanuel M, Rene M, Joseph H, Elias B (2023) Development of AI-based maize storage monitoring system using IoT. In: Proceedings of the international conference on intelligent computing and control. IEEE, pp 38–42. https://doi.org/10.1109/ICCC56789.2023.10087456
Vellidis G, Liakos V, Andreis JH, Perry CD, Porter WM, Barnes EM, Morgan KT, Fraisse C& Migliaccio KW (2016) Development and assessment of a smartphone application for irrigation scheduling in cotton. Computers and Electronics in Agriculture, 127, 249–259
Wahab I, Hall O, Jirström M (2018) Remote sensing of yields using UAV-derived NDVI for estimating maize vigor and yields in complex farming systems in sub-saharan africa.. Drones 2(3):28. https://doi.org/10.3390/drones2030028
Wang L, Ma X, Li Z, Chen Y (2022) Deep learning for automated diagnosis of maize diseases. Comput Electron Agric 198:107024. https://doi.org/10.1016/j.compag.2022.107024
Yamac SS, Kurtuluş B, Memon AM, Alomair G, Todorovic M (2024) Suitability of supervised learning methods for estimating crop water consumption. Agronomy 14(3):532. https://doi.org/10.3390/agronomy14030532
Yang S, Kang R, Xu T, Guo J, Deng C, Zhang L, Kaufmann HJ (2024) Improving satellite-based retrieval of maize leaf chlorophyll content. Drones 8(12):783. https://doi.org/10.3390/drones8120783
Zhang D, Kovacs JM (2019) The application of small UAVs for precision agriculture. Remote Sens 11(2):151. https://doi.org/10.3390/rs11020151

Fodder Maize and Silage: Enhancing Fodder Security and Dairy Productivity

17

Pardeep Kumar, Mukesh Choudhary, Bhupender Kumar, N. Sunil, Bharat Bhushan, Bahadur Singh Jat, Ankush Sharma, Sandesh Kumar Mishra, B. Jyoti, Subhank Dixit, and Hanuman Sahay Jat

Abstract

Maize is a vital cereal crop serving as a key source of high-quality green fodder and silage for livestock, due to its high biomass yield, nutritional value, palatability, and adaptability across diverse agro-climatic zones. With the projected rise in global demand for milk and meat, and persistent shortages of green and dry fodder in countries like India, maize offers a sustainable solution for enhancing fodder and feed security. Moreover, maize silage, produced from whole-plant maize through anaerobic fermentation, ensures year-round feed availability, particularly during lean periods. This chapter highlights the role of maize as green fodder and silage, outlines desirable forage and silage quality traits, and explores innovations such as hydroponic maize fodder for space- and water-efficient production. Different breeding strategies are employed, ranging from conventional selection and utilization of genetic resources, to molecular breeding, genomic selection, and genetic engineering to improve biomass yield, forage quality, and digestibility. The integration of traits such as *brown midrib* mutants, high-lysine, high-oil, leafy, and waxy endosperm types for silage and green fodder ideotype development in maize. Advances in QTL mapping, genome-wide association studies, and genome editing tools like CRISPR-Cas are likely to accelerate the improvement of maize for fodder purposes, offering promising avenues for sustainable livestock production and feed security.

Keywords

Maize · Fodder · Silage · Quality · Mutants · Breeding

P. Kumar (✉) · M. Choudhary · B. Kumar · N. Sunil · B. Bhushan · B. S. Jat · A. Sharma · S. K. Mishra · B. Jyoti · S. Dixit · H. S. Jat
ICAR-Indian Institute of Maize Research, Ludhiana, Punjab, India
e-mail: mukesh.choudhary1@icar.org.in

M. Choudhary et al. (eds.), *Next-Generation Maize Research*,
https://doi.org/10.1007/978-981-92-0722-0_17

17.1 Introduction

The global food demand is going to increase by about 50–60% by 2050 from the current level for food security (Falcon et al. 2022). Similarly, the livestock sector also plays an important role in ensuring world food security. The demand for livestock-based milk and meat production is rising globally due to increasing population, changing food habits, and rising income status (Ahamed et al. 2023). Meat and milk consumption in developing countries grow at 2.8% and 3.3% per annum, respectively (Delgado et al. 1999). Furthermore, the demand for milk and meat consumption is projected to reach 1077 and 455 million tons, respectively, by 2050 (Alexandratos and Bruinsma 2012). Besides the genetic makeup of the livestock, higher milk productivity, feed, and nutrition hold significant importance. Maize is one of the important fodder crops widely cultivated across the globe as a green fodder crop due to its high biomass yield, good palatability, and rich nutritional profile (Choudhary et al. 2020). It plays a crucial role in supporting ruminant livestock systems and sustainable animal feed production in both tropical and temperate regions. The utilization pattern of maize is 61% for feed, 17% for food, and 22% for industry over the global consumption (Prasanna and Hoisington 2003). Maize can be utilized as green fodder in various forms, as it can be used as a whole plant and individual parts like green tassel, green cob husk, leaf stem, and stover (Kumar et al. 2025).

Although green fodder serves as a good source of roughage and energy for livestock, its deficit (more so during summer) and year-round unavailability are major limitations. Silage is the best alternative to address these limitations. The global production and market for maize silage have seen substantial growth over the past two decades. In the United States alone, corn silage production reached approximately 143.78 million tons from 6.71 million acres in 2020 (Karnatam et al. 2023). Maize silage serves as a critical feed resource, particularly for ruminants, as it typically contains 60–70% moisture and is rich in starch and protein derived from the grains. This composition provides high energy content and enhances digestibility, making it an efficient and reliable source of nutrition for livestock. Maize varieties developed specifically for silage need to be bred for traits such as high biomass yield, superior digestibility, and enhanced energy content, making them ideal for livestock feeding (Karnatam et al. 2023). Owing to its high green mass yield per hectare, efficient fermentation potential, and excellent animal palatability, maize silage remains a reliable and cost-effective supplementation strategy, especially during dry seasons (Kuehn et al. 1999). "Furthermore, maize offers wide opportunities for utilization as green fodder and silage on a large scale. However, substantial genetic improvement is still required to enhance traits such as high biomass yield, ideotype development, and nutritional/digestibility characteristics to ensure future livestock security. To achieve these improvements, approaches including conventional breeding, molecular breeding, genomic selection, fodder mutants, and genetic engineering are playing critical roles and will be discussed in detail in this chapter (Fig. 17.1).

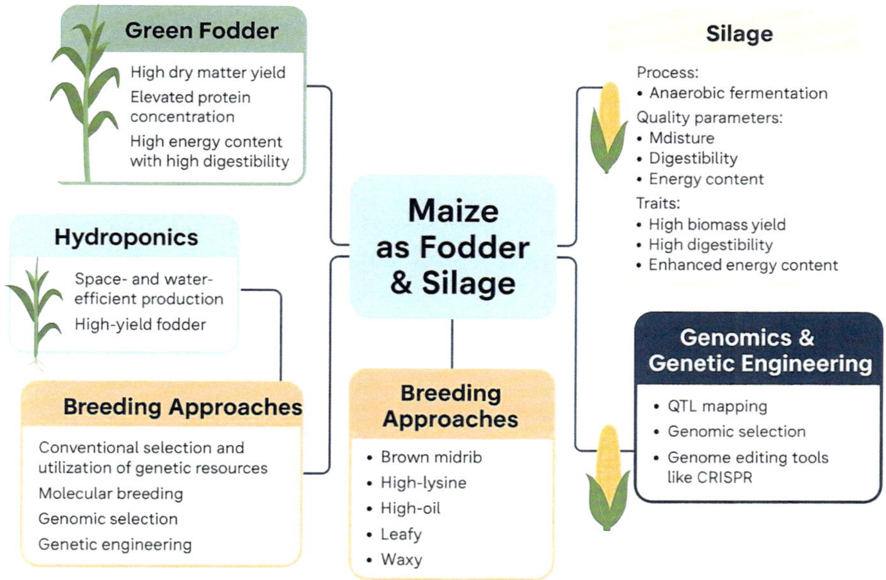

Fig. 17.1 Overview of utilization techniques of maize as fodder and genetic interventions

17.2 Maize as Green Fodder

Desirable characteristics of an ideal forage fodder include high green fodder yield, elevated protein concentration, high energy content with superior digestibility, strong intake potential due to low fibre content, and an optimal dry matter concentration at harvest (Lewis et al. 2004). Leguminous crops generally provide more nutritious fodder because of their protein richness, but they produce comparatively less biomass. Maize, on the other hand, is a promising crop for large-scale biomass production; as its cultivation and productivity increase, the availability of biomass also expands (Chaudhary et al. 2014). Maize straw has long been used as animal fodder, and the quality of green maize is regarded as the best among non-legume forage crops and superior among cereals. It is considered an ideal forage due to its rapid growth, high yields, palatability, rich nutrient content, and its ability to enhance body weight and milk quality in cattle (Sattar et al. 1994). As livestock feed, maize is both highly nutritious and sustainable (Hukkeri et al. 1977; Chaudhary et al. 2014) reported that green fodder maize contains 7–10% crude protein and 52–68% in vitro matter digestibility when harvested at 55–65 days. Alongside high biomass, quality traits such as dry matter, crude protein (CP), crude fat (CF), neutral detergent fibre (NDF), acid detergent fibre (ADF), ash content, lignin content, and in-vitro dry matter digestibility (IVDMD) must also be considered. To meet the growing demand for green fodder, high-biomass maize cultivars have been developed in India. Popular examples include African Tall (Mahatma Phule Krishi Vishwavidyalaya, Rahuri, 1983), J 1006 (Punjab Agricultural University, Ludhiana,

1992), and PMC 6 (Maharana Pratap University of Agricultural & Technology, Udaipur, 2009).

Specialty corn, such as baby corn and sweet corn, is a good source of green fodder, since the plant remains green after cob harvesting. Specialty corns are particularly popular in peri-urban areas due to their dual utility in both food and fodder systems. Baby corn is highly nutritious, classified as a vegetable, and considered a high-value agricultural product in national and international markets. Its strong export potential further encourages widespread cultivation. Similarly, sweet corn is widely consumed in hotels, restaurants, and food service industries, with numerous value-added products derived from it gaining popularity for human consumption. Beyond their market value, specialty corn cultivation generates substantial quantities of green biomass after picking the cobs, which can be efficiently utilized as high-quality animal fodder, thereby strengthening both food and feed security (Chaudhary et al. 2014). Additionally, the cob husks of baby corn and sweet corn also serve as a nutrient source of green fodder (Kumar et al. 2025).

17.3 Maize as Silage

Fodder conservation is essential for efficient livestock farming, ensuring a steady supply of quality feed during scarcity periods and preserving surplus fodder. It prevents over-maturity and maintains forage availability throughout the year. Ensilage, the controlled anaerobic fermentation of forage by natural microorganisms (primarily lactic acid-producing bacteria), effectively preserves nutrients, suppresses harmful microbes, and minimizes losses, making it the most preferred conservation method (Guo et al. 2023). The preserved fodder becomes ready for feeding after 40–45 days. Silage represents the stored form of harvested crops, prepared through anaerobic fermentation in which soluble carbohydrates are converted into lactic acid, similar to the process used in pickle or achaar. Once prepared, silage can be stored for long durations and fed to animals during lean periods when green fodder is scarce. Maize silage is particularly versatile, suitable not only for cattle but also for sheep, goats, and camels. Among available fodder crops, maize is considered the most suitable for silage due to its high biomass, dry matter yield, and nutritional value (Chaudhary et al. 2014). Maize silage is produced by harvesting whole maize plants, including cobs, a few weeks before physiological maturity, chopping the crop, and ensiling it in pits, silage bags, or bales. During fermentation, soluble carbohydrates in developing grains and other plant parts are converted into lactic acid, which preserves nutrients and ensures high-quality feed for livestock. An ideal maize silage hybrid should combine high biomass yield, elevated energy content, and superior fibre digestibility. Such genotypes can be developed by integrating diverse traits—including high starch, high lysine, high oil, floury, waxy, low lignin, and leafy mutants—which are otherwise available in isolation (Dolstra and Miedema 1986). In general, dual-purpose or normal corn hybrids possess high grain production potential but only moderate fibre digestibility, limiting their suitability for silage. Therefore, targeted breeding for silage-specific hybrids is essential. This approach has already been successfully implemented in countries such as Canada, the USA, and China, but remains underexplored in India, highlighting the need for focused research and breeding programs to develop maize hybrids optimized for silage production.

High fodder biomass in maize plays a crucial role in livestock feeding; however, quality must be equally emphasized to ensure higher digestibility and net energy availability. Alongside biomass-related traits such as plant height, number of leaves, leaf length and width, and stem girth, several quality traits are critical. These include pH, crude protein content, dry matter yield, moisture content, crude fibre, silica content, starch content, aflatoxin content, ash content, and in-vitro digestibility parameters such as neutral detergent fibre (NDF), acid detergent fibre (ADF), lignin content, dry matter intake, dry matter digestibility, and overall feed value. Good-quality maize silage is assessed by nutrient composition and fermentation characteristics. Ideally, dry matter (DM) content should be 30–35%, allowing the material to pack tightly and ferment properly. Crude protein (CP) content generally ranges from 7.2 to 10%, depending on the hybrid and harvest stage. Fibre components are equally important: ADF should fall between 23.6 and 33.2%, while NDF is optimal in the range of 41.3–54.1%, ensuring efficient digestibility. Energy availability is reflected through total digestible nutrients (TDN), ideally 66.8–70.9% in high-quality silage. Fermentation quality is another key indicator: well-fermented silage typically shows a pH of 3.7–4.2, low ammonia-nitrogen (5–7% of total N), and a lactic acid concentration of 4–7% of DM, with acetic acid maintained at 1–3%. When maize silage meets these ranges, it is considered nutritious, well-fermented, and suitable for sustaining high livestock performance.

17.4 Hydroponics Maize Fodder

Maize hydroponics is gaining popularity due to its low space and water requirements and its high nutritional value. Hydroponics is a modern plant cultivation technique that eliminates the need for soil, instead relying on water or nutrient-rich solutions. This approach minimizes water usage, reduces labour inputs, and enables year-round production independent of climatic conditions. The term "hydroponics" originates from the Greek words hydro (water) and ponos (work), literally meaning "water working" (Arif et al. 2023). In this system, a single kilogram of maize grain can yield 4–8 kilograms of fresh green fodder within 8–9 days, with plants forming a compact root mat reaching 25–28 cm in height (Rajkumar et al. 2018). Hydroponic fodder, also referred to as sprouted fodder or sprouted grains, is produced by germinating seeds in controlled environments without soil (Dung et al. 2010). The technology has remarkable production capacity, capable of generating up to 1 ton of maize feed per day from an area of just 45–50 m^2 (Ma et al. 2024; Naik et al. 2013). This makes hydroponic maize a sustainable and efficient option for meeting livestock feed demands under resource-limited conditions.

17.5 Approaches for Genetic Improvement of Fodder Maize

17.5.1 Traditional Breeding Approaches

The primary objective of plant breeders is to broaden the genetic base of maize, thereby enhancing the potential for crop improvement through heterotic breeding strategies. Valuable genetic variation can be explored in maize landraces, wild relatives, and improved populations, providing desirable alleles associated with agronomic and quality traits. One key challenge in maize fodder is the high lignin content, which adversely affects overall digestibility. Conventional breeding approaches have been effective in modifying lignin composition to improve cell wall digestibility through selection and backcross breeding. For example, *brown midrib* (*bmr*) maize, characterized by reduced lignin content, has been developed in the USA (population WGRCOMP C2), while the *gt1* (*grassy tiller 1*) gene has been identified to enhance forage production (Coors et al. 1994). In India, GBPUAT, Pantnagar, released the first interspecific hybrid DFH 2, derived from a cross between wild *Zea mays* ssp. *parviglumis* and a cultivated maize inbred line, which exhibited tillering and high biomass production. Selective breeding via recurrent selection can progressively increase forage yield and improve quality attributes. Moreover, several mutants are available that contribute to high biomass production and enhanced nutritional quality in maize silage and green fodder, offering promising resources for future breeding programs.

17.5.2 Genetic Mutants for Maize Fodder and Silage Enhancement

17.5.2.1 Brown Midrib

Several low-lignin genetic mutants have been identified in maize, offering significant potential for improving silage quality by reducing lignin content, which is inversely correlated with digestibility (Coors et al. 1994). The first *brown-midrib* (*bm*) mutant was discovered at the University of Minnesota in 1924 (Jorgenson 1931) and later designated as *bm1*. Subsequently, three additional mutants—*bm2*, *bm*3, and *bm4*—were identified (Burnham and Brink 1932; Emerson et al. 1935; Burnham 1934). These mutations arose naturally and exhibit simple recessive Mendelian inheritance. Among them, *bm3* has proven to be the most widely utilized in maize fodder breeding programs, owing to its consistent ability to reduce lignin content and enhance cell wall digestibility, thereby improving the nutritional quality of maize silage.

17.5.2.2 High-Quality Protein Maize

The *opaque2* (*o2*) and *floury2* (*fl2*) alleles are known to enhance lysine and tryptophan concentrations in maize kernels (Mertz et al. 1964), thereby improving the protein quality of corn. These alleles are particularly beneficial for animal feed, especially for monogastric animals such as poultry and swine, which require higher levels of essential amino acids. Beyond grain utilization, there has also been growing interest in incorporating these alleles into corn silage breeding programs, as improved amino acid profiles could enhance the nutritional value of silage and contribute to better livestock performance. Both alleles follow a simple recessive inheritance pattern but are influenced by genetic modifiers (Vasal 2001).

17.5.2.3 High Oil

The oil content in maize hybrids generally ranges from 3.5% to 5.0%, whereas varieties with more than 6% oil are classified as high-oil maize (Watson 1987). Owing to its elevated oil concentration, high-oil corn possesses greater energy density, since oil provides approximately 2.5 times more calories than carbohydrates (Alexander 1988). As a result, feeding high-oil corn has been shown to enhance daily weight gain and improve feed efficiency in poultry, swine, and dairy cattle, making it a valuable resource for livestock nutrition.

17.5.2.4 Leafy

The leafy mutant in maize is a co-dominant mutation widely utilized in the development of silage-specific hybrids. It is particularly characterized by the presence of 8–12 leaves above the ear, compared to only 4–6 leaves in normal cultivated maize (Shaver 1983). These additional leaves enhance photosynthate production, thereby increasing both grain yield and whole-plant biomass. Another advantage of the leafy mutant is its ability to reduce lignin content in the lower part of the plant, as the majority of the green biomass is concentrated above the cob. This trait contributes to improved digestibility and overall quality of maize silage.

17.5.2.5 Waxy Endosperm

Maize starch is typically composed of approximately 75% amylopectin, consisting of highly branched chains of glucose molecules, and 25% amylose, which is formed by straight chains of glucose molecules (Coors et al. 1994). The waxy phenotype, controlled by the recessive wx1 allele, results in the almost complete replacement of amylose with nearly 100% amylopectin (Alexander 1988). In theory, the increased branching of amylopectin enhances starch digestibility, making waxy maize particularly valuable for both food and feed applications.

17.5.3 Genomics-Assisted Breeding

Conventional breeding remains a valuable tool but is often time-consuming in the current era. To accelerate genetic improvement, Collard and Mackill (2008) demonstrated the use of molecular markers for efficiently identifying and transferring desirable traits into crop plants. This approach is particularly useful for marker-assisted backcrossing as well as advanced strategies such as marker-assisted recurrent selection and genomic selection. Marker-assisted selection (MAS) enables precise detection of specific genes or alleles within plant lines, cultivars, and breeding populations, substantially reducing the time required for selection compared to conventional phenotypic methods (Wani et al. 2018). Advancements in marker discovery and gene annotation have accelerated due to powerful data-mining resources such as POPcorn, MaizeGDB, and Panzea. Modern genotyping platforms, including SNPs and high-resolution mapping panels like the Nested Association Mapping panel, Multi-parent Advanced Generation Inter-Cross, offer great potential for strengthening genome-wide association studies on biomass yield and quality in maize.

Molecular markers also help dissect genomic regions underlying complex traits. For instance, several studies have identified *bmr* mutants, which exhibit altered lignin composition and reduced lignin concentration, thereby improving cell wall digestibility (Sattler et al. 2010). However, lignin content alone does not fully determine digestibility, highlighting the need for comprehensive QTL analyses to capture the genomic regions influencing both biomass yield and forage quality. Both major and minor quantitative trait loci (QTLs) can be introgressed into elite cultivars through molecular breeding, thereby facilitating the development of superior fodder maize hybrids with enhanced biomass and bioenergy potential (Table 17.1). For example, numerous QTLs have been reported for traits such as plant height (Chen et al. 2011; Ku et al. 2010; Leipner et al. 2008; Liu et al. 2017: Kaeppler et al. 2000; Qiu et al. 2007; Rahman et al. 2011) and shoot dry weight Dolferus et al. 2011; Osman et al. 2013; Liu et al. 2017). Similarly, QTLs associated with quality traits (Table 17.2), including crude protein content, lignin, acid detergent fibre (ADF), neutral detergent fibre (NDF), cell wall digestibility, and starch digestibility, have been identified (Barrière et al. 2008; Wei et al. 2009). Fodder maize can also be exploited for stay-greenness, tillering ability, and regrowth ability (ratooning), ensuring prolonged availability of fresh green fodder. Wild species such as *Zea mays* ssp. *parviglumis* and *Zea mays* ssp. *diploperennis* possess stay-green characteristics that can be introgressed into cultivated maize to develop multi-tiller, stay-green genotypes with enhanced fodder supply (Choudhary et al. 2020).

17.5.4 Genome-Wide Association and Genomic Selection

Genome-wide association studies (GWAS) are highly effective for identifying marker–trait associations, thereby facilitating the selection of desirable traits in maize. Using GWAS, markers associated (Table 17.3) with plant height, stalk diameter, and stover yield have been reported (Mazaheri and Heckwolf 2019; Vinayan et al. 2013; Montesinos-López et al. 2021). Similarly, markers linked to quality parameters such as crude protein, acid detergent fibre (ADF), and in-vitro organic matter digestibility have also been identified (Vinayan et al. 2013).

In the current era, genomic selection (GS) has emerged as a promising approach for improving polygenic traits where marker-assisted selection (MAS) alone is insufficient. Riedelsheimer et al. (2012) demonstrated the utility of genomic prediction in estimating heterosis for biomass-related traits across 570 test crosses derived from 285 diverse dent inbreds and two flint testers. SNP-based genomic prediction achieved a maximum accuracy of 0.81, while metabolite prediction reached 0.80,

Table 17.1 QTLs for green fodder and silage traits in maize using different types of mapping population

S. No.	Fodder traits	Population type	References
1.	Biomass	$F_{2:3}$	Chen et al. (2011), Rahman et al. (2011)
2.	Above ground dry biomass	RILs	Trachsel et al. (2016)
3.	Total biomass and shoot dry weight	RILs	Laza et al. (2006)
4.	Dry matter yield and fresh stover yield	$F_{2:4}$	Wei et al. (2009)
5.	Dry weight	$F_{2:3}$	Rodríguez et al. (2014)
6.	Relative shoot fresh and dry weight	RILs	Liu et al. (2017)
7.	Full fresh and shoot fresh weight	DH	Luo et al. (2019)
8.	Shoot dry weight	$F_{2:3}$	Qiu et al. (2007), Osman et al. (2013)
9.	Shoot dry weight	BC_4F_3	Dolferus et al. (2011)
10.	Shoot dry weight	DH	Li et al. (2021)
11.	Shoot fresh weight	RILs	Cui et al. (2017)
12.	Shoot weight	RILs	Kaeppler et al. (2000)
13.	Straw dry weight	$F_{2:3}$	Wei et al. (2009), Leipner et al. (2008)
14.	Straw dry weight	F_2	Jompuk et al. (2005)
15.	Plant height	$F_{2:3}$	Qiu et al. (2007), Chen et al. (2011), Ku et al. (2010), Leipner et al. (2008), Rahman et al. (2011)
16.	Plant height and ear height	$F_{2:3}$	Wei et al. (2009)
17.	Plant height	RILs	Liu et al. (2017), Kaeppler et al. (2000)
18.	Plant height	DH	Li et al. (2021)
19.	Leaf length	$F_{2:3}$	Ku et al. (2010)
20.	Leaf width	$F_{2:3}$, RILs	Ku et al. (2010; Liu et al. (2017)
21.	Leaf length and width	$F_{2:3}$	Chen et al. (2011), Wei et al. (2009)
22.	Leaf length and width	RILs	Liu et al. (2017)
23.	Leaf area	RILs	Cui et al. (2017)
24.	Leaf area	$F_{2:3}$	Osman et al. (2013)

(continued)

Table 17.1 (continued)

S. No.	Fodder traits	Population type	References
25.	Leaf area	DH	Luo et al. (2019)
26.	Leaf area and angle	RILs	Kaeppler et al. (2000)
27.	Leaf senescence and chlorophyll content	DH	Li et al. (2021)
28.	Leaf area and senescence	RILs	Trachsel et al. (2016)

RILs recombinant inbred lines, *DH* Doubled haploids

underscoring the effectiveness of GS in screening diverse inbreds for superior hybrid development. Furthermore (Massman et al. 2013) highlighted the advantages of genomic selection over marker-assisted recurrent selection, showing that GS can achieve greater genetic gains for fodder biomass improvement in maize. Collectively, these studies demonstrate that GWAS and GS are powerful tools for dissecting complex traits and accelerating the development of high-yielding, high-quality fodder maize hybrids.

17.5.5 Genetic Engineering and Modifications

Genetic engineering has been successfully employed in maize to enhance digestibility and biomass-related traits (He et al. 2003) utilized an antisense gene strategy to develop transgenic maize lines resembling the naturally occurring bm3 (brown midrib) mutant phenotype, characterized by reduced lignin content. This was achieved by down-regulating the activity of caffeic acid O-methyltransferase, a key enzyme in lignin biosynthesis, resulting in lower lignin concentration and improved in vitro enzymatic digestibility. Similarly, RNA interference targeting cinnamyl alcohol dehydrogenase, another critical enzyme in lignin formation, has proven effective in reducing lignin levels and enhancing digestibility (Fornalé et al. 2012). Beyond transgenic approaches, Targeting Induced Local Lesions in Genomes (TILLING), a reverse genetics technique, introduces random point mutations to identify functional alleles. More recently, CRISPR-Cas genome editing has emerged as a powerful tool, enabling precise mutations in bm genes to reduce lignin content (Doudna and Charpentier 2014). Such targeted interventions can facilitate the development of high-biomass, low-lignin hybrids with superior digestibility. Collectively, these molecular tools not only improve lignin composition but also enable accurate trait mapping, thereby accelerating the development of maize hybrids with enhanced forage quality and bioenergy potential.

17.6 Conclusion

Maize improvement specifically for fodder purposes is critical to ensuring sustainable livestock productivity and long-term feed security. While extensive genetic research has focused on maize grain, relatively less attention has been directed

Table 17.2 QTLs for quality traits for green fodder and silage in maize using different types of mapping population

S. No.	Fodder quality traits	Population type	References
1.	In-vitro dry matter digestibility (IVDMD)	$F_{2:3}$, RILs	Wei et al. (2009), Barrière et al. (2008)
2.	Crude protein (CP)	$F_{2:3}$, RILs	Wei et al. (2009)
3.	Acid detergent fibre (ADF)	$F_{2:3}$, RILs	Wei et al. (2009)
4.	Crude protein content, lignin and neutral detergent fibre (NDF)	RILs	Barrière et al. (2001)
5.	Crude protein content, Lignin, ADF, NDF, Cell wall digestibility, starch digestibility, and content	RILs	Barrière et al. (2008)
6.	Lignin, ADF, NDF, and Starch digestibility	$F_{2:3}$	Chen et al. (2011)

Table 17.3 Genome-wide association studies for fodder and silage traits

S. No.	Traits	Association panel size with name	References
1.	Cellulose	AM368	Li et al. (2016)
2.	Hemicellulose		
3.	Lignin		
4.	Plant height	IAP453	Mazaheri and Heckwolf (2019)
5.	Stalk diameter		
6.	CP, ADF, IVOMD	DTMA panel (276)	Vinayan et al. (2013)
7.	p-coumaric acid, ferulic acid, total diferulates, diferulic acid	IAP (276)	López-Malvar et al. (2019)
8.	Stover yield	MAGIC (408)	Lopez-Malvar et al. (2021)

toward fodder improvement. To meet the rising global demand for dairy and meat products, there is an urgent need to employ advanced genetic tools alongside targeted mutants to develop high-biomass maize with superior quality traits that can enhance milk yield and overall livestock performance. Maize, with its exceptional biomass yield, nutritive value, and versatility for both green fodder and silage, remains a cornerstone of modern livestock production systems. An integrated breeding approach—combining conventional selection with molecular tools, genomic prediction, and targeted genetic modifications—offers the most promising pathway for improvement. Exploiting genetic diversity from landraces, wild relatives, and identified mutants can strengthen key traits such as digestibility, stay-green capacity, energy density, and reduced lignin content. Emerging technologies, including CRISPR-based genome editing and genomic selection, provide unprecedented opportunities to develop high-biomass, nutritionally superior maize hybrids tailored specifically for fodder use. Future strategies should emphasize ideotype breeding, region-specific cultivar development, and innovative production systems such as hydroponics, which ensure year-round availability of fresh green fodder. Together, these approaches will secure a consistent supply of high-quality feed, thereby supporting sustainable livestock productivity in the face of increasing global food demand.

References

Ahamed MS, Sultan M, Shamshiri RR, Rahman MM, Aleem M, Balasundram SK (2023) Present status and challenges of fodder production in controlled environments: A review. Smart Agric Technol 3:100080. https://doi.org/10.1016/j.atech.2022.100080

Alexander DE (1988) Breeding special nutritional and industrial types. In: Sprague GF and Dudley JW (Eds) Corn and corn improvement, 3rd edn,, vol 18. American Society of Agronomy, pp 869–880 https://doi.org/10.2134/agronmonogr18.3ed.c14

Alexandratos N, Bruinsma J (2012) World agriculture towards 2030/2050: The 2012 revision. ESA Working Paper No. 12-03. https://doi.org/10.22004/ag.econ.288998

Allen MS, Coors JG, Roth GW (2003) Buxton DR, Muck RE and Harrison JH (Eds) Silage science and technology, ASA, CSSA and SSSA, Madison, WI, vol 42. American Society of Agronomy, pp 547–608

Arif M, Khalaf QA, ur Rehman, A. (2023) Effects of feeding maize hydroponic fodder on growth performance, nitrogen balance, nutrient digestibility, hematology, and blood metabolites of water buffalo calves. Open Veterin J 13(12):1607. https://doi.org/10.5455/OVJ.2023.v13.i12.10

Barrière Y, Gibelin C, Argillier O, Méchin V (2001) Genetic analysis in recombinant inbred lines of early dent forage maize. I. QTL mapping for yield, earliness, starch and crude protein contents. Maydica 46:253

Barrière Y, Thomas J, Dénoue D (2008) QTL mapping for lignin content and cell wall digestibility in maize. Plant Sci 175(4):585–595

Barrière Y, Guillaumie S, Pichon M, Emile JC (2009) Breeding for silage quality traits. In: Cereals. Springer, New York, pp 367–394

Bin L, Fei G, Bai-zhao R, Shu-ting D, Peng L, Bin Z (2021) Lignin metabolism regulates lodging resistance of maize hybrids. J Integr Agric 20(8):2077–2089. https://doi.org/10.1016/S2095-3119(20)63346-9

Burnham CR (1934) Cytogenetic studies of an interchange between chromosomes 8 and 9 in maize. Genetics 19(5):430–447

Burnham CR, Brink RA (1932) Linkage relations of a second brown midrib gene (bm2) in maize. Agron J 24(12):960–963. https://doi.org/10.2134/agronj1932.00021962002400120003x

Chaudhary DP, Jat SL, Kumar R, Kumar A, Kumar B (2014) Fodder quality of maize: Its preservation. In: Chaudhary DP, Kumar S and Singh S (Eds) Maize: Nutrition dynamics and novel uses, Springer India

Chen J, Cai Y, Xu L, Wang J (2011) Identification of QTLs for biomass production in maize under different phosphorus levels. Front Agric China 5(2):152–161. https://doi.org/10.1007/s11703-011-1077-3

Chen J, Xu W, Velten J, Xin Z, Stout J (2012) Characterization of maize inbred lines for drought and heat tolerance. J Soil Water Conserv 67(5):354–364

Choudhary M, Kumar P, Kaswan S, Jat SL (2020) Harnessing tillering ability of *Zea mays* ssp. *parviglumis*. Indian J Agric Sci 90:2317–2322

Collard BC, Mackill DJ (2008) Marker-assisted selection. Philos Trans R Soc B 363(1491):557–572. https://doi.org/10.1098/rstb.2007.2170

Coors JG, Carter PR and Hunter RB (1994) Silage corn. In: Hallauer AR (Ed) Specialty corns, CRC Press, Boca Raton, FL, pp 305–340

Coors JG, Eilert DT, Flannery PJ (2010) Plants and seeds of corn comprising brown midrib and gt1 genes. U.S. Patent No. 7,723,584

Cui TT, He KH, Chang LG, Zhang XH, Xue JQ, Liu JC (2017) QTL mapping for leaf area in maize. J Integr Agric. https://doi.org/10.1016/S2095-3119(16)61524-1

Delgado CL, Rosegrant MW, Steinfeld H, Ehui S, Courbois C (1999) The growing place of livestock products in world food. MSSD Discussion Paper No. 28

Dolferus R, Ji X, Richards RA (2011) Abiotic stress and control of grain number in cereals. Plant Sci 181(4):331–341. https://doi.org/10.1016/j.plantsci.2011.05.015

Dolstra O, Miedema P (1986) Breeding of silage maize. In: Proceedings of the 13th EUCARPIA congress, pp 61–71

Doudna JA, Charpentier E (2014) Genome engineering with CRISPR-Cas9. Science 346(6213):1258096. https://doi.org/10.1126/science.1258096

Dung DD, Godwin IR, Nolan JV (2010) Nutrient content of hydroponic barley sprouts. J Anim Vet Adv 9(18):2432–2436

Emerson RA, Beadle GW, Fraser AC (1935) A summary of linkage studies in maize. Cornell Univ Agr Exp Sta Memoir 180

Falcon WP, Naylor RL, Shankar ND (2022) Rethinking global food demand for 2050. Popul Dev Rev 48(4):921–957

Fornalé S, Capellades M, Encina A (2012) Altered lignin biosynthesis improves bioethanol production. Mol Plant 5(4):817–830. https://doi.org/10.1093/mp/ssr097

Guo X, Xu D, Li F, Bai J, Su R (2023) Roles of lactic acid bacteria in crop silage. Microb Biotechnol 16(1):67–87. https://doi.org/10.1111/1751-7915.14184

He X, Hall MB, Gallo-Meagher M, Smith RL (2003) Improvement of forage quality by downregulation of maize O-methyltransferase. Crop Sci 43(6):2240–2251. https://doi.org/10.2135/cropsci2003.2240

http://www.fao.org/fileadmin/user_upload/lon/HLEF2050_Global_Agriculture.pdf

Hukkeri SB, Shukla NP, Rajput RK (1977) Effect of soil moisture and nitrogen on fodder yield of oat. Indian J Agric Sci 47:204

Jompuk C, Fracheboud Y, Stamp P, Leipner J (2005) Mapping QTLs for chilling tolerance in maize. J Exp Bot 56(414):1153–1163. https://doi.org/10.1093/jxb/eri108

Jorgenson LR (1931) Brown midrib in maize and its linkage relations. Agron J 23:507–510. https://doi.org/10.2134/agronj1931.00021962002300070005x

Kaeppler SM, Parke JL, Mueller SM, Senior L, Stuber C, Tracy WF (2000) Variation among maize inbred lines. Crop Sci 40(2):358–364. https://doi.org/10.2135/cropsci2000.402358x

Karnatam KS, Mythri B, Un Nisa W, Sharma H, Meena TK, Rana P (2023) Silage maize for sustainable animal husbandry. Front Genet 14:1150132. https://doi.org/10.3389/fgene.2023.1150132

Ku LX, Zhao WM, Zhang J, Wu LC (2010) QTL mapping of leaf angle in maize. Theor Appl Genet 121(5):951–959. https://doi.org/10.1007/s00122-010-1364-z

Kuehn CS, Linn JG, Johnson DG, Jung HG, Endres MI (1999) Feeding silages from corn hybrids to dairy cattle. J Dairy Sci 82(12):2746–2755. https://doi.org/10.3168/jds.S0022-0302(99)75531-1

Kumar P, Hundal JS, Bhushan B, Choudhary M, Kumar B, Sharma A (2025) Nutritional quality evaluation of maize by-products. RMA 46:170–175

Kumari S, Hiloidhari M, Naik SN, Dahiya RP (2019) Methane emission assessment from Indian livestock and its role in climate change using climate metrics. In: Hussain S (Ed). In: Climate change and agriculture IntechOpen, pp 1–16

Laza MR, Kondo M, Ideta O, Barlaan E, Imbe T (2006) Identification of QTLs for productivity in rice. Crop Sci 46(2):763–773. https://doi.org/10.2135/cropsci2005.05.0084

Leipner J, Jompuk C, Camp KH, Stamp P, Fracheboud Y (2008) QTL studies of chilling-related traits in maize. Theor Appl Genet 116(4):555–562. https://doi.org/10.1007/s00122-007-0690-2

Lewis AL, Cox WJ, Cherney JH (2004) Hybrid and cutting height interactions in corn forage. Agron J 96(1):267–274. https://doi.org/10.2134/agronj2004.2670

Li Z, Li K, Yang X, Hao H, and Jing HC (2021) Combined QTL mapping and association study reveals candidate genes for leaf number and flowering time in maize. Theor. Appl. Genet. 134, 3459–3472. https://doi.org/10.1007/s00122-021-03907-x

Li K, Wang H, Hu X, Liu Z, Wu Y, Huang C (2016) GWAS of stalk cell wall components in maize. PLoS One 11(8):e0158906. https://doi.org/10.1371/journal.pone.0158906

Liu G, Hou P, Xie R, Ming BO, Wang K, Xu W, Li S (2017) Canopy characteristics of high-yield maize with yield potential of 22.5 Mg ha^{-1}. Field Crop Res 213:221–230. https://doi.org/10.1016/j.fcr.2017.08.011

López-Malvar A, Butrón A, Samayoa LF, Figueroa-Garrido DJ, Malvar RA, Santiago R (2019) GWAS for maize stem cell wall traits. BMC Plant Biol 19(1):519. https://doi.org/10.1186/s12870-019-2135-x

Lopez-Malvar A, Malvar RA, Butron A, Revilla P, Pereira-Crespo S, Santiago R (2021) Genetic dissection of forage digestibility traits. Agronomy 11(1):104. https://doi.org/10.3390/agronomy11010104

Luo M, Zhang Y, Chen K, Kong M, Song W, Lu B, Zhao J (2019) Mapping QTLs for salt tolerance in maize. Mol Breed 39(5):64. https://doi.org/10.1007/s11032-019-0974-7

Ma Y, Guo T, Zhang Z, Amat G, Jing Y, Tuo Y, Hou L (2024) Feeding hydroponic barley seedlings to ewes. Front Veterin Sci 10:1280544. https://doi.org/10.3389/fvets.2023.1280544

Massman JM, Gordillo A, Lorenzana RE, Bernardo R (2013) Genomewide predictions from maize single-cross data. Theor Appl Genet 126(1):13–22. https://doi.org/10.1007/s00122-012-1955-y

Mazaheri M, Heckwolf M (2019) Genome-wide association analysis of stalk biomass and anatomical traits in maize. BMC Plant Biol 19(1):45. https://doi.org/10.1186/s12870-019-1653-x

Mertz ET, Bates LS, Nelson OE (1964) Mutant gene increasing lysine in maize endosperm. Science 145(3629):279–280. https://doi.org/10.1126/science.145.3629.279

Montesinos-López A, Runcie DE, Ibba MI, Pérez-Rodríguez P, Montesinos-López OA, Crespo LA (2021) Multi-trait genomic prediction in wheat. G3 11(10):jkab270. https://doi.org/10.1093/g3journal/jkab270

Naik PK, Dhuri RB, Karunakaran M, Swain BK, Singh NP (2013) Hydroponics technology for green fodder production. Indian Dairyman 65:54

Osman KA, Tang B, Wang Y, Chen J, Yu F, Li L (2013) Dynamic QTL analysis for waterlogging tolerance in maize. PLoS One 8(11):e79305. https://doi.org/10.1371/journal.pone.0079305

Prasanna BM, Hoisington DA (2003) Molecular breeding for maize improvement. Indian J Biotechnol 2(1):85–98

Qiu F, Zheng Y, Zhang Z, Xu S (2007) Mapping QTLs for waterlogging tolerance. Ann Bot 99(6):1067–1081. https://doi.org/10.1093/aob/mcm055

Rahman H, Pekic S, Lazic-Jancic V, Quarrie SA, Shah SM, Pervez A, Shah MM (2011) Molecular mapping of drought tolerance QTLs in maize. Genet Mol Res 10(2):889–901

Rajkumar G, Reddy BS, Rao KR (2018) Hydroponics as an alternative fodder production method. Int J Livestock Res 8(2):1–12

Riedelsheimer C, Czedik-Eysenberg A, Griedetar C et al (2012) Genomic prediction of heterotic traits in maize. Nat Genet 44(2):217–220. https://doi.org/10.1038/ng.1033

Rodríguez VM, Butrón A, Rady MO, Soengas P, Revilla P (2014) QTLs for cold stress response in maize. Mol Breed 33(2):363–371. https://doi.org/10.1007/s11032-013-9955-4

Sattar MA, Haque MF, Rahman MM (1994) Intercropping maize with broadcast rice. Bangladesh J Agric Res 19:159–164

Sattler SE, Funnell-Harris DL, Pedersen JF (2010) Brown midrib mutations in cereals. Plant Sci 178(3):229–238. https://doi.org/10.1016/j.plantsci.2010.01.001

Shaver DL (1983) Genetics and breeding of maize with extra leaves. In: Proceedings of the annual Corn and Sorghum research conference, vol 38, pp 161–180

Trachsel S, Sun D, SanVicente FM (2016) Identification of QTLs for early vigor and stay-green in maize. PLoS One 11(3):e0149636. https://doi.org/10.1371/journal.pone.0149636

Vasal SK (2001) High quality protein corn. In: Hallauer AR (Ed) Specialty corn, 2nd edn, CRC Press, pp 85–12

Vinayan MT, Babu R, Jyothsna T, Zaidi PH, Blümmel M (2013) Candidate genomic regions for maize stover quality. Field Crop Res 153:102–106. https://doi.org/10.1016/j.fcr.2013.03.018

Wani SH, Choudhary M, Kumar P, Akram NA, Surekha C, Ahmad P, Gosal SS (2018) Marker-assisted breeding for abiotic stress tolerance. In: Biotechnologies of crop improvement, vol 3. Springer, pp 1–23. https://doi.org/10.1007/978-3-319-94746-4_1

Watson SA (1987) Structure and composition. In: Corn chemistry and technology. American Association of Cereal Chemists (AACC), St. Paul

Wei M, Li X, Li J, Fu J, Wang Y, Li Y (2009) QTL detection for stover yield and quality traits in maize. Plant Physiol Biochem 47(10):886–894. https://doi.org/10.1016/j.plaphy.2009.06.001

18 Maize to Ethanol: Pathways to Sustainable Biofuels

Alla Singh, Mamta Gupta, Gurpinder Singh, and Krishan Kumar

Abstract

Maize is an important crop for bioethanol production due to its high productivity, wide adaptability, and relatively low demand for water and other resources. Globally, maize-based ethanol has contributed to energy security, the reduction of greenhouse gas emissions, and additional income for farming communities. In India, recent policy initiatives under the Ethanol Blended Petrol (EBP) Programme and the National Policy on Biofuels have expanded the use of maize to support the accelerated E20 ethanol blending target. This chapter provides an overview of the role of maize in bioethanol production at global and national levels. It discusses the advantages of maize over other cereal crops and reviews the major pathways for converting maize into bioethanol. First-generation (1G) ethanol production from maize starch and second-generation (2G) ethanol production from maize stover are explained, along with their respective benefits and limitations. Recent advances in enzyme technology, biomass pretreatment, and integrated biorefinery approaches that allow the production of ethanol along with other value-added biochemicals are highlighted. Overall, maize is a climate-resilient and versatile feedstock with strong potential to support long-term bioethanol programs, provided that food security, residue management, and policy support are carefully balanced.

Keywords

Maize · Starch · Enzyme · Bioethanol · Enzyme · Feedstock

A. Singh (✉) · M. Gupta · G. Singh · K. Kumar
ICAR-Indian Institute of Maize Research, Ludhiana, Punjab, India
e-mail: alla.singh@icar.org.in

M. Choudhary et al. (eds.), *Next-Generation Maize Research*,
https://doi.org/10.1007/978-981-92-0722-0_18

18.1 Introduction

Maize or corn (*Zea mays* L.), commonly known as the "Queen of cereals," is a versatile crop with wide-ranging industrial applications. Several value-added products derived from maize hold significant industrial relevance. In recent years, there has been growing interest in its use for bioenergy generation. The establishment of the Global Biofuel Alliance in 2023 highlights the commitment of the Global South toward achieving energy security and sustainability. Extensive research has focused on the conversion of maize residues into biofuels and biochemicals through various bioprocessing technologies. The conversion of maize to bioethanol represents one of the most significant developments in renewable energy over the past few decades. Maize-based ethanol has emerged as a viable pathway toward sustainable transportation fuels and as a viable alternative to fossil fuels globally. Recent advancements have significantly improved production efficiency and environmental performance, with maize-based ethanol demonstrating considerable potential to reduce greenhouse gas emissions while providing economic benefits to agricultural communities. This chapter discusses the multifaceted aspects of maize-to-ethanol conversion, examining its importance, historical development, technological innovations, and future prospects.

18.2 Importance of Maize and Its Relevance for Long-Term Bioethanol Programs

Maize is an established crop in global bioethanol production, particularly in countries seeking to reduce petroleum dependence while supporting agricultural communities. The strategic value of maize for bioethanol production is due to several factors that position it favourably compared to alternative feedstocks. When evaluated against other common bioethanol feedstocks like sugarcane and rice, maize offers several advantages. First, maize requires significantly less water than either sugarcane or rice cultivation, making it a more sustainable option in regions facing water scarcity. Globally in 2019, maize consumed 705 billion m^3 of green water and only 59 billion m^3 of blue water, with 87.1% of its total water footprint coming from rainfall—compared to rice, which required 424 billion m^3 of green and 266 billion m^3 of blue water (only 61.4% green), and sugarcane, which used 222 billion m^3 green and 75 billion m^3 blue (about 74.7% green) (Mialyk et al. 2024). In India, maize's crop water use (CWU) is around 363 mm/year for rainfed and 460 mm/year for irrigated systems, significantly lower than rice (up to 562 mm/year) and sugarcane (up to 1192 mm/year) (Mialyk et al. 2024). This makes maize particularly well-suited for Kharif-season cultivation, where it capitalizes on monsoonal rainfall and minimizes reliance on overdrawn groundwater resources. Its high green water share, lower irrigation burden, and compatibility with ethanol production position maize as a critical crop for advancing climate-smart agriculture and ensuring sustainable biofuel pathways in India and beyond.

In a controlled field study in the Philippines, flooded rice emitted up to 121.9 kg CH_4-C ha^{-1}, resulting in the highest seasonal global warming potential (GWP) of 4.26 Mg CO_2-eq ha^{-1} (zero-N treatment), predominantly due to methane emissions from anaerobic soils (Weller et al. 2015). In contrast, maize systems emitted only 0.96 kg CH_4-C ha^{-1} and 3.95 kg N_2O-N ha^{-1}, with a significantly lower GWP of 1.76 Mg CO_2-eq ha^{-1} under conventional nitrogen fertilization. Most notably, yield-scaled GWP was lowest in maize (<300 kg CO_2-eq Mg^{-1} grain) compared to flooded rice (450–1250 kg CO_2-eq Mg^{-1} grain) and aerobic rice (~400 kg CO_2-eq Mg^{-1} grain), due to maize's higher yield potential (>10 Mg ha^{-1}) and reduced CH_4 emissions (Weller et al. 2015). These results support the promotion of rice–maize rotations and conservation agriculture-based maize systems as climate-smart alternatives in South and Southeast Asia, particularly under increasing irrigation water constraints. To maximize sustainability, policies should simultaneously address soil carbon conservation, as long-term aerobic conditions may increase the risk of Soil Organic Carbon depletion if residue management is inadequate. From a production flexibility standpoint, maize offers considerable advantages. It can be grown across diverse climatic conditions and has shorter crop cycles than sugarcane, allowing for more responsive production adjustments based on market demands. Additionally, maize provides greater flexibility in processing timing since harvested grain can be stored for extended periods, unlike sugarcane, which must be processed promptly after harvest to prevent sugar degradation. Schauberger et al. (2022) analysed more than a hundred years of data of ten crops being cultivated in France, and showed that unlike other crops, maize has not reached a yield plateau, implying scope for further improvements in productivity. This is essential to further reduce the carbon footprint of the bioethanol produced from it. Deutsch et al. (2018) have simulated a 2 °C rise in temperature and the crop yield losses due to increase pest infestation scenario. The authors mentioned that in the case of South & South East Asia, the crop losses for maize are expected to be around 7%, much less than rice, whose yield is expected to drop by as much as 59%. All the above characteristics position maize as a preferred feedstock for a long-term bioethanol program.

18.3 The Global Evolution of Maize Bioethanol

Erenstein et al. (2022) have reviewed the production, consumption, and trade trends of maize globally. The authors mentioned that in comparison to rice and wheat, maize is more versatile due to its multiple uses. The percent relative change in the production of maize from 1995 to 2019 is 118%, while the same is 41% and 39% for paddy and wheat, respectively (Fig. 18.1).

The use of ethanol as a fuel has a surprisingly long history. As early as 1826, ethanol powered the first internal combustion engine. In 1876, Nicolaus Otto used ethanol in his pioneering four-cycle engines. Henry Ford's iconic Model T, introduced in 1908, was designed to run on ethanol, demonstrating early recognition of alcohol fuels' potential. However, petroleum's emergence as an abundant, inexpensive fuel source largely sidelined ethanol development for much of the twentieth

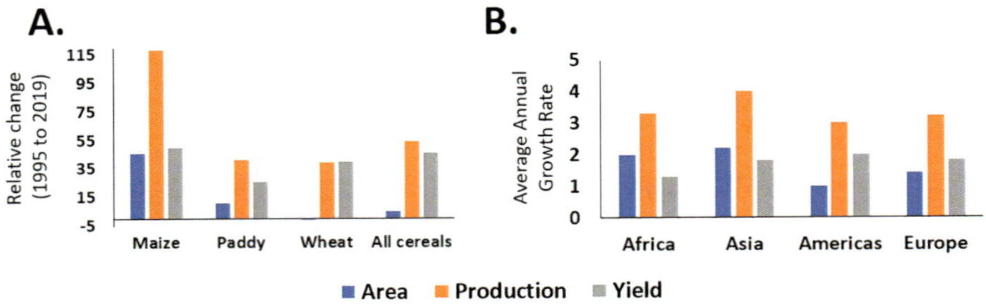

Fig. 18.1 Area, production, and yield trends. (**a**) Relative change in percentage in maize, paddy, wheat and all cereals combined, from 1995 to 2019. (**b**) Average annual growth rate in different regions of the world. Area, production, and yield figs are in million hectares, million tonnes, and ton/hectare, respectively (FAOStat, 2021)

century. The modern maize-ethanol industry took root during the 1970s energy crisis, when petroleum shortages and price spikes renewed interest in alternative fuels. In 1974, Archer Daniels Midland (ADM) began actively promoting ethanol as a gasoline additive, seeing an opportunity to market the ethanol by-product generated during corn syrup production. This corporate initiative coincided with growing national interest in reducing dependency on imported oil, creating favourable conditions for policy support. The 1978 Energy Tax Act provided the first significant tax exemption for ethanol blends, marking the beginning of sustained policy support for the industry. Throughout the 1980s, federal and state subsidies helped maintain ethanol production despite fluctuating oil prices, establishing the economic foundation for industry growth. During this period, the "Minnesota Model" emerged, in which farmer cooperatives developed ethanol plants to create markets for their corn and keep profits within rural communities. Ethanol's role expanded significantly in the 1990s when it emerged as an oxygenate to improve gasoline combustion and reduce carbon monoxide emissions. The phase-out of competing oxygenate MTBE due to groundwater contamination concerns further bolstered ethanol demand. However, the transformative policy development came with the Renewable Fuel Standard (RFS), established in 2005 and expanded in 2007, which mandated increasing volumes of renewable fuels in the U.S. transportation supply. These policies led to a dramatic rise in the production of corn ethanol in the United States.

Brazil's foundational ethanol policy, Proalcool, launched in 1975, initially focused exclusively on sugarcane. Since 2017, with the opening of the country's first maize-based bioethanol plant in Mato Grosso, a joint venture between Tapajós Participações S.A. and Summit Agricultural Group, the share of maize in bioethanol has increased. This integration has been particularly successful in "flex" plants that process sugarcane during its harvest season and switch to maize during the off-season, optimizing capacity utilization throughout the year. China initiated its bioethanol program in 2001, initially targeting surplus grain stocks in the national reserve. The 10th Five-Year Plan (2001–2005) established pilot programs for blending ethanol with gasoline, expanding to cover 18 cities and 5 counties across 11 provinces by 2004. However, China's approach has been notably cautious regarding food security implications. During the global food price spike of 2007–2008, China implemented the "Limitation Policy" restricting grain processing to 26% of domestic demand and strictly controlling the development of grain-based ethanol facilities. The European Union has developed a distinctive approach characterized by stringent sustainability criteria. While the EU's ethanol program utilizes multiple feedstocks, including wheat, barley, and sugar beets alongside limited maize, its regulatory framework emphasizes lifecycle greenhouse gas reductions and strict protections against indirect land use change. This sustainability-focused approach has influenced global standards and certification systems while driving research into advanced biofuels with superior environmental performance.

18.4 India's Ethanol Blending Program and Maize Integration

The National Policy on Biofuels of the Government of India is the guiding document for the development of the bioethanol complex in the country. In 2020, the National Biofuel Coordination Committee gave approval for the utilization of maize for ethanol production in the country. In addition, the interest subvention scheme was extended to grain-based distilleries for the strengthening of ethanol production capacity. Efforts were undertaken by Oil and Marketing Companies to enhance their ethanol storage capacity. Due to the efforts made, the country achieved the E10 target and advanced the E20 target from 2030 to 2025. Another document, Roadmap for Ethanol Blending in India 2020–25, prominently mentions the potential role of maize for meeting the upcoming bioethanol demand of the country, including second- generation ethanol. The Ethanol Blended Petrol (EBP) Program was launched in January 2003, initially targeting 5% ethanol blending in nine states and four union territories. By 2006, the Ministry of Petroleum and Natural Gas extended this 5% target to 20 states, though implementation proceeded slowly due to multiple challenges, including "non-inclusion of conversion of grain to ethanol, high taxation of ethanol, limited availability of feedstock and infrastructure."

A transformative shift began in the last decade with multipronged reforms to boost Ethanol usage aligned with energy security, climate change mitigation, and rural economic development goals. The program initially focused on sugarcane molasses and juice as primary feedstocks, reflecting India's position as the world's second-largest sugarcane producer. Maize ethanol will benefit from the increasing biorefineries network of the country, which includes grain-based and dual-purpose ethanol plants. The already installed distilleries are concentrated in the North Western Plain Zone and central India, especially Maharashtra. The upcoming capacity is witnessing good potential in Eastern India as well (Fig. 18.2).

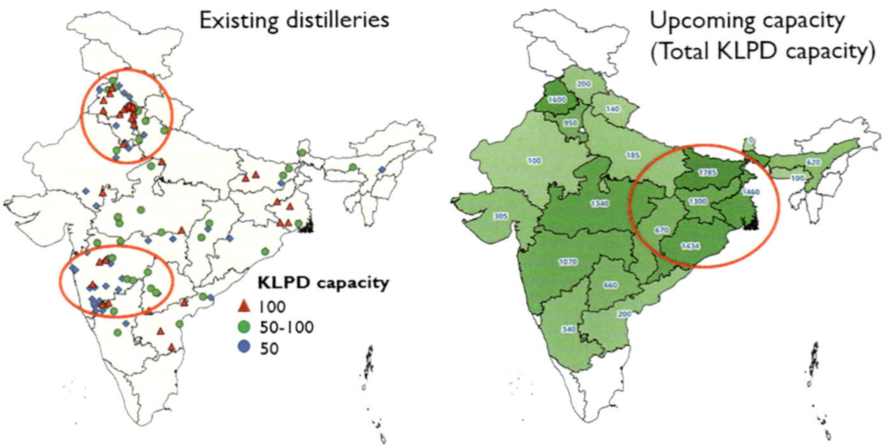

Fig. 18.2 Existing and upcoming capacity of distillation in India (NAAS 2024 Strategy Paper No. 19)

18.5 Conversion of Maize to Ethanol and Other Biochemicals

18.5.1 First-Generation Bioethanol (1G): From Maize Grain (Starch)

Figure 18.3a shows the first-generation route, where maize flour, rich in starch (a polymer), is used as the primary feedstock. The starch is hydrolyzed into fermentable sugars (monomers) using amylase enzymes under high-temperature conditions (around 90 °C for 1–3 h). This thermal pretreatment step necessitates the use of plate heat exchangers for cooling, adding to the capital expenditure (Capex) and increasing the carbon footprint of the process. The starch hydrolysis process has undergone significant optimization, with modern enzymatic approaches replacing traditional thermal methods. Advanced enzymatic systems now employ optimized α-amylase and glucoamylase combinations that can achieve 96–98% glucose yields within 8 h at temperatures of 68–72 °C (Khvedelidze et al. 2019). The optimal enzyme ratios involve 60 units of glucoamylase and 30 units of α-amylase activity per gram of starch, with mesophilic α-amylase (0.6 g/kg), glucoamylase (1.0 g/kg), and raw amylase (0.8 g/kg) providing the highest filtration rates and conversion efficiency. Following hydrolysis, the released sugars are subjected to fermentation under mesophilic conditions (28–37 °C, 150 rpm, 1–3 days) to yield bioethanol. This approach is commercially mature and widely adopted due to high ethanol yields and established infrastructure, but it competes with food and feed uses of maize, raising concerns about food security.

Starch structure optimization reveals that waxy corn varieties demonstrate superior conversion efficiency compared to normal corn (Yangcheng et al. 2013). Waxy corn achieves starch-ethanol conversion efficiencies of 93.0–93.2%, substantially

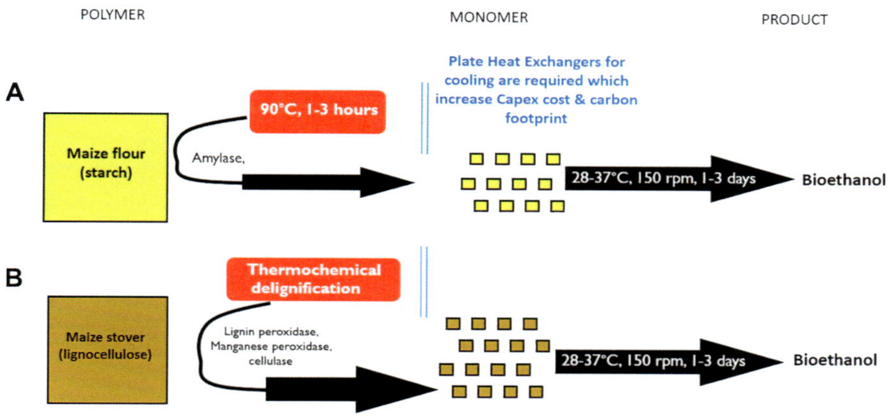

Fig. 18.3 Conversion of maize to ethanol using grains (first generation) and non-grain lignocelluloses (second generation) methodologies

higher than normal corn's 88.0–88.4%, primarily due to greater starch hydrolysis rates where >90% of waxy corn starch is hydrolyzed compared to <80% for normal corn. This improved performance results from the absence of amylose–lipid complexes and reduced starch retrogradation in waxy varieties. Corn fibre utilization within first-generation systems has shown remarkable potential for yield improvement (Gao et al. 2022). Novel lignocellulolytic enzyme cocktails with strong substrate specificity for high-viscosity, high-sugar corn mash can increase ethanol concentration by approximately 12% compared to conventional processing. The in-situ conversion of cellulose in corn fibre during SSF decreases fermentation broth viscosity by 40.9% while reducing residual sugars by 56.3%, with 75% of the ethanol yield increase attributed to cellulose hydrolysis and 25% to enhanced starch utilization.

18.5.2 Second-Generation Bioethanol (2G): From Maize Stover (Lignocellulose)

The second-generation pathway, which utilizes maize stover—the non-edible, lignocellulosic residue left after grain harvest (Fig. 18.3). This feedstock comprises complex polymers like cellulose, hemicellulose, and lignin, making it more recalcitrant to enzymatic breakdown. To release fermentable sugars, thermochemical delignification is performed using a cocktail of enzymes such as lignin peroxidase, manganese peroxidase, and cellulase. These help in breaking down lignin and freeing up cellulose for subsequent enzymatic saccharification. After delignification, the monomeric sugars are fermented under the same conditions (28–37 °C, 150 rpm, 1–3 days) to produce bioethanol. Although 2G bioethanol is more sustainable—since it utilizes agricultural waste and avoids food-feed conflict—it requires more complex pretreatment steps and enzymatic formulations, making it technologically intensive and currently less cost-competitive. Deep eutectic solvents represent a breakthrough in sustainable pretreatment technology, offering high delignification efficiency (>90% lignin removal) under mild conditions (Wang and Lee 2021). Natural deep eutectic solvents (NADESs) provide renewable and biodegradable alternatives to conventional ionic liquids, with choline chloride: glycerol (ChCl:Gly) systems enabling continuous high-solid pretreatment at loadings up to 50%. These systems achieve glucose and xylose yields >85% from enzymatic saccharification while preserving lignin structural characteristics and avoiding significant chemical composition changes.

Ionic liquid pretreatment using biomass-derived compounds offers a "closed-loop" biorefinery approach. Ionic liquids derived from vanillin, p-anisaldehyde, and furfural, major by-products of lignocellulosic processing, achieve 90–95% glucose and 70–75% xylose yields after 72-h incubation (Wang et al. 2017). Similarly, pentose utilization optimization addresses the critical challenge of efficiently converting C5 sugars (xylose and arabinose) present in hemicellulose. Engineered *Saccharomyces cerevisiae* strains incorporating heterologous pentose pathways can utilize both oxidoreductase and isomerase routes for xylose and arabinose conversion (Young et al. 2010). The oxidoreductase pathway employs *Pichia stipitis* xylose reductase (XYL1) and xylitol dehydrogenase (XYL2), while cofactor imbalances limiting theoretical yields can be addressed through metabolic engineering of the pentose phosphate pathway.

Apart from this, many other biochemicals can be produced from maize. Residues comprising maize stover (stalks, leaves, husks) and cobs—are an abundant lignocellulosic feedstock containing approximately 30% cellulose, 30% hemicellulose, and 17% lignin, and can be sustainably converted into second-generation bioethanol through a sequence of pretreatment, hydrolysis, fermentation, and distillation steps (Da Rosa et al. 2024; Sun et al. 2021). Pretreatment is essential to overcome biomass recalcitrance: dilute acid or alkaline hydrolysis, steam explosion, hydrothermal treatments, and enzymatic/biological approaches disrupt the lignin–carbohydrate matrix and reduce cellulose crystallinity, improving enzyme accessibility during saccharification (Joshi and Manjare 2024; Pan et al. 2006). Following pretreatment, a tailored cocktail of cellulases and hemicellulases cleaves cellulose into glucose and hemicellulose into pentoses; for example, simultaneous saccharification and fermentation of corn husk by a coculture strategy has been shown to streamline these steps, enhancing ethanol titres and mitigating inhibitory byproduct accumulation (Sharma et al. 2018; Shafiei Alavijeh et al. 2020). Typical ethanol yields range from 200 to 300 L per tonne of dry stover, with Shafiei Alavijeh et al. (2020) reporting 214 L ethanol from a dry 1 t maize stover basis after acid pretreatment and enzymatic hydrolysis. Beyond standalone ethanol production, integrated biorefinery designs leverage all major biomass fractions to improve process economics and environmental performance: in one model, acid pretreated and enzymatically hydrolyzed stover sugars are fermented by the fungus *Mucor indicus* to produce bioethanol, glycerol and fungal biomass; lipids extracted from that biomass are then converted to biodiesel, the residual solids generate biogas via anaerobic digestion, and lignin streams provide process steam or higher value aromatics (Alavijeh et al.

2020; Arias et al. 2023). Circular valorization approaches also recover xylooligosaccharides and other specialty chemicals from hemicellulose, while reductive catalytic fractionation of lignin yields phenolic monomers suitable for polymers and resins, ensuring near complete utilization of the feedstock (Da Rosa et al. 2024).

Quantitative sustainability assessments via life cycle analysis (LCA) and techno-economic assessment (TEA) are vital before industrial deployment: the basic LCA framework encompasses maize cultivation, residue collection at sustainable removal rates, transport, pretreatment, hydrolysis, fermentation, product recovery, and waste management (Nicolae et al. 2018; Siol et al. 2023). Case studies show that acid-based pretreatments carry the largest environmental burdens, but integrating renewable energy into downstream processing can substantially reduce fossil resource depletion and greenhouse gas emissions (Hussin et al. 2023; Sun et al. 2021). Comparative LCA of biochemical ethanol versus thermochemical routes (pyrolysis, gasification) demonstrates that microbial fermentation yields up to 50% lower CO_2 emissions than pyrolysis, and that advanced pyrolysis electrocatalytic hydrogenation pathways powered by renewables can even outperform cellulosic ethanol in global warming potential (Sun et al. 2021).

TEA highlights how co-product diversification—such as biomethane, biodiesel, and biopolymers—distributes capital costs across multiple revenue streams and improves internal rates of return under realistic market scenarios (Siol et al. 2023; Joelsson et al. 2016). Key challenges to sustainable scale up include the high costs of enzymes and pretreatment chemicals, complex feedstock logistics (collection, drying, grinding and transport), and the need to manage fermentation inhibitors such as acetic acid and phenolics; mathematical models quantifying inhibitor impacts on microbial kinetics guide pretreatment optimization to balance maximal sugar release with minimal inhibitor formation (Pan et al. 2006; Hussin et al. 2023). Looking forward, advances in deep eutectic solvent pretreatment, consolidated bio-processing (where one engineered microorganism both hydrolyzes biomass and ferments sugars), and pentose utilizing yeast strains promise further cost and energy savings, while coupling precision agriculture for optimized residue availability and blockchain for supply chain traceability will be critical to realizing the full potential of corn residues as a sustainable, low carbon source of bioethanol and allied biofuels (Arias et al. 2023; Da Rosa et al. 2024).

The conversion of corn residue into value-added chemicals as three principal categories—furans, polyols, and organic acids—each derived from the sugar monomers released during lignocellulosic deconstruction (Fig. 18.4). Both hexose (e.g., glucose) and pentose (e.g., xylose) sugars undergo acid-catalyzed dehydration to yield furfural (FF) and 5-hydroxymethylfurfural (HMF), whose established markets and versatile reactivity underpin a variety of downstream oxidation and reduction pathways. Oxidative routes transform FF and HMF into maleic anhydride (MAH) and maleic acid (MA), with FF alone giving rise to fumaric acid (FA), while both intermediates can be further oxidized to 2,5-furandicarboxylic acid (FDCA) and, in the presence of low-molecular-weight alcohols, to furan-2,5-dimethylcarboxylate (FDMC). Conversely, catalytic hydrogenation of FF yields furfural alcohol (FFA), which can be sequentially reduced to tetrahydrofurfuryl alcohol (THFA),

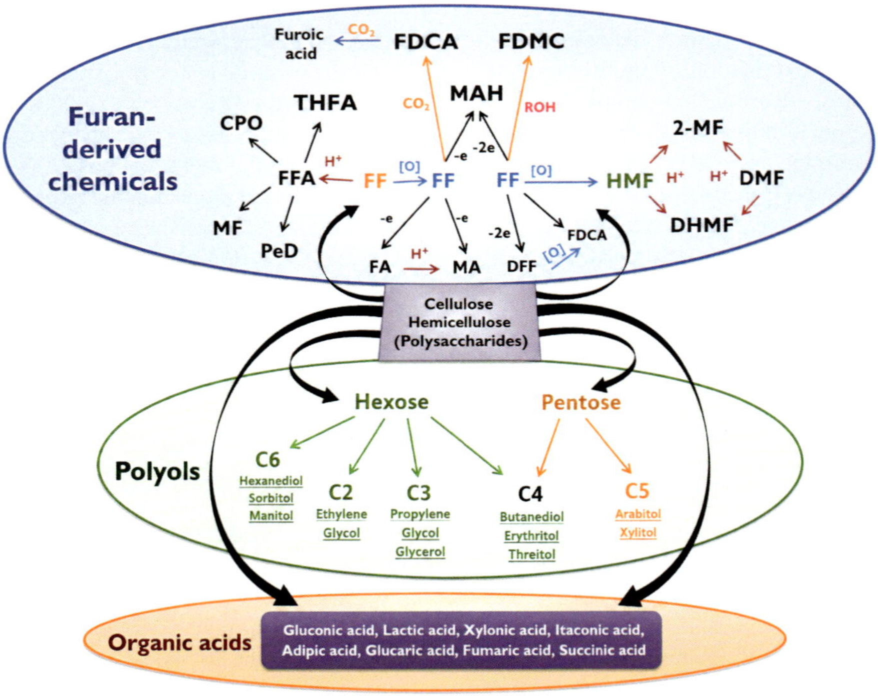

Fig. 18.4 Conversion of corn residue to chemicals

cyclopentanone (CPO), methylfuran (MF), and 1,5-pentanediol (PeD), whereas HMF reduction affords 2-methylfuran (2-MF), 2,5-dihydroxymethylfuran (DHMF), and 2,5-dimethylfuran (DMF). Beyond furan chemistry, a spectrum of C_2–C_6 polyols—including ethylene glycol, propylene glycol, glycerol, sorbitol, mannitol, hexanediol, erythritol, threitol, butanediol, arabitol, and xylitol—can be biosynthetically or catalytically produced from hexoses and pentoses. Finally, both classes of sugars serve as feedstocks for numerous organic acids such as gluconic, lactic, xylonic, adipic, glucaric, levulinic, fumaric, and succinic acids, thereby positioning corn residue as a versatile platform for sustainable chemical manufacture.

Furans, viz., furfural and 5-hydroxymethylfurfural; polyols, viz., C2-C6 alcohols and organic acids can be produced from corn residue. [O] and H^+ represent oxidation and acid-catalyzed transformations, respectively. The addition of other chemicals like carbon dioxide or acid is indicated. In the case of C4 alcohols, 1,2-butanediol can be synthesized from hexose only, while erythritol and threitol can be synthesized by both hexoses and pentoses. Similarly, organic acids can generally be produced by both hexoses and pentoses. FF: Furfural, HMF: 5-hydroxymethylfurfural, MAH: Maleic anhydride, MA: Maleic acid, FA: Fumaric Acid, FDCA: 2,5-furandicarboxylic acid, DFF: 2,5-Diformylfuran, FDMC:

Furan-2,5-dimethylcarboxylate, FFA: Furfural alcohol, MF: Methylfuran, 2-MF: 2-methylfuran, DHMF: 2,5-dihydroxymethylfuran, DMF: 2,5-dimethylfuran, THFA: Tetrahydrofurfuryl alcohol, CPO: Cyclopentanone, PeD: 1,5-pentanediol. Modified from Deng et al. (2023) and enhanced using AI tools.

18.6 Conclusions

The transition from fossil fuels to bio-based alternatives is critical for achieving long-term energy sustainability, and maize stands out as a promising feedstock in this global shift. Both first-generation (1G) and second-generation (2G) bioethanol production pathways from maize offer unique advantages and face specific challenges. The 1G approach utilizes maize flour (starch), benefiting from mature technologies and higher ethanol yields, but it competes with food supply and incurs higher capital and energy costs due to thermal processing and cooling requirements. On the other hand, the 2G pathway leverages maize stover (lignocellulose)—a non-edible agricultural residue—aligning more strongly with sustainability and waste valorization principles. However, it requires complex pretreatment steps and advanced enzymatic processes, making it currently less cost-effective. The launch of the Global Biofuel Alliance 2023 underscores the geopolitical and developmental importance of sustainable biofuels, especially for the Global South. India's inclusion of maize in its ethanol blending roadmap, supported by policy interventions and infrastructure development, highlights the crop's growing significance in national energy security frameworks. Looking ahead, integrated biorefinery models that convert both starch and lignocellulosic components into a suite of biofuels and biochemicals present the most holistic pathway forward. Innovations in enzyme engineering, consolidated bioprocessing, and circular valorization can help overcome existing economic and technological barriers. Ultimately, maize offers a strategic advantage in both food and fuel sectors, and its dual role must be optimized through smart policy, scientific innovation, and sustainability assessments.

References

Alavijeh RS, Karimi K, van den Berg C (2020) An integrated and optimized process for cleaner production of ethanol and biodiesel from corn stover by *Mucor indicus*. J Clean Prod 249:119321

Arias A, Feijoo G, Moreira MT (2023) Biorefineries as a driver for sustainability: key aspects and future prospects. J Clean Prod 418:137925

Da Rosa RG, Ferreira VC, Barroso TLC, Castro LEN, Forster Carneiro T (2024) Corn biomass as a feedstock for biorefinery. Biofuels Bioprod Biorefin 19:206

Deng W, Feng Y, Fu J, Guo H, Guo Y, Han B, Jiang Z, Kong L, Li C, Liu H, Nguyen PT (2023) Catalytic conversion of lignocellulosic biomass into chemicals and fuels. Green Energy & Environment. Feb 1;8(1):10–14

Deutsch CA, Tewksbury JJ, Tigchelaar M, Battisti DS, Merrill SC, Huey RB, Naylor RL (2018) Increase in crop losses to insect pests in a warming climate. Science 361(6405):916–919

Erenstein O, Jaleta M, Sonder K, Mottaleb K, Prasanna BM (2022) Global maize production, consumption and trade: trends and R&D implications. Food Secur 14:1295–1319

Food and Agriculture Organization (FAO) (2021) FAOSTAT Statistical Database. Rome, Italy: FAO
Gao L, Zhang D, Wu X (2022) In situ corn fiber conversion for ethanol improvement by the addition of a novel lignocellulolytic enzyme cocktail. J Fungi 8(3):221
Hussin F, Hazani NN, Khalil M, Aroua MK (2023) Environmental life cycle assessment of biomass conversion using hydrothermal technology: a review. Fuel Process Technol 246:107747
Joelsson E, Dienes D, Kovacs K, Galbe M, Wallberg O (2016) Combined production of biogas and ethanol at high solids loading from wheat straw impregnated with acetic acid: experimental study and techno-economic evaluation. Sustain Chem Proc 4:1–19
Joshi M, Manjare S (2024) Chemical approaches for biomass valorisation: a comprehensive review of pretreatment strategies. Environ Sci Pollut Res 31:48928–48954
Khvedelidze R, Kutateladze L, Sadunishvili T, Zakariashvili N, Darbaidze Z, Aleksidze T (2019) Heat-stable α- and glucoamylase performing deep enzymatic hydrolysis of starch to fermentable glucose. Eur Sci J 15(21):35–44
Mialyk O, Zagaria C, Berger M, Sala S (2024) Water scarcity impacts of major crops: a global green and blue water footprint assessment. J Clean Prod 427:139060
NAAS (2024) Maize for bioethanol production in India: prospects and strategy. Strategy Paper No. 19. New Delhi: National Academy of Agricultural Sciences, 20 pp
Nicolae EC, Camia A, Erenstein O (2018) Sustainable removal rates of maize stover: soil health implications. Waste Manag 30(10):1889–1899
Pan X, Gilkes N, Saddler JN (2006) Effect of acetyl groups on enzymatic hydrolysis of cellulosic substrates. Biotechnol Biofuels 99:423–430
Schauberger B, Kato H, Kato T, Watanabe D, Ciais P (2022) French crop yield, area and production data for ten staple crops from 1900 to 2018 at county resolution. Sci Data 9:38
Shafiei Alavijeh R, Shahvandi A, Okoro OV, Denayer JF, Karimi K (2020) Two-stage acid pretreatment and saccharification of maize stover: ethanol and biodiesel yields. Biomass Bioenergy 161:106447
Sharma S, Sharma V, Kuila A (2018) Simultaneous saccharification and fermentation of corn husk by co-culture strategy. J Pet Environ Biotechnol 9:1000360
Siol C, Thrän D, Majer S (2023) Utilizing residual biomass from agriculture and forestry: LCA and TEA boundaries. Biomass Bioenergy 174:106839
Sun H, Luo Z, Li S, Xue S, Zhou Q, Wei T, Du L (2021) Comparative LCA of bioethanol versus thermochemical biofuel production from corn stover. Biomass Convers Biorefinery 13:1–13
Wang W, Lee DJ (2021) Lignocellulosic biomass pretreatment by deep eutectic solvents on lignin extraction and saccharification enhancement: a review. Bioresour Technol 339:125587
Wang FL, Li S, Sun YX, Han HY, Zhang BX, Hu BZ, Gao YF, Hu XM (2017) Ionic liquids as efficient pretreatment solvents for lignocellulosic biomass. RSC Adv 7(76):47990–47998
Weller S, Kraus D, Butterbach-Bahl K, Wassmann R, Tirol-Padre A, Kiese R (2015) Diurnal patterns of methane emissions from paddy rice fields in The Philippines. J Plant Nutr Soil Sci 178(5):755–767
Yangcheng H, Jiang H, Blanco M, Jane JL (2013) Characterization of normal and waxy corn starch for bioethanol production. J Agric Food Chem 61(2):379–386
Young E, Lee SM, Alper H (2010) Optimizing pentose utilization in yeast: the need for novel tools and approaches. Biotechnol Biofuels 3(1):24

Value-Added Maize Products as Healthy Future Choices for Fast Food

19

Sukhmani Dhaliwal, Mamta Gupta, Ankita Kataria,
Ph. Romen Sharma, Mukesh Choudhary,
and Manesh Chander Dagla

Abstract

The increasing demand for healthier fast-food alternatives has prompted exploration of maize as a versatile, nutritious, and sustainable ingredient for value-added products. Traditional fast foods are often criticized for their high content of saturated fats, refined carbohydrates, and salt, contributing to obesity, diabetes, and cardiovascular diseases. In contrast, maize offers complex carbohydrates, dietary fibre, essential amino acids, vitamins, minerals, and bioactive compounds with antioxidant, anticancer, and anti-inflammatory properties. Its gluten-free nature makes it particularly suitable for individuals with celiac disease or gluten intolerance, while its slow-digesting starch provides sustained energy and satiety. Advanced processing techniques such as extrusion, fortification, fermentation, nixtamalization, and minimal-oil cooking enable the transformation of maize into diverse fast-food options that retain flavour and convenience while enhancing nutritional value. Popular maize-based fast foods include popcorn, nachos, tortillas, cornbread, polenta, tamales, arepas, pretzels, pancakes, waffles, burgers, pizzas, baby corn fries, pasta, cheese spreads, corn nuts, and fritters. These products not only cater to evolving consumer preferences for functional, allergen-free, and low-glycaemic foods but also support agricultural sustainability and economic growth. By integrating maize into mainstream fast-food formulations, the food industry can address public health concerns while offering innovative, culturally diverse, and appealing recipes. This chapter highlights the nutritional benefits of maize, its role in promoting smart functional foods, and its potential to redefine fast food as a healthier, convenient, and sustainable choice for future generations.

S. Dhaliwal · M. Gupta (✉) · A. Kataria · Ph. R. Sharma · M. Choudhary · M. C. Dagla
ICAR-Indian Institute of Maize Research, Ludhiana, Punjab, India
e-mail: mukesh.choudhary1@icar.org.in

M. Choudhary et al. (eds.), *Next-Generation Maize Research*,
https://doi.org/10.1007/978-981-92-0722-0_19

Keywords

Maize · Value-added products · Fast-food · Health · Quality

19.1 Introduction

In the present scenario of global health trends, people are moving towards quick, convenient, tasty, and healthier food options. Traditional fast foods are criticized for saturated fats, refined carbohydrates, sugar, and salt, causing cardiovascular diseases, obesity, diabetes, and certain non-communicable diseases (PIB 2021). This has led to a growing interest in healthier alternatives with better taste. Therefore, as a staple crop in many countries, maize-based value-added fast-food recipes can be a promising solution for versatile, nutritious, and economically viable products. The innovative ideas for maize-based fast foods and advanced processing techniques like extrusion, fortification, fermentation, nixtamalization, and air frying/minimal frying can transform maize into a wide range of food items such as gluten-free wraps, tortillas and tacos, high-fibre snacks and baked products, minimal oil-extruded items, fortified and fermented cereals, maize burgers and nuggets, etc. The maize-based fast-food products is the first choice of consumers due to nutritional benefits because it contains dietary fibres, complex carbohydrates, phytochemicals/vitamins (A, B, E, and K), minerals (Mg, K, P), essential amino acids like lysine and tryptophan (in quality protein maize), a range of antioxidants, carotenoids, phytosterols and phenolic compounds (syringic acid, coumaric acid and ferulic acid) (Siyuan et al. 2018; Félix-Medina et al. 2021; Kumar et al. 2019, 2024). Maize fibre is an excellent source of dietary fibre, and harnessing its prebiotic effects on health has become a promising approach for developing high-fibre foods (Delcour et al. 2016). The complex carbohydrates in maize digest slowly, which provides the sustained energy for the day and maintains fullness for longer periods (Kazerooni et al. 2019). Maize contains B vitamins (thiamine, niacin, and folate), which support energy metabolism, as well as minerals like magnesium and potassium, essential for muscle and nerve function. It is reported that maize has anti-HIV activity due to the presence of *Galanthus nivalis agglutinin* lectin (*GNA* lectin) (Rouf Shah et al. 2016).

Maize has been recognized as a valuable source of bioactive peptides, with potential applications in health and nutrition. These peptides have demonstrated a range of bioactive properties, including anticancer activity (Díaz-Gómez et al. 2017). Maize provides a viable alternative for individuals with gluten intolerance or celiac disease. Polyphenols from pigmented corn help to regulate the inflammatory and oxidative conditions in individuals with celiac disease at the intestinal level (Piazza et al. 2024). In addition to celiac disease, there is a condition called "gluten intolerance," which does not cause damage to the intestinal mucosa but leads to symptoms like bloating, diarrhoea, fatigue, headache, confusion, and joint pain. People with this condition experience a significant decline in their quality of life when consuming gluten-containing products and have to avoid gluten (Jericho et al. 2017). Maize grains, consumed either in their raw or cooked form, provide significant health benefits by supporting the proper functioning of the heart, brain, and digestive system. Additionally, they contribute to making a strong immune system and offer protective effects against elevated blood pressure and cholesterol levels (Saritha et al. 2020). These products can meet the demand of consumers who require foods which have low-glycaemic index, high resistant starch and vitamin B-complex, improve micronutrient and antioxidant concentration, high-satiety value, and are gluten or allergen-free (Meena et al. 2022; Rouf Shah et al. 2016). These benefits offer the food industry to adopt maize for the re-formulation of fast foods to promote them as convenient and healthier alternatives. The step towards popularization of these fast-food products will not only address the health concerns about traditional fast foods, but it will also support agricultural sustainability and the economy.

With this idea, we can promote next-generation healthy maize-based fast-food products labelled as smart functional food, and a sustainable choice of fast food without sacrificing flavour and convenience. This chapter will focus on fast-food products and their recipes, which can be prepared from maize.

19.2 Popular Value-Added Maize-Based Fast-Food Options

19.2.1 Popcorn

Popcorn is a whole-grain food and stands as one of the most widely consumed snack foods across various regions of the world (Soylu and Tekkanat 2007). It is a snack made from kernels of specialty corn, i.e. popcorn, that expand and puff up after heating (Fig. 19.1). This specialty corn pops because its kernels contain a small amount of water inside a soft starch layer, surrounded by a hard outer shell. When heated, the water becomes steam, building up pressure until the shell bursts. This causes the kernel to explode and turn inside out, releasing the steam (Hoseney et al.

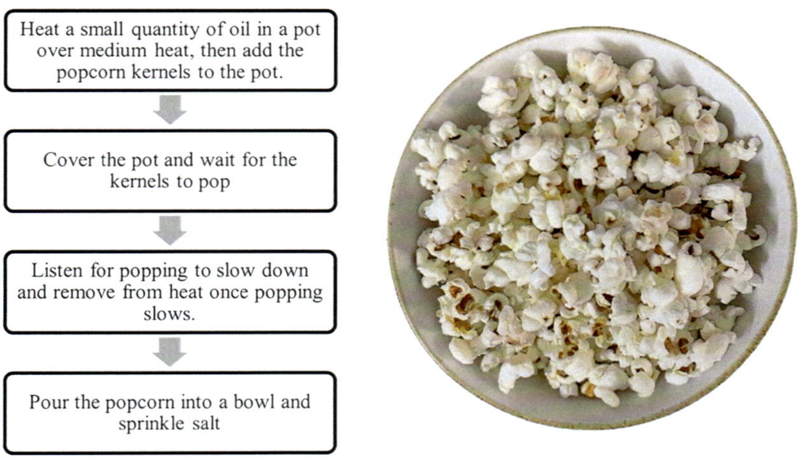

Fig. 19.1 Protocol for popcorn preparation and the resultant product

1983). Popcorn is a rich source of phytonutrients, minerals, and vitamins. Furthermore, it is considered both flavourful and an excellent source of dietary fibre (Chawla et al. 2022). It has been reported that including popcorn as a regular snack in the diet can result in notable enhancements to cardiovascular health indicators (Nguyen et al. 2012).

The two primary types of popcorn commonly consumed are:

(i) *Butterfly (or snowflake)* This variety features irregular, light, and fluffy shapes, making it ideal for movie theatres or pre-packaged popcorn (Fig. 19.1).
(ii) *Mushroom* This variety is round and compact, making it ideal for caramel or coated popcorn due to its durability.

The basic recipe to prepare popcorn has been mentioned (Fig. 19.1) with basic ingredients: popcorn kernels, oil, and salt.

Popcorn has evolved from a simple snack into a highly customizable and diverse food product. While traditional types such as butter and cheese remain popular, the expansion of the gourmet snack market has introduced a wide array of inventive flavour combinations. These include caramel, piña colada, Swiss onion, sriracha lime, cherry almond, and many others. This wide range of flavour profiles reflects the growing consumer demand for innovative snack options, as well as popcorn's versatility as a base for both sweet, savoury, and spicy flavour combinations. This innovation has made popcorn not just a traditional snack but also a medium for creative cooking and specialty foods.

- *Caramel popcorn*: The process of making caramel popcorn closely resembles the preparation of plain popcorn, with the key difference being the addition of melted caramel. After the popcorn is popped, it is coated with the caramel mixture, and the coated popcorn is then baked at 120 °C for 1 h to achieve a crispy texture (Food Crumbles 2021).
- *Piña colada popcorn*: It is a tropical version of regular popcorn, mixing the sweet and creamy flavours of a piña colada with the crunch of popcorn. The process begins by popping the popcorn, followed by coating it with a mixture of melted butter, shredded coconut, powdered sugar, and pineapple juice powder (Johnson 2019).
- *Swiss onion popcorn:* The preparation involves popping the popcorn and then coating it with a mixture of melted butter, Swiss cheese powder, and onion seasoning. This blend of flavours gives the popcorn a creamy, savoury taste with a subtle onion flavour (Kitchen Concoctions 2017).
- *Sriracha lime popcorn*: It is a spicy, tangy snack that mixes the heat of sriracha sauce with the fresh flavour of lime. The process begins by popping the popcorn, after which it is evenly coated with melted butter, sriracha sauce, lime juice, and a pinch of salt. The heat from the sriracha is perfectly balanced with the zesty lime, resulting in a delicious snack (Popcorn.org 2025).

19.2.1.1 Nachos

It is a popular Mexican dish consisting of tortilla chips topped with cheese and other ingredients. They are often served as a snack or appetizer. They can either be baked or fried (Manasa et al. 2024). The nachos have been transformed into a widely popular comfort food, offering countless variations that can be found at restaurants, fast food chains, and even in home kitchens. Their adaptability and ability to cater to diverse tastes have made them a staple on menus around the globe. Traditional nachos feature tortilla chips made from corn tortillas that are fried until crispy. The chips can be topped with melted cheese.

Maize nachos come in various types, generally categorized by the flavour and style of the chips:

(i) *Classic nachos/Traditional nachos:* Simple tortilla chips made from corn tortillas that are fried until crispy. They are mostly topped with melted cheese, typically cheddar or a cheese blend.
(ii) *Loaded nachos*: Chips topped with an assortment of ingredients, such as ground meat, beans, sour cream, jalapeños, guacamole, and salsa, and olives, etc.
(iii) *Baked nachos*: Tortilla chips are baked rather than fried, typically offering a healthier alternative (Fig. 19.2).

The basic recipe to prepare nachos has been mentioned (Fig. 19.2) with basic ingredients: corn flour, oil, salt, and spices.

19.2.1.2 Corn Smoothies

They are a nutritious way to enjoy the natural sweetness of corn. They can be made with fresh or cooked corn kernels and combined with fruits and flavouring agents (Paolicchi 2013). The idea is to create a creamy, slightly sweet drink with the

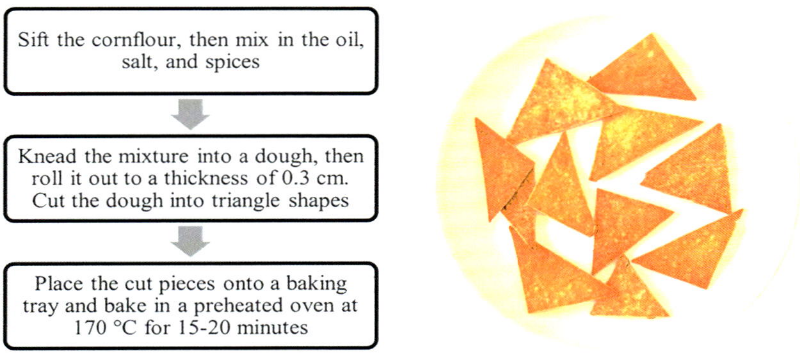

Fig. 19.2 Protocol for preparation of nachos and the resultant product

Fig. 19.3 Protocol for preparation of corn smoothie and the resultant product

natural flavour of corn (Fig. 19.3). Corn smoothies can be prepared in a variety of ways:

(i) *Tropical Maize Smoothie*: A blend of maize, pineapple, mango, and coconut milk as shown in Fig. 19.3.
(ii) *Green Maize Smoothie:* Maize combined with spinach, kale, and other leafy greens for a nutrient-rich drink.
(iii) *Spiced Maize Smoothie:* Maize paired with cinnamon, nutmeg, and ginger for a warm, flavourful twist.

The basic recipe to prepare corn smoothie has been mentioned (Fig. 19.3) with basic ingredients: corn kernels, fruits, lime juice, salt, and flavouring agents.

(i) *Tortillas*

They are flat, round pieces of bread made from corn, commonly used in Mexican and Latin American cuisine to wrap or hold fillings like beans, meat, vegetables, and cheese. They are soft and have a slightly earthy flavour. The process of transforming maize into tortillas, called nixtamalization, starts by cooking the grain in lime, which produces nixtamal. The moist nixtamal is then stone-ground into a dough (masa), which is shaped into disks and cooked on a hot griddle or in a gas-fired oven to make tortillas (Fig. 19.4). Notably, nixtamalization enhances the nutritional profile of tortillas, especially by supplying the essential calcium that is often lacking in the diet (Serna-Saldivar 2015).

There are generally two types of tortillas:

(i) *Traditional Tortillas:* They are made with fresh masa through the nixtamalization process and are considered to have a superior taste to the modern tortillas (Fig. 19.4). These tortillas are made with basic ingredients, without relying on modern shortcuts or processed additives. They are prepared using traditional methods passed down through generations, especially in Mexico and Central America.
(ii) *Modern Tortillas:* Modern tortillas typically refer to commercially produced versions of the traditional tortilla, made using automated processes and syn-

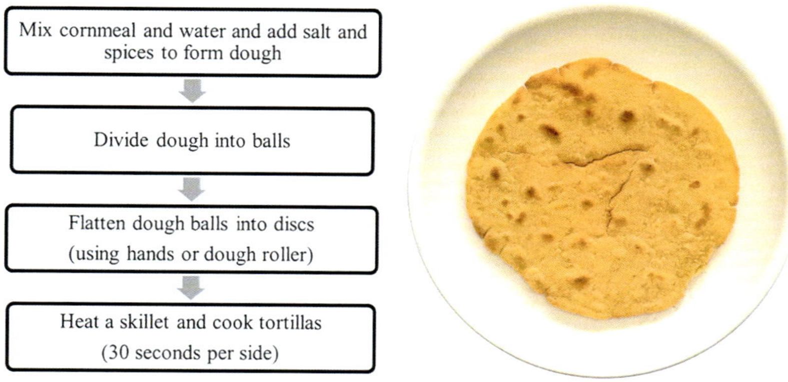

Fig. 19.4 Protocol for preparation of tortilla and the resultant product

thetic ingredients. These tortillas contain additives, preservatives, and processed ingredients to extend shelf life and enhance texture.

The basic recipe to prepare corn tortillas has been mentioned (Fig. 19.4) with basic ingredients: corn flour treated with lime, water, and salt.

19.3 Cornbread

It is primarily made from cornmeal (ground dried corn), typically leavened with baking powder to give it a light and fluffy texture (Serna-Saldivar and Perez Carrillo 2019). It has a slightly sweet, crumbly texture and can be baked, fried, or even steamed. It is a popular dish in southern, southwestern, and some Latin American cuisines and is commonly served as a side dish, often with chili, barbecue, or stews. The basic recipe to prepare corn bread has been mentioned (Fig. 19.5) with basic ingredients: cornmeal, flour, leavening agents, buttermilk, oil, and eggs.

There are generally four main types of cornbread that are commonly consumed:

(i) *Mexican Cornbread:* It is commonly eaten in Mexican cuisine, combining the basic ingredients of classic cornbread, like cornmeal and flour, with additional elements that add moisture, richness, and a touch of spice (Fig. 19.5). It is a twist on traditional cornbread, blending with ingredients often found in Mexican cuisine to create a unique flavour.

(ii) *Sweet Cornbread:* Sweet cornbread is a variation of traditional cornbread that includes added sugar or other sweeteners, giving it a sweeter flavour. It is frequently served as a side dish with meals or as a dessert, especially in the Southern United States. It may also include ingredients such as milk, butter, or honey to amplify its sweetness and richness.

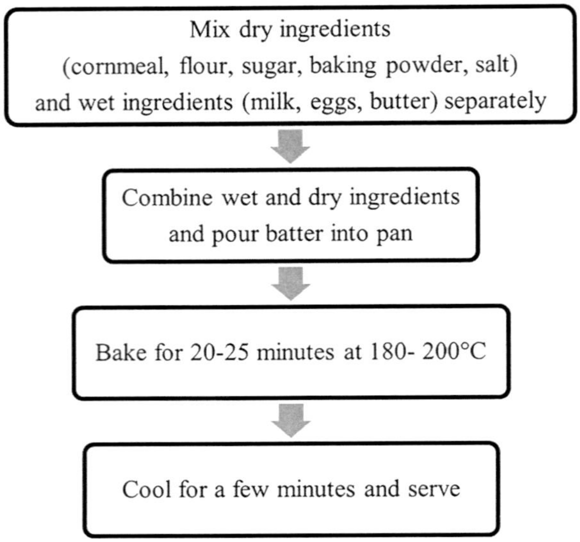

Fig. 19.5 Protocol for cornbread preparation

(iii) *Northern Cornbread:* It is a type of cornbread that tends to be less sweet and more savoury compared to its Southern counterpart. It is typically baked in a pan or baking dish and is often served as a side dish with savoury meals such as stews, soups, or roasted meats.

(iv) *Southern Cornbread:* It is made with buttermilk and often includes ingredients such as eggs, butter, and bacon drippings, which provide moisture and enhance its flavour. It is traditionally baked in a cast-iron skillet, which helps create a golden-brown crust while keeping the interior moist and tender. It is commonly served as a side dish with meals like fried chicken, barbecue, or beans, and can also serve as a base for dishes such as cornbread stuffing.

19.3.1 Polenta

Polenta is a versatile, porridge-like dish traditionally made by simmering cornmeal with water and salt over low heat, constantly stirring to achieve a smooth, creamy consistency (Fig. 19.6). Polenta can be enjoyed on its own or enhanced with varieties of toppings such as cheese, meat, sausages, or fish (Marchetti 2023). While it is most closely associated with northern Italy, polenta is also widely consumed in other parts of Europe and South America. It can be served in various ways, either soft and creamy or allowed to set and then sliced. It can either be used as a side dish or as a base for sauces or stews. Polenta comes in different types depending on the grinding and preparation method of the cornmeal. Here are the main types:

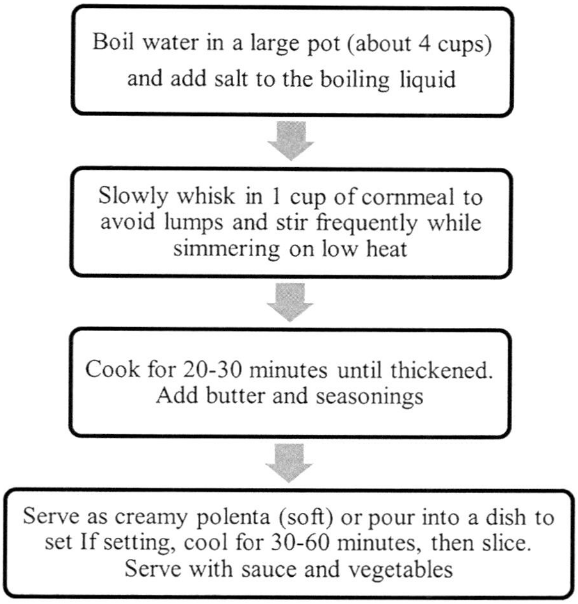

Fig. 19.6 Protocol for Polenta preparation

(i) *Coarse Polenta:* It is produced from cornmeal that is ground more coarsely, giving it a grainy texture. It takes longer to cook compared to finer polenta varieties and produces a more earthy dish.
(ii) *Medium Polenta:* This has a finer grind than coarse polenta. It cooks quicker and has a smoother texture, but still preserves some grit.
(iii) *Fine Polenta:* It is made from finely ground cornmeal, which cooks quickly and creates a creamy texture. It is commonly used to prepare delicate dishes.
(iv) *Instant Polenta:* It is pre-cooked and dried, which results in a significantly faster cooking time of just a few minutes. Instant polenta has a smoother texture but lacks the rich depth of flavour present in traditional polenta.
(v) *Yellow Polenta:* It is the most widely consumed variety of polenta made from yellow cornmeal. It has a characteristic golden colour and is typically available in fine, medium, and coarse grinds.
(vi) *White Polenta:* It is made from white cornmeal and has a subtler flavour than yellow polenta.

The basic recipe to prepare polenta has been mentioned (Fig. 19.6) with the essential ingredients: cornmeal, water, salt, and butter.

19.3.2 Tamales

This is a traditional Mexican dish, steam-cooked and typically made from nixtamalized cornmeal dough and salt, all wrapped in corn husks. (Rodríguez-Huezo et al.

2017). Due to their simple and neutral flavour, tamales serve as a staple diet in everyday meals, much like bread in French cuisine or rice in many Eastern cultures. Tamales can also be filled with a variety of sauces to enhance their flavour and texture, turning them into a main course (Cruz-Vazquez et al. 2019).

Tamales differ significantly throughout Mexico and other regions of Latin America. Every region has its unique take on tamales, featuring distinct fillings, cooking methods, and masa preparations.

(i) *Mexico*: Popular varieties include tamales de pollo (chicken), tamales de puerco (pork), and tamales de elote (corn).
(ii) *Central America*: In countries such as Guatemala and Honduras, tamales are typically stuffed with savoury fillings, including vegetables (Fig. 19.7).
(iii) *Southern United States*: In states like Texas, tamales are a popular dish, usually filled with beef or pork and covered in red chilli sauce.

The basic recipe to prepare tamales has been mentioned (Fig. 19.7) with basic ingredients: masa harina (corn flour for tamales), vegetable oil, chicken or vegetable broth, baking powder, salt, and corn husks.

19.3.3 Arepas

Arepa is an unleavened bread made from cornmeal, mixed with salt and water. It is a staple in Venezuelan cuisine and is commonly enjoyed either baked or fried (Siciliano-Rosen 2025). Arepas are made from cornmeal dough and resemble flatbreads, though they have a slightly thicker texture. They can be prepared

Fig. 19.7 Protocol for Tamales preparation

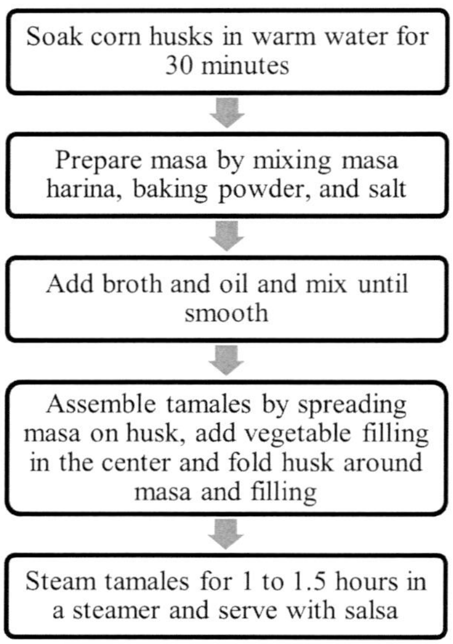

Fig. 19.8 Protocol for the preparation of Arepas

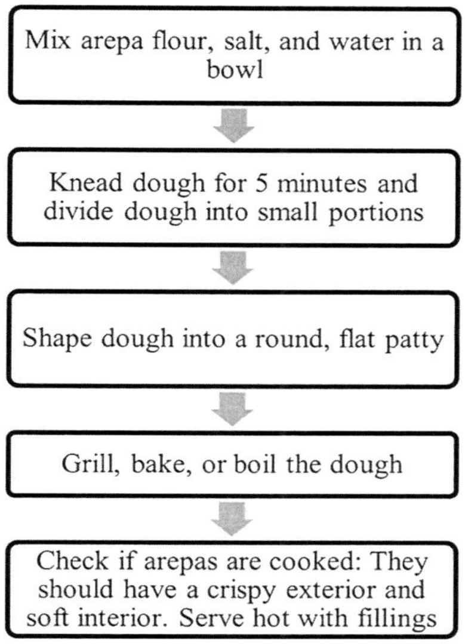

through various methods, including grilling, baking, frying, or boiling, providing a range of culinary options (Fig. 19.8). Arepas come in various types, each with unique characteristics influenced by the country or region. Venezuelan arepas are commonly packed with chicken, avocado, and cilantro, offering a flavourful combination (Puyana 2017). In contrast, Colombian arepas are known for their simple yet delicious filling of cheese, highlighting the rich dairy flavours of the region (Dinho 2023). Each type of arepa differs not only in its fillings but also in the preparation methods and ingredients, showcasing the extensive variety of this treasured dish across different regions. The basic recipe to prepare arepa has been mentioned (Fig. 19.8) with basic ingredients: Arepa flour, water, salt, and oil.

19.3.4 Maize Pretzels

These maize products are prepared using cornmeal as a primary ingredient. Pretzels can be prepared through a straightforward process involving a limited number of ingredients, such as flour, oil, sugar, and yeast. During the production process, the dough is shaped with an extruder at mild pressure, immersed in a hot-alkali solution, salted for flavour, and then baked (Fig. 19.9). The baking process generally occurs in two phases: an initial quick bake at a high temperature, followed by a gradual drying stage at a lower temperature (Lusas and Rooney 2001).

Maize pretzels come in a range of flavours, both sweet and savoury. The sweet varieties are coated with cinnamon, sugar, or chocolate, while the savoury ones are seasoned with flavours like cheese, garlic, ranch, or spicy chilli. Additionally, these pretzels can be filled with indulgent options like peanut butter or chocolate, offering

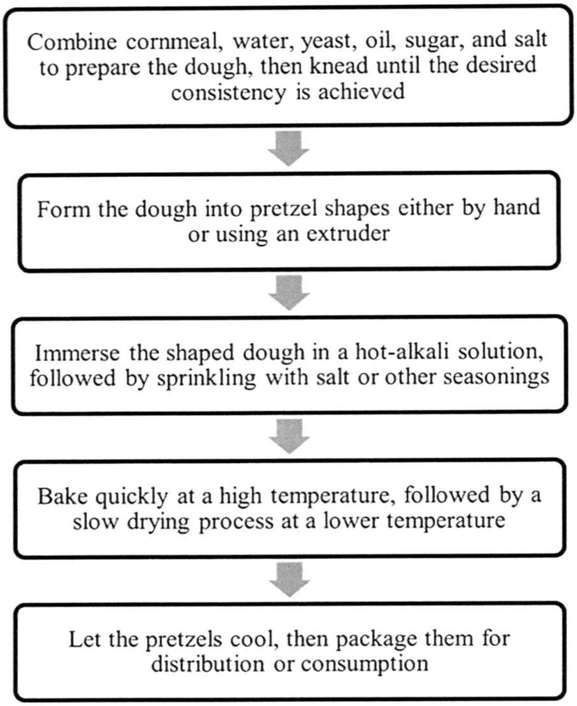

Fig. 19.9 Protocol for preparation of maize pretzels

a richer snack experience. The basic recipe to prepare maize pretzels has been mentioned (Fig. 19.9) with basic ingredients: cornmeal, yeast, oil, sugar, and salt.

19.3.5 Maize Pancakes

These pancakes are made with cornmeal as the primary ingredient instead of traditional wheat flour (Fig. 19.10). Pancakes are among the oldest bread products, traditionally made from a batter-based mixture (Taylor 2023). These pancakes are a popular choice for breakfast or brunch across many cultures. Maize pancakes can be made either sweet or savoury. Sweet pancakes are typically flavoured with sugar or sweet toppings, as shown in Fig. 19.10, while savoury pancakes highlight saltier or spicier ingredients, often complemented with savoury accompaniments. The basic recipe to prepare the maize pancakes has been mentioned (Fig. 19.10) using the essential ingredients: cornmeal, sugar, baking powder, milk, egg, butter, and salt.

19.3.6 Maize Waffles

There are variations of traditional waffles made using cornmeal as the primary ingredient instead of wheat flour. This imparts a unique texture and flavour to the waffles,

Fig. 19.10 Protocol for pancake preparation and the resultant product (without and with topping)

Fig. 19.11 Protocol for the preparation of maize waffles

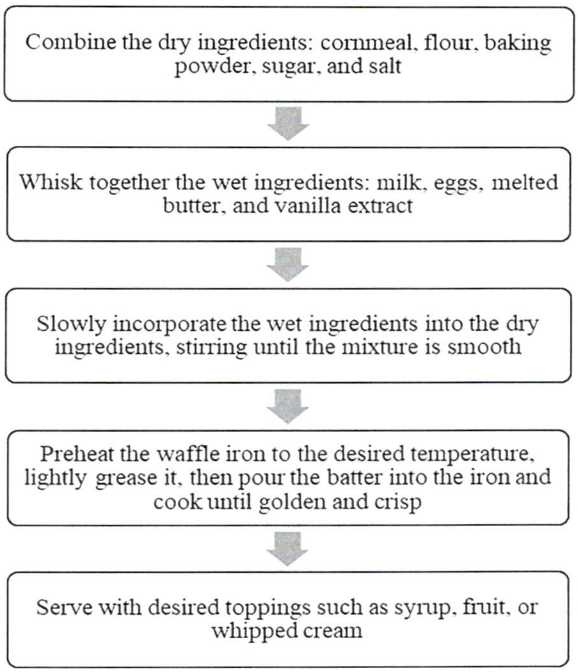

with a noticeably granular consistency and a mild, sweet, and nutty undertone. Waffles are convenience foods made by baking batters or doughs in indented waffle irons, which impart a cake-like texture (Fig. 19.11). They are leavened through aeration, chemical leavening agents, or yeast. Maize waffles are often enjoyed with a variety of toppings, such as syrup, butter, fruit, or whipped cream, and are popular for breakfast or as a snack. Waffles can be either ready-to-eat (RTE) or frozen. Frozen waffles are partially baked to retain approximately 40% moisture content and are promptly frozen. Their cooking is completed in a toaster or microwave oven just before consumption. These frozen waffles are primarily consumed in North America.

The basic recipe to prepare the maize waffles has been mentioned (Fig. 19.11) with basic ingredients: cornmeal, baking powder, sugar, milk, eggs, butter, vanilla extract, and salt (Sharma 2022).

19.3.7 Maize Burgers

These burgers are variations of traditional burgers, where maize is used as a primary ingredient in the bun and patty. The bun is primarily prepared from cornmeal, with additional ingredients such as flour, yeast, salt, and occasionally sugar or honey. It is then baked to produce a dense yet soft texture (Kubert 2014). The maize patty can be prepared using cornmeal, whole corn kernels, or a combination of both, typically mixed with additional ingredients such as vegetables, spices, and binding agents like egg or flour to hold the patty together (Kapoor 2016a). Maize burgers are widely consumed in different regions and offer a vegetarian or vegan alternative to traditional meat-based burgers. They can be enjoyed with various toppings such as lettuce, tomato, pickles, and sauces. The basic recipe to prepare the maize bun has been mentioned (Fig. 19.12) with basic ingredients: cornmeal, flour, salt, sugar, yeast, butter, and egg. Similarly, the basic recipe to prepare the maize patty has been mentioned (Fig. 19.13) with basic ingredients: corn kernels, vegetables, oil, spices, egg, and salt.

19.3.8 Maize Pizza

It is a variation of traditional pizza that incorporates maize as the primary ingredient in the crust or dough. Unlike conventional pizza dough, which is made from wheat flour, the dough for maize pizza is prepared using cornmeal or corn flour, resulting in a distinct texture and flavour profile (Kapoor 2016b). Maize pizza is commonly enjoyed in regions where maize is a staple food, such as Mexico, Central America, and South America. Its popularity extends across various global regions, with numerous variations including traditional toppings like grilled meats, fresh herbs, and hot sauces. In some African and Asian countries, variations of maize pizza may feature indigenous vegetables, spices, or legumes. The basic recipe to prepare the maize pizza has been mentioned (Fig. 19.14) with basic ingredients: cornmeal, corn flour, salt, yeast, cheese, vegetables, and fresh herbs.

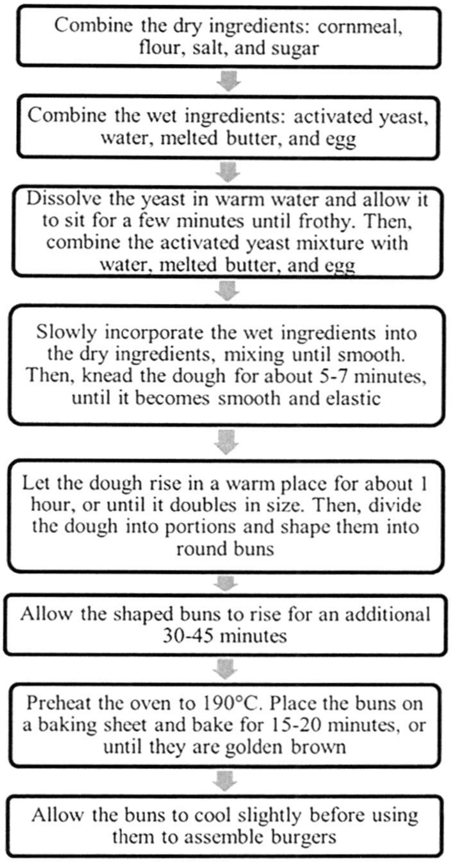

Fig. 19.12 Protocol for maize bun preparation

19.4 Corn on the Cob

It refers to the whole ear of corn that is still attached to its husk, with the kernels left intact. It is a popular method of enjoying corn, often prepared by boiling, steaming, or grilling (Allrecipes 2025). Once cooked, the corn is commonly served on the cob and seasoned with ingredients such as butter, salt, or spices (Serna-Saldivar and Perez Carrillo 2019). The corn is usually eaten directly from the cob by holding it with hands and biting off the kernels, making it a fun and casual dish. This preparation is popular at social gatherings such as barbecues and picnics, and is frequently served as a side dish alongside the main course. Its simplicity and versatility make it a favoured option for both casual and festive meals.

The various preparation methods include:

(i) *Boiling:* This is the most common method to enjoy corn on the cob. The corn is usually boiled in water with a pinch of salt for about 5–10 min until tender.

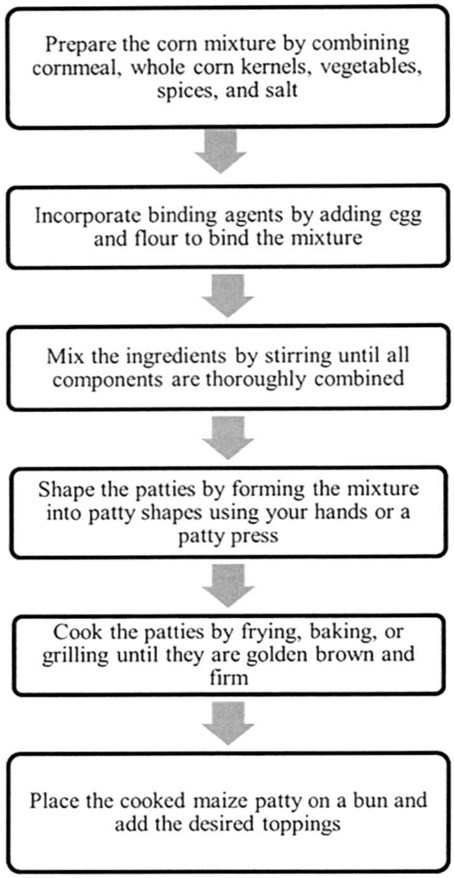

Fig. 19.13 Protocol for maize patty preparation

(ii) *Grilling:* This cooking method imparts a unique smoky flavour to the corn. It's often grilled with the husk on to retain moisture or husked and grilled directly for a smoky, charred flavour.
(iii) *Roasting:* Corn can also be roasted in the oven or air fryer, creating a deliciously caramelized flavour (Fig. 19.15).
(iv) *Microwaving:* An efficient method involves wrapping corn on the cob in a damp paper towel and microwaving it.

19.4.1 Baby Corn Fries

These are a variation of traditional fries made using baby corn instead of regular potatoes. Baby corn is the young, immature ear of specialty corn, harvested before it reaches full maturity, and it has a small, tender texture. Baby corn is a flavourful vegetable that contains no cholesterol. It is a low-calorie vegetable rich in fibre

Fig. 19.14 Protocol for maize pizza preparation

Prepare the cornmeal dough by combining cornmeal, corn flour, salt, sugar, water, and yeast. Knead the dough until it becomes smooth and elastic

↓

Allow the dough to rise in a warm location for approximately 1 hour

↓

Roll the dough to the desired size and thickness. Spread a layer of tomato sauce over the dough, then top with cheese and other desired toppings

↓

Bake the pizza in the preheated oven for 10-15 minutes at 190°C

↓

Take the pizza out of the oven, slice it, and serve with optional garnishes such as fresh herbs

Fig. 19.15 Air-fried corn on the cob

content (Rani et al. 2019). To make baby corn fries, the baby corn is usually sliced into smaller pieces, coated in batter or breadcrumbs, and then either deep-fried or baked until crispy (Fig. 19.16). They are often served as a snack or appetiser and can be seasoned with various spices or accompanied by dipping sauces, much like regular fries. The unique flavour and texture of baby corn offer a different and lighter alternative to traditional potato fries.

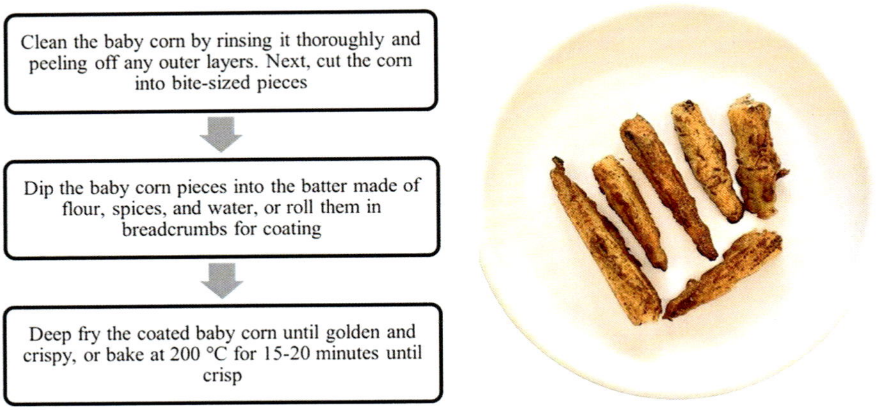

Fig. 19.16 Protocol for preparation of baby corn fries and the resultant product

There are several variations of baby corn fries:

(i) *Deep or Air Fried Baby Corn Fries* A more flavourful variation, where the batter or coating is enhanced with spices such as chilli powder, turmeric, or garam masala (Richa 2022) (Fig. 19.16).
(ii) *Baked Baby Corn Fries* A healthier alternative that involves baking baby corn in the oven instead of frying, resulting in a version with reduced fat content (Anand 2021).
(iii) *Asian Baby Corn Fries* These fries are coated with soy sauce, sesame seeds and ginger to create a savoury and tangy flavour (Rawat 2019).

The basic recipe to prepare the baby corn fries has been mentioned (Fig. 19.16) with the basic ingredients: baby corn, flour, spices, and oil.

19.5 Maize Pasta

Since maize is a gluten-free cereal, maize pasta is a significant healthy option for celiac patients. It is also a technological challenge to develop quality pasta since gluten plays a key role in influencing the texture, structure, and shelf-life of the resultant product. However, pre-processing, which can reorganize the starch into new structures or the addition of certain additives, may be required so that the product can withstand the cooking stress and retain the structural integrity. Therefore, additives like egg albumin, milk proteins, and gums can be added, or pregelatinization of starch may be carried out (Bongianino et al. 2023; Marti and Pagani 2013). Figure 19.17 gives the flow chart for the preparation of maize pasta.

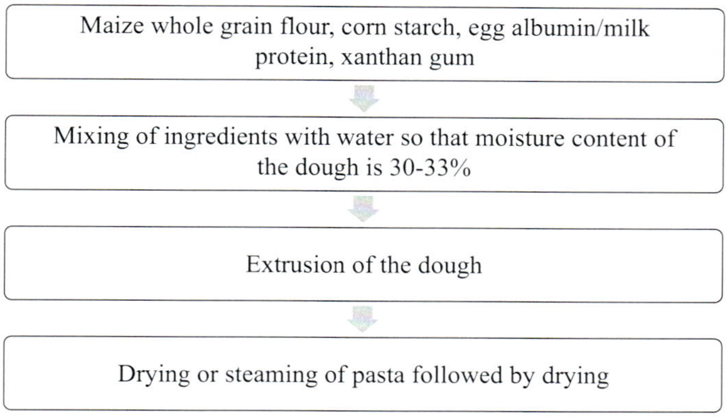

Fig. 19.17 Preparation of maize pasta

Fig. 19.18 Plate of maize cheese spread

19.5.1 Maize Cheese Spread

It is a savoury spread composed of maize and cheese, typically blended with seasonings to form a creamy, flavourful paste. The spread usually combines cooked or pureed maize with cheese, such as cheddar or cream cheese, along with other ingredients like butter, cream, or spices to enhance its flavour and texture (Fig. 19.18). Maize cheese spread is commonly consumed as a quick snack, often paired with bread or crackers, or used as a filling in sandwiches (Khushboo 2022). Its creamy texture, subtle sweetness from the corn, and savoury cheese flavour contribute to its popularity in casual dining and comfort food contexts.

19.5.2 Corn Nuts

Corn nuts are a crunchy snack made from corn kernels that are soaked and then cooked until crisp. While deep-frying was common in the past, modern methods

Fig. 19.19 Protocol for preparation of corn nuts and the resultant product

often use the oven or air fryer to make them healthier. Oven roasting gives them an even, toasted flavour, while air frying creates a crunchy texture in less time (Fig. 19.19). They are often seasoned with salt, chili, lime, or ranch. These oven-baked and air-fried versions show how traditional snacks can be adapted to healthier, modern cooking methods (Jampel 2015).

19.6 Corn Fritters

Corn fritters are savoury, fried snacks commonly prepared from a batter that typically consist of sweet corn kernels, flour, eggs, milk, and a variety of seasonings. These fritters are usually deep-fried or pan-fried until golden and crispy. The batter is moulded into small patties and then fried until the exterior becomes golden and crispy, while the interior remains tender and filled with corn (Zaffary 2025). These are widely enjoyed in various cuisines globally and can be adapted with ingredients such as cheese, herbs, or other vegetables. They are commonly served as appetizers, side dishes, or light meals, and are often paired with condiments such as sour cream, salsa, or chutney. They are commonly found at casual dining establishments, food trucks, and quick-service restaurants.

19.6.1 Crispy Corn

It is a flavourful and crunchy snack commonly served as an appetizer or party food. It is prepared by deep-frying or pan-frying corn kernels, typically coated with a mixture of spices and flour (Sharmilee 2022). The kernels develop a crispy texture, resulting in a highly delectable snack. It is frequently featured in various Asian cuisines, particularly in Chinese and Indo-Chinese restaurants, where it is often combined with sauces such as sweet chilli sauce, soy sauce, or other seasonings to enhance its flavour.

19.7 Cheese Corn Balls

This maize product is a popular and widely consumed snack prepared by combining sweet corn kernels and cheese, which are then coated in breadcrumbs and deep-fried or pan-fried to develop a crispy exterior. The filling is usually prepared from mashed corn, cheese varieties such as mozzarella or cheddar, and various spices, creating a creamy, gooey, and flavourful centre (Anand 2024). These bite-sized balls are commonly served as appetizers or snacks, often accompanied by dipping sauces such as ketchup, mayonnaise, or spicy condiments.

19.7.1 Maize Sandwich

A maize sandwich is a variation of a sandwich in which corn kernels is the central ingredient, often combined with other components such as vegetables, and various condiments. It typically includes cooked or canned corn kernels, which are blended with a variety of seasonings and, in some cases, cheese or mayonnaise. The mixture is spread between slices of bread to create a simple yet flavourful snack (Swasthi 2022). Variations of the maize sandwich may include additional ingredients such as onions, tomatoes, lettuce, and herbs, and some versions may be grilled or toasted to enhance the texture. The maize sandwich is often served as a quick, light meal or appetizer in informal dining settings.

19.8 Conclusion

The maize-based value-added fast-food products discussed in this chapter offer a promising path towards nutritious fast-food options. The maize can be processed into snacks and food products rich in complex carbohydrates, fibre, antioxidants, essential vitamins and minerals via innovative approaches and fortifications. Their nutritional benefits and ability to incorporate them into a wide range of tasty and convenient food make them ideal for reshaping the healthy fast-food sector and quick service. They also appeal the modern tastes and contribute to good dietary habits, including plant-based gluten-free diets. Moreover, the fast-food industry can adopt these products to help consumers eat healthy food without compromising their taste. These products will contribute to the health, food security and agricultural sustainability. As consumer awareness and adaptation by the food industry are increasing, these maize-based products can be a viable solution to bridge the gap between health and the fast-food sector.

References

Allrecipes (2025) Jamie's sweet and easy corn on the cob. Available at: https://www.allrecipes.com/recipe/222352/jamies-sweet-and-easy-corn-on-the-cob/. Accessed 11 Dec 2025

Anand R (2021) Tandoori style baby corn recipe, in oven. Rak's Kitchen. Available at: https://rakskitchen.net/tandoori-style-baby-corn-recipe/. Accessed 11 Dec 2025

Anand R (2024) Cheese corn balls recipe. Rak's Kitchen. Available at: https://rakskitchen.net/corn-cheese-balls-recipe. Accessed 11 Dec 2025

Bongianino NF, Steffolani ME, Morales CD, Biasutti CA, León AE (2023) Technological and sensory quality of gluten-free pasta made from flint maize cultivars. Foods 12(14):2780. https://doi.org/10.3390/foods12142780

Chawla S, Kerna N, Roberson R, Carsrud V, Ngwu D, Florence O, Obiamaka P, Okereke A, Nwokorie U, Holets H, Ayozie C, Nkwopara V (2022) Revisiting an old friend: the gastrointestinal benefits of regular popcorn consumption. Ecol Cult Gastron Diet Sci 9:73–85

Cruz-Vazquez C, Villanueva-Carvajal A, Estrada-Campuzano G, Dominguez-Lopez A (2019) Tamales texture properties as a function of corn endosperm type. Int J Gastron Food Sci 16:100153. https://doi.org/10.1016/j.ijgfs.2019.100153

Delcour JA, Aman P, Courtin CM, Hamaker BR, Verbeke K (2016) The health effects of whole grain consumption: from epidemiological evidence to mechanistic insights. Adv Nutr 6(2):320–328. https://doi.org/10.3945/an.115.010546

Díaz-Gómez JL, Castorena-Torres F, Preciado-Ortiz RE, García-Lara S (2017) Anti-cancer activity of maize bioactive peptides. Front Chem 5:44. https://doi.org/10.3389/fchem.2017.00044

Dinho E (2023) Arepas de queso (cheese arepas). My Colombian Recipes. Available at: https://www.mycolombianrecipes.com/cheese-cornmeal-patties-arepas-de-queso/. Accessed 11 Dec 2025

Félix-Medina JV, Gutiérrez-Dorado R, López-Valenzuela JA, López-Ángulo G, Quintero-Soto MF, Perales-Sánchez JXK, Montes-Ávila J (2021) Nutritional, antioxidant and phytochemical characterization of healthy ready-to-eat expanded snack produced from maize/common bean mixture by extrusion. LWT 142:111053

Food Crumbles (2021) The science of caramel popcorn. Available at: https://foodcrumbles.com/the-science-of-caramel-popcorn/. Accessed 11 Dec 2025

Hoseney RC, Zeleznak K, Abdelrahman A (1983) Mechanism of popcorn popping. J Cereal Sci 1(1):43–52. https://doi.org/10.1016/S0733-5210(83)80007-1

Jampel S (2015) Fulfill every snacking need with homemade corn nuts, two ways. Food52. Available at: https://food52.com/story/14738-fulfill-every-snacking-need-with-homemade-corn-nuts-two-ways. Accessed 11 Dec 2025

Jericho H, Sansotta N, Guandalini S (2017) Extraintestinal manifestations of celiac disease: effectiveness of the gluten-free diet. J Pediatr Gastroenterol Nutr 65(1):75–79. https://doi.org/10.1097/MPG.0000000000001420

Johnson L (2019) Pina colada popcorn. Cooking with Curls. Available at: https://cookingwithcurls.com/2019/10/11/pina-colada-popcorn/. Accessed 11 Dec 2025

Kapoor S (2016a) Spicy corn patties. Available at: https://www.sanjeevkapoor.com/Recipe/Spicy-Corn-Patties.html. Accessed 11 Dec 2025

Kapoor S (2016b) Cornmeal crust pizza. Available at: https://www.sanjeevkapoor.com/Recipe/Cornmeal-Crust-Pizza.html. Accessed 11 Dec 2025

Kazerooni EG, Atia Sharif HN, Rehman R, Nisar S (2019) Maize (corn): a useful source of human nutrition and health—a critical review. Int J Chem Biochem Sci 15:35–41

Khushboo (2022) Cheese masala corn dip. Carve Your Craving. Available at: https://carveyourcraving.com/cheese-masala-corn-dip/. Accessed 11 Dec 2025

Kitchen Concoctions (2017) Sour cream and onion popcorn. Available at: https://www.kitchen-concoctions.com/2017/02/sour-cream-and-onion-popcorn/. Accessed 11 Dec 2025

Kubert K (2014) Cornmeal burger buns recipe. Comfortably Domestic. Available at: https://comfortablydomestic.com/2014/09/cornmeal-burger-buns/. Accessed 11 Dec 2025

Kumar P, Choudhary M, Hossain F, Singh NK, Choudhary P, Gupta M, Singh V, Chikappa GK, Kumar R, Kumar B, Jat SL, Rakshit S (2019) Nutritional quality improvement in maize (Zea mays): Progress and challenges. Indian J Agric Sci 89:895–911. https://doi.org/10.56093/ijas.v89i6.90756

Kumar P, Longmei N, Choudhary M, Gupta M, Kumar B, Jat BS, Bhushan B, Dagla MC, Aggarwal SK (2024) Enhancement of nutritional quality in maize grain through QTL-based approach. Cereal Res Commun 52(1):39–55

Lusas EW, Rooney LW (eds) (2001) Snack foods processing. CRC Press, Boca Raton, Florida

Manasa R, Farooqui M, Deepika M, Sofia R, Naik S, Shivananjappa M (2024) Formulation of nutrient-rich nachos using little millet (*Panicum sumatrense*) flour. Ann Geriatr Educ Med Sci 10:51–54. https://doi.org/10.18231/j.agems.2023.012

Marchetti D (2023) Creamy polenta. The Mediterranean Dish. Available at: https://www.themediterraneandish.com/polenta-recipe/. Accessed 11 Dec 2025

Marti A, Pagani MA (2013) What can play the role of gluten in gluten-free pasta? Trends Food Sci Technol 31(1):63–71. https://doi.org/10.1016/j.tifs.2013.03.001

Meena KK, Taneja NK, Jain D, Ojha A, Saravanan C, Mudgil D (2022) Bioactive components and health benefits of maize-based fermented foods: a review. Biointerface Res Appl Chem 13(4):338. https://doi.org/10.33263/BRIAC134.338

Nguyen V, Cooper L, Lowndes J, Melanson K, Angelopoulos TJ, Rippe JM, Reimers K (2012) Popcorn is more satiating than potato chips in normal-weight adults. Nutr J 11(1):71. https://doi.org/10.1186/1475-2891-11-71

Paolicchi S (2013) Sweet corn smoothie. Family Focus Blog. Available at: https://familyfocusblog.com/sweet-corn-smoothie/. Accessed 11 Dec 2025

Piazza S, Bani C, Colombo F, Mercogliano F, Pozzoli C, Martinelli G, Petroni K, Pilu SR, Sonzogni E, Fumagalli M, Sangiovanni E, Restani P, Dell'Aglia M, Di Lorenzo C (2024) Pigmented corn as a gluten-free source of polyphenols with anti-inflammatory and antioxidant properties in Caco-2 cells. Food Res Int 168:114640. https://doi.org/10.1016/j.foodres.2024.114640

Popcorn.org (2025) Sriracha lime popcorn. Available at: https://www.popcorn.org/Recipes/Details/sriracha-lime-popcorn. Accessed 11 Dec 2025

Press Information Bureau (PIB) (2023) HFW/PQ/Adverse effect of junk fast foods/30th July 2021/9, Release ID: 1740750

Puyana A (2017) Arepas are conquering the world- but dying at home in Venezuela. NPR. Available at: https://www.npr.org/sections/thesalt/2017/07/26/538515889/arepas-are-conquering-the-world-but-dying-at-home-in-venezuela. Accessed 11 Dec 2025

Rani R, Sheoran R, Soni P (2019) Baby corn: a wonderful vegetable. Int J Sci Environ Technol 6(2):1407–1412

Rawat S (2019) Crispy chilli garlic baby corn. That Delicious Dish. Available at: https://thatdeliciousdish.com/chilli-garlic-babycorn/. Accessed 11 Dec 2025

Richa (2022) Crispy baby corn fry. My Food Story. Available at: https://myfoodstory.com/baby-corn-fry/. Accessed 11 Dec 2025

Rodríguez-Huezo M, Flores P, García S, Meraz M, Vernon-Carter EJ, Alvarez-Ramirez J (2017) Effect of fat type on starch and protein digestibility of traditional tamales. Starch/Stärke 70:201700286. https://doi.org/10.1002/star.201700286

Rouf Shah T, Prasad K, Kumar P (2016) Maize-A potential source of human nutrition and health: a review. Cogent Food Agric 2(1):1166995. https://doi.org/10.1080/23311932.2016.1166995

Saritha A, Ramanjaneyulu AV, Sainath N, Umarani E (2020) Nutritional importance and value addition in maize. Biotica Res Today 2(9):974–977

Serna-Saldivar S (2015) History of corn and wheat tortillas. In: Rooney LW, Serna-Saldivar SO (eds) Tortillas, wheat flour and corn products. AACC International Press, Minnesota, pp 10–20

Serna-Saldivar SO, Perez Carrillo E (2019) Food uses of whole corn and dry-milled fractions. In: Serna-Saldivar SO (ed) Corn, 3rd edn. AACC International Press, pp 435–467. https://doi.org/10.1016/B978-0-12-811971-6.00016-4

Sharma A (2022) Maize flour-based bakery products. MSc dissertation. Dr. Yashwant Singh Parmar University of Horticulture and Forestry, Solan, India

Sharmilee J (2022) Crispy corn recipe: Barbeque Nation style crispy corn. Sharmis Passions. Available at: https://www.sharmispassions.com/crispy-corn-recipe-barbeque-nation-style/. Accessed 11 Dec 2025

Siciliano-Rosen L (2025) Arepa. Encyclopaedia Britannica. Available at: https://www.britannica.com/topic/arepa. Accessed 11 Dec 2025

Siyuan S, Tong L, Liu R (2018) Corn phytochemicals and their health benefits. Food Sci Hum Wellness 7(3):185–195. https://doi.org/10.1016/j.fshw.2018.09.003

Soylu S, Tekkanat A (2007) Interactions amongst kernel properties and expansion volume in various popcorn genotypes. J Food Eng 80(1):336–341. https://doi.org/10.1016/j.jfoodeng.2006.06.001

Swasthi (2022) Corn cheese sandwich recipe. Indian Healthy Recipes. Available at: https://www.indianhealthyrecipes.com/corn-cheese-sandwich-recipe/. Accessed 11 Dec 2025

Taylor JRN (2023) Flatbreads and pancakes. In: Shewry PR, Koksel H, Taylor JRN (eds) ICC handbook of 21st century cereal science and technology. Academic Press, United Kingdom. pp 291–302

Zaffary J (2025) Corn fritters. Allrecipes. Available at: https://www.allrecipes.com/recipe/18040/corn-fritters/. Accessed 11 Dec 2025

20 Linking Innovations and Farmers: Upscaling Maize Production and Understanding Market Trends in India

Ph. Romen Sharma, Alok Kumar Sahoo, Laishram Priscilla, Shankar Lal Jat, Prakash Goraksha Athare, Manish Kakraliya, and Shaik Rehana

Abstract

Maize has emerged as a strategic cereal in global and Indian agriculture due to its multifunctional role across food, feed, fodder, fuel, and industrial value chains. In India, maize has evolved from a subsistence-oriented coarse cereal to the third most important cereal, driven by technological change, including single-cross hybrids, improved agronomy, mechanization, and expanding demand from feed, industrial, and bioethanol markets. Despite this progress, productivity gains remain uneven, with large yield gaps persisting in rainfed kharif maize systems due to climatic risks, pest pressures, and institutional and market constraints. Strengthening innovative and demand-responsive pathways that align technology, markets, and institutions is therefore central to upscaling maize production.

Keywords

Food–feed–fuel nexus · India · Maize · Market integration · Technological innovation · Yield gaps

---

Ph. R. Sharma (✉) · A. K. Sahoo · S. L. Jat · P. G. Athare · M. Kakraliya · S. Rehana
ICAR-Indian Institute of Maize Research, Ludhiana, Punjab, India

L. Priscilla
Punjab Agricultural University, Ludhiana, Punjab, India
e-mail: priscilla@pau.edu

M. Choudhary et al. (eds.), *Next-Generation Maize Research*,
https://doi.org/10.1007/978-981-92-0722-0_20

20.1 Introduction

Maize (*Zea mays* L.) is among the most dynamic crops in global agriculture, supporting food, feed, and fuel systems. Unlike rice and wheat, which are consumed largely as staple foods, maize functions as a multifunctional crop linking food, feed, fodder, energy, and industrial value chains. This versatility has elevated maize to a strategic global commodity, with production expanding rapidly over the past three decades and surpassing that of both rice and wheat. Continued advances in hybrid breeding, agronomic intensification, mechanization, and sustained demand from livestock, industrial, and bioenergy sectors are expected to further strengthen maize's position as the world's most widely cultivated and traded cereal in the coming decade (Erenstein et al. 2022).

By 2024, maize occupied roughly 210 million hectares worldwide, ranking second only to wheat in harvested area and already surpassing rice. This structural shift is evident in long-term triennium-ending (TE) trends for area, yield, and output of major cereals (Table 20.1). Between TE-1964 and TE-2024, global maize area nearly doubled, productivity increased almost threefold, and production expanded by nearly sixfold, substantially outpacing rice and wheat. In contrast, growth in rice and wheat output has been driven mainly by yield gains on relatively stable cultivated areas, highlighting maize's distinctive dual pathway of both area expansion and yield intensification. The expansion of maize is driven primarily by its multifunctional nature, serving simultaneously as food, feed, fodder, industrial raw material, and bioenergy feedstock. Unlike rice and wheat, this diversified use profile, combined with strong responsiveness to technological and agronomic intensification, has enabled maize to expand in both cultivated area and productivity, underpinning its sustained growth over time.

Within this global context, India's maize sector has undergone a significant transformation, mirroring broader shifts in global maize production and commercialization (Ganguly et al. 2017). Once regarded as a secondary coarse cereal, maize has evolved into the country's third most important cereal after rice and wheat, cultivated year-round across diverse agro-ecological regions and production systems. Technological innovation—particularly the shift to single-cross hybrids, improvements in crop and resource management, mechanization, and plant protection—has played a central role in driving productivity growth (Dass et al. 2015; Sivamurugan et al. 2025). At the same time, expanding demand from poultry feed, starch, ethanol, and emerging specialty markets has strengthened maize's commercial orientation.

However, evidence consistently shows that the benefits of maize innovations are unevenly realized. Productivity gains, adoption of improved technologies, and income outcomes vary widely across regions, seasons, farm sizes, and socio-economic groups (Kumar and Savadatti 2023). Rainfed *kharif* maize systems, which account for the largest share of area, continue to exhibit substantial yield gaps and vulnerability to climatic and biotic stresses, while irrigated rabi and spring maize systems achieve higher and more stable yields (Yadav et al. 2015, 2016). Access to quality seed, mechanization, advisory services, post-harvest infrastructure, and

Table 20.1 Change in area, productivity, and production of major cereals, TE-1964 to TE-2024

Crop	Area (m ha)			Productivity (t/ha)			Production (Mt)		
	TE-1964	TE-2024	Increase (×)	TE-1964	TE-2024	Increase (×)	TE-1964	TE-2024	Increase (×)
Maize	106.6	208.6	2.0	2.00	5.78	2.9	213	1206	5.7
Rice	121.6	168.8	1.4	2.02	4.75	2.35	245	801	3.3
Wheat	209.4	218.6	1.04	1.19	3.66	3.1	251	800	3.2

Note: TE-1964 refers to the average of 1962–1964 and TE-2024 to the average of 2022–2024. Increase (×) Source: FAOSTAT markets strongly conditions farmers' ability to translate technological potential into realized gains (Ganguly et al. 2017; Sivamurugan et al. 2025). As maize demand expands across food, feed, and fuel uses, these disparities raise critical questions about who benefits from maize growth, through which pathways, and under what institutional and market conditions. As maize demand expands across food, feed, and fuel uses, these disparities raise critical questions about who benefits from maize growth, through which pathways, and under what institutional and market conditions.

20.2 Scenario of Maize in India

In India, maize has emerged as the third most important cereal after rice and wheat, cultivated year-round across diverse agro-ecological regions and production systems. Between 1966–67 and 2024–25, maize area expanded from about 5.1–12.1 million hectares, while production increased sharply from 4.9 to 43.4 million tonnes. Over the same period, average productivity rose from less than 1.0 t/ha to nearly 3.59 t/ha, representing a 3.7-fold increase. Overall, maize production increased almost ninefold, with productivity gains accounting for the bulk of growth, complemented by moderate area expansion (Fig. 20.1).

A defining feature of India's maize trajectory is the structural break around 2003–04, marking a transition from slow, subsistence-oriented growth to a more dynamic, market-driven phase. Post-2003–04, maize productivity consistently exceeded 2 t/ha, production growth accelerated, and area expansion resumed, reflecting the combined effects of single-cross hybrid adoption, expanding feed and industrial demand, and improvements in crop management. This transition highlights maize's evolution from a secondary coarse cereal to a strategic, multifunctional crop increasingly integrated into commercial value chains.

Maize's expansion has been enabled by its cultivation across *kharif*, *rabi*, and spring/summer seasons, allowing it to fit into both rainfed and irrigated systems. *Kharif* maize dominates the national area with about 60% and is primarily rainfed, while rabi maize, grown under assured irrigation, consistently achieves higher yields. Spring/summer maize, though limited in area, is expanding rapidly in intensive systems due to its flexibility, high returns, and strong linkage with feed and silage markets. In regions such as the north-western India, maize is increasingly promoted as a diversification alternative to the rice–wheat system, motivated by groundwater depletion, rising energy costs, and environmental externalities (Rakshit et al. 2021; Jat et al. 2025). Empirical evidence from large-scale farmer participatory experiments in the northwestern Indo Gangetic Plains shows that replacing rice with maize substantially reduces irrigation water use, energy consumption, labour demand, and global warming potential, while delivering higher net returns under both subsidized and non-subsidized electricity regimes (Jat et al. 2025). These findings reinforce maize's suitability as a diversification alternative to resource-intensive rice in appropriate agro-ecological contexts. Similarly, in eastern India, low-productivity upland paddy systems face persistent yield constraints and income stagnation, presenting a significant opportunity for maize as a viable crop replacement alternative. Moreover, maize is better suited to upland ecologies due to its shorter duration, lower water requirement, and greater responsiveness to improved management practices (NASS 2024). Despite these gains, India's national average maize productivity remains well below attainable potential. Yield comparisons highlight a substantial gap between India and global leaders such as the United States (>11 t/ha), as well as countries like China, Ukraine, and Argentina (6–7 t/ha). The expansion of single-cross hybrid coverage with the adoption of mechanization and best crop management practices will be crucial in productivity enhancement (NASS 2024). Importantly, national averages mask pronounced spatial, seasonal, and socio-economic heterogeneity. Evidence from ICAR-IIMR Frontline Demonstrations during 2024–25 shows relatively moderate yield gaps in rabi and spring maize (16–18%), indicating reasonably well-managed systems with scope for incremental gains. In contrast, kharif maize exhibits much wider gaps (around 31% on average), with site-specific gaps exceeding 65–80% in eastern, central, hill, and rainfed regions.

Programme-wise analysis further reveals that yield gaps are systematically larger in socio-economically vulnerable social classes and regions, including tribal areas, Schedule caste, and the North Eastern Hill region. These patterns confirm that maize yield gaps in India are not primarily genetic, but are driven by management quality, access to inputs and services, risk exposure, and market connectivity. Consequently, the largest untapped productivity potential is concentrated in *kharif* maize cultivated in rainfed and marginal ecologies, where targeted dissemination of proven technologies and services can deliver substantial gains.

Kharif maize, therefore, constitutes the principal binding constraint in India's maize productivity growth. Its dominance in area, combined with high exposure to climatic variability, biotic stresses, and institutional constraints, results in the widest yield gaps and the greatest yield variability. High incidence of fall armyworm and other pests, along with delayed or inappropriate control measures, further exacerbates losses. Addressing constraints in kharif maize thus represents the most critical leverage point for raising national maize productivity while simultaneously advancing equity and resilience.

20.3 Technological Innovations in Maize Systems

Technological innovation has been the principal driver of productivity enhancement in maize systems worldwide, and India is no exception (Yadav et al. 2015). Genetic improvement, agronomic management, mechanization, conservation agriculture, and pest management interact with farmers' capacities, institutions, and markets, so productivity gains arise from bundles of innovations rather than isolated technologies (Parihar et al. 2016; Ghosh et al. 2020).

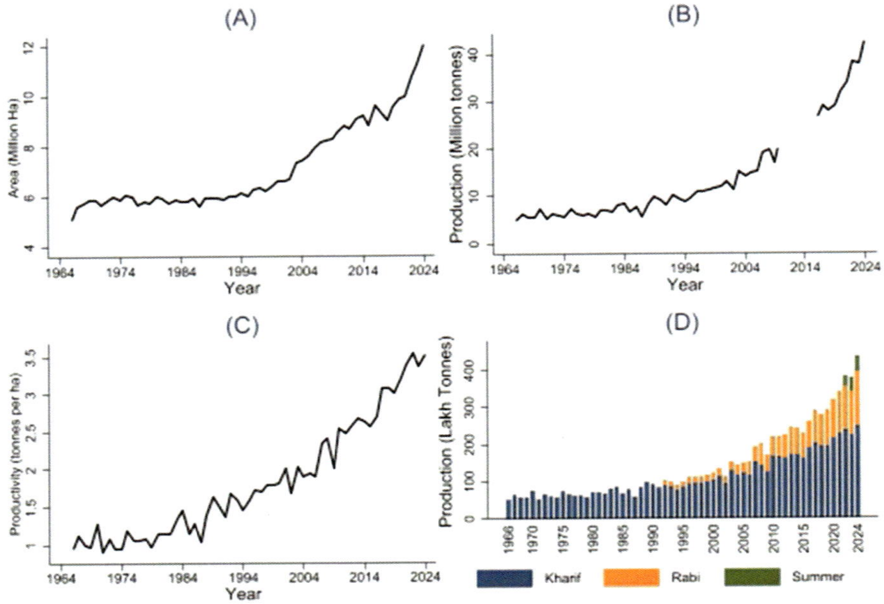

Fig. 20.1 Trends in maize area, production, productivity, and season-wise production in India

20.3.1 Genetic Improvement, Hybrid Development, and Seed Systems

Genetic improvement has been a central driver of maize yield growth in India, although its benefits have been unevenly realized across regions and production ecologies. Since its establishment in 1957, the All India Coordinated Research Project (AICRP) on Maize has developed a large and diverse varietal pipeline, with 657 cultivars released to date, including 166 open-pollinated varieties, 305 public hybrids, and 186 private hybrids. Over time, breeding efforts have shifted decisively from composites and double-cross hybrids to single-cross hybrids tailored to specific agro-ecologies and market requirements, a transition that has underpinned national productivity gains from less than 0.5 t/ha in the early 1950s to about 3.59 t/ha in recent years (Yadav et al. 2015).

Improved hybrids increasingly combine tolerance to drought, heat, waterlogging, and major diseases, improving yield levels and stability, especially in irrigated, input-rich, and market-linked systems (Dhillon and Jat 2024). At the same time, public breeding programmes deliberately maintain a dual varietal pathway: single-cross hybrids for high-input, commercial systems, and improved OPVs, composites or synthetics for risk-prone, resource-constrained ecologies where seed recyclability and low cost are valued (Yadav et al. 2015). The public sector has now also extended to nutritional quality with Quality Protein Maize (QPM) and other biofortified hybrids and end-use specificity. However, adoption is constrained by misalignment with farmers' risk and trait preferences, weak market incentives for resilience and nutritional attributes, and perceptions of lower yield relative to commercial checks. Scaling is further limited by seed-system bottlenecks, as these segments are dominated by public-sector products with restricted seed systems and last-mile delivery. Specialty types such as sweet corn, popcorn, and baby corn have also expanded rapidly, though popcorn breeding in particular remains constrained by a narrow genetic base and reliance on imported or proprietary germplasm (Dhillon and Jat 2024). In baby corn systems, male-sterile varieties have eliminated the need for manual detasseling, substantially reducing labour requirements and production costs. Variety releases during 2021–25 indicate that across field corn, most releases were of the *Kharif* season, with 90 releases and an average yield of 7.65 t/ha. Rabi had fewer releases (32), but recorded a higher average yield of 9.49 t/ha. Among biofortified maize, *Kharif* accounted for 37 releases with an average yield of 6.82 t/ha, while two *Rabi* releases achieved a higher mean yield of 7.95 t/ha. However, realization of genetic gains is mediated by the seed system: despite roughly 65% of maize area being under improved cultivars, adoption of stress-tolerant and biofortified types remains limited, especially in *kharif*, eastern, tribal and hill regions, where local varieties and old composites dominate, and seed replacement rates are low (Yadav et al. 2015). Around 15–20% of maize area remains under low-yielding composites and landraces, indicating substantial scope for replacement with hybrids.

Private seeds dominate both normal grain hybrids and speciality corn seed systems in commercially oriented belts, driven by strong marketing and feed demand, while many promising public cultivars remain undersupplied due to weak, decentralized seed production and last-mile delivery (Dhillon and Jat 2024). Strengthening pluralistic, decentralized seed systems, linking public breeding, local producers, and private distributors, and ensuring better targeting of climate-resilient and cultivars to marginal ecologies, is therefore essential to translate breeding progress into more inclusive and climate-resilient maize productivity gains (Maqbool et al. 2021).

20.3.2 Agronomic and Crop Management

Optimizing sowing date, plant geometry, and resource management consistently increases maize yields and yield stability by aligning crop growth with favourable climate, improving canopy structure, and enhancing input-use efficiency (Wu et al. 2024; Lakhena et al. 2025). Across diverse environments, timely sowing maximizes radiation capture and avoids heat and drought around flowering and grain filling, with late sowing commonly reducing yields by about 15–30% (Lu et al. 2016; Cao et al. 2024; Bamboriya et al. 2025; Jiang et al. 2023). Timely sowing improves post-anthesis dry matter accumulation, nutrient uptake and use efficiency, and yield stability indices compared with delayed planting (Lu et al. 2016; Guo et al. 2022). Improved maize hybrids perform best at row spacings of 60–75 cm and intra row spacing of 15–22 cm, corresponding to plant populations of about 70,000–100,000 plants per hectare, which optimize light interception and nutrient uptake under adequate water and fertility. Higher plant densities may raise yields in irrigated, high-input systems but can reduce productivity under moisture or nutrient stress due to increased competition (Jat et al. 2013; Goyal et al. 2024). Configurations that improve canopy light distribution and space per plant (e.g. paired or strip rows) enhance photosynthesis, delay senescence, and support high yields (Guo et al. 2022; Deng et al. 2022; Wu et al. 2025).

Improved establishment on ridges, beds, or permanent beds, especially under conservation-agriculture-type practices, increases maize grain yield and water-use efficiency by roughly 5–25% over conventional tillage and enhances system productivity in maize-based rotations (Das et al. 2017; Jat et al. 2018; Lakhena et al. 2025). Conservation agriculture with residue retention and diversification improves yield stability, soil carbon, and resilience in long-term maize systems (Das et al. 2017; Jat et al.).

Integrated nutrient management—mineral fertilizer combined with organic inputs and micronutrients—raises yields, nutrient-use efficiency and profitability while lowering environmental footprints compared with mineral fertilizer alone (Dhaliwal et al. 2021; Kamal et al. 2025). Precision or model-based nutrient and water management further improves yields and water/nutrient productivity relative to blanket recommendations (Jat et al. 2018; Wu et al. 2023). Crop diversification through rotations and intercropping, especially with legumes, improves nutrient cycling, land use efficiency, profitability, and pest and mycotoxin management (Meena et al. 2024; Drakopoulos et al. 2021).

20.3.3 Mechanization, Precision Agriculture, and Post-Harvest Technologies

Maize mechanization in India increasingly covers the seed-to-seed production cycle, improving timeliness, labour efficiency, and productivity; however, manual planting and manual cob picking still dominate among small and marginal farmers, especially in rainfed, tribal, and eastern regions, due to fragmented landholdings, small plot sizes, and limited access to machinery services. Pre-sowing operations are largely mechanized through tractor- and power-tiller–mounted implements such as ploughs, harrows, rotavators, and laser levellers, while conservation-agriculture–oriented options—zero-till, strip-till, and permanent-bed planters—reduce tillage intensity, fuel use, and cultivation costs, with strip-till multi-crop planters showing yield and cost advantages over conventional tillage (Johansen et al. 2012; Bell et al. 2017; Kumar and Karmakar 2025). Mechanized sowing using seed-cum-fertilizer drills, inclined-plate, vacuum, and pneumatic planters improves seed singulation, spacing, and depth control, enhancing crop establishment and enabling precise single-seed placement (Kus 2021; Pareek et al. 2023). Nevertheless, manual dibbling remains widespread, particularly where access to planters through Custom Hiring Centres (CHCs) is limited. During crop growth, inter-row weeders, power weeders, sprayers, and fertigation systems support timely weed, nutrient, and pest management, although manual weeding continues to be common in labour-abundant areas (Nyamangara et al. 2014; Sundaram et al., 2021).

Harvest mechanization is expanding through maize pickers, combine harvesters, and forage harvesters for silage, yet manual harvesting of grain maize remains predominant, constrained by uneven maturity, lodging, and small field size. Post-harvest mechanization—shellers, dryers, cleaners, and graders—significantly reduces losses and improves grain. Maize drying reduces grain moisture to safe levels (≈13–14%), limiting storage losses, mould and mycotoxin risks, and insect damage. Access to mechanical dryers is therefore a market-enabling technology rather than merely a post-harvest convenience, as it allows farmers to delay sales, aggregate produce, and negotiate better prices instead of accepting discounts imposed by traders for wet grain. By improving quality compliance and timing of sales, dryers strengthen farmers' bargaining power and enhance income realization. The market benefits of mechanization are most evident where machinery access is embedded within collective institutions such as Custom Hiring Centres and Farmer Producer Organizations. These platforms link mechanized services with aggregation, grading, drying, and bulk marketing, enabling smallholders to benefit from economies of scale and participate in formal value chains. Government supported Custom Hiring Centres play a crucial role in promoting inclusive farm mechanization by providing affordable machinery services, thereby reducing cultivation costs and improving the net income of small and resource-poor farmers (Kumar et al. 2026). In the absence of such institutional linkages, mechanization remains confined to production efficiency gains without translating into improved market outcomes. Emerging precision tools such as sensor-based nutrient management and digital decision support show promise, but their scaling will depend on service-based delivery models that complement existing manual practices rather than replace them outright (Jat et al. 2018; Hasanain et al. 2025).

20.3.4 Integrated Pest Management and the Fall Armyworm Challenge

Fall armyworm (FAW), *Spodoptera frugiperda*, first reported on maize in Karnataka in 2018, has rapidly spread across all major maize-growing regions of India and emerged as a major yield constraint (Sharanabasappa et al. 2018; Suby et al. 2020). Surveys in rainfed maize record widespread infestation (44–100%) and substantial plant damage (22–47%), with FAW largely displacing native stem borers (Navik et al. 2021). Severe infestations occur from early vegetative to tasselling stages in both *kharif* and *rabi* crops, with damaging populations persisting in areas where maize is grown year-round (Varshney et al. 2020; Firake and Behere 2020). FAW management in India has been dominated by chemical control, which can be effective when applied early and accurately but has led to high costs, variable results, and rising resistance risks (Deshmukh et al. 2020; Ahir et al. 2022; Sharma and Kumar

2024). In contrast, integrated pest management (IPM) approaches—combining pheromone-based monitoring, biological and botanical options, and need-based insecticide use—consistently outperform calendar spraying, reducing pest pressure and increasing yields at lower cost (Varshney et al. 2020; Rajashekhar et al. 2024; Sireesha et al. 2024).

Low-cost farmer practices such as applying soil, sand, or ash into the maize whorl, and whorl-targeted application of control agents using carriers, further improve control when timed at early larval stages, while reducing pesticide use and off-target effects (Babendreier et al. 2020; Divya et al. 2021; Kalleshwaraswamy et al. 2022). Large-scale IPM demonstrations confirm that surveillance-led, threshold-based interventions reduce FAW infestation, increase maize yields by 12–25%, and improve benefit–cost ratios relative to sole insecticide-based strategies (Ramadevi et al. 2024).

20.3.5 Farmers' Innovations on Maize

Farmer innovation in maize extends beyond field practices to encompass seed choice, agronomy, post-harvest management, finance, and marketing along the value chain (Kar et al. 2025). These farmer-led, co-created innovations are typically low-cost, context-specific, and closely aligned with local constraints, making them more acceptable than standardized technologies, though they often lack formal recognition and scaling pathways (Perinelle et al. 2021). Through on-farm experimentation and participatory trials, farmers routinely modify recommendations on spacing, varietal choice, and pest management, generating locally viable "self-innovations" that formal extension frequently overlooks (Kwapong and Ankrah 2023).

Adoption is highest for practices with clear, immediate benefits—such as timely harvesting, rapid drying, and basic pest control—while more input- and knowledge-intensive innovations lag due to resource and access constraints (Jones-Garcia and Krishna 2021). In India, practices such as zero-tillage maize and optimized seasonal packages have emerged through farmer experimentation supported by on-farm trials, resulting in higher productivity and profitability (Pilli et al. 2025). Innovation also extends to markets, where digital platforms, aggregation models, and informal finance reduce losses and improve price discovery, though weak aggregation, poor storage, and persistent inclusion gaps—especially for women and poorer farmers—continue to limit returns (Reddy et al. 2025). Overall, durable gains occur when farmer co-creation is embedded within supportive innovation systems that integrate extension, finance, ICT tools, and market institutions, enabling local knowledge to translate into scalable, quality-oriented maize value chains (Banda 2022; Paleologo et al. 2025).

20.4 Market Trends and Industrial Demand

In India, maize has rapidly evolved into a multi-use, market-responsive crop, with demand dominated by poultry and livestock feed, expanding industrial uses, emerging bioethanol programs, and small but lucrative specialty segments.

20.4.1 Changing Utilization Patterns and Demand Growth

Recent utilization patterns indicate a pronounced shift in maize use in India toward feed and industrial markets. Currently, about 44% of total maize production is consumed as poultry feed and a further 11% as livestock feed, making the poultry–livestock sector the dominant sink for maize. In addition, there is a growing trend of maize cultivation specifically for silage, particularly in the western Indo-Gangetic Plains (IGPs), to meet the rising demand from the dairy sector. Industrial uses are also significant: roughly 12% of maize is absorbed by the starch industry, while about 18% is utilized by ethanol and other industrial users. In contrast, direct food consumption and food processing together account for only about 15% of total maize use.

Maize is the preferred energy ingredient in poultry rations, constituting 50–60% of broiler feed and 25–35% of layer feed, which makes feed manufacturers highly sensitive to maize prices and quality attributes. Rapid growth of organized poultry production and feed milling has therefore been a key driver of maize demand expansion. Bioethanol has emerged more recently as an additional demand source, reflecting global trends. India has achieved the 20% ethanol blending (E20) target for 2025–26 and now aims to reach 30% blending (E30) by 2030. Achieving this target will require a substantial expansion in annual maize production, implying a sustained annual growth rate of about 8–9% (NASS 2024). Empirical evidence points to strong spatial and vertical integration between maize and poultry markets. Prices in major maize-producing and consuming states, such as Andhra Pradesh, Tamil Nadu, and Karnataka, are cointegrated with egg and broiler prices, indicating efficient price transmission along the maize–poultry value chain. While this integration provides relatively stable long-run demand for producers, it also transmits price volatility to farm-gate levels when feed companies adjust ration formulations or substitute maize with alternative ingredients.

Looking ahead, maize demand for ethanol is projected to rise sharply over the next decade. Based on current utilization patterns and blending targets, maize demand for ethanol alone could reach 71.5–76.3 million tonnes by 2035–36, excluding export requirements (Jat et al. 2024). This trajectory necessitates strengthening maize productivity, post-harvest quality management, and market institutions to balance competing food, feed, and fuel demands, while ensuring that rising industrial demand translates into stable and remunerative outcomes for farmers.

20.4.2 Export Markets and Global Integration

India's ability to grow maize year-round, its relatively low freight costs, a mature seed system, multiple seaports, and its non-GMO status give it strong potential as a reliable maize exporter, especially to nearby Asian markets such as Nepal, Bangladesh, Vietnam, Malaysia, and Indonesia (Tsering and Shende 2022; Kolhe et al. 2022; Patel et al. 2025; Chauhan et al. 2021). However, exports remain largely surplus-driven and volatile, and are increasingly constrained by the need to comply with stringent and segmented food and feed safety standards—particularly for mycotoxins, moisture content, traceability, and GMO identity preservation—in global maize trade (Shiferaw et al. 2011; Ngwenya et al. 2025; Chen and Villoria 2019; Bren d'Amour and Anderson 2020). Inadequate drying, storage, and post-harvest management increase contamination risks, which restrict access to high-value markets and depress price realization (Ngwenya et al. 2025; Shiferaw et al. 2011; Tsering and Shende 2022; Murdia et al. 2016; Chauhan et al. 2021).

Sustained growth in maize exports, therefore, hinges on strengthening quality-assurance systems and accredited laboratory capacity, together with investments in improved drying, storage, and bulk-handling infrastructure to consistently meet the regulatory thresholds of importing countries (Tsering and Shende 2022; Kolhe et al. 2022; Chauhan et al. 2021; Nirmal et al. 2024). Although exports are unlikely to dominate India's maize demand in the near term, they can act as a strategic outlet for periodic surpluses and help stabilize prices as domestic feed, industrial, and bioenergy demand expands, provided that export quality and safety standards are progressively aligned with domestic market requirements (Shiferaw et al. 2011; Deepak et al. 2025; Beckman et al. 2025; Opadokun et al. 2025).

20.5 Linking Innovation, Markets, and Scaling Pathways

Evidence from India and other maize-growing regions shows that productivity, farmer adoption, and market demand are tightly interlinked rather than operating as separate domains (Pingali 2012; Joshi et al. 2019). Sustainable upscaling of maize in India, therefore, requires a shift from fragmented, technology-centric interventions towards integrated pathways that couple technological change with market incentives and supportive institutions. In this perspective, breeding priorities, agronomic practices, value-chain organization, and policy instruments must be co-designed so that innovations respond to end-user demand while remaining accessible to diverse categories of farmers, including smallholders and marginalized groups (Agarwal et al. 2021).

20.5.1 Trait–Market Alignment and Demand-Responsive Innovation

Trait market alignment is a central mechanism through which innovation systems connect to evolving demand, ensuring that breeding investments translate into varieties valued by both producers and buyers. Different market segments priorities distinct grain and plant characteristics such as energy density, protein quality, kernel hardness, starch content, colour, and mycotoxin resistance, which in turn shape breeding pipelines, seed system strategies, and management recommendations (Prasanna et al. 2020; Rakshit et al. 2014). Feed markets, particularly poultry, reward high metabolizable energy, good pelletability, and low aflatoxin contamination, while starch and ethanol industries demand high and stable starch content, good fermentability, and uniform grain quality. In contrast, specialty food markets emphasize taste, texture, nutritional attributes, and appearance (Murthy et al. 2016; Kumar et al. 2020). In this context, waxy maize exhibits higher starch to ethanol conversion efficiency (approximately 93%) than normal maize (approximately 88.2%) (Yangcheng et al. 2013), supporting the case for developing waxy maize hybrids with high amylopectin and rapidly degradable starch for bioethanol applications (Jat et al. 2024). Similarly, QPM addresses both human nutritional needs and high-quality poultry and pig feed requirements, illustrating how trait-specific breeding can serve multiple demand segments simultaneously (Dei 2017; Kaul et al. 2019; Bidi et al. 2019). Demand responsive innovation requires these downstream requirements to be explicitly incorporated into breeding programmes, varietal design, agronomic advisories, and extension curricula through market-informed trait prioritization and target product profiles (Morris et al. 2001; Setimela et al. 2017). When clear and credible price signals and buyer requirements are communicated, producers are more likely to invest in suitable hybrids, fertilizer regimes, and post-harvest practices (Agarwal et al. 2021). However, weak and inconsistent price premiums for quality and end-use traits at the farm gate continue to limit adoption of demand-responsive hybrids and management practices, thereby constraining effective trait market alignment.

20.5.2 Extension and Institutional Pathways for Scaling Maize Innovations

Effective scaling of maize innovations in India depends on integrated extension systems that link technical advice with institutional and market access. Face-to-face approaches—such as on-farm demonstrations, Farmer Field Schools (FFS), lead farmers, and participatory trials—remain essential for trust building, local adaptation, and farmer-led learning under climate variability. These approaches gain scale and timeliness when complemented by digital advisories delivering weather, pest, nutrient, and price information, which have been shown to improve adoption of climate-smart practices and enhance yields and incomes in maize systems.

Collective arrangements, including Custom Hiring Centres (CHCs), cooperatives, and Farmer Producer Organizations (FPOs), play a pivotal role by embedding extension within access to mechanization, inputs, storage, and markets, thereby reducing costs and enabling quality-oriented maize production (Prabhavathi et al. 2021; Basavaraj 2022; Ganavi et al. 2024; Kotyal 2025). CHCs facilitate timely land preparation, planting, harvesting, and drying, while FPOs strengthen aggregation, quality control, and price negotiation, aligning technical recommendations with economic incentives (Das and Patil 2024a, b; Anand et al. 2025; Singh 2023).

Market-oriented and digitally enabled platforms covering grading, traceability, contract farming, and bundled weather and agronomy services can further improve coordination and risk sharing, although their effectiveness depends on connectivity, digital literacy, trust, and appropriate regulation (Jose et al. 2024). Extension priorities must therefore extend beyond production to include integrated pest management, particularly Fall Armyworm surveillance and threshold-based control, post-harvest quality management including drying, storage, and mycotoxin prevention, and emerging market-linked maize systems.

In this context, scaling specialty corn such as sweet corn, baby corn, and popcorn requires focused extension on varietal choice, crop scheduling, harvest timing, and post-harvest handling, supported by contract arrangements and linkages with organized processors for assured quality and price realization. Likewise, scaling silage maize calls for focused extension and market integration to strengthen feed and fodder security in India.

Scaling maize in India is therefore best understood as a policy and institutional challenge. Productivity gains from hybrids, mechanization, and improved agronomy depend on supportive price policies, market reforms, infrastructure investment, and adaptive extension. Building stronger FPOs, shared mechanization and service-provider models, and robust public–private partnerships, together with pluralistic, digitally enabled extension systems, is crucial to convert technological and market opportunities into resilient, inclusive maize growth. In this context, large-scale Frontline Demonstrations across diverse agro-climatic zones demonstrate that integrated maize technology packages can deliver substantial yield gains under real farm conditions, highlighting the pivotal role of field-based demonstrations in effective scaling.

20.6 Future Outlook and Policy Interventions

Maize in India is poised to expand at the food feed fuel interface amid rising climate stress and rapid market diversification. Future progress will depend on balancing productivity growth with evolving market demand and environmental sustainability. Breeding priorities should increasingly focus on multi-stress-tolerant and end use specific hybrids for feed, silage, bioethanol, and specialty markets, with particular emphasis on improving kharif productivity where growth potential is highest. At the same time, affordable, timely, and quality seed must be delivered through efficient and responsive seed systems that are aligned with regional market requirements.

Adoption of improved technologies remains constrained by interacting climatic, market, and institutional risks. Weak and volatile farm gate prices, limited quality differentiation, and asymmetric bargaining power often discourage investment in improved seed and inputs. Policies must therefore strengthen risk mitigation through insurance, extension, contract arrangements, and farmer organizations, while improving market access and price discovery. Strategic investments in maize dryers, storage infrastructure, grading facilities, integrated e-market platforms, and expanded procurement by NAFED and NCCF can strengthen market integration, reduce distress sales, and stabilize producer prices.

Capacity building and effective technology demonstration are essential to translate innovation into impact. Strengthening extension systems, scaling on-farm demonstrations of integrated maize packages, expanding access to mechanization through Custom Hiring Centres, and linking these services with aggregation and marketing platforms will enable timely operations, reduce labour bottlenecks, and lower production risk. Coordinated interventions across seed systems, irrigation, mechanization services, market infrastructure, and ethanol-linked value chains, supported by targeted research incentives for industrially suitable maize, are necessary to build a resilient, competitive, and equitable maize economy. At the same time, rapid expansion of bioethanol demand introduces potential trade-offs with food security and price stability, reinforcing the need for coordinated policy frameworks that balance competing market uses while safeguarding producer incentives.

20.7 Conclusion

Maize in India has moved beyond its traditional status as a secondary coarse cereal and has become a strategically important crop linking food, feed, fodder, fuel, and industrial systems. Sustained growth in maize production over the past two decades reflects the combined effects of genetic improvement, better agronomic management, mechanization, and expanding demand from poultry, starch, and bioethanol industries. However, this growth has been uneven across regions, seasons, and socio-economic groups. Large and persistent yield gaps, particularly in rainfed kharif maize, indicate that constraints to productivity are driven more by management quality, institutional access, and market linkages than by genetic potential alone.

The evidence shows that isolated technological interventions are insufficient to realize maize's full potential. Productivity and income gains are achieved when innovations are applied as integrated packages that combine improved cultivars, timely crop management, balanced nutrient and pest management, mechanization services, and effective post-harvest practices such as drying and storage. Equally important is aligning production practices with evolving market requirements for quality, safety, and end use. Future progress in the maize sector will depend on integrated scaling pathways that connect research, extension, markets, and institutions.

Prioritizing maize systems by strengthening seed and extension delivery, expanding shared mechanization and post-harvest infrastructure, and improving price discovery through collective and digital platforms are central to building a resilient, competitive, and equitable maize economy in India.

References

Agarwal B, Anthwal P, Verma R (2021) Smallholder agriculture, markets and gendered value chains in India. J Agrar Chang 21(3):456–479

Ahir KC, Bharpoda TM, Patel HC, Patel MG (2022) Bio-efficacy of newer insecticides against fall armyworm, *Spodoptera frugiperda* (J.E. Smith) infesting maize. J Entomol Zool Stud 10(1):173–178

Anand R, Burman RR, Saini S (2025) Farm mechanization for smallholder farmers in India: prospects and challenges. Agric Eng Today 49(1):1–12

Babendreier D, Agboyi LK, Beseh P, Osae M, Nboyine J, Ofori-Sekyere E, Gatsi B, Buah S, Clottey V, Adama I, Osei-Mensah E, Kenis M, Kwadjo KE (2020) The efficacy of alternative, environmentally friendly plant protection measures for control of fall armyworm, *Spodoptera frugiperda*, in maize. Insects 11(4):240

Bamboriya S, Sepat S, Kumar R, Jat SL, Rakshit S (2025) Optimization of sowing time to mitigate heat stress in spring maize (*Zea mays*) in Indo Gangetic plains of India. Indian J Agric Sci 95(2):136–143

Banda G (2022) Evolution of Zimbabwe's maize innovation ecosystems: building an institutional innovation infrastructure that supports food security. Afr Dev 47(4):107–134

Basavaraj G (2022) A review on custom hiring services under Indian conditions: farmer's perception, associated factors, constraints, and suggestions. Asian J Agric Exten Econ Sociol 40(9):1–12

Beckman J, Johnson ME, Ajewole K, Kaufman J, Sabala E (2025) The growing demand for animal products and feed in India: future prospects for production, trade, and technology innovation. https://doi.org/10.32747/2025.9015821

Bell RW, Johansen C, Haque ME, Hossain MM, Rashid MH (2017) Mechanised minimum soil disturbance seeding systems for smallholder conservation agriculture in South Asia. In: Reicosky DC (ed) Conservation agriculture. CABI, Wallingford, UK

Bidi B, Nsahlai IV, Apaloo J (2019) Performance of broilers and pigs fed quality protein maize: a meta-analysis. Anim Feed Sci Technol 251:66–78

Cao ZY, Chen ZH, Tang B, Zeng Q, Guo HL, Huang WH, Luo Y, Shen S, Zhou SL (2024) The effects of sowing date on maize: phenology, morphology, and yield formation in a hot subtropical monsoon region. Field Crop Res 303:109244

Chauhan D, Walia R, Singh C, Deivakani M, Kumbhkar M (2021) Detection of maize disease using random forest classification algorithm. Turkish J Comput Maths Educ 12(9):715–720

Chen B, Villoria NB (2019) Climate shocks, food price stability and international trade: evidence from 76 maize markets in 27 net-importing countries. Environ Res Lett 14(1):014007

D'Amour CB, Anderson W (2020) International trade and the stability of food supplies in the Global South. Environ Res Lett 15(7):074005

Das C, Patil S (2024a) Optimizing farm machinery utilization through Custom Hiring Centers: a farmers' perspective. Ecol Environ Conserv 30(1):120–126

Das C, Patil SB (2024b) Economic analysis of custom hiring centres in Vijayapur district: viability and impact. Int J Res Agron 7(3):45–55

Das A, Ghosh PK, Lal R, Saha R, Ngachan S (2017) Soil quality effect of conservation practices in maize–rapeseed cropping system in Eastern Himalaya. Land Degrad Dev 28(6):1862–1874. https://doi.org/10.1002/ldr.2325

Dass S, Chikkappa GK, Kumar B, Kaul J, Ashok Kumar R, Kumar V, Singh V (2015) Genetic enhancement and crop management lead maize revolution in India. In: Gupta SK, Singh AK, Gautam NC (eds) Maize for food, nutrition and environmental security. Indian Society

of Genetics and Plant Breeding/ICAR–Indian Institute of Maize Research, New Delhi, pp 267–283
Deepak TN, Vasudevan SN, Palanimuttu, Mahadevu P, Shobharani M, Siddaraju R, Thimmegowda (2025) Influence of modified atmospheric storage conditions on seed longevity and health of maize genotypes (*Zea mays* L.). Cereal Res Commun 53(4):2335–2342
Dei HK (2017) Quality protein maize as animal feed: a review. Anim Nutr 3(1):1–7
Deng X, Bai H, Liu D, Tang J, Wang B, Shen Y, Cao J, Feng P (2022) Projecting future changes in extreme climate for maize production in the North China Plain and the role of adjusting the sowing date. Mitig Adapt Strateg Glob Chang 27(5):35
Deshmukh SS, Pavithra H, Kalleshwaraswamy CM, Shivanna B, Maruthi M, Mota-Sanchez D (2020) Field efficacy of insecticides for management of invasive fall armyworm, *Spodoptera frugiperda* (J.E. Smith) (Lepidoptera: Noctuidae) on maize in India. Fla Entomol 103(2):221–227
Dhaliwal SS, Sharma S, Sharma V, Shukla AK, Walia SS, Alhomrani M, Hossain A (2021) Long-term integrated nutrient management in the maize–wheat cropping system in alluvial soils of North-Western India: influence on soil organic carbon, microbial activity and nutrient status. Agronomy 11(11):2258
Dhillon BS, Jat SL (2024) Developing synthetics as a by-product of hybrid maize breeding. Indian J Genet Plant Breed 84(2):281–289
Divya K, Kumar A, Singh R (2021) On-farm evaluation of eco-friendly management practices against fall armyworm, *Spodoptera frugiperda* (J.E. Smith) in maize. J Entomol Zool Stud 9(1):114–120
Drakopoulos D, Kägi A, Six J, Zorn A, Wettstein FE, Bucheli TD, Vogelgsang S (2021) The agronomic and economic viability of innovative cropping systems to reduce Fusarium head blight and related mycotoxins in wheat. Agric Syst 192:103198
Erenstein O, Jaleta M, Sonder K, Mottaleb K, Prasanna BM (2022) Global maize production, consumption and trade: trends and R&D implications. Food Secur 14:1295–1319
Firake DM, Behere GT (2020) Natural mortality of invasive fall armyworm, *Spodoptera frugiperda* (J.E. Smith) (Lepidoptera: Noctuidae) in maize agro ecosystems of Northeast India. Biol Control 148:104290
Ganavi NR, Kumar NR, Ravikumar S, Kambale P (2024) Impact of custom hiring centers on cost, profitability, and efficiency in crop production in Anantapur district of Andhra Pradesh, India. J Sci Res Rep 30(5):1–14
Ganguly K, Gulati A, von Braun J (2017) Innovations spearheading the next transformations in India's agriculture. India Policy Forum 13(2016–17):27–79
Ghosh A, Singh P, Singh R, Jat ML, Kumar D (2020) Crop productivity, soil health and energy dynamics of Indian Himalayan intensified organic maize-based systems. Int Soil Water Conserv Res 8(4):457–468
Goyal M, Kaur R, Umar M (2024) Assessment of sowing dates and spacing for indogangetic plain zone kharif maize (*Zea mays*) cultivars to achieve higher productivity in Western IGP. Indian J Agron 69(4):384–390
Guo D, Chen C, Li X, Wang R, Ding Z, Ma W, Zhou B (2022) Adjusting sowing date improves the photosynthetic capacity and grain yield by optimizing temperature condition around flowering of summer maize in the North China Plain. Front Plant Sci 13:934618
Hasanain M, Singh R, Jat ML (2025) Digital and precision technologies for climate-smart maize systems in South Asia: opportunities and pathways for scale. In: Prasanna BM (ed) Climate-smart maize systems for the global tropics. CIMMYT, Mexico
Jat SL, Parihar CM, Singh AK, Jat ML, Sinha AK, Mishra BN, Singh RN (2013) Integrated nutrient management in Quality Protein Maize (*Zea mays*) planted in rotation with wheat (*Triticum aestivum*): effect on productivity and nutrient use efficiency under different agro-ecological conditions. 83(4) https://epubs.icar.org.in/index.php/IJAgS/article/view/28836
Jat HS, Datta A, Choudhary M, Sharma PC, Yadvinder-Singh, Sidhu HS, Jat ML (2018) Assessing soil properties and crop productivity in a maize–wheat rotation under precision conservation agriculture in north-west India. Soil Tillage Res 180:240–253

Jat SL, Hossain F, Zaidi PH, Muthusamy V, Prasanna BM (2024) Waxy maize: opportunities and breeding strategies for bioethanol and specialty starch markets in India. Curr Sci 126(7):980–988

Jat SL, Jat HS, Rakshit S, Sharma PR, Kumar B, Kakraliya M, Gathala MK, Bijarniya D, Kalwania KC, Singh Y, Choudhary M, Jat ML (2025) Maize as an alternative to resource-intensive rice: empirical insights from on-farm participatory study under diverse agricultural scenarios in the Indo-Gangetic Plains of Northwestern India. Front Sustain Food Syst 9:1700854

Johansen C, Haque ME, Bell RW, Thierfelder C, Esdaile RJ (2012) Conservation agriculture for small holder rainfed farming: opportunities and constraints of new mechanized seeding systems. Field Crop Res 132:18–32

Jones-Garcia E, Krishna VV (2021) Farmer adoption of sustainable intensification technologies in the maize systems of the Global South. A review. Agron Sustain Dev 41(1):8

Jose J, Balkrishna A, Pathak R, Kumar S, Singh S (2024) A comprehensive analysis of the advances in Indian digital agricultural architecture. Smart Agric Technol 5:100249

Joshi VR, Thorp KR, Coulter JA, Johnson GA, Porter PM, Strock JS, Garcia y Garcia A (2019) Improving site-specific maize yield estimation by integrating satellite multispectral data into a crop model. Agronomy 9(11):719

Jiang L, Wang M, Chu Z, Gao Y, Guo L, Ji S, Jiang L, & Gong L (2023) Effects of Temperature on Growth and Grain Maturity of Spring Maize in Northeast China: A Study of Different Sowing Dates. Atmosphere. https://doi.org/10.3390/atmos14121755

Kalleshwaraswamy CM, Sharanabasappa D, Pradeep P, Sannathimmappa HG (2022) Low-cost and farmer-friendly approaches for management of fall armyworm, *Spodoptera frugiperda*, in maize in India. Pest Manag Horticult Ecosyst 28(1):1–10

Kamal N, Saeed M, Sultan R, Khanum S, Abbasi SH, Saleem S, Shah SH (2025) Performance of spring wheat lines for agronomic and disease resistance traits at Wheat Research Sub-Station, Murree, Pakistan. Planta Anim 4(6):67–71

Kar P, Sharma R, Jat SL, Nigam S, Debnath S, Dar ZA, Sravani D, Sen D, Bhuyan N, Rakshit S (2025) Tracking the emergence and usage of farmer led innovations (FLIs) in maize ecosystem: a pan India exploration. Indian J Tradit Knowl 24(2):177–184

Kaul J, Kumar R, Hossain F, Jat SL (2019) Quality protein maize: breeding, adoption and prospects in India. Indian J Genet Plant Breed 79(4):625–635

Kolhe P, Perke DS, Chande JA (2022) Growth and export performance of maize from India. Asian J Agric Exten Econ Sociol 40(9):140–149

Kotyal K (2025) Economic impact of Farmer Producer Company membership on maize cultivation in Kollegala, Karnataka, India. Adv Res 25(3):1–15. https://doi.org/10.9734/air/2025/v26i21286

Kumar, N. R., Athare, P. G., Kumar, A., & Kumar, S. (2026). Impact of custom hiring services on paddy cultivation: A propensity score matching approach. Agricultural Research. https://doi.org/10.1007/s40003-025-00933-9

Kumar A, Karmakar S (2025) Mechanization of maize cultivation in India: status, challenges and opportunities. Agric Mechaniz Asia Afr Latin Am 56(3):15579

Kumar P, Savadatti PM (2023) Role of maize in accelerating agricultural growth and structural transformation in India. J Rural Dev 42(2):145–166

Kumar R, Hossain F, Muthusamy V, Jat SL, Yadava P (2020) Grain quality traits and end-use diversification in maize: priorities for breeding and value chain development in India. Indian J Genet Plant Breed 80(4):510–520

Kus E (2021) Evaluation of some operational parameters of a vacuum single-seed planter in maize sowing. J Agric Sci 27(3):327–334

Kwapong NA, Ankrah D (2023) Understanding innovation process within an interactive social network: empirical insights from maize innovations in southern Ghana. Cogent. Soc Sci 9(1):2185771. 4

Lakhena KK, Pooniya V, Biswakarma N, Zhiipao R, Singh A, Kumar D, Shivay Y, Govindasamy P, Bana R, Rathore SS, Upadhayay PK, Barman A, Swarnalakshmi K, Choudhary R, Jat R, Ramakrishnan S, Jat SL, Litskas VD, Goulao LF, Hirzel J (2025) Impact of nine years of conservation tillage and precise nutrient management on equivalent yields, soil microbial dynamics, and water–energy footprints of the maize–mustard rotation. Front Sustain Food Syst 9:1407061

Lu HD, Xue J, Guo D (2016) Efficacy of planting date adjustment as a cultivation strategy to cope with drought stress and increase rainfed maize yield and water use efficiency. Agric Water Manag 177:357–366

Maqbool MA, Beshir AR, Iqbal M (2021) Strengthening seed systems for climate-resilient and nutritionally enriched maize in South Asia. In: Prasanna BM (ed) Breeding and deploying multiple stress-tolerant maize varieties in the tropics. CIMMYT, Mexico, D.F., pp 145–166

Morris ML, Tripp R, Dankyi AA (2001) Adoption and impacts of improved maize production technology: a case study of the Ghana Grains Development Project. CIMMYT, Mexico, D.F.

Murdia LK, Wadhwani R, Wadhawan N, Bajpai P, Shekhawat S (2016) Maize utilization in India: an overview. Am J Food Nutr 4(6):169–176

Murthy CH, Hossain F, Muthusamy V, Zaidi PH (2016) Grain quality requirements of feed, food and industrial users of maize in India. In: Prasanna BM, Pixley KV, Hodson D et al (eds) Proceedings of the 13th Asian maize conference: maize for food, feed and nutrition security. CIMMYT, Mexico

Meena L, Meena S, Kumar L, Pillai N, & Ram T (2024) Evaluation of food and fodder based cropping systems for sustaining productivity, resource use efficiency and profitability in western plain zone of Uttar Pradesh. Indian Journal of Agronomy. https://doi.org/10.59797/ija.v68i4.5462

Nyamangara J, Mashingaidze N, Masvaya EN, Nyengerai K, Kunzekweguta M, Tiivavi R, & Mazvimavi K (2014) Weed growth and labor demand under hand hoe based reduced tillage in smallholder farmers' fields in Zimbabwe. Agriculture, Ecosystems & Environment, 187, 146–154. https://doi.org/10.1016/j.agee.2013.10.005

NAAS (2024) Maize for bioethanol production in India: prospects and strategy. Strategy Paper No. 19, National Academy of Agricultural Sciences, New Delhi: 20 pp

Navik O, Shylesha AN, Patil J, Venkatesan T, Lalitha Y, Ashika T (2021) Damage, distribution and natural enemies of invasive fall armyworm *Spodoptera frugiperda* (J. E. Smith) under rainfed maize in Karnataka, India. Crop Prot 143:105465

Ngwenya K, Luckert MK, Mohapatra S (2025) Gendered determinants of adaptation in growing maize in smallholder agriculture in East Africa. Environ Dev Sustain 27(1):1905–1932

Nirmal KR, Reddy CM, Tulasi R, Maheswari GU, Sapthami I (2024) Advancements in environmental data analysis for climate-resilient agriculture using remote sensing and deep learning. Remote Sens Earth Syst Sci 7(4):670–680

Opadokun WO, Ajewole T, Babatunde M, Olorunmaiye KS (2025) Influence of S-Metolachlor and Nicosulfuron on Weed Flora, density, and yield performance of maize. Al-Hikmah J Pure Appl Sci 5(1):18096802

Paleologo M, Acampora M, Barello S, Graffigna G (2025) Uncovering the landscape of participatory research in agricultural innovation: a scoping review. Citizen Sci Theory Pract 10(1):1–18

Pareek R, Singh D, Meena BS (2023) Performance evaluation of tractor operated inclined plate planter for maize crop. Int J Agric Eng 16(1):56–62

Parihar CM, Jat SL, Singh AK, Datta A, Parihar MD, Jat ML, Saharawat YS, Kakraliya SK, Yadav OP (2016) Conservation agriculture in irrigated intensive maize-based systems of north-western India: effects on crop productivity, water use and economics. Field Crop Res 193:104–116

Patel P, Bhanderi DR, Tank RV, Gadhiya A, Panchal B (2025) Impact of integrated nutrient management strategies on yield and economic returns of baby corn (*Zea mays* l.) production under south Gujarat conditions, India. Food Sci Agric Res Highligh 5:1–19

Perinelle A, Meynard J-M, Scopel E (2021) Combining on-farm innovation tracking and participatory prototyping trials to develop legume-based cropping systems in West Africa. Agric Syst 186(102952):7

Pilli K, Venkanna Y, Rao BB, Srinivas A, Kumar TV, Navya B, Vijaya D (2025) Impact of on-farm trials and front-line demonstrations on zero tillage maize in Peddapalli District, Telangana, India. J Sci Res Rep 31(4):251–259

Pingali PL (2012) Green revolution: impacts, limits, and the path ahead. Proc Natl Acad Sci USA 109(31):12302–12308

Prabhavathi Y, Kishore N, Charishma C (2021) Penetration of farm mechanization services through FPOs: rural dynamics-based business modelling. Int J Agric Sci 13(6):10837–10841

Prasanna BM, Cairns JE, Zaidi PH, Beyene Y, Makumbi D, Gowda M, Magorokosho C, Zaman-Allah M, Olsen MS, Das A, Worku M, Gethi J, Vivek BS, Pixley KV (2020) Harnessing maize genetic resources for multiple end uses: challenges and opportunities. In: Prasanna BM (ed) Maize in the developing world: trends, challenges and opportunities. CIMMYT. Quality, post-harvest handling, and mycotoxins, Mexico, D.F.

Rajashekhar M, Rajashekar B, Reddy TP, Chandrashekara KM, Vanisree K, Ramakrishna K, Sunitha V, Shaila O, Sathyanarayana E, Shahanaz S, Reddy S, Shankar A, Jahan A, Kumar P, Reddy MJM (2024) Evaluation of farmers-friendly IPM modules for the management of fall armyworm, *Spodoptera frugiperda* (J.E. Smith) in maize in the hot semi-arid region of India. Sci Rep 14:5972

Rakshit S, Sekhar JC, Guleria SK, Singode A, Dass S (2014) Maize research in India—status and future prospects. Maize J 3(2):1–16

Rakshit S, Singh NP, Khandekar N, Rai PK (2021) Diversification of cropping system in Punjab and Haryana through cultivation of maize, pulses and oilseeds. Policy paper. ICAR-Indian Institute of Maize Research, Ludhiana, p 37

Ramadevi K, Reddy DDR, Prasad NVVSD, Srinivasa Rao V (2024) Large-scale demonstrations of IPM modules against fall armyworm, *Spodoptera frugiperda* (J.E. Smith) in maize under farmers' fields. J Biol Control 38(1):18–26

Reddy KJ, Chandrashekara CP, Rajyalakshmi B, Biradar DP, Rajakumar GR, Kanannavar PS (2025) Development of soil test crop response (STCR) and STCR-integrated plant nutrition supply (IPNS) models to achieve targeted yields of maize (*Zea mays* L.) in vertisols of Northern Karnataka. Indian. J Agric Res 59(1):1

Sundaram PK, Rahman A, Singh AK, & Sarkar B (2021) A novel method for manual weeding in row crops. Indian J. Agric. Sci., 91(6), 946–948. https://doi.org/10.56093/ijas.v91i6.114307

Setimela PS, Magorokosho C, Lunduka R, Gasura E, Makumbi D, Tarekegne A, Cairns JE, Mwangi W (2017) On-farm productivity and farmer preferences for tropical maize hybrids in sub-Saharan Africa. Food Secur 9(3):593–608

Sharanabasappa D, Kalleshwaraswamy CM, Asokan R, Mahadeva Swamy HM, Maruthi MS, Pavithra HB, Hegde K, Navi S, Prabhu ST, Goergen G (2018) First report of the fall armyworm, *Spodoptera frugiperda* (J.E. Smith) (Lepidoptera: Noctuidae), an alien invasive pest on maize in India. Pest Manag Horticult Ecosyst 24(1):23–29

Sharma R, Kumar P (2024) Economic impact and resistance risk of insecticide-based management of fall armyworm (*Spodoptera frugiperda*) in Indian maize. Indian J Entomol 86(1):45–52

Shiferaw B, Prasanna BM, Hellin J, Bänziger M (2011) Crops that feed the world 6. Past successes and future challenges to the role played by maize in global food security. Food Secur 3(3):307–327

Singh A (2023) Agricultural extension reforms and institutional innovations for inclusive outreach in India. In: Burman RR (ed) Agricultural extension reforms in South Asia. Springer, New Delhi, pp 45–72

Sireesha K, Madhumathi T, Srinivas A (2024) Validation of integrated pest management modules against fall armyworm, *Spodoptera frugiperda* (J.E. Smith) in maize. Indian. J Entomol 86(2):312–319

Sivamurugan A, Surendrakumar A, Bharathi C, Karthikeyan R, Pazhanivelan S, Manivannan V, Shanmugapriya P (2025) Growth and yield of irrigated maize (*Zea mays* L.) as influenced by mechanization and nutrient management practices. Plant Sci 12:1–6

Suby SB, Mooventhan P, Ravichandran G, Kannan M (2020) First report of fall armyworm, *Spodoptera frugiperda* (J.E. Smith) (Lepidoptera: Noctuidae) on maize in Tamil Nadu, India. J Entomol Zool Stud 8(3):1707–1710

Tsering R, Shende NV (2022) Export dimension of maize in India, pp 473–479

Varshney R, Poornesha B, Raghavendra A, Lalitha Y, Apoorva V, Ramanujam B, Rangeshwaran R, Subaharan K, Shylesha AN, Bakthavatsalam N, Chaudhary M, Pandit V (2020) Biocontrol-based management of fall armyworm, *Spodoptera frugiperda* (J.E. Smith) (Lepidoptera: Noctuidae) on Indian maize. J Plant Dis Protect 127:325–337

Wu Y, Zhou G, Song Y, Ren S, Geng J, Zhao H, Song X (2023) A simulation study on optimization of sowing time of maize (*Zea mays* L.) for maximization of growth and yield in the present context of climate change under the North China Plain. Agronomy 13(2):385

Wu W, Yue W, Bi J, Zhang L, Xu D, Peng C, Chen X, Wang S (2024) Influence of climatic variables on maize grain yield and its components by adjusting the sowing date. Front Plant Sci 15:1406634

Wu L, Wang B, Quan H, Li Liu D, Feng H, Chen F, Wu L (2025) Designing virtual maize cultivars with optimal planting date and density can improve yield and water use efficiency under plastic mulching conditions. Field Crop Res 322:109723

Yadav OP, Hossain F, Karjagi CG, Kumar B, Zaidi PH, Jat SL, Chawla JS, Kaul J, Hooda KS, Kumar P, Yadava P, Dhillon BS (2015) Genetic improvement of maize in India: retrospect and prospects. Agric Res 4(4):325–338

Yadav OP, Prasanna BM, Yadava P, Jat SL, Kumar D, Dhillon BS, Solanki IS, Sandhu J (2016) Doubling maize (*Zea mays*) production of India by 2025—challenges and opportunities. Indian J Agric Sci 86(4):427–436

Yangcheng H, Jiang H, Blanco M, Jane J (2013) Characterization of normal and waxy corn starch for bioethanol production. J Agric Food Chem 61(4):932–939

Public-Private Synergy: The Success Story Behind India's Hybrid Maize Seed Production

21

Bhupender Kumar, Rakhi Singh, Shrikant Yankanchi, Chayanika Lahkar, Pushpendra, Pardeep Kumar, Bahadur Singh Jat, Shankar Lal Jat, Abhijit Kumar Das, Mukesh Choudhary, and Hanuman Sahay Jat

Abstract

Maize is presently the leading cereal in terms of production volume, and is poised to become the most widely grown industrial crop in the coming decade. It is a versatile and multivalent crop that finds its major application as animal feed, biofuel, and a nutritious food source. India ranks as the second most significant maize-producing country in Asia and is the fifth largest consumer and fifth largest producer in the world. India's hybrid maize seed production is characterized by distinct roles of public and private sectors. Despite dynamism shared by both the public and private sectors, significant gaps exist that hinder the effectiveness of the hybrid maize seed supply system, ultimately compromising optimal performance and equitable outcomes. The key gaps include inefficient translation of public research findings into commercial products, restricted availability of high-quality hybrid seeds and support for smallholder farmers, and region-centric maize seed production sites in the country. Collaboration between the two sectors is often hindered by issues such as a lack of mutual trust, Intellectual Property Rights complexities and benefit sharing, policy and regulatory ambiguities, divergent institutional goals, and operational challenges. Therefore, in this chapter, we will discuss the gaps between the public and private sectors and how to bridge them in order to achieve more efficient and resilient hybrid maize production in India.

B. Kumar (✉) · R. Singh · S. Yankanchi · Pushpendra · P. Kumar · B. S. Jat · S. L. Jat · A. K. Das · M. Choudhary · H. S. Jat
ICAR-Indian Institute of Maize Research Ludhiana, Ludhiana, Punjab, India
e-mail: mukesh.choudhary1@icar.org.in

C. Lahkar
Assam Agricultural University- Zonal Research Station (AAU-ZRS), Gossaigaon, Assam, India
e-mail: chayanika.lahkar@aau.ac.in

M. Choudhary et al. (eds.), *Next-Generation Maize Research*,
https://doi.org/10.1007/978-981-92-0722-0_21

Keywords

Maize · Seed supply · Hybrid · Public and private sectors · Gap

21.1 Introduction

Maize (*Zea mays* L.) stands as a critical cereal crop within the Indian agricultural landscape, demonstrating remarkable growth in area, production, and productivity over recent decades. Maize is also known as the "Queen of Cereals" due to its highest yield potential among the cereals (Kumar et al. 2025). As a C4 crop, it is comparatively more climate-resilient and can be cultivated across tropical, subtropical, and temperate regions, ranging from sea level up to 3000 metres in elevation. Maize plays a vital role in ensuring food and nutritional security, particularly in regions where it is traditionally cultivated. However, its significance extends beyond direct human consumption, playing a vital role as a primary ingredient in poultry and livestock feed, and serving as a crucial raw material for various industries, including starch and ethanol production (Kumar et al. 2025). Maize with high energy value and less anti-nutritional properties has tremendous demand in the feed industry. The most important use and demand driver of maize is poultry feed, which accounts for 47% of total maize consumption. Food consumption accounts for 20% of maize production, with 13% being direct consumption and 7% being processed food (YES Bank and FICCI 2024). It is the most versatile crop being grown on 12.0 million hectares with a production of 42.3 million tonnes (2024–25) (Fig. 21.1). Among Indian states, Madhya Pradesh and Karnataka are the largest producers of maize in India (Fig. 21.2) (https://upag.gov.in/).

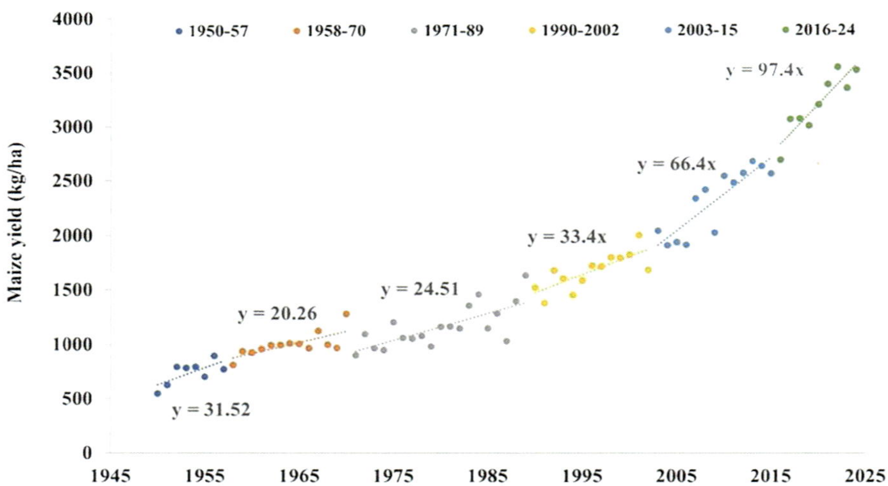

Fig. 21.1 All India maize area, production, and productivity trends (1950–2025)

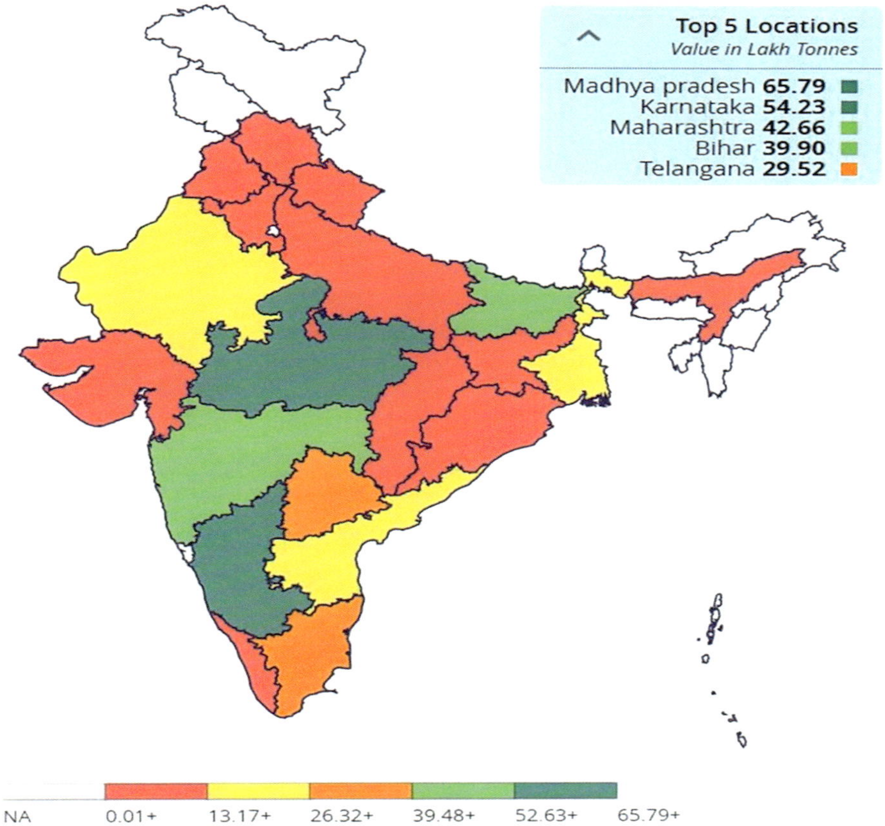

Fig. 21.2 Distribution of maize production in India 2024–25 (https://upag.gov.in/)

India's maize production has increased significantly, with yields growing faster than those of other major food grains. However, despite this progress, projections indicate a potential demand-supply gap emerging in the coming decades, necessitating strategies focused on productivity-led growth. High-quality seed, especially hybrids, plays a pivotal role in boosting agricultural productivity. The Indian hybrid maize seed supply system is characterized by distinct but interdependent roles played by the public and private sectors. Public organizations/institutions like the Indian Council of Agricultural Research (ICAR) and State Agricultural Universities (SAUs) provide the foundational research, germplasm resources, and policy frameworks. Historically dominant, their role shifted following the 1988 New Seed Development Policy, which catalysed substantial private investment, particularly in hybrid maize R&D, production, and marketing. Despite this dynamism, there are significant gaps and coordination weaknesses between the public and private sectors. Therefore, in this chapter, we will discuss the key gaps between the public and private sectors and how to address the existing gaps for a more efficient and resilient hybrid maize seed system in India.

21.2 The Maize Economy of India

As the third most important cereal crop after rice and wheat, maize plays a crucial part in India's economy. Maize makes a substantial contribution to Indian agricultural production. The rising demand from diverse sectors, particularly feed and fuel, underscores maize's growing importance for farmers' income and the overall agricultural economy (Erenstein et al. 2022). The maize economy has shifted, and production has increased drastically since the introduction of hybrid varieties. However, in recent years, major changes have taken place as a result of the growing commercial focus of the agricultural economy and heightened demand for diversified end uses, particularly in feed and industrial applications. Concurrently, significant investments in maize research have led to the development of advanced production technologies that have provided farmers with the means to respond to changes in demand. Approximately 2.0 lakh tons of hybrid maize seeds are currently needed in our nation to cover about 10 million hectares of land under maize cultivation.

21.3 Policy Dynamics and Their Impact on India's Hybrid Maize Seed Industry

The Royal Commission on Agriculture, established in 1928, was the first official organization to acknowledge the importance of high-quality seeds. In seeking to enhance India's agricultural sector, the Royal Commission focused on the production and distribution of quality seed. Subsequently, the National Seeds Corporation, Ltd (NSC) was established in 1963 for the production and distribution of seeds. This was followed in 1966 by the enactment of the Central Seed Act, which offered a legal framework for regulating and ensuring seed quality. In 1967, the National Seed Review Team was formed to make recommendations to the National Commission on Agriculture. These recommendations called for strengthening the authority of the NSC and the establishment of new government bodies to produce and distribute seed (Singh et al. 1995). Policy reforms introduced in the late 1980s increased the involvement of the private sector in the Indian seed industry, leading to a significant rise in investments in plant breeding research and seed production. The maize seed industry, in particular, experienced significant growth as a result of these reforms. In the early 1990s, there was a notable increase in the number of private companies involved in maize seed production in India, along with a significant rise in private sector funding for maize research (Singh et al. 1995). Concurrently, the growing accessibility of proprietary hybrids developed by private seed companies has sped up the diffusion of hybrid maize seeds. While recent developments have undoubtedly benefited many of India's maize farmers, concerns remain in some circles that private seed companies are primarily targeting assured ecologies, large-scale commercial growers, often overlooking small-scale, subsistence-oriented farmers who are seen as less profitable markets for commercial seed (Morris 2002). It is challenging to determine whether or not this concern is justified without comprehensive information about the actual conditions under which maize seed is utilized.

Both the public and private sectors are crucial to India's hybrid maize seed supply (Fig. 21.3). Public sector institutions mainly focus on research and development, while the private sector places greater emphasis on the development and commercialization of hybrid seeds (Mishra et al. 2025). Until the late 1980s, India's seed industry was largely under public sector control. This changed with the introduction of two major policy reforms that enabled private sector participation. The Industrial Licensing Policy of 1987, for the first time, allowed large Indian companies, including those with up to 40 percent foreign equity-to produce and market seeds in the country. This was followed by the New Policy for Seed Development in 1988, which liberalized domestic seed trade, reduced restrictions on the import of germplasm, seeds, and seed-processing equipment, and encouraged the entry of foreign firms into the seed sector. These reforms were further strengthened by the New Industrial Policy of 1991, which permitted foreign direct investment in the seed industry. Collectively, these policy changes led to a rapid expansion of private sector participation in the seed industry and generated valuable insights for public policymakers and corporate decision-makers in South Asia on the linkages between seed policy reforms, private investment in research and development, and productivity growth in agriculture (Fig. 21.3).

21.4 Public Sector Contributions

The public sector in India serves a foundational role in agriculture, with a broad mandate encompassing research, education, extension, and policy formulation. At the apex is the ICAR, based in New Delhi, which coordinates, directs, and manages agricultural research, education, and extension activities throughout the nation (Pal 2017). ICAR functions through an extensive network of over 100 research institutes, which include deemed universities, supports Central and SAUs, and manages Krishi Vigyan Kendras (KVKs). SAUs mainly focus on conducting agricultural research and education that is specific to each state, while the State Departments of Agriculture primarily handle agriculture extension. Public funding originates from both the Union Government and State governments, with some funds transferred from ICAR to SAUs. Public institutions viz ICAR-Indian Institute of Maize Research (IIMR) and the SAUs, have historically played a crucial role in breeding initiatives, resulting in the development of numerous improved maize varieties and hybrids. The All India Coordinated Research Project on Maize (AICRPs), a collaborative multi-location evaluation system managed by ICAR, has been instrumental in developing and releasing varieties that are well adapted to diverse agro-climatic environments (Kumar et al. 2012).

IIMR specifically mandates basic and strategic research for maize productivity enhancement, coordination of multi-location research, managing germplasm, developing different types of corn hybrids/varieties, and disseminating technology (Kumar et al. 2012). A critical role is played by the ICAR-National Bureau of Plant Genetic Resources (NBPGR), which serves as the nodal agency for the management of plant genetic resources (Tyagi et al. 2021). The NBPGR is involved in the

Fig. 21.3 An overview of public–private synergy in hybrid maize seed production, depicting complementary roles of public research and private enterprise in germplasm development, policy and technical support, seed production, and distribution. The partnership drives accelerated hybrid development, high-quality seed availability, improved farmer incomes, and strengthened food security of the nation

exploration, collection, evaluation, maintenance, and documentation of germplasm. It facilitates the import, export, and domestic distribution of germplasm for research, adhering to national regulations and international agreements (Singh et al. 2020). The public sector's historical policy of allowing open access to its germplasm has played a crucial role in the development of the private seed industry, particularly after the 1988 policy reforms. This access provided private companies with the necessary foundational genetic material to initiate and enhance their own breeding programs, particularly for hybrid development, effectively reducing their entry barriers and accelerating their contribution to the market (Pal et al. 2002). This represents a significant, though perhaps under-appreciated, contribution of public research infrastructure to private sector development.

The public sector's role in seed multiplication mainly focuses on the initial generations, namely Breeder Seed (BS). Public seed agencies such as the National Seed Corporation (NSC) and State Seed Corporations (SSCs) are responsible for the multiplication and marketing of Certified Seed (CS). The NSC, being the largest public agency, possesses considerable infrastructure, including farms and an extensive dealer network, managing numerous varieties of diverse crop species. SSCs usually concentrate on varieties that are popular within their respective states. Public sector-bred maize hybrids such as DMRH 1308, DMRH 1301, DMH 117, DMH 121, and HQPM 1 have also made a significant impact in the country, with a considerable share in DAC breeder seed indents. Many private seed companies have taken up these hybrids by signing MoUs with the respective institutes. The maize seed sector in India is largely dominated by the private sector, which accounts for about 65–70% of the market, while the public sector contributes approximately 25–30% (Kumar et al. 2025).

21.5 Private Sector Contribution

It was the liberalization of the new seed development policy in the year 1988 that seems to be the key player in remodelling the scenery of the Indian seed industry, especially for maize. It was the easement of policy that offered subsidized entry barriers and relieved restrictions; this has contributed immensely to increased involvement and financial assistance from the private sector. Huge investments were made by both domestic and international firms (MNCs), which have contributed well to stretching their actions and establishing their R&D centres. Under the impact of the 1988 policy and trade liberalization, the private sector made substantial investments in research and development, mainly concentrated on plant breeding in generating potent maize hybrids. While establishing their own breeding programs, private companies have consistently relied on public organizations (ICAR, SAUs) and international centres (CIMMYT) for access to germplasm developed by these organizations. One of the major objectives shared by private R&D remains with the establishment of hybrids having features like enhanced yield, pest, and disease resistance.

Innovation in this field has advanced through the adoption of cutting-edge technologies, including the integration of biotech traits and the application of novel techniques such as double haploid, genomic selection, and genome editing. Several large companies have established agro-climatic testing zones across India to develop and evaluate region-specific varieties. The hybrid maize seed sector is dominated by numerous private players that are leading in popularity, as reflected by the widespread adoption of their varieties across cultivated areas (Pavithra et al. 2018). The profit-driven nature of the private sector, combined with farmers' substantial annual investment in purchasing hybrid seeds due to their high yield potential, has laid the foundation for notable expansion of the hybrid seed market (Pavithra et al. 2018). This has achieved private business in gaining firm command in the hybrid maize seed market, and has also aided in accessing a large share of the market. The private sector is predominantly engaged in the production of certified seeds, primarily hybrids, as well as truthfully labelled seeds (Loch and Boyce 2003). Employment of contract farming for seed production by private companies involves the dissemination of technical knowledge and parental seed materials to the farmers for cultivation and vending it back to companies at a predetermined premium cost over the commercial grain cost (Eaton and Shepherd 2001). Whilst this model seems to elevate earning opportunities for the farmers, the diffused nature of production across the fields of various small farmers could present varied complications in modulating robust quality monitoring and varietal purity. The considerable, revolutionary marketing and distribution network of the private sector stands out as one of its key strengths. Their network is multi-layered, involving distributors, dealers, and retailers that manage to reach farmers across varied locations. Companies often utilize comprehensive marketing strategies, including portfolio branding, promotional campaigns, farmer engagement initiatives and brand loyalty programs, which can play a decisive role in farmers' purchasing decisions. Additionally, online platforms have emerged as significant channels for seed distribution.

21.6 Bridging the Gap Between Public and Private Sector

Despite the dynamism of both the public and private sectors, there exist potential flaws that affect their synergy within the hybrid maize seed supply chain. These include diverse gaps, possible coincidences, and inefficient synchronization, hampering ideal performance and equitable outcomes. Although India has a huge seed industry, driven by a strong public research foundation, there still remains scope for coordination between public and private stakeholders that has hindered the sector's full potential. One of the potential disconnects remains ineffective translation of scientific outcomes from the public sector into commercial products (Singh et al. 2002). The contributing barriers include complicated access procedures, inappropriate knowledge dispersal, and distrust around Intellectual property rights (IPR) (Tonapi et al. 2015). This leads to the underutilization of public assets and significant overlap between the two sectors. In order to overcome this, there arises a need for appropriate synchronization in research and development, with the public sector concentrating precisely on upstreaming research and manufacturing intermediate products (Pal and Tripp 2002). Public research institutions possess robust R&D initiatives and have developed numerous improved maize varieties over the decades. However, these institutions lack marketing strategies and distribution networks to promote their products effectively. Private seed companies have successfully captured significant market share through aggressive marketing, reliable distribution, and contract-based seed production (Kumar et al. 2012).

Public seed agencies must dedicate resources to branding and visibility, including field demonstrations and partnerships with Farmer Producer Organizations (FPOs) to build reliability in public-bred hybrids and develop efficient "Seed Villages" or "Seed Hubs". Public research institutions should also evolve measures to take their products to the growers, demonstrate the potential of the seeds developed, and produce the seed materials adequately to fill the gap left open by the private players. Without strong outreach, even well-developed varieties fail to achieve widespread adoption (YES Bank and FICCI 2024). Most importantly, the development of an efficient and integrated maize grain market is essential to drive the sector hand-in-hand with the maize seed sector in the regions. Another potential gap lies in the smooth progression of the seed proliferation process. Supply of suitable maize hybrid seeds at the right time or season renders farmers with the right yield, lacking which farmers lose trust in the varietal seed (Upadhyay et al. 2018). For effective progression of the process, there arises the need for punctual and optimum supply of high-quality Breeder Seed (BS) by public sectors and public agencies (NSC/ SSCs) or private companies that carry out the proliferation of seeds under license. Any sort of inadequacy here can predominantly hamper the accessibility of Foundation and Certified/ Truthfully Labelled (TL) seeds. Some of the inadequacies include the narrow interest of the private sectors in the proliferation and commercialization of Open Pollinated Varieties (OPV) due to the limited profit margins they offer over hybrids (Ranjit et al. 2012). As a result, farmers are often dependent on potentially old varieties instead of OPVs, thus restricting their access to superior genetics. However, conservation of optimum seed quality through the scattered production system comprising public agencies, Multinational Companies (MNCs), SMEs, and numerous small-scale farmers tends to pose an additional challenge (Vorley et al. 2009).

Quality assurance is another challenge known to strengthen this gap. India possesses a potential framework for certification and testing of seeds along with voluntary certification (Prasad et al. 2017). However, hybrids from the private sector are marketed under the name of truthfully labelled (TL) seeds, where quality control remains with the production company (Loch and Boyce 2003). Undoubtedly, the majority of the production companies aim at optimum quality assurance, relying primarily on self-regulation, which can introduce potential vulnerabilities into the system. Therefore, the doping of the market with poor-quality seeds despite the recognized superiority of hybrid seeds is a matter of concern that generates an urgent need for a robust monitoring and enforcement system, strengthened with collaborative efforts across the sectors. Restricted distribution and limited market access also contribute to bridging the gap between the public and private sectors. Seed distribution network of private companies is often limited to commercially feasible regions, that devoid small-scale farmers of high-quality seeds. It is not just the limited availability but also the higher cost that restricts farmers' access to these seeds. The efforts must be put to diversify the maize seed production sites beyond the peninsular India. There do exist public distribution channels that may be associated with various government schemes, which also need to be explored. There is no government regulation on the price of hybrid maize seed. Thus, instead of a restrictive price policy, the state governments may provide a subsidy directly to the marginal farmers on the purchase of hybrid seeds to increase maize productivity in the region (YES Bank and FICCI 2024).

Promoting lower-input, climate-resilient maize hybrids along with input support can improve larger adoption among the smallholders. These efforts must be accompanied by direct benefit transfers for seed purchases instead of restrictive price controls, thereby preserving market incentives while making hybrid seeds affordable (Hanchinal 2017). Another contributing challenge is the weak coordination between public agencies and private companies in providing agronomic guidance and accurately assessing the diverse needs for the wide range of marketed hybrids. This results in farmers receiving fragmented and inconsistent information. Financial and human resource limitations are one such cause that is known to raise the gap, which often hinders public extension systems such as KVKs and ATMAs from providing precise and extensive guidance on the management of private sector hybrid varieties (Poonam and Surya Rathore 2017; Panda et al. 2023). However, private agencies also aim to support farmers with technical education, which is basically associated with the promotion of their own products. This prevents smallholders from adopting new hybrids, from perceiving the essential knowledge needed for proper crop management. However, a well-coordinated approach is missed that integrates the public sector's product-neutral guidance with the private sector's product-specific insights and agronomic recommendations (Bruwer 2023). Another area of concern is the limited exchange of germplasm between the two players due to a lack of trust and mutual benefit frameworks. To resolve this, transparent material transfer agreements (MTAs) should be encouraged with defined benefit-sharing clauses and intellectual property protections. A centralized coordination can reduce duplication and strengthen the seed value chain (Kumar et al. 2013). In order to overcome these potential flaws, a mutual collaboration of public and private sectors, i.e., Public Private Partnership (PPPs), can be beneficial in attaining management reforms. This partnership is based on mutual understanding, equality, trust, sharing knowledge, and focusing on combined goals and mutual welfare. This enhances information dissemination, fosters innovation, and accessibility to improved technology. This collaboration has an essential role in the improvement of research technologies for seed proliferation and quality assurance of commercial products. One of the major advantages of PPP is sharing innovation and information that offers a significant reduction in research cost (Ayyappan et al. 2007).

The introduction of policies and programs can help bridge the gap between the public and private sectors. Such policies and programs should aim at offering equitable access to all farmers, especially small-scale and marginal farmers in remote or underserved areas (Qureshi et al. 2015). Implementation gaps in seed policy and legislation also affect quality assurance and fair market practices. A decentralized model of enforcement, including state-level seed certification bodies, should be strengthened with digital monitoring tools. Regular market sampling and surprise audits can improve compliance and transparency across the supply chain (Gauchan et al. 2014). Infrastructural gaps in remote regions further limit seed access. To counter this, mobile seed processing units and seed storage facilities should be promoted through public funding and private investment incentives. Such interventions can bring seed services closer to marginalized regions, including hilly and tribal areas. Another strategy to overcome the gap between the two players is the involvement of farmers in R&D, as it can aid in developing suitable and farmer-friendly hybrid varieties of maize (Scoones and Thompson 2009). The public and private partnership to produce, multiply, and distribute seed is critical to maintain the seed cycle and provide it to the ultimate users. Therefore, there exists an urgent need to bridge the prevailing gap that would not only contribute to helping small-scale farmers but would also aid in agriculture by the introduction of improved seed varieties.

21.7 Policy Implications

Investing in R & D and strengthening the maize seed system are critical to improving hybrid performance, seed replacement rate (SRR), and farmer adoption in India. By diversifying seed production beyond peninsular India, improving seed infrastructure, promoting public-private partnerships, rationalizing IPR and pricing policies, and integrating male sterility systems, seed costs can be reduced, and labour constraints will be addressed. Additionally, increased investments in developing climate-resilient, high-yielding, and high-starch hybrids will contribute to bridging yield gaps, particularly under rainfed conditions and for ethanol production. A resilient, profitable, and sustainable maize production system supporting India's food, feed, and fuel security requires promoting affordable mechanization, strengthening extension and field demonstrations, and scaling up sustainable practices like drip irrigation, zero tillage, and integrated pest management.

21.8 Conclusion

The maize industry in India holds immense potential for growth and transformation, driven by technological innovations, strategic partnerships, supply chain optimization, and supportive policy frameworks. The Government of India has introduced a range of initiatives, such as various programs led by the Indian Institute of Maize Research, Sub-Mission on Feed and Fodder Development under the National Livestock Mission, the Sub-Mission on Agricultural Mechanization, the National Agriculture Market (e-NAM), and the Agriculture Infrastructure Fund, to strengthen the maize ecosystem. To fully realize this potential, it is essential for the private sector to work collaboratively with the government in designing and implementing strategies that create shared value for farmers and stakeholders across the maize value chain. Public and private sector collaboration can boost national maize productivity, enhance food and nutritional security, foster climate resilience, and stimulate economic growth. However, strategic interventions are necessary to foster this synergy due to the prevailing gaps between the two sectors. Some of the potential strategies include developing a dedicated National PPP Framework for the seed sector, revising IPR guidelines to promote transparency and fairness, establishing neutral platforms for dialogue and joint project development, policy reforms involving price and quality control, enhancing mechanisms for technology transfer, designing PPP models to incentivize private participation, integrating public and private extension efforts, diversify seed production sites, developing seed storage and processing infrastructure, regulating maize seed price and quality monitoring, and investing in capacity building for all stakeholders. Addressing these gaps and fostering genuine partnership is crucial for building a more efficient, equitable, and resilient hybrid maize seed system capable of meeting India's future needs.

References

Ayyappan S, Chandra P, Tandon SK (2007) Agricultural transformation through public-private partnership: an Interface. Indian Council of Agricultural Research, New Delhi, pp 1–145

Bruwer PW (2023) Establishing effective communication between the agricultural researcher and the farmer with special reference to the large-scale sugarcane, soybeans and maize industry-a comparative case study in South Africa. Doctoral dissertation, University of the Free State

Eaton C, Shepherd A (2001) Contract farming: partnerships for growth (No. 145). Food & Agriculture Org

Erenstein O, Jaleta M, Sonder K, Mottaleb K, Prasanna BM (2022) Global maize production, consumption and trade: trends and R&D implications. Food Secur 14(5):1295–1319

Gauchan D, Thapa Magar DB, Gautam S, Singh S, Singh US (2014) Strengthening seed system for rice seed production and supply in Nepal. IRRI-NARC collaborative EC-IFAD funded project on Seed Net Development. Socioeconomics and Agricultural Research Policy Division, Nepal Agricultural Research Council, Nepal. 40p. Khumaltar: NARC, Socio-Economics and Agricultural Research Policy Division (SARPOD). p 40

Hanchinal RR (2017) The Indian seed industry: achievements and way forward. Souvenir published during XIV national seed seminar on "Food Security Through Augmented Seed Supply Under Climate Change" organized during 28–30 January, 2017, pp 9–20

Kumar R, Alam K, Krishna VV, Srinivas K (2012) Value chain analysis of maize seed delivery system in public and private sectors in Bihar. Agric Econ Res Rev 25(conf):387–398

Kumar R, Srinivas K, Sivaramane N (2013) Assessment of the maize situation, outlook and investment opportunities in India. Country Report—Regional Assessment Asia (MAIZE-CRP), National Academy of Agricultural Research Management, Hyderabad, India. Available at https://naarm.org.in/wp-content/uploads/2016/07/Assessment-of-the-Maize-Situation-outlook-and-investment-opportunities-in-India.pdf. Accessed 15 Jan 2026

Kumar B, Kumara Kirna TM, Jat SL, Jat HS (2025) Economic impacts of maize hybrids. ICAR-National Institute of Agricultural Economics and Policy Research. Policy Brief No. 70. p 4

Loch DS, Boyce KG (2003) Balancing public and private sector roles in an effective seed supply system. Field Crop Res 84(1–2):105–122

Mishra GP, Dadlani M, Vijay D, Sripathy KV, Kumar S, Dikshit HK (2025) Development of seed sector in India. In: Indian seed sector: evolution, technology, trade and impact. Springer Nature Singapore, Singapore, pp 21–57

Morris ML (2002) The development of the seed industry under globalization. In: Globalization and the developing countries: emerging strategies for rural development and poverty alleviation. CABI Publishing, Wallingford UK, pp 201–217

Pal S (2017) Agricultural R&D policy in India. ICAR-National Institute of Agricultural Economics and Policy Research

Pal S, Tripp R (2002) India's seed industry reforms: prospects and issues. Indian J Agric Econ 57(3):443–458

Panda CK, Dei S, Priya R, Paswan A, Patel DK, Mandal PK, Kumari M (2023) Study on constraints of ATMA functionaries in Bihar. Econ Aff 68(3):1757

Pavithra S, Boeber C, Shah SA, Subash SP, Birthal PS, Mittal S (2018) Adoption of modern maize varieties in India: insights based on expert elicitation methodology. Agric Res 7(4):391–401

Poonam P, Surya Rathore SR (2017) An analysis of constraints as perceived by KVK managers for effective functioning of Krishi Vigyan Kendras. Indian J Agric Res 51(3):277–281

Prasad SR, Chauhan JS, Sripathy KV (2017) An overview of national and international seed quality assurance systems and strategies for energizing seed production chain of field crops in India. Indian J Agric Sci 87(3):287–300

Qureshi ME, Dixon J, Wood M (2015) Public policies for improving food and nutrition security at different scales. Food Secur 7:393–403

Ranjit K, Khurshid A, Vijesh K, Srinivas K (2012) Value chain analysis of maize seed delivery system in public and private sectors in Bihar. Agric Econ Res Rev 25:387–398

Scoones I, Thompson J (2009) Farmer first revisited: innovation for agricultural research and development, 1st edn. Practical Action, Rugby

Singh RP, Pal S, Morris ML (1995) Maize research, development, and seed production in India: contributions of the public and private sectors. CIMMYT economics working paper 95–03. CIMMYT, Mexico, D.F

Singh RP, Kumar R, Singh NP (2002) Transitioning maize seed industry in India: sectoral dimensions. Indian J Agric Econ 57(3):430–442

Singh K, Gupta K, Tyagi V, Rajkumar S (2020) Plant genetic resources in India: management and utilization. Vavilov J Genet Breed 24(3):306

Tonapi VA, Rao YY, Chowdary PN, Mahala RS, Elangovan M (2015) Need for innovative aptitude and mechanisms for Public Private Partnership and germplasm exchange. Seed Times. National Seed Association of India. https://www.millets.res.in/books/Vilas_A_Tonapi_Need_for_innovation_aptitude.pdf

Tyagi V, Brahmi P, Yadav SK, Singh SP, Singh S, Singh K (2021) Germplasm Access from ICAR-NBPGR and use within India. Indian J Plant Genet Resour 34(02):216–220

Upadhyay N, Ghimire YN, Sharma B, Acharya Y, Gairhe S, Sapkota S (2018) Factors affecting adoption of maize varieties in Nepal. J Inst Agric Anim Sci 35(1):39–45

Vorley B, Lundy M, MacGregor J (2009) Business models that are inclusive of small farmers. In: Agro-industries for development. CABI, Wallingford UK, pp 186–222

YES Bank & FICCI (2024) The Indian Maize Sector: trends, challenges & imperatives for sustainable growth. Available at The Indian Maize Sector- Trends, Challenges & Imperatives for Sustainable Growth. Accessed 15 Jan 2026

The Future of Maize: Trends, Challenges, and Opportunities 22

Pardeep Kumar, Bhupender Kumar, Mukesh Choudhary, Pravin Kumar Bagaria, and Hanuman Sahay Jat

Abstract

Maize (*Zea mays L.*) has emerged as the most produced cereal over the globe, surpassing rice and wheat, with expanding roles in livestock nutrition, industrial starch, and bioethanol. In India, maize ranks third in acreage after rice and wheat, yet its demand trajectory is far more dynamic, driven by poultry feed requirements, starch industries, and renewable energy programs. Current projections from NITI Aayog indicate a looming demand–supply imbalance of 6 MT between 2035 and 2047, underscoring the urgency of systemic interventions. Key constraints include limited irrigation coverage, uneven adoption of improved hybrids, vulnerability to climate variability, post-harvest losses, and fragmented market structures. At the same time, significant opportunities exist: hybrid seed development aligned with end-use industries, integration of digital agriculture and AI-enabled breeding pipelines, investment in post-harvest infrastructure, and Farmer Producer Organization (FPO)-led aggregation, expansion of maize-based bioethanol production, and enhanced private sector participation. Case studies such as Brazil's *safrinha* model, India's poultry sector, and pilot bioethanol initiatives illustrate practical pathways for transformation. Viewed through the lens of *Viksit Bharat 2047*, maize is positioned not merely as a cereal but as a lever of resilience, innovation, and prosperity. Its integration into climate-resilient farming systems, renewable energy frameworks, and global trade networks will be central to India's transition toward developed-nation status.

Keywords

Maize · Future · Demand · Energy · Food

P. Kumar · B. Kumar · M. Choudhary · P. K. Bagaria · H. S. Jat (✉)
ICAR-Indian Institute of Maize Research, Ludhiana, India
e-mail: mukesh.choudhary1@icar.org.in

M. Choudhary et al. (eds.), *Next-Generation Maize Research*,
https://doi.org/10.1007/978-981-92-0722-0_22

22.1 Introduction

Maize, known scientifically as *Zea mays* L. and originating in Central America nearly 9000 years ago, spread across continents through trade, conquest, and cultural exchange (Kistler et al. 2020; Li 2024). Maize is cultivated across nearly 170 countries worldwide, covering approximately 208 million hectares. Global maize production has now reached about 1241 million MT (FAOSTAT 2025). Its journey from the ancient milpas of Mexico to the intensively cultivated fields of Punjab and Bihar reflects not only maize's remarkable adaptability but also its capacity to integrate seamlessly into diverse food systems and expanding industrial value chains. India's maize story is one of transformation in terms of production and productivity. Traditionally, it was grown in tribal belts and consumed as staple food in states like Bihar, Madhya Pradesh, and Rajasthan, but now become a commodity crop driven by industrial demand. Further, poultry feed, starch industries, and ethanol blending programs have redefined maize's role in the national economy (NAAS 2024). In 2024–25, India produced 43.40 million MT of maize across 12.09 million hectares with a productivity of 3.59 MT per hectare, ranking fourth in acreage and fifth in production globally (Ministry of Agriculture and Farmers Welfare 2025, available at https://upag.gov.in/). Currently, energy industries are expanding rapidly in response to changing consumer preferences and energy policies (Murdia et al. 2016).

The introduction of maize into India's agricultural landscape also reflects broader historical and cultural shifts. Unlike rice and wheat, which are deeply embedded in traditional diets and procurement systems, maize has often been seen as a "secondary" crop. Yet its rise in recent decades challenges this perception. The poultry revolution, driven by rising incomes and protein demand, has made maize indispensable. Similarly, the starch industry, supplying inputs to textiles, paper, pharmaceuticals, and processed foods, relies heavily on maize (Kaul et al. 2019; Kumar et al. 2012). The ethanol blending program, targeting 20 percent blending by 2025–26, positions maize as a cornerstone of India's renewable energy strategy (NITI Aayog & Ministry of Petroleum and Natural Gas 2021). The label *"queen of cereals"* is not a mere rhetorical flourish, as maize's dominant role in global food security, high productivity, and multipurpose utility (Selvi et al. 2014; Shiferaw et al. 2011). Maize's versatility is unmatched: it can be consumed directly as food, transformed into animal feed, processed into starch and sweeteners, and converted into ethanol for fuel (Ranum et al. 2014). Few crops straddle the domains of food, feed, and energy with such ease. In India, maize occupies a unique position as the third most important cereal after rice and wheat, yet its growth trajectory is far more dynamic (Ayalew and Sekar 2016). While rice and wheat are heavily dependent on irrigation and government procurement systems, maize thrives in rainfed conditions, offering resilience against monsoon vagaries. This resilience is crucial in a country where climate change increasingly threatens agricultural stability (Mall et al. 2017).

The strategic importance of maize is accentuated by India's vision of Viksit Bharat 2047. As the nation aspires to become a developed economy by its centenary of independence, ensuring food, feed, and energy security has gained prominence within agricultural and economic planning (NITI Aayog *Strategic Imperatives for*

Viksit Bharat 2047). Projections indicate that the demand for maize in India is expected to rise sharply in the coming decades, with estimates suggesting that total maize demand could reach around 86 million MT by 2047–48 (NAAS 2024) under a business-as-usual scenario due to population growth and expanding use in food, feed, and industrial applications. However, production under current productivity trends may fall short of this demand, highlighting the need for enhanced productivity and supportive policy measures to bridge the gap. The narrative of maize is also deeply intertwined with farmer livelihoods. Case studies from Andhra Pradesh and Punjab illustrate how farmers transitioning from water-intensive paddy to maize have improved incomes and reduced vulnerability to groundwater depletion. In Bihar, however, rainfed maize farmers struggle with erratic rainfall and pest outbreaks, highlighting the unevenness of maize's promise. These stories underscore the need for targeted interventions that address regional challenges while leveraging maize's inherent strengths. As India marches toward *Viksit Bharat 2047*, maize will be more than a crop; it will be a symbol of resilience, innovation, and prosperity. Its future lies not only in higher yields but in smarter systems-digital agriculture, stress-tolerant hybrids, efficient post-harvest infrastructure, and integrated value chains. The challenge is formidable, but the opportunity is equally compelling. Maize can anchor India's food, feed, and energy security, making it a true queen of cereals in the twenty-first century.

22.2 Global and Indian Trends

The global trajectory of maize reflects an interplay of unprecedented production scale, continuous technological innovation, and evolving demand driven by livestock expansion, industrial uses, and biofuel policies. In 2024–25, the world produced 1.23 billion MT of maize, surpassing rice and wheat by a wide margin (https://www.fas.usda.gov/). This dominance is not accidental; it reflects decades of investment in hybrid breeding, mechanization, and policy support. The United States remains the undisputed leader, with yields exceeding 10 MT per hectare, resulting from single-cross hybrids, precision nutrient management, and advanced post-harvest systems. The second largest producer is China, which has expanded maize acreage to feed its burgeoning livestock sector, while Brazil's "safrinha" revolution has transformed the country into a maize powerhouse (Júnior and Sentelhas 2019; USDA 2023). By planting maize as a second crop after soybeans, Brazil now produces nearly three-quarters of its maize in this system, demonstrating how cropping innovation can unlock productivity without expanding land (Gurgel et al. 2024). Further, in Africa and Latin America, white maize remains a staple food, ground into flour for tortillas, ugali, or roti. In North America and Asia, yellow maize dominates, serving primarily as animal feed. Globally, over 60% of maize is consumed by livestock, underpinning the poultry, dairy, and pork industries (Grote et al. 2021). Industrial uses are expanding: maize starch feeds into textiles, paper, and pharmaceuticals; sweeteners like high-fructose corn syrup dominate processed foods; and ethanol has become a cornerstone of renewable energy strategies. In 2022, the

United States alone diverted 40% of its maize crop to ethanol, reshaping global supply-demand balances.

Approximately 28% of the maize produced in India is consumed directly as food, largely in processed forms such as popcorn, baby corn, sweet corn, maize flour, and corn flakes. According to Murdia et al. (2016), maize utilization is dominated by the feed and industrial sectors, with poultry feed accounting for 48% of total consumption, animal feed for 11%, starch for 11%, and brewing and seed each contributing about 1%. Additionally, nearly one million tonnes of maize are absorbed annually by the processed food industry for manufacturing products such as bread, muffins, doughnuts, pancake mixes, infant foods, biscuits, wafers, breakfast cereals, and pet foods. The majority fuels poultry feed, starch industries, and increasingly, ethanol blending programs. India's poultry revolution, driven by rising incomes and protein demand, has made maize indispensable. The starch industry, supplying inputs to textiles, paper, and pharmaceuticals, relies heavily on maize.

Trade dynamics further illustrate India's niche in maize exports. In the 2023–24 period, India exported about 1.44 million tonnes of maize (corn) valued at approximately USD 443.5 million, with major destinations including Vietnam, Nepal, Bangladesh, Malaysia, and Thailand. Nearly 90% of India's maize exports continue to be directed to neighbouring markets such as Nepal, Bangladesh, Sri Lanka, and Vietnam, reflecting strong regional demand for Indian maize (APEDA maize dashboard, 2025). India's non-GM maize earns a premium in Asian markets because many countries restrict imports of GM grain. This gives India an export advantage, but it also limits options during domestic shortages, as GM maize cannot be imported easily, causing supply pressure and price fluctuations for processors. Looking ahead, NITI Aayog projects domestic maize demand to reach 86 million MT by 2047–48, while production may only touch 80 million MT (Fig. 22.1).

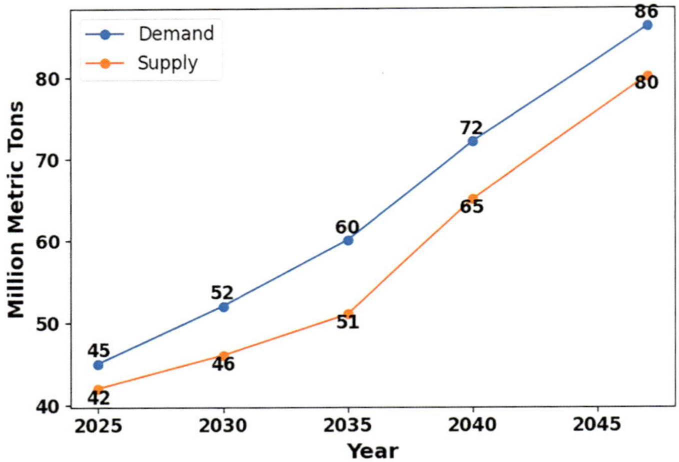

Fig. 22.1 Projected maize demand vs. supply (2025–2047) in India. (Source: NITI Aayog Report on Working Group Report on Crop Husbandry, Agriculture Inputs, Demand & Supply, 2024)

Bridging this projected gap of six million MT will require productivity-led growth, seasonal diversification (expanding rabi maize), and supply chain efficiencies.

22.3 Challenges

The modern agricultural production sector is under significant pressure due to constant, fluctuating market conditions. Specifically, the combined forces of increasing global demand and unpredictable prices for staple commodities, like maize grain, are driving producers to actively seek out more resilient and profitable operational strategies.

Maize is mainly a rainfed crop in India and hence dependence on monsoon rains exposes farmers to climate volatility, resulting sometimes in crop delay or erratic rainfall and heat stress devastating yields. In Bihar, for instance, farmers often lose entire crops to mid-season droughts, highlighting the vulnerability of rainfed maize. Similarly, sowing is dependent on the onset of the monsoon season. Agriculture faces an existential threat from climate change, a phenomenon expected to intensify its adverse effects on cultivation environments globally. Direct evidence of this change includes the recorded increase in the global average surface temperature by 0.99 °C during the first two decades of the twenty-first century compared to the 1850–1900 pre-industrial baseline, alongside a concurrent, dramatic rise in atmospheric concentrations of potent greenhouse gases, CO_2, CH_4, and N_2O. These shifts contribute to more frequent and severe extreme weather events, altered growing seasons, and increased pest and disease pressures, profoundly challenging the stability and productivity of food systems worldwide (Martínez-Ortiz et al. 2024). These changes have led to a higher frequency and intensity of extreme climatic events, including droughts and altered precipitation patterns (Sharma et al. 2022), floods and tropical cyclones (Murray-Tortarolo et al. 2018), as well as heatwaves and cold waves. Collectively, these stressors exacerbate crop vulnerability by increasing damage from biotic factors in maize, driving shifts in pest and disease distributions, and negatively affecting beneficial soil microbial communities (Martínez-Ortiz et al. 2024). Climate change compounds these vulnerabilities, such as rising temperatures, shifting rainfall patterns, and pest outbreaks, such as fall armyworm (FAW), which threaten productivity. FAW, first reported in India in 2018, now affects maize across multiple states, causing yield losses of up to 30% (Prakash et al. 2024). Over-reliance on chemical sprays risks resistance, while integrated pest management remains underutilized. On the other hand, soil degradation and declining organic matter further erode resilience, reducing maize's ability to withstand stress (Shahbazi et al. 2025).

Quality seed adoption of hybrids is another constraint. While high-yielding hybrids exist, their penetration is uneven, and seed replacement rates remain low. Farmers often rely on saved seed, perpetuating yield drag and limiting resilience (Alemu 2011). Regulatory hurdles further slow innovation: multi-year trials and complex approval processes delay the release of proven hybrids, discouraging private investment. Without streamlined pathways, India risks falling behind in deploying stress-tolerant, end-use specific varieties (Mekonnen et al. 2021).

Post-harvest losses are significant, as inadequate drying, storage, and logistics infrastructure lead to quality deterioration and mycotoxin risks. Studies estimate that 10–15% of maize is lost post-harvest, reducing farmer incomes and compromising food safety. Lack of scientific bulk storage and efficient aggregation exacerbates the problem (Bala et al. 2010). Farmers often sell maize at harvest under distress, receiving lower prices due to moisture and quality discounts.

Despite impressive growth, maize faces daunting challenges across the value chain. Market inefficiencies further hinder growth, including fragmented supply chains, weak farmer linkages, and dependence on intermediaries, which reduce price realization. Farmers often receive only 50–60% of consumer prices, with the rest absorbed by middlemen and logistics costs (https://naarm.org.in/). Volatility in poultry demand and ethanol procurement creates price cycles, leaving farmers exposed (Hellin and Erenstein 2009). Without hedging mechanisms, warehouse receipts, or standardized grading, farmers lack tools to manage risk (Awasthi et al. 2025).

Policy hurdles add complexity, like import restrictions on GM maize that limit flexibility during shortages, while export obligations constrain processors (Kimani and Gruère 2010). Quarantine norms and fumigation requirements create bottlenecks. Mechanization finance remains limited, with smallholders unable to access affordable credit for harvesters, planters, dryers, and storage units. Current subsidies do not fully bridge adoption risks, slowing mechanization. These challenges are not insurmountable, but they require coordinated action. Without addressing irrigation, seed adoption, climate resilience, post-harvest infrastructure, and market integration, maize's potential will remain unrealized.

22.4 Opportunities and Futuristic Imperatives

The challenges facing India's maize sector are formidable, but they also open the door to transformative opportunities. If maize is to become a cornerstone of *Viksit Bharat 2047*, the sector must embrace innovation, infrastructure, and inclusivity. Opportunities span from seed technology to digital agriculture, from post-harvest systems to biofuel revolutions, and from farmer empowerment to global positioning. Each of these imperatives is not just about incremental gains but about reimagining maize as a strategic crop for food, feed, and energy security in a rapidly changing world (Pathak et al. 2025).

22.4.1 Hybrid Seed Adoption and End-Use Alignment

The first and most immediate opportunity lies in accelerating the adoption of high-yielding, stress-tolerant hybrids. India's maize productivity has grown from 2.6 MT/ha in 2012–13 to 3.59 MT/ha in 2024–25 (Ministry of Agriculture and Farmers

Welfare 2025, available at www.upag.gov.in), yet this remains far below global leaders such as the USA, which routinely achieves 10 MT/ha. The gap reflects the limited penetration of advanced hybrids and low seed replacement rates. By 2047, India must aim to double its average productivity to at least 7 MT/ha, a target achievable through widespread hybrid adoption. End-use alignment is critical. Feed industries require maize with high energy density and uniform grain size; starch processors demand varieties with elevated starch fractions and consistent protein ranges; food processors seek hybrids with milling quality, pop ability, and taste attributes. Breeding pipelines must therefore be demand-driven, not supply-driven (Lee et al. 2007). Case studies from private seed companies show how targeted hybrids, such as those designed for poultry feed in Karnataka or starch processing in Gujarat, have delivered yield gains of 20–30% while meeting industrial specifications. Public breeding programs, too, must pivot toward end-use specificity, supported by strong regulation under the Plant Variety Protection and Farmers' Rights Act to safeguard innovation and encourage private investment (Ganesh et al. 2022).

22.4.2 Mechanization, Digital Agriculture, and Precision Management

The second frontier is mechanization and digital agriculture. Mechanized cultivation, precision planting, GPS-guided tractors, drones, and IoT-based nutrient management will improve efficiency, reduce labour dependency, and minimize post-harvest losses. Precision farming, drones, sensors, and AI-driven advisory platforms can revolutionize maize cultivation. In Andhra Pradesh, pilot projects using drones for targeted pesticide application reduced chemical use by 25% while maintaining yields (Singh and Singh 2025). In Karnataka, soil moisture sensors linked to mobile advisories helped farmers optimize irrigation, saving water and improving productivity. By 2047, such technologies must become mainstream, not niche.

Digital platforms can democratize access to advanced tools. Equipment rental and sharing services, facilitated through mobile apps, can make harvesters, planters, and dryers accessible to smallholders. AI-driven advisory systems can deliver real-time information on weather, pests, and nutrient management, tailored to specific geographies. Imagine a farmer in Bihar receiving a mobile alert that predicts a fall armyworm outbreak 3 days in advance, along with recommended biological treatments and market price forecasts. Such integration of data, technology, and extension can transform farmer decision-making. The futurist vision is of AI-driven breeding pipelines that deliver hybrids tailored to micro-climates (Khan et al. 2022). By 2047, India could deploy machine learning models that analyze genomic data, climate projections, and farmer preferences to design hybrids optimized for resilience and profitability. Precision agriculture will not just be about managing inputs but about orchestrating entire cropping systems in harmony with climate and market dynamics.

22.4.3 Post-Harvest Infrastructure and Market Integration

Post-harvest losses remain a major constraint, with estimates suggesting that 10–15% of maize is lost due to inadequate drying, storage, and logistics (Nath et al. 2024). This is both an economic and a food security issue. Investments in drying units, bulk storage facilities, and maize clusters can reduce losses and improve quality. Farmer Producer Organizations (FPOs) can play a pivotal role in aggregation, enabling farmers to negotiate better prices and connect directly with markets. Case studies from Madhya Pradesh show how FPOs equipped with community dryers and silos reduced post-harvest losses by 12% and increased farmer incomes by 18%. In Bihar, cooperative models linking farmers to starch processors ensured stable demand and fair pricing. By 2047, India must envision carbon-neutral maize clusters equipped with solar-powered dryers, smart silos with temperature and moisture sensors, and blockchain-enabled traceability systems. Such infrastructure will not only reduce losses but also enhance export competitiveness by meeting stringent quality standards. Market integration is equally important. E-marketplaces can enhance transparency and competitive pricing, ensuring farmers receive fair value. Warehouse receipt systems can allow farmers to store maize and sell when prices are favorable, reducing distress sales. Futures and options markets can provide hedging mechanisms against volatility. By 2047, India must build a maize market ecosystem that is efficient, transparent, and farmer-centric.

22.4.4 Bioethanol Revolution and Energy Security

The bioethanol program represents perhaps the most transformative opportunity. India's mandate for 20% ethanol blending by 2025–26 and 30% by 2030 positions maize as a key feedstock. While sugarcane has traditionally dominated ethanol production, its water intensity and regional concentration limit scalability. Maize offers a more sustainable alternative, with lower water requirements and broader geographic adaptability. Pilot biorefineries in Maharashtra and Karnataka have demonstrated the viability of maize-based ethanol, producing fuel that reduces carbon emissions and enhances energy security. By 2047, maize could anchor India's renewable energy ambitions, supplying not only ethanol for blending but also biojet fuels for aviation (https://ciire.in/). Imagine Indian airlines powered by maize-derived biofuels, reducing dependence on imported crude and aligning with net-zero goals. Such a vision is not far-fetched; global companies are already investing in biojet fuel technologies, and India can position itself as a leader in this domain.

Balancing food, feed, and fuel priorities will be critical. Policies must ensure that maize diverted to ethanol does not undermine food or feed security. Scientific quota-setting, flexible import regulations, and industry-specific varieties can help manage this balance. By 2047, India must achieve a harmonious integration of maize into its energy strategy, ensuring sustainability across sectors.

22.4.5 Private Sector Investment and Global Positioning

Private sector investment across the maize value chain can catalyze change. Processing units, silage training, and integrated value chains can attract capital and expertise (FICCI and YES BANK. 2024). Partnerships with FPOs and cooperatives can strengthen farmer linkages and ensure equitable growth. Case studies from Gujarat show how private starch companies partnered with FPOs to supply certified seed, provide extension services, and guarantee buy-back, resulting in yield gains and stable incomes. India's position as a reliable exporter of non-GM maize to Asian markets can be further consolidated. Certification systems and industry-specific varieties can enhance competitiveness. By 2047, India could emerge as a global hub for specialty maize—biofortified, non-GM, and climate-resilient serving both domestic and international markets. Futuristic view that India is exporting maize fortified with Vitamin A to Africa, non-GM maize to Europe, and high-starch maize to Southeast Asia. Such diversification can enhance India's global standing and farmer incomes.

22.4.6 Advancing Climate-Smart Agriculture and Regional Expansion

The future of maize cultivation lies in the widespread adoption of climate-smart agriculture (CSA) practices and expansion into non-traditional areas. CSA interventions such as drought-tolerant hybrids, zero tillage, precision irrigation, and agroforestry have already demonstrated yield gains of 15–30% while enhancing resilience and reducing emissions. With maize yields declining by 1.7% for every 1 °C rise above optimum flowering temperatures, research into CRISPR-based breeding, genomic selection, and soil health restoration will be critical. India's National Mission on Sustainable Agriculture (NMSA) is promoting CSA through funding and awareness, but scaling adoption requires targeted innovations in crop genetics, water use efficiency, and farmer-centric digital advisory tools. Expansion into non-traditional regions such as Jammu & Kashmir, Himachal Pradesh, Tamil Nadu, and the North-East further strengthens food security and rural economies, with Assam alone recording a 60% increase in maize area between 2015 and 2022 due to improved varieties and rising demand for poultry feed and ethanol.

22.4.7 Vision for Viksit Bharat 2047

The vision for *Viksit Bharat 2047* places maize at the heart of India's transformation. Food security will be enhanced through diversified diets and biofortified/fortified maize products, reducing malnutrition. Feed security will be ensured through resilient supply chains supporting poultry and dairy, meeting the protein needs of a growing population. Energy security will be anchored by maize-based ethanol and biofuels, reducing carbon intensity and import dependence. Climate resilience will be achieved through AI-driven breeding and regenerative practices, ensuring stability under extreme weather. Rural prosperity will be driven by farmer success stories scaled nationally, with maize clusters and FPOs powering collective bargaining and industrial linkages. Policies supporting gender-responsive approaches, digital extension, crop insurance, and FPO-led aggregation will be essential to ensure equitable adoption. With these innovations, maize can evolve from a staple cereal into a driver of resilience, sustainability, and prosperity, aligning with India's vision of *Viksit Bharat 2047*.

22.5 Conclusion

Maize stands at the crossroads of India's agricultural transformation. Its trajectory from a traditional food crop to a strategic industrial commodity reflects both the dynamism of Indian agriculture and the challenges of balancing food, feed, and fuel priorities. The demand–supply gap projected for 2035 and 2047 is a warning signal, but also an opportunity to reimagine maize cultivation and value chains. The imperatives are clear: accelerate hybrid seed adoption, deploy digital agriculture, invest in post-harvest infrastructure, and integrate maize into India's renewable energy strategy. Farmer Producer Organizations, private sector partnerships, and government policy must converge to build resilient, efficient, and inclusive maize ecosystems. By 2047, maize can anchor India's vision of *Viksit Bharat*—ensuring food security through fortified diets, feed security through resilient supply chains, and energy security through bioethanol and biojet fuels. The future of maize is not just about higher yields; it is about smarter systems, climate resilience, and global competitiveness. With coordinated action, maize can truly become the queen of cereals in the twenty-first century, symbolizing resilience, innovation, and prosperity in India's march toward developed-nation status.

References

Alemu D (2011) The political economy of Ethiopian Cereal Seed Systems: state control, market liberalization and decentralization. IDS Bull 42(4):69–77

Awasthi NP, Singh SK, Yadav PK (2025) Technical challenges and opportunities in export of food from India: food regulatory compliance and future prospects to prevent food frauds using analytical techniques. Int Adv Res J Sci Eng Technol 12(7):178–194

Ayalew B, Sekar I (2016) Trends and regional disparity of maize production in India. J Dev Agric Econ 8(9):193–199

Bala BK, Haque MA, Hossain MA, Majumdar S (2010) Post harvest loss and technical efficiency of rice, wheat and maize production system: assessment and measures for strengthening food security. Final Report CF 6(08)

FAOSTAT (2025). Maize production statistics. Rome: Food and Agriculture Organization of the United Nations

FICCI and YES BANK (2024) The Indian Maize Sector: trends, challenges & imperatives for sustainable growth. Federation of Indian Chambers of Commerce and Industry/YES BANK Ltd., New Delhi/Mumbai

Ganesh R, Kumar A, Pilania DS, Singh AK, Choudhury DR, Nagarathna TK, Agarwal DK, Prabhu KV (2022) Farmers' rights and plant breeders' rights under Indian Regulations of Plant Variety Protection: a TRIPS compliant harmonised coexistence. Indian J Plant Genet Resour 35(03):13–17

Grote U, Fasse A, Nguyen TT, Erenstein O (2021) Food security and the dynamics of wheat and maize value chains in Africa and Asia. Front Sustain Food Syst 4:617009

Gurgel AC, Seabra JE, Arantes SM, Moreira MM, Lynd LR, Galindo R (2024) Contribution of double-cropped maize ethanol in Brazil to sustainable development. Nat Sustainability 7(11):1429–1440

Hellin J, Erenstein O (2009) Maize-poultry value chains in India: implications for research and development. J New Seeds 10(4):245–263

Júnior RDSN, Sentelhas PC (2019) Soybean-maize succession in Brazil: impacts of sowing dates on climate variability, yields and economic profitability. Eur J Agron 103:140–151

Kaul J, Jain K, Olakh D (2019) An overview on role of yellow maize in food, feed and nutrition security. Int J Curr Microbiol App Sci 8(02):3037–3048

Khan MHU, Wang S, Wang J, Ahmar S, Saeed S, Khan SU, Xu X, Chen H, Bhat JA, Feng X (2022) Applications of artificial intelligence in climate-resilient smart-crop breeding. Int J Mol Sci 23(19):11156

Kimani V, Gruère GP (2010) Implications of import regulations and information requirements under the Cartagena Protocol on Biosafety for GM commodities in Kenya. AgBioforum 13(3):222–241

Kistler L, Thakar HB, VanDerwarker AM, Domic A, Bergström A, George RJ, Harper TK, Allaby RG, Hirth K, Kennett DJ (2020) Archaeological Central American maize genomes suggest ancient gene flow from South America. Proc Natl Acad Sci 117(52):33124–33129

Kumar RS, Kumar BH, Kaul JY, Karjagi CG, Jat SL, Parihar CM, Kumar AS (2012) Maize research in India-historical prospective and future challenges. Maize J 1(1):1–6

Lee KM, Herrman TJ, Rooney L, Jackson DS, Lingenfelser J, Rausch KD, McKinney J, Iiams C, Byrum L, Hurburgh CR Jr, Johnson LA (2007) Corroborative study on maize quality, dry-milling and wet-milling properties of selected maize hybrids. J Agric Food Chem 55(26):10751–10763

Li JS (2024) The Columbian exchange: maize's global journey and ecological impact. Maize Genom Genet 15(3):102–110

Mall RK, Gupta A, Sonkar G (2017) Effect of climate change on agricultural crops. In: Current developments in biotechnology and bioengineering. Elsevier, pp 23–46

Martínez-Ortiz MA, Salinas-Moreno Y, Ramírez-Díaz JL, Ledesma-Miramontes A, Alemán de la Torre I (2024) Challenges and opportunities in the specialization of maize cultivation. Agro Product 17(9):81–92

Mekonnen DK, Abate GT, Yimam S, Benfica R, Spielman DJ, Place F (2021) The impact of Ethiopia's direct seed marketing approach on smallholders' access to seeds, productivity, and commercialization. IFPRI Discussion Paper, vol 1998. International Food Policy Research Institute

Ministry of Agriculture & Farmers Welfare (2025) All India: crop-wise area, production & yield. Available at: https://upag.gov.in/dash-reports/allindiaapy. Accessed 7 Oct 2025

Murdia LK, Wadhwani R, Wadhawan N, Bajpai P, Shekhawat S (2016) Maize utilization in India: an overview. Am J Food Nutr 4(6):169–176

Murray-Tortarolo GN, Jaramillo VJ, Larsen J (2018) Food security and climate change: the case of rainfed maize production in Mexico. Agric Forest Meteorol 253:124–131

NITI Aayog and Ministry of Petroleum and Natural Gas (2021). Roadmap for Ethanol Blending in India 2020–25. India.

NAAS (2024) Maize for bioethanol production in India: prospects and strategy. Strategy Paper No. 19. National Academy of Agricultural Sciences, New Delhi. 20

Nath B, Chen G, O'Sullivan CM, Zare D (2024) Research and technologies to reduce grain post-harvest losses: a review. Foods 13(12):1875

Pathak H, Joshi PK, Lakra WS, Singh AK, Baranwal VK, Jain RK (2025) Indian agriculture by 2047: a roadmap for research, education and extension. National Academy of Agricultural Sciences, New Delhi, p xxiv+371

Prakash KN, Venkataramana MN, Gaddi GM, Umesh KB, Muralimohan K, Gururaj B, Sharif M, Khandibagur V (2024) Impact of invasive Fall Army Worm on maize yield and return pattern in India: a comprehensive analysis. J Exp Agric Int 46(5):1–11

Ranum P, Peña-Rosas JP, Garcia-Casal MN (2014) Global maize production, utilization, and consumption. Ann N Y Acad Sci 1312(1):105–112

Selvi RP, Gopal SM, Swaminathan B (2014) Structural changes in the growth dimensions of maize (*Zea mays* L.) during post-green revolution and post-liberalization periods of Indian agriculture. Trends Biosci 7(20):3247–3254

Shahbazi F, Shahbazi S, Nadimi M, Paliwal J (2025) Losses in agricultural produce: a review of causes and solutions, with a specific focus on grain crops. J Stored Prod Res 111:102547

Sharma RK, Kumar S, Vatta K, Bheemanahalli R, Dhillon J, Reddy KN (2022) Impact of recent climate change on corn, rice, and wheat in southeastern USA. Sci Rep 12(1):16928

Shiferaw B, Prasanna BM, Hellin J, Bänziger M (2011) Crops that feed the world 6. Past successes and future challenges to the role played by maize in global food security. Food Secur 3(3):307–327

Singh R, Singh S (2025) A review of Indian-based drones in the agriculture sector: issues, challenges, and solutions. Sensors 25(15):4876

USDA (2023). Grain and Feed Annual. Brasilia. Public Distribution Report, April 10. Report No. BR2023-0008. Prepared by Formiga, Marcela.

Kolibri 360 GmbH
Ferdinand-Jühlke-Straße 7
99095 Erfurt, Deutschland
produktsicherheit@kolibri360.de